Mosquitoes, Communities, and Public Health in Texas

Mosquitoes, Communities, and Public Health in Texas

Edited by

Mustapha Debboun
Martin Reyna Nava
Leopoldo M. Rueda

ACADEMIC PRESS
An imprint of Elsevier

Academic Press is an imprint of Elsevier
125 London Wall, London EC2Y 5AS, United Kingdom
525 B Street, Suite 1650, San Diego, CA 92101, United States
50 Hampshire Street, 5th Floor, Cambridge, MA 02139, United States
The Boulevard, Langford Lane, Kidlington, Oxford OX5 1GB, United Kingdom

Notices
Knowledge and best practice in this field are constantly changing. As new research and experience broaden our
understanding, changes in research methods, professional practices, or medical treatment may become necessary.
Practitioners and researchers must always rely on their own experience and knowledge in evaluating and using any
information, methods, compounds, or experiments described herein. In using such information or methods they
should be mindful of their own safety and the safety of others, including parties for whom they have a professional
responsibility.

To the fullest extent of the law, neither the Publisher nor the authors, contributors, or editors, assume any liability for
any injury and/or damage to persons or property as a matter of products liability, negligence or otherwise, or from
any use or operation of any methods, products, instructions, or ideas contained in the material herein.

Library of Congress Cataloging-in-Publication Data
A catalog record for this book is available from the Library of Congress

British Library Cataloguing-in-Publication Data
A catalogue record for this book is available from the British Library

ISBN: 978-0-12-814545-6

For information on all Academic Press publications
visit our website at https://www.elsevier.com/books-and-journals

Publisher: Charlotte Cockle
Acquisition Editor: Anna Valutkevich
Editorial Project Manager: Susan Ikeda
Production Project Manager: Punithavathy Govindaradjane
Cover Designer: Alan Studholme

Typeset by SPi Global, India

Working together
to grow libraries in
developing countries

www.elsevier.com • www.bookaid.org

Contents

5. Key to Genera of Fourth Instar Mosquito Larvae of Texas

Leopoldo M. Rueda

6. Key to the Species of Fourth Stage Mosquito Larvae of Texas

Leopoldo M. Rueda

Part II
Communities

7. Mosquito Surveillance

*Nina M. Dacko, Martin Reyna Nava,
Christopher Vitek and Mustapha Debboun*

8. Mosquito Control

*Kyndall Dye-Braumuller, Chris Fredregill
and Mustapha Debboun*

13. Functional Relationship Between Public Health and Mosquito Abatement

Robert Vanderslice, Nicholas Porter, Samantha Williams and Abraham Kulungara

14. Vaccines for Mosquito-Borne Human Viruses Affecting Texas

Peter J. Hotez

15. Personal Protective Measures Against Mosquitoes

Stephen P. Frances and Mustapha Debboun

Contributors

Numbers in parentheses indicate the pages on which the authors' contributions begin.

Alan D.T. Barrett (339), Sealy Institute for Vaccine Sciences, University of Texas Medical Branch, Galveston, TX, United States

Nina M. Dacko (221), Tarrant County Public Health Environmental Health Division, Fort Worth, TX, United States

Mustapha Debboun (3, 9, 221, 249, 319, 387), Mosquito and Vector Control Division, Harris County Public Health, Houston, TX, United States

Dagne Duguma (319), Mosquito and Vector Control Division, Harris County Public Health, Houston, TX, United States

Kyndall Dye-Braumuller (249), Mosquito and Vector Control Division, Harris County Public Health, Houston, TX, United States

Stephen P. Frances (387), Australian Defence Force Malaria & Infectious Disease Institute, Enoggera, QLD, Australia

Chris Fredregill (249), Mosquito and Vector Control Division, Harris County Public Health, Houston, TX, United States

Peter J. Hotez (381), Texas Children's Hospital Center for Vaccine Development, National School of Tropical Medicine, Baylor College of Medicine, Houston, TX, United States

Abraham Kulungara (359), Association of State and Territorial Health Officials, Arlington, VA, United States

Aldo I. Ortega-Morales (279), Parasitology Department, Agricultural Autonomous University Antonio Narro, Laguna Unit, Torreón, Coahuila, México

Nicholas Porter (359), Association of State and Territorial Health Officials, Arlington, VA, United States

Martin Reyna Nava (9, 221, 279), Mosquito and Vector Control Division, Harris County Public Health, Houston, TX, United States

Leopoldo M. Rueda (3, 169, 177, 193, 199, 319), Walter Reed Biosystematics Unit, Department of Entomology, Smithsonian Institution, Suitland, MD, United States

Daniel Strickman (307), Global Health Program, Bill & Melinda Gates Foundation, Seattle, WA, United States

Robert Vanderslice (359), Environmental Health Consultant, The Vanderslice Group, North Kingstown, RI, United States

Christopher Vitek (221), The University of Texas Rio Grande Valley, Center for Vector-Borne Diseases, Edinburg, TX, United States

Scott C. Weaver (339), Institute for Human Infections and Immunity, Western Gulf Center of Excellence for Vector-borne Diseases, University of Texas Medical Branch, Galveston, TX, United States

Samantha Williams (359), Association of State and Territorial Health Officials, Arlington, VA, United States

Preface

Mosquitoes have caused great mortality and misery to humans and animals and sometimes even influencing our economy and history. Thus, their surveillance and control are very important to public health. The recent introduction of West Nile, Zika, and chikungunya viruses into North America and the continuing threat of dengue viruses from their reservoirs in Latin America are immediate concerns for people in Texas. The subtropical climate of much of the state, large-scale transport of goods and people with Latin America, and an extensive border with Mexico make Texas particularly vulnerable to additional invasive pathogens and mosquitoes. Climate change makes mosquito species invasion even more likely as habitats become more disturbed and as conditions become potentially more conducive for tropical species. *Mosquitoes, Communities, and Public Health in Texas* is a comprehensive treatment of these pests and vectors.

This book was conceived by Dr. Mustapha Debboun in 2017 while serving as the Director of Mosquito & Vector Control Division at Harris County Public Health in Houston, Texas. The idea for this book came about shortly after publishing a checklist of mosquitoes in Harris County and the city of Houston. Since that time, it has been supported by the coeditors and authors who contributed their expertise to produce a useful, high-quality volume. We are greatly thankful to the Walter Reed Biosystematics Unit (WRBU) at the Walter Reed Army Institute of Research (WRAIR) and the Smithsonian Institution for the use of the national mosquito collection during the development of the adult and larval keys. High praise goes to our photographer, Gene White, for his excellent work on preparing the photographs of mosquitoes. We are also grateful to our colleagues who participated in the review process and contributed their expertise to improve the book.

Mosquitoes, Communities, and Public Health in Texas is organized into three parts. The first part provides the means to identify adult females and larvae of all the species of the state and maps of their distribution at a county level in Texas and in adjacent states of Mexico. The chapters in the second part address the interests in mosquitoes at a community level. These topics include surveillance, control, mosquito species in adjacent states of Mexico, and considerations on potentially invasive mosquitoes. The third and final part of the book concentrates on the importance of mosquitoes in public health, recent expansion of mosquito-borne pathogens into Texas, potential roles of vaccines for prevention of mosquito-borne diseases, functional relationship between public health and mosquito abatement, and personal protective measures against mosquitoes.

This is the first book providing a comprehensive treatment of the 87 species of mosquitoes known from Texas. It combines information on identification, distribution, biology, veterinary and public health importance, and control of mosquitoes in the state. Identification through fully illustrated dichotomous keys to larvae or adult females leads the reader to species-by-species treatment of geographic distribution, habitat associations, and medical importance.

The recent award from the US Centers for Disease Control and Prevention to the University of Texas Medical Branch at Galveston, Texas A&M University, and Harris County Public Health Mosquito & Vector Control Division resulted in the establishment of the Western Gulf Center of Excellence for Vector-Borne Diseases. The establishment of this institution is an indication of the importance given at a national level to the challenges faced in the region from mosquitoes and other arthropods that transmit pathogens. *Mosquitoes, Communities, and Public Health in Texas* provides a good framework for that work and a significant training tool. Up until now, information on Texas mosquitoes has been scattered in the scientific literature, state publications, and any records of individual presentations. In fact, the last manual on mosquitoes of Texas was published over 60 years ago in 1944 by George W. Cox of the state health department.

As the editors of this book, we thank the chapter authors and coauthors who generously donated their time and decades of accumulated professional experience. We also thank Susan Ikeda of Elsevier who guided us through the publication process. We hope this book will be useful and interesting to entomologists, public health professionals, and all those concerned with mosquitoes in Texas. It is our goal that this book will contribute to the protection of the state's citizens and provide a framework for students of medical entomology.

Mustapha Debboun
Martin Reyna Nava
Leopoldo M. Rueda

Part I

Mosquitoes

Chapter 1

Taxonomy, Identification, and Biology of Mosquitoes

Leopoldo M. Rueda[a], Mustapha Debboun[b]

[a]Walter Reed Biosystematics Unit, Department of Entomology, Smithsonian Institution, Suitland, MD, United States, [b]Mosquito and Vector Control Division, Harris County Public Health, Houston, TX, United States

Chapter Outline

1.1 Taxonomy and identification

Mosquitoes belong to the family Culicidae, order Diptera, class Insecta (Hexapoda), and phylum Arthropoda. There are two recognized subfamilies, the Anophelinae and Culicinae. Subfamily Culicinae has 11 tribes worldwide, and in Texas, 7 tribes (with 11 genera) are found: Aedini (*Aedes*, *Haemagogus*, and *Psorophora*), Culicini (*Culex* and *Deinocerites*), Culisetini (*Culiseta*), Mansoniini (*Coquillettidia* and *Mansonia*), Orthopodomyiini (*Orthopodomyia*), Toxorhynchitini (*Toxorhynchites*), and Uranotaeniini (*Uranotaenia*). There are about 3600 described mosquito species and subspecies, under 188 subgenera in 41 genera of mosquitoes worldwide (WRBU, 2018). In Texas, there are 87 known species, under 22 subgenera in 12 genera of mosquitoes (see Chapter 2). Although Carpenter and LaCasse (1955) treated *Psorophora (Janthinosoma) varipes* (Coquillett) as a valid species, it was not included in the current list of Texas mosquitoes since its distribution and morphological identity are still questionable (Darsie and Ward, 2005). Harrison et al. (2008) provided new diagnostic morphological features of seven species of *Psorophora (Janthinosoma)*. They also followed the suggestion of Belkin and Heineman (1975) about the restricted distribution of *Psorophora varipes* to an area along the Pacific slope from Mexico to Nicaragua, and therefore it is not found in the United States. Additional studies, however, particularly the use of molecular techniques (deoxyribonucleic acid barcoding), are needed to ascertain the true identity, morphological variations, and regional distribution of *Psorophora varipes*.

Mosquitoes, like other arthropods, are bilaterally symmetrical. The adult mosquito (Fig. 1.1) is covered with an exoskeleton and with three distinct body regions: head, thorax, and abdomen. The head is ovoid in shape, with two large compound eyes and five appendages such as two antennae, two maxillary palpi, and the proboscis. The thorax has three segments: the prothorax, mesothorax, and metathorax. Each segment has a pair of jointed legs, while the mesothorax has a pair of functional wings and the metathorax, a pair of knobbed structures called halters. The abdomen has 10 segments, of which the three terminal segments are specialized for reproduction and excretion. Mosquito adults resemble Chironomidae, Dixidae, Chaoboridae, and other Nematocera, which like mosquitoes have aquatic immature stages. Mosquitoes, however, are distinguished from such similar looking dipterous insects by the presence of scales on the wing veins and wing margins and by their long proboscis that is adapted for piercing and sucking. In contrast to an adult, the larva (Fig. 1.2) is mainly composed of soft, membranous tissues in thorax and abdomen and hardened, sclerotized plates in the head.

Detailed morphological descriptions and glossaries of the adult, pupa, larva, and egg of mosquitoes are found in several publications (Harbach and Knight, 1980, 1982; Darsie and Ward, 2005). Most taxonomic keys to identify mosquitoes are based on morphological characters (Carpenter and LaCasse, 1955; Belkin, 1962; Rueda et al., 1998b; Rueda, 2004; Rattanarithikul et al., 2005; Huang and Rueda, 2017, 2018). Additional list of identification keys and references can be found in the Walter Reed Biosystematics Unit website (WRBU, 2018).

Mosquitoes, Communities, and Public Health in Texas. https://doi.org/10.1016/B978-0-12-814545-6.00001-8

FIG. 1.1 Mosquito adult female, lateral view.

FIG. 1.2 Mosquito larva, dorsal view.

In addition to morphological techniques, several approaches to identify mosquitoes have been used, including cytogenetics (polytene chromosomes with discernible banding patterns); electrophoresis (allozymes); and other molecular methods involving deoxyribonucleic acid, particularly microsatellites, markers, randomly amplified polymorphic deoxyribonucleic acid, and polymerase chain reactions and amplification of ribosomal deoxyribonucleic acid (Walton et al., 1999).

The diversity of the mosquito species varies among different regions of the world. The greatest diversity of mosquito species is found in the Neotropical region (31% of total known species), followed by the Oriental (30%), Afrotropical (22%), and Australasian (22%) regions. The Nearctic region (5%), including the United States and Canada, has the lowest species diversity. In the Nearctic region the greatest number of species in Culicinae is found in the tribe Culicini, followed by Aedini (Rueda, 2008). In Texas the greatest number of species in Culicinae occurs in the tribe Aedini (40 species) followed by Culicini (21 species).

Mosquito taxonomists use new methods of computerized and molecular analyses and comprehensive data sets to address phylogeny and classification of mosquitoes. Numerous problems, however, usually arise to change the mosquito classification and nomenclature. Reinert et al. (2004, 2006, 2008, 2009) published their extensive morphology phylogenetic studies on tribe Aedini, where they created 74 new, elevated, or resurrected genera from the single genus *Aedes*, and thus increasing about three times the number of genera in the family Culicidae. The tribe Aedini contains almost 25% of known species of mosquitoes, including vectors of human disease agents. Wilkerson et al. (2015) reanalyzed Reinert et al.'s data on generic groupings and found that their phylogeny was obviously weakly supported and their taxonomic rankings failed priority and other taxon-naming criteria. Wilkerson et al. (2015) proposed simplified Aedine generic designations to restore a more useful classification for the operational community and to correct the unstable classification of tribe Aedini. They reduced the ranks of the genera and subgenera of Reinert et al. to subgenera or informal groups, respectively, and followed the generic structure of tribe Aedini to its status prior to the year 2000. For more details, visit the online taxonomic catalog of the Walter Reed Biosystematics Unit at www.mosquitocatalog.org.

1.2 Biology

Mosquitoes have a holometabolous type of development, with four distinct stages in their life cycle: egg, larva, pupa, and adult (Fig. 1.3). Larvae and pupae of mosquitoes require aquatic habitats (mainly standing or flowing water) for proper development. The female adult lays either single eggs (*Aedes* and *Anopheles*) or in clusters (*Culex* and *Culiseta*) up to several hundred at a time, on the surface of water, on the surface of floating plants, along margins of water pools and streams, on the walls of artificial containers, or in moist habitat subject to flooding. The larvae (or *wrigglers*) undergo shedding (or *molting*) of the skin (or *exuviae*) four times before becoming pupae. Larvae of most species usually feed on organic matter and other microorganism in the water for about 1–3 weeks or longer depending on the water temperature. Larvae of mosquito predators (*Toxorhynchites*) feed on larvae of other mosquitoes. In some predatory species the first instar is a filter feeder, while the second to fourth instars have well-developed predaceous feeding structures. The pupae (or *tumblers*) develop after the fourth instar. Unlike larvae, pupae do not feed and may live from 1 to 3 days before becoming adults. Only adult female mosquitoes bite humans and animals and feed on blood. Males feed primarily on flower nectars, while the females require the blood meal to produce viable eggs. Anthropophilic species prefer to feed on humans, while zoophilic species feed in nature on animals (including mammals and birds) other than humans. Females of predators (*Toxorhynchites*) and other mosquitoes do not feed on blood. Some autogenous females can also produce viable eggs, even without a blood meal. Females usually feed every 3–5 days, and in a single feeding a female usually engorges more than its own weight of blood. Some species of mosquitoes bite mostly during daytime (*Aedes*), while others prefer to feed at dusk, twilight, and nighttime (*Anopheles*). Some species exhibit seasonal switching of hosts that resulted to the transmission of diseases from animals to humans (or *zoonotic disease transmission*). Diapause (i.e., hibernation, aestivation, or overwintering) is found in various life stages, that is, as eggs in some *Aedes* and *Psorophora*; as larvae in *Coquillettidia*, some *Culiseta*, *Mansonia*, *Orthopodomyia*, *Toxorhynchites*, and some *Aedes*; as adults, often fertilized females in *Uranotaenia*, most *Culex*, some *Anopheles*, and other *Culiseta*; and as either eggs or larvae in *Culiseta morsitans* (Theobald) (Stojanovitch and Scott, 1997).

Mosquitoes, like other biting arthropods, use visual, thermal, and olfactory stimuli to locate hosts. Olfactory cues may be important as mosquito nears the host, while visual stimuli appear important for in-flight orientation, mainly over wide ranges. For daytime biters, host movement may initiate orientation toward a human or animal. Carbon dioxide is released primarily from the breath and the skin. Carbon dioxide and octenol are common attractants that are used in monitoring and surveillance of mosquitoes in various habitats (Rueda et al., 2001). The antennae of mosquitoes have chemoreceptors that

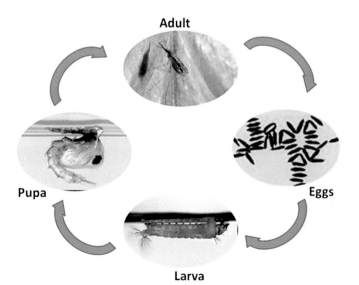

FIG. 1.3 Life cycle of the mosquito, with four stages: egg, larva, pupa, and adult.

are stimulated by lactic acid but inhibited by repellents such as deet (*N,N*-diethyl-3-methyl-benzamide), which are effective personal protective measures against biting insects to reduce or prevent transmission of vector-borne diseases (Rueda et al., 1998a).

Mosquitoes have diverse habitats that allow them to colonize various kinds of environments. The immature stages of mosquitoes are found in different aquatic habitats, that is, ditches, swamps, streams, ponds, marshes, temporary and permanent ground pools, rock holes, tree holes, crab holes, lake margins, artificial containers (tin cans, tires, bird feeders, and flower vases), natural or plant containers (fruits, leaves, husks, bamboo nodes, and tree holes), and other habitats (Laird, 1998, Rueda et al., 2005, 2006; Rueda, 2008). Knowledge of mosquito larval habitats is essential in designing effective mosquito control programs and disease prevention strategies. The most practical method to reduce a local population of pestiferous mosquitoes is to eliminate their habitats, such as discarded tires, clogged gutters, and stumped tree holes. Larval habitats that are not possibly eliminated can be modified (cleaning clogged ditches, open water management in salt marshes, etc.). Several appropriate techniques can be applied such as using biological control agents (predatory fish and insects and microbials), selected larvicides for permanent breeding habitats (lakes, ponds, and ground pools), or other environmental modification methods to control mosquito breeding.

1.3 Mosquito-borne diseases

Mosquito bites affect the health of humans and animals more than any other arthropod pest throughout the world. Mosquito-borne diseases are transmitted to humans by the bite of an infected mosquito by a parasite as in the case of malaria or by a virus as in the case of dengue, yellow fever, chikungunya, West Nile virus, Zika virus, etc. In addition, mosquitoes not only can transmit diseases that afflict humans but also can transmit serious pathogens that dogs and horses are susceptible to, such as dog heartworm, eastern equine encephalitis, and western equine encephalitis. Mosquito-borne diseases in Texas will be discussed in detail in Chapter 11.

References and further reading

Belkin, J.N., 1962. The Mosquitoes of the South Pacific (Diptera, Culicidae). University of California Press, Berkeley, CA. vol. 1, 608 pp., vol. 2, 412 pp.

Belkin, J.N., Heineman, S.J., 1975. *Psorophora (Janthinosoma) mathesoni*, sp. nov. for "varipes" of the Southeastern USA. Mosq. Syst. 7, 363–366.

Carpenter, S.J., LaCasse, W.J., 1955. Mosquitoes of North America (North of Mexico). University of California Press, Berkeley, CA. 495 p.

Darsie, R.F., Ward, R.A., 2005. Identification and Geographical Distribution of the Mosquitoes of North America, North of Mexico. University Press of Florida, Gainesville, FL. 383 pp.

Harbach, R.E., Knight, K.L., 1980. Taxonomists' Glossary of Mosquito Taxonomy. Plexus Publishing, Inc., Marlton, NJ. 415 pp.

Harbach, R.E., Knight, K.L., 1982. Corrections and additions to taxonomists' glossary of mosquito taxonomy. Mosq. Syst. (for 1981) 13, 201–217.

Harrison, B.A., Varnado, W., Whitt, P.B., Goddard, J., 2008. New diagnostic characters for females of *Psorophora (Janthinosoma)* species in the United States, with notes on *Psorophora mexicana* (Bellardi) (Diptera: Culicidae). J. Vector Ecol. 33, 232–237.

Huang, Y.M., Rueda, L.M., 2017. Pictorial keys to the sections, groups, and species of the *Aedes (Finlaya)* in the Afrotropical Region (Diptera: Culicidae). Zootaxa 4221 (1), 131–141.

Huang, Y.M., Rueda, L.M., 2018. A pictorial key to the species of the *Aedes (Mucidus)* in the Afrotropical Region (Diptera: Culicidae). Proc. Entomol. Soc. Wash. 120 (4), 798–806.

Laird, M., 1998. The Natural History of Larval Mosquito Habitats. Academic Press, London, UK. 555 pp.

Rattanarithikul, R., Harrison, B.A., Panthusiri, P., Coleman, R.E., 2005. Illustrated keys to the mosquitoes of Thailand. I. Background; geographic distribution; lists of genera, subgenera, and species; and a key to the genera. Southeast Asian J. Trop. Med. Public Health 36, 1–80. Suppl. 1.

Reinert, J.F., Harbach, R.E., Kitching, I.J., 2004. Phylogeny and classification of Aedini (Diptera: Culicidae), based on morphological characters of all life stages. Zool. J. Linnean Soc. 142, 289–368.

Reinert, J.F., Harbach, R.E., Kitching, I.J., 2006. Phylogeny and classification of *Finlaya* and allied taxa (Diptera: Culicidae: Aedini) based on morphological data from all life stages. Zool. J. Linnean Soc. 148, 1–101.

Reinert, J.F., Harbach, R.E., Kitching, I.J., 2008. Phylogeny and classification of *Ochlerotatus* and allied taxa (Diptera: Culicidae: Aedini) based on morphological data from all life stages. Zool. J. Linnean Soc. 153, 29–114.

Reinert, J.F., Harbach, R.E., Kitching, I.J., 2009. Phylogeny and classification of tribe Aedini (Diptera: Culicidae). Zool. J. Linnean Soc. 157, 700–794.

Rueda, L.M., 2004. Pictorial keys for the identification of mosquitoes (Diptera: Culicidae) associated with dengue virus transmission. Zootaxa 589, 1–60.

Rueda, L.M., 2008. Global diversity of mosquitoes (Insecta: Diptera: Culicidae) in freshwater. Hydrobiologia 595, 477–487.

Rueda, L.M., Rutledge, L.C., Gupta, R.K., 1998a. Effect of skin abrasions on the efficacy of the repellent deet against *Aedes aegypti*. J. Am. Mosq. Control Assoc. 14, 178–182.

Rueda, L.M., Stockwell, S.A., Pecor, J.E., Gaffigan, T., 1998b. Key to the Mosquito Genera of the World (INTKEY Module). Walter Reed Biosystematics Unit, Smithsonian Institution, Washington, DC. [CD]. Also, in The Diptera Dissemination Disk, vol. 1. North American Dipterists Society, Washington, DC.

Rueda, L.M., Harrison, B.A., Brown, J.S., Whitt, P.B., Harrison, R.Y., Gardner, R.C., 2001. Evaluation of 1-octen-3-ol, carbon dioxide, and light as attractants for mosquitoes associated with two distinct habitats in North Carolina. J. Am. Mosq. Control Assoc. 17, 61–66.

Rueda, L.M., Iwakami, M., O'Guinn, M., Mogi, M., Prendergast, B.F., Miyagi, I., Toma, T., Pecor, J.E., Wilkerson, R.C., 2005. Habitats and distribution of *Anopheles sinensis* and associated *Anopheles* Hyrcanus group in Japan. J. Am. Mosq. Control Assoc. 21, 458–463.

Rueda, L.M., Kim, H.C., Klein, T., Pecor, J., Li, C., Sithiprasasna, R., Debboun, M., Wilkerson, R.C., 2006. Distribution and larval habitat characteristics of *Anopheles* Hyrcanus Group and related mosquito species (Diptera: Culicidae) in South Korea. J. Vector Ecol. 31, 199–206.

Stojanovitch, P., Scott, H.G., 1997. Illustrated Key to the Adult Male Mosquitoes of America (North of Mexico). Harold G. Scott, Louisiana. 121 pp.

Walton, C., Sharpe, R.G., Pritchard, S.J., Thelwell, N.J., Butlin, R.K., 1999. Molecular identification of mosquito species. Biol. J. Linnaean Soc. 68, 241–256.

Wilkerson, R.C., Linton, Y.-M., Fonseca, D.M., Schultz, T.R., Price, D.C., Strickman, D.A., 2015. Making mosquito taxonomy useful: a stable classification of tribe Aedini that balances utility with current knowledge of evolutionary relationships. PLoS One 10 (7), e0133602.

WRBU, 2018. Walter Reed Biosystematics Unit. Mosquito Identification Resources. Smithsonian Institution, Washington, DC. http://www.wrbu.org/VecID_MQ.html. (Accessed September 12, 2018).

Chapter 2

Mosquito Species of Texas

Martin Reyna Nava, Mustapha Debboun

Mosquito and Vector Control Division, Harris County Public Health, Houston, TX, United States

Chapter Outline

Abbreviations

Ae.	Aedes
AMCA	American Mosquito Control Association
An.	*Anopheles*
CALV	California serogroup viruses
CDC	Centers for Disease Control and Prevention
CEV	California Encephalitis virus
CHIKV	chikungunya virus
Cq.	*Coquillettidia*
Cs.	*Culiseta*
CVV	Cache Valley virus
Cx.	*Culex*
De.	*Deinocerites*
DENV	dengue virus
DSHS	Department of State Health Services

Mosquitoes, Communities, and Public Health in Texas. https://doi.org/10.1016/B978-0-12-814545-6.00002-X

EEEV	Eastern equine encephalitis
Hg.	*Haemagogus*
HJV	Highland J virus
JCV	Jamestown Canyon virus
JEV	Japanese encephalitis virus
KEYV	Keystone virus
LACV	La Crosse encephalitis virus
Ma.	*Mansonia*
MAYV	Mayaro virus
Or.	*Orthopodomyia*
P.	*Plasmodium*
POTV	Potosi virus
Ps.	*Psorophora*
RVFV	Rift Valley fever
SAV	San Angelo virus
SLEV	Saint Louis encephalitis
TAHV	Tahyna virus
TENV	Tensaw virus
TMCA	Texas Mosquito Control Association
TVTV	Trivittatus virus
TX	Texas
Tx.	*Toxorhynchites*
Ur.	*Uranotaenia*
VEEV	Venezuelan equine encephalitis
WEEV	Western equine encephalitis
WNV	West Nile virus
WRBU	Walter Reed Biosystematics Unit
WYOV	Wyeomyia virus
YFV	yellow fever virus
ZIKV	Zika virus

2.1 Introduction

The physiography of Texas (TX) is of utmost importance in understanding the ecology of mosquitoes and their distribution patterns. The physiographic regions in Texas are the Atlantic Plain (Gulf Coastal Plain), the Interior Plains (Central Lowland or North Central Plains and Great Plains), and Intermontane Plateaus (basin and range or mountains and basins). These three divisions can be further divided into 26 regions (Wikipedia, 2018a,b). In addition, Texas is divided into 254 geographic regions for administrative purposes or jurisdictions denominated counties. They are the intermediate tier of state government, between the statewide tier and the immediately local government tier (typically a city, town/borough or village/township) (Wikipedia, 2018c).

About 60 submitters from approximately 40 counties report or submit mosquito specimens to the Texas Department of State Health Services (DSHS) for arbovirus testing. At present, there are 19 organized mosquito control districts in Texas and 15 according to the Texas Mosquito Control Association (TMCA, 2001a).

The total number of mosquito species recognized in the United States (US) has increased from 170 (Darsie and Ward, 2005) to approximately 190 including Alaska and Hawaii (WRBU, 2019). This number includes the discoveries of *Anopheles grabhamii* Theobald (Darsie et al., 2002), and *Aedes pertinax* Grabham (Shroyer et al., 2015). However, this number increased with the recent discoveries of *Culex* (*Melanoconion*) *panocossa* Dyar (Blosser and Burkett-Cadena, 2017), *Aedeomyia squamipennis* (Lynch Arribalzaga) (Burket-Cadena and Blosser, 2017), and the possibility of *Aedes notoscriptus* (Skuse) in California (Daily Breeze, 2018) to 193. The American Mosquito Control Association (AMCA) mentions that there are *about 200 species* (AMCA, 2019a) and then states on a subsequent webpage a total of 176, including *An. grabhamii* (AMCA, 2019b). Eighty-seven species occur in Texas and are contained in 12 genera: *Anopheles* Meigen, *Aedes* Meigen, *Culex* Linnaeus, *Culiseta* Felt, *Coquillettidia* Dyar, *Deinocerites* Theobald, *Haemagogus* Williston, *Mansonia* Blanchard, *Orthopodomyia* Theobald, *Psorophora* Robineau-Desvoidy, *Toxorhynchites* Theobald, and *Uranotaenia* Lynch Arribalzaga. (Family Culicidae Meigen: subfamilies Anopheline Grassi and Culicinae Meigen [Tribes Aedini, Culicini, Culisetini, Mansoniini, Orthopodomyiini, Toxorhynchitini, and Uranotaeniini] (Table 2.1).)

TABLE 2.1 Mosquitoes of Texas: 87 species (Diptera/Culicidae).

I) Subfamily anopheline Grassi

    I. Genus **ANOPHELES** Meigen

        • Subgenus *Anopheles* Meigen

          1) *atropos* Dyar and Knab

          2) *barberi* Coquillett

          3) *bradleyi* King

          4) *crucians* Wiedemann

          5) *franciscanus* McCraken

          6) *freeborni* Aitken

          7) *judithae* Zavortink

          8) *pseudopunctipennis* Theobald

          9) *punctipennis* (Say)

          10) *quadrimaculatus* Say

          11) *smaragdinus* Reinert

          12) *walkeri* Theobald

        • Subgenus *Nyssorhynchus* Blanchard

          13) *albimanus* Wiedemann

II) Subfamily Culicinae Meigen

    • Tribe Aedini Neveu-Lemaire

    II. Genus **AEDES** Meigen

        • Subgenus *Aedimorphus* Theobald

          14) *vexans* (Meigen)

        • Subgenus *Georgecraigius* Reinert, Harbach and Kitching

          15) *epactius* Dyar & Knab

         • Subgenus *Lewnielsenius* Reinert, Harbach and Kitching

          16) *muelleri* Dyar

        • Subgenus *Ochlerotatus* Lynch Arribalzaga

          17) *atlanticus* Dyar and Knab

          18) *bimaculatus* (Coquillett)

          19) *campestris* Dyar and Knab

          20) *canadensis* (Theobald)

          21) *dorsalis* (Meigen)

          22) *dupreei* (Coquillett)

          23) *fulvus pallens* Ross

          24) *grossbecki* Dyar and Knab

          25) *infirmatus* Dyar and Knab

          26) *mitchellae* (Dyar)

          27) *nigromaculis* (Ludlow)

          28) *scapularis* (Rondani)

          29) *sollicitans* (Walker)

          30) *sticticus* (Meigen)

          31) *taeniorhynchus* (Wiedemann)

          32) *thelcter* Dyar

          33) *thibaulti* Dyar and Knab

          34) *tormentor* Dyar and Knab

          35) *trivittatus* (Coquillett)

        • Subgenus *Protomacleaya* Theobald

          36) *brelandi* Zavortink

          37) *hendersoni* Cockerell

          38) *triseriatus* (Say)

          39) *zoosophus* Dyar and Knab

        • Subgenus *Stegomyia* Theobald

          40) *aegypti* (Linnaeus)

          41) *albopictus* (Skuse)

III. Genus **HAEMAGOGUS** Williston

        • Subgenus *Haemagogus* Williston

          42) *equinus* Theobald

IV. Genus **PSOROPHORA** Robineau-Desvoidy

        • Subgenus *Grabhamia* Theobald

          43) *columbiae* (Dyar and Knab)

          44) *discolor* (Coquillett)

          45) *signipennis* (Coquillett)

        • Subgenus *Janthinosoma* Lynch Arribálzaga

          46) *cyanescens* (Coquillett)

          47) *ferox* (von Humboldt)

          48) *horrida* (Dyar and Knab)

          49) *longipalpus* Randolph and O'Neill

          50) *mathesoni* Belkin and Heinemann

          51) *mexicana* (Bellardi)

        • Subgenus *Psorophora* Robineau-Desvoidy

          52) *ciliata* (Fabricius)

          53) *howardii* Coquillett

Continued

TABLE 2.1 Mosquitoes of Texas: 87 species (Diptera/Culicidae).—cont'd

- Tribe Culicini Meigen

  V. Genus **CULEX** Linnaeus

  - Subgenus *Culex* Linnaeus

    54) *chidesteri* Dyar

    55) *coronator* Dyar and Knab

    56) *declarator* Dyar and Knab

    57) *erythrothorax* Dyar

    58) *interrogator* Dyar and Knab

    59) *nigripalpus* Theobald

    60) *quinquefasciatus* Say

    61) *restuans* Theobald

    62) *salinarius* Coquillett

    63) *stigmatosoma* Dyar

    64) *tarsalis* Coquillett

    65) *thriambus* Dyar

  - Subgenus *Melanoconion* Theobald

    66) *abominator* Dyar and Knab

    67) *erraticus* (Dyar and Knab)

    68) *peccator* Dyar and Knab

    69) *pilosus* (Dyar and Knab)

  - Subgenus *Neoculex* Dyar

    70) *apicalis* Adams

    71) *arizonensis* Bohart

    72) *territans* Walker

  VI. Genus **DEINOCERITES** Theobald

    73) *mathesoni* Belkin and Hogue

    74) *pseudes* Dyar and Knab

- Tribe Culisetini Belkin

  VII. Genus **CULISETA** Felt

  - Subgenus *Climacura* Howard, Dyar and Knab

    75) *melanura* (Coquillett)

  - Subgenus *Culiseta* Felt

    76) *incidens* (Thomson)

    77) *inornata* (Williston)

- Tribe Mansoniini Belkin

  VIII. Genus **COQUILLETTIDIA** Dyar

  - Subgenus *Coquillettidia* Dyar

    78) *perturbans* (Walker)

  IX. Genus **MANSONIA** Blanchard

    - Subgenus *Mansonia* Blanchard

    79) *titillans* (Walker)

- Tribe Orthopodomyiini Belkin, Heinemann and Page

  X. Genus **ORTHOPODOMYIA** Theobald

    80) *alba* Baker

    81) *kummi* Edwards

    82) *signifera* (Coquillett)

- Tribe Toxorhynchitini Lahille

  XI. Genus **TOXORHYNCHITES** Theobald

  - Subgenus *Lynchiella* Lahille

    83) *moctezuma* (Dyar and Knab)

    84) *rutilus septentrionalis* (Dyar and Knab)

- Tribe Uranotaeniini Lahille

  XII. Genus **URANOTAENIA** Lynch Arribalzaga

  - Subgenus *Pseudoficalbia* Theobald

    85) *anhydor syntheta* Dyar and Shannon

  - Subgenus *Uranotaenia* Lynch Arribalzaga

    86) *lowii* Theobald

    87) *sapphirina* (Osten-Sacken)

Traditional classification (Knight and Stone, 1977; Knight, 1978; Harbach and Kitching, 1998; Wilkerson et al., 2015; WRBU, 2019).

Culex quinquefasciatus Say tops the list in Texas with 59% followed by *Cx. tarsalis* Coquillett (55%), *Cx. coronator* Dyar and Knab (50%), *Ae. vexans* (Meigen) (46%), *An. quadrimaculatus* Say (44%), *An. pseudopunctipennis* Theobald and *Ps. columbiae* (Dyar and Knab) with 43% each, *Ae. aegypti* (Linnaeus) (42.5%), *An. punctipennis* (Say) (41.7%), *Ps. cyanescens* (Coquillett) (40.5%), *Cx. salinarius* Coquillett (40%), *Cx. erraticus* (Dyar and Knab) (39%), and *Cs. inornata* (Williston) and *Ps. signipennis* (Coquillett) (37.8%) and *Ae. albopictus* (Skuse) with 36%.

There are 19 species of *Culex* mosquitoes in Texas. However, when the species from the subgenus *Culex (Melanoconion)* Theobald that occur in Texas (*Cx. abominator* Dyar and Knab, *Cx. erraticus* (Dyar and Knab), *Cx. peccator* Dyar and Knab, and *Cx. pilosus* Dyar and Knab) are not identified to species (due to the complex taxonomy, incomplete descriptions of many species, and misidentification of specimens in the subgenus *Melanoconion*), they are lumped and are reported as "*Culex (Melanoconion)* sp." as is the case in the Texas A&M AgriLife Extension Agricultural and Environmental Safety Department website (Texas A&M, 2013). Moreover, a newly discovered species in Texas, *Cx. arizonensis* Bohart, is not included in the list as this species was not discovered until 2002 (Reeves and Darsie, 2003), which brings the total to 19 not including "*Culex (Melanoconion)* sp."

Furthermore, TMCA lists the number of species (TMCA, 2001b) recorded in Texas based on Fournier et al. (1989). However, TMCA list is also missing *Cx. arizonensis* and *An. smaragdinus* Reinert. The discrepancies between both lists continue with the number of *Psorophora* species found in Texas as both include *Ps. mathesoni* Belkin and Heinemann and *Ps. varipes* (Coquillett), wherein the only species present in the Southeastern United States is *Ps. mathesoni* (Belkin and Heinemann, 1975; Harrison et al., 2008), thus bringing the total number of *Psorophora* species to 11 and not 12 as listed on both web pages.

Another difference between both lists of species is that TMCA includes *Orthopodomyia kummi* Edwards, a species that is not included in Fournier's list, which is the basis for TMCA's records. This species is neither listed by Texas A&M nor by Darsie and Ward (2005) where the latter documents 85 species in Texas. Zavortink (1972) documents the presence of *Or. kummi* in the Chisos Mountains Basin in association with *Ae. brelandi* Zavortink. The recent collection and identification of the larvae of *Or. kummi* collected within the Chisos Mountain Basin, TX in 2007 (Byrd et al., 2009), further confirms its presence as documented previously, and the recent report of *Tx. moctezuma* in Brownsville, TX (Uejio et al., 2014) increases the total number of species in Texas to 87. There are publications of other species not included in any of the lists or publications previously mentioned: *Deinocerites spanius* (Dyar and Knab) in Brownsville, TX in 1939 reported by Fisk (1941), which was also reported by McGregor and Eads (1943), although Belkin and Hogue (1959) recognized those specimens reported by the former as *mathesoni* Belkin and Hogue. *Deinocerites epitedeus* (Knab) was collected in Harlingen, Cameron County, TX by Rueger and Druce (1950), but it was believed to be an intrusion from Central America or a misidentification (Carpenter and LaCasse, 1955).

The distribution of the mosquito species in Texas has been compiled by those records in literature and the provision of records from vector control personnel. As a result of comprising these data for Texas, nearly 87% of the counties have contributed at least one specimen in recording species that occur in those counties, while the other 13% have not reported any specimens to the abundance and diversity of mosquito species of Texas (Fig. 2.88 and Appendix 2.1). In addition, some counties may have seen an increase in the number of species they had initially provided, as some species reported in literature had not been accounted for previously on their own records. Most of the lists submitted by different districts, counties, or municipalities provided what they have collected from their trapping efforts throughout the years or from an old list they had been operating with at their job location. For example, there is a record of *Ae. thelcter* Dyar in Harris County, TX that is not accounted for in the recent checklist of species in Harris County (Nava and Debboun, 2016), which would bring the total to 57 instead of the previous 56 published from Harris County. Such record is based on an unpublished 1971–75 mosquito light trap summary conducted at the Brooks Air Force Base, TX (Fournier et al., 1989).

Only 10 counties report more than half of the total number of species documented in Texas: Bexar (59), Harris (57), Cameron (54), Dallas (50), Travis (48), Galveston (47), Tarrant (47), Jefferson (45), Wichita (45), and Montgomery (44). In addition to these 10 counties, only five more report above 37 number of species, that is, Hidalgo (39), Lamar (39), Orange (39), San Patricio (37), and Val Verde with 37.

2.1.1 How to read this chapter

The mosquito species of Texas are listed taxonomically in Table 2.1 following the traditional classification of Knight and Stone (1977), Knight (1978), Harbach and Kitching (1998), Wilkerson et al. (2015), and Walter Reed Biosystematics Unit (WRBU, 2019). Brief general notes are provided for some genera, followed by general morphological characters of each species not intended to be used as identification keys, their bionomics and medical importance, and their distribution in Texas by county represented by their corresponding distribution maps. In addition, Appendix 2.1 is provided depicting the presence or absence of the species in each Texas county.

2.2 *Anopheles* Meigen

Mosquitoes belonging to this genus are distinct from other mosquitoes in the family *Culicidae* Meigen especially in the immature stages where eggs have lateral floats and larvae have palmate setae on their abdomen and lack the presence of the respiratory siphon characteristic of other mosquito species, thus allowing them to lie parallel underneath the water surface to obtain their oxygen via the spiracles on the eight abdominal segment. Anopheline adults have palps that are as long as the proboscis, and their abdomen is raised at an angle when resting.

There are over 400 species of *Anopheles* worldwide and 13 have been recorded in Texas including some of the *Anopheles crucians* and the *An. quadrimaculatus* complexes. They are crepuscular and nocturnal; some are anthropophilic and others are zoophilic. Some are responsible for the transmission of the protozoans causing malaria in humans, animals, and birds.

2.2.1 Subgenus *Anopheles* Meigen

Anopheles atropos (Dyar and Knab, 1906)

A medium-sized mosquito with a long, dark proboscis and entirely dark palps or with pale rings at the apices of the distal segments. The scutum is dark brown to dark reddish brown with short golden-brown hairs and larger brown setae. The abdomen is dark brown to black with dark hairs. Wing scales are dark and narrow with four indistinct dark spots, and the legs are entirely dark.

Bionomics. This is a saltwater mosquito that develops in permanent saltwater pools or in marshes where salt content varies from brackish to undiluted seawater. Adults are abundant near salt marshes where they breed. They overwinter in the egg stage and have multiple broods per year. Females are fierce biters active during the day, even under direct sunlight, and at night. They feed on large mammals including humans and enter dwellings to feed. It has been collected in large numbers in light traps in June and July in Galveston County, TX (Texas State Health Department, 1944). Although it has been infected with *Plasmodium vivax*, it is not a natural vector of malaria due to its restricted distribution to saltmarshes (Carpenter et al., 1946).

Distribution. In Texas, it has been found in 10 counties (Fig. 2.1).

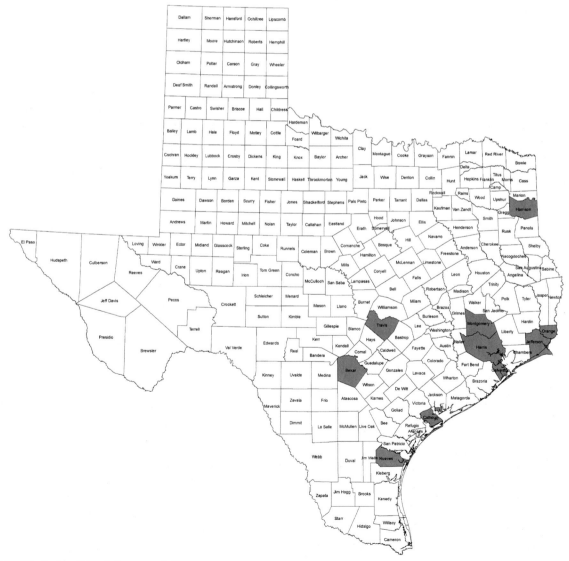

FIG. 2.1 Distribution of *Anopheles atropos* in Texas.

Anopheles barberi (Coquillett, 1903) and Anopheles judithae (Zavorthink, 1969)

These are closely related allopatric species belonging to one of three primary groups of tree hole breeding *Anopheles*. *Anopheles barberi* is a small, grayish to shiny brown mosquito with gray palps and proboscis. The mesonotum has dark setae, half as long as the width of the scutum. The abdomen is brown with dark brown hairs. Their wing scales are slightly broadened, gray and without the typical spots of other *Anopheline* species. Their legs are entirely dark. The acrostichal setae of *An. judithae* are amber, and it has less setae on the proepisternum and forecoxa than *An. barberi*.

Bionomics. *Anopheles barberi* is a tree hole breeding mosquito. Eggs are found singly on the surface of water in containers. Larvae have been collected from stump holes and found living in cavities lined with wood, tree holes, and artificial containers. They are multivoltine and found any time of the year when habitats are flooded. Larvae have been observed to prey on other mosquito larvae. Adults usually emerge in June and are found resting underneath bridges, culverts, and buildings in or near wooded areas. Females are persistent biters and may enter houses around their breeding sites. *Anopheles judithae* occurs at elevations between 2200 and 6000 m in evergreen forests at higher altitude and restricted to narrow tree lines along watercourses. They overwinter in the larval stage and are frequently found in permanent tree holes rather than temporary ones. Although *An. barberi* has been infected with *P. vivax*, it is not known to be involved in the transmission of any human pathogens (Carpenter et al., 1946).

Distribution. In Texas, *An. barberi* has been found in 11 counties (Fig. 2.2), and *An. judithae* has been collected in the Southern Trans-Pecos area at the Big Bend National Park in Brewster County (Fig. 2.3).

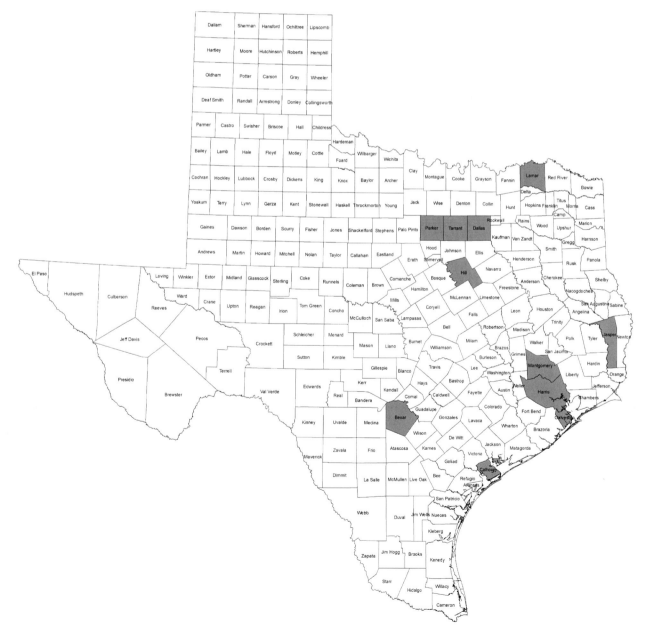

FIG. 2.2 Distribution of *Anopheles barberi* in Texas.

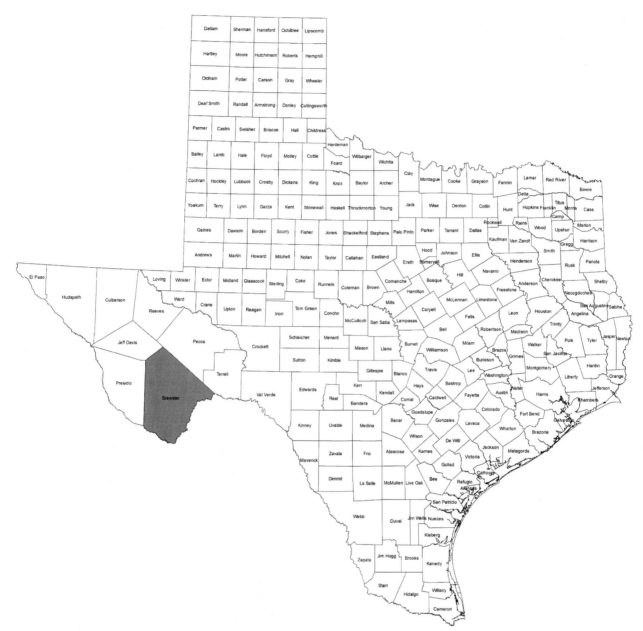

FIG. 2.3 Distribution of *Anopheles judithae* in Texas.

Anopheles crucians complex: *Anopheles bradleyi* (King, 1939) and *Anopheles crucians* (Wiedemann, 1828)

Anopheles bradleyi and *An. crucians* are two of the seven mosquito species (*An. crucians* sp. A, B, C, D, E; *An. georgianus*; and *An. bradleyi*) belonging to the *An. crucians* complex. These species are very difficult to tell apart, and four of them have not been completely described. However, *An. bradleyi* and species C are separate and distinct species (Wilkerson et al., 2004). Adults are medium to large with wings having distinctive alternate patterns of light and dark scales and patches of white scales on the distal end of the wing. Their proboscis is long and dark as are the palps where segment 3 has white scales basally, segment 4 basally and apically, and segment 5 is entirely white. The stem of vein 5 of *An. bradleyi* is entirely white scaled.

    Bionomics. *Anopheles bradleyi* larvae have been found in pools, margin of ponds and marshes in brackish water near the coast, pools containing emergent vegetation, and along with *An. crucians* in fresh water. Larvae of *An. crucians* can also be found in ponds, lakes, swamps, and permanent and semipermanent pools associated with aquatic vegetation and prefer habitats with acid water and found throughout the winter in the Southern United States. Adults of *An. bradleyi* can fly considerable distances from their breeding sites and are active at night, dusk, and dawn feeding on mammals including humans. They overwinter in the adult stage and have multiple broods per year. The biology of *An. crucians* is similar to that of *An. bradleyi* and can also feed on birds. Adults can enter houses to feed but are mainly exophagic. They rest during the day underneath houses, bridges, culverts, and other similar shelters. This is more commonly a fall and winter species in the southern United States. *Anopheles bradleyi* has been infected successfully with *P. falciparum* in the laboratory, but is not believed to be a natural vector of malaria. *Anopheles crucians*, on the other hand, has been found naturally infected with malaria, although it is not considered an important vector (Carpenter et al., 1946). It has also been found to be infected with EEE and WNV (Varnado et al., 2012).

   Distribution. In Texas, *An. bradleyi* has been found in 6 counties (Fig. 2.4) and *An. crucians* in 84 counties (Fig. 2.5).

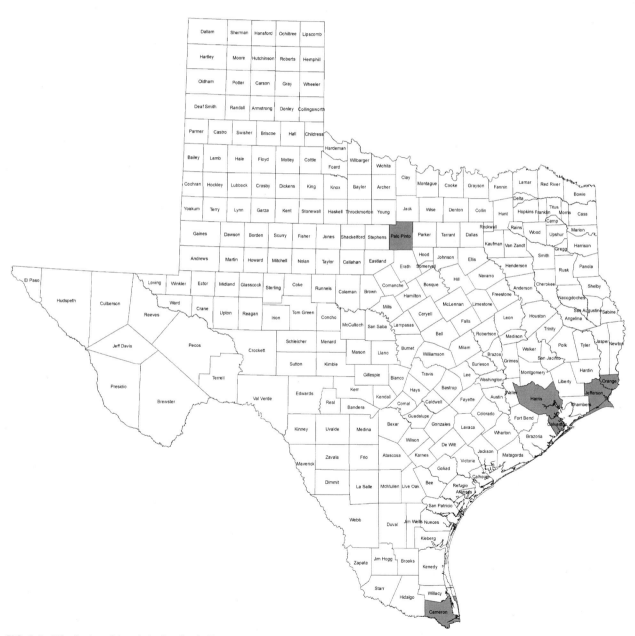

FIG. 2.4   Distribution of *Anopheles bradleyi* in Texas.

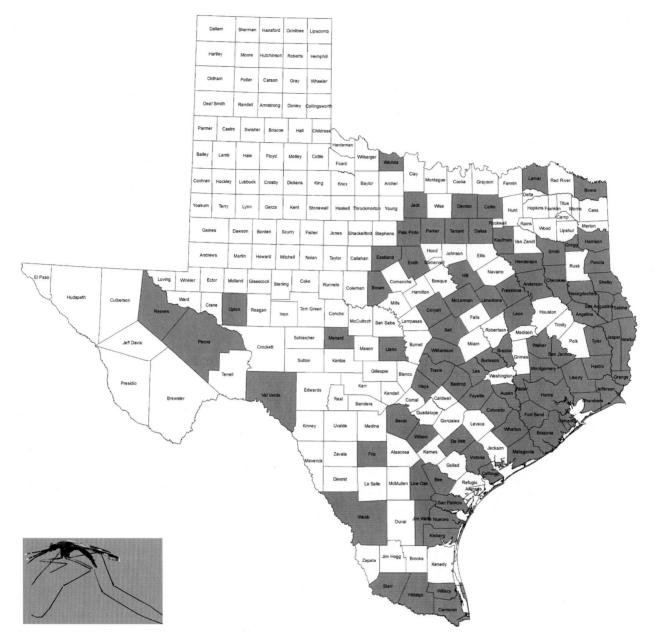

FIG. 2.5 Distribution of *Anopheles crucians* in Texas.

Anopheles franciscanus (McCracken, 1904)

A medium-sized mosquito similar to *An. pseudopunctipennis* Theobald with a dark, long proboscis and palps that are black with apices of segments 2, 3, and 4 white ringed, and the tip of last segment is dark, which differentiates it from *An. pseudopunctipennis*.

Bionomics. It is common in arid regions. Eggs have floats and can rest in the water vertically and horizontally. Larvae are found in shallow pools during the dry season and in drainage ditches of irrigated areas on the western tip of Texas. Sites are associated with green algae and in pools open to sunlight. They are also found in brackish water and artificial containers. Adults are crepuscular and feed on mammals and fowl. They are exophagic, seldom enter dwellings, and rarely feed on humans. They fly from their larval sites to find resting places after a blood meal.

Distribution. In Texas, it has been found in seven counties (Fig. 2.6).

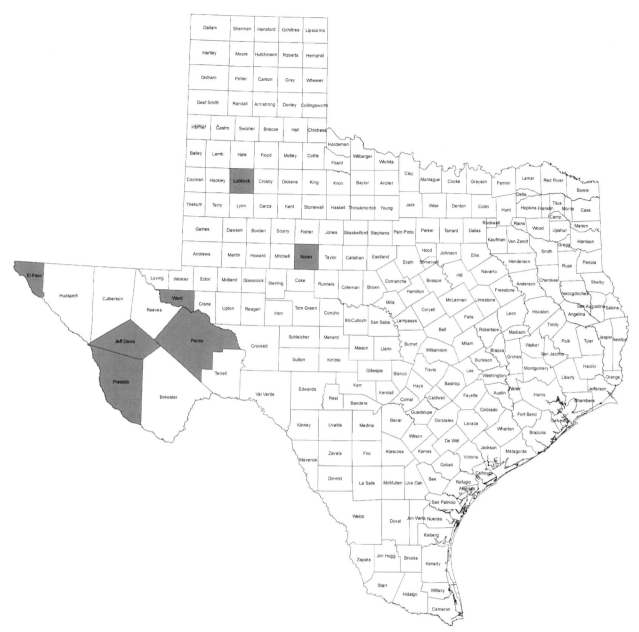

FIG. 2.6 Distribution of *Anopheles franciscanus* in Texas.

Anopheles freeborni (Aitken, 1939)

It is a member of the North American *maculipennis* complex. A medium-sized mosquito similar to *An. maculipennis* Meigen and the female is indistinguishable from *An. quadrimaculatus* Say. Its larvae are also indistinguishable from *An. punctipennis* (Say). It has a dark proboscis and palps with raised scales on the basal part. The scutum is brown to black with grayish-brown median and submedian stripes. The abdomen is brown to black; legs are dark with femora and tibiae with pale scales at the tip. Wing scales are narrow and dark and sometimes forming four darker spots.

    Bionomics. This is a multivoltine species with generations starting in April to November. Eggs are laid singly in shallow pools. Larvae are found in clear seepage water with grass or cattails, roadside pools, rice fields, drainage ditches rich in algae, cold to hot springs, and lakes in clear shallow grassy water. Adults can disperse significant distances from their larval locations during the fall season. They rest during the day in dark and cool places inside dwellings, culverts, and underneath bridges. Females are active at dusk and at night but can feed during the day as well. They feed readily on humans inside houses but are opportunistic in nature feeding on any animal host available. It is a main vector of malaria in arid and semiarid regions of the United States. WEE has also been isolated in collected mosquitoes (Carpenter and LaCasse, 1955).

Distribution. In Texas, it has been found in four counties (Fig. 2.7).

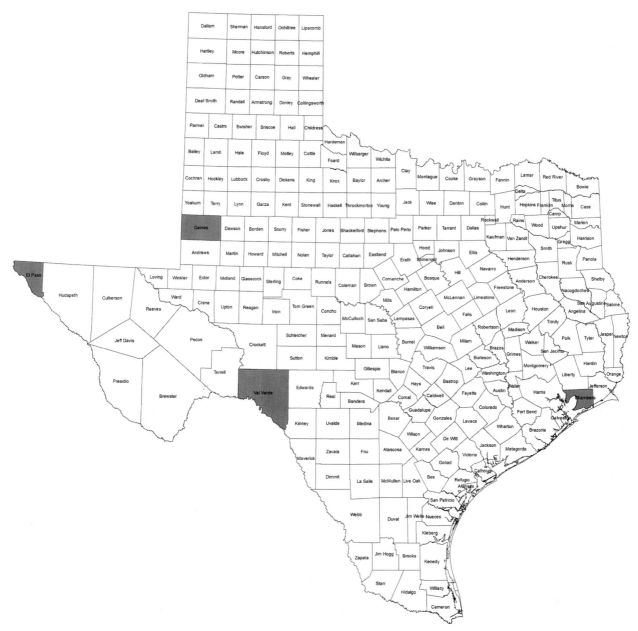

FIG. 2.7 Distribution of *Anopheles freeborni* in Texas.

Anopheles pseudopunctipennis (Theobald, 1901)

A medium-sized mosquito with palps as long as the proboscis and with pale rings on palp tips. The scutum has a median frosted stripe with yellowish-white scales and pale yellow hairs, dark brown laterally with long, dark setae. Blunt or rounded dark brown to black abdomen with dark legs, where the tips of femora and tibiae are pale and the hind tarsi are entirely dark. The wings have two pale patches on the anterior margin and pale areas along the posterior fringe of the wing. It is morphologically similar to *An. punctipennis* but with more pale scales on their wings.

Bionomics. Larvae are found in clear sunlit water rich in algae and in slow moving streams where they rest on floating debris during the dry season. The lakes and springs at San Marcos below the Balcones Fault are good breeding places where clear water seeps and where streams and ponds with limestone bottom and emergent vegetation are found. They have multiple generations that are continuous and occur at the end of the dry season. They overwinter in the adult stage. They are domestic, entering houses to shelter early in the mornings and go outdoors at dusk to feed on humans, cattle, pig, sheep, fowl, and horses but can also feed indoors. It has been experimentally infected with *P. vivax* and *P. falciparum* and has been found naturally infected with malaria parasites. It is an important vector of malaria in Mexico and parts of South America. It is not known to play an important role in the transmission of malaria in the United States (Carpenter et al., 1946).

  Distribution. In Texas, it has been found in 110 counties (Fig. 2.8).

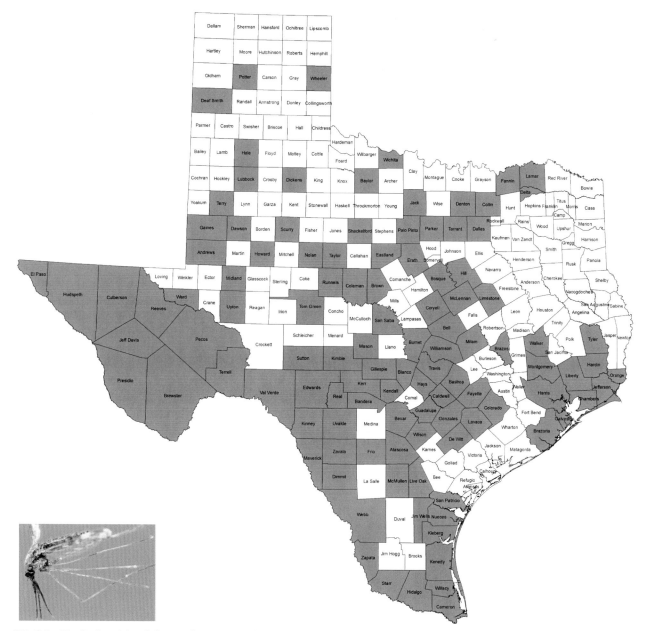

FIG. 2.8 Distribution of *Anopheles pseudopunctipennis* in Texas.

Anopheles punctipennis (Say, 1823)

A medium-sized mosquito with palps that are dark and as long as the proboscis. The scutum has a broad median frosted stripe with short pale yellow hairs; dark brown laterally with larger dark setae. The abdomen is blunt or rounded with pale and dark setae. The legs are dark scaled with the femora and tibiae having white scales at the tips and the hind tarsi are entirely dark. There are two pale patches on the anterior wing margin and vein 1 displays a patch of white scales between two patches of dark scales.

Bionomics. Larvae are found in clear, cool water in a variety of habitats including ponds, pools, streams, borrow pits, and margins of flowing streams, and in artificial containers. They have multiple generations per year and overwinter in the adult stage hibernating in well-protected shelters and are regarded as an outdoor species that very seldom enters houses. Adult females feed on humans and large mammals after dusk and at night but may bite during daytime in woodlands and daytime resting places. In spite of the successful experimental infection with malaria parasites (*P. vivax*, *P. falciparum*, and *P. malariae*), it is not known to be a natural vector for malaria (Carpenter et al., 1946).

Distribution. In Texas, it has been found in 106 counties (Fig. 2.9).

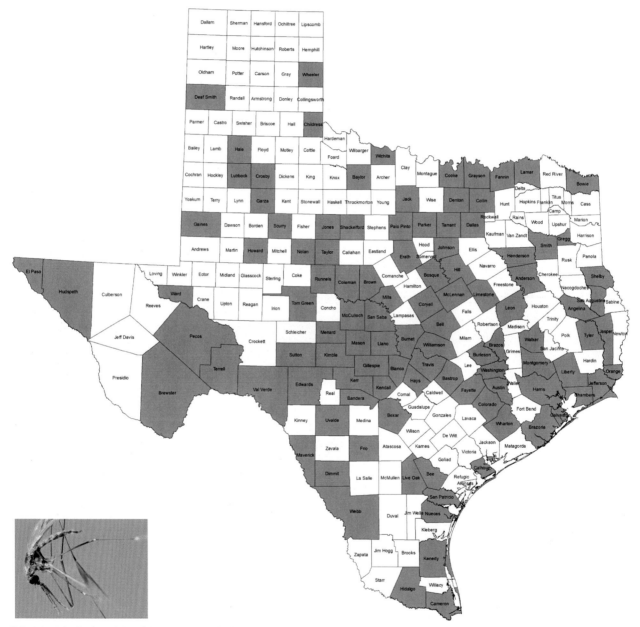

FIG. 2.9 Distribution of *Anopheles punctipennis* in Texas.

Anopheles quadrimaculatus complex: *Anopheles quadrimaculatus* (Say, 1824) and *Anopheles smaragdinus* (Reinert, 1997)

The *Anopheles quadrimaculatus* complex comprises five species that overlap in range, and only *An. quadrimaculatus* and *An. smaragdinus* have been recorded in Texas. All species are very similar and can only be distinguished by slight morphological differences. *Anopheles quadrimaculatus* is a medium-sized mosquito that can be identified by the four darker spots of dense scales on their wings where vein 1 is entirely dark. Their legs are entirely dark, and the femora and tibiae are tipped with pale scales. It has more setae on the scutal fossa, prealar area, and the interocular area than *An. smaragdinus*.

Bionomics. Eggs are deposited singly on the surface of water in patterns of vegetation chosen more frequently than others. Larvae occur in fresh water, slow moving streams, canals, ponds, and lakes containing emerging vegetation or floating debris, rice fields, and occasionally in temporary pools and permanent water swamps. Adults have multiple generations and overwinter in the hibernating adult stage. They are very active at dusk and at night in search of a blood meal and before dawn where they go into daytime resting places. They feed on humans, wild, and domestic animals. The species in this group have been infected successfully with *P. vivax*, *P. falciparum*, and *P. malariae*, and infection in wild-caught mosquitoes has been well established. *Anopheles quadrimaculatus* was regarded as the most important vector prior to the eradication of malaria in the United States (Carpenter and LaCasse, 1955) and a potential vector of other viruses (Reinert et al., 1997).

Distribution. In Texas, *An. quadrimaculatus* has been found in 113 counties (Fig. 2.10), while *Anopheles smaragdinus* has in 7 counties (Fig. 2.11).

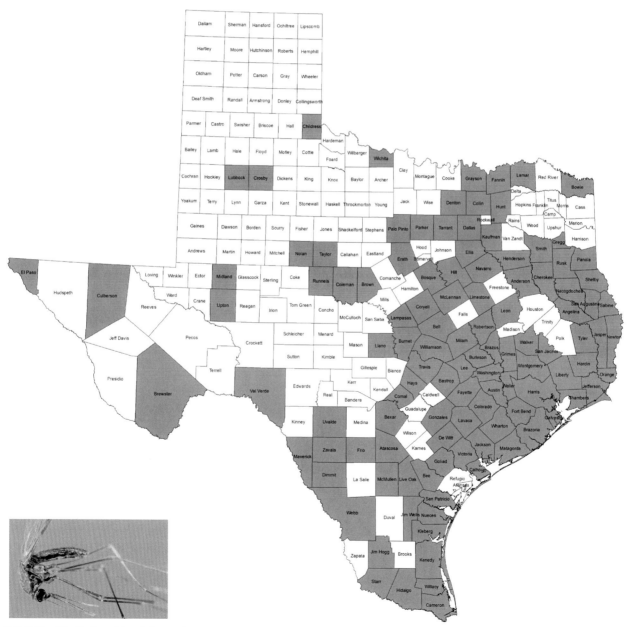

FIG. 2.10   Distribution of *Anopheles quadrimaculatus* in Texas.

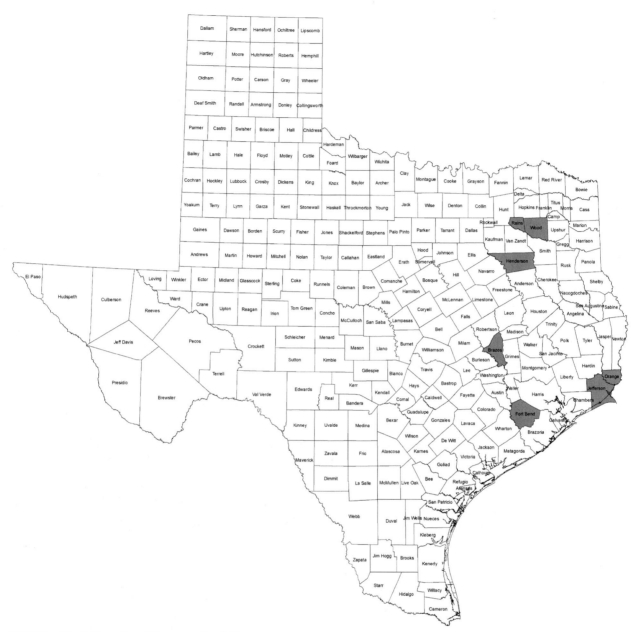

FIG. 2.11   Distribution of *Anopheles smaragdinus* in Texas.

Anopheles walkeri (Theobald, 1901)

A medium-sized, dark brown mosquito with dark palps that are white ringed apically. Its legs are entirely dark with the femora and tibiae tipped with pale scales, and their wing scales are narrow and dark where some of the scales can form four darker spots. It resembles *An. atropos* and others in the *quadrimaculatus* complex. The halteres in *An. walkeri* are pale scaled, while those of *An. atropos* are dark scaled.

Bionomics. *Anopheles walkeri* has multiple generations per year from May to October. Eggs are produced in the winter and hibernate in the northern part of its range. Larvae are found in freshwater swamps and lake margins with emerging and floating vegetation or floating debris. Adults feed on humans and large mammals and are active biters at night entering houses. During the day, they are found in cool, humid, shaded, resting places outside dwellings and have been found 2 miles from their breeding sites. In Texas, it breeds in giant bulrush and switch-cane marshes along the Sabine River. It has been infected with malaria, experimentally, and also in a single wild-collected mosquito (Bang et al., 1943).

Distribution. In Texas, it has been found in two counties (Fig. 2.12).

FIG. 2.12 Distribution of *Anopheles walkeri* in Texas.

2.2.2 Subgenus *Nyssorhynchus* Blanchard

Anopheles albimanus (Wiedemann, 1821)

This is the only member of the subgenus *Nyssorhynchus* Blanchard found in Texas. A medium-sized mosquito with a black, long proboscis and dark palps that are as long as the proboscis with white markings on the apices of each segment except the first one, and the last segment is all white. The scutum is dark to dark brown with a pair of small submedian black spots in the middle. The abdomen is dark brown to black with pale scales on the middle of the segments. The legs are dark except for the terminal segments of the hind tarsi. First hind tarsi are dark, second are white apically, third and fourth are white, and fifth are dark basally and white apically. Wings are white spotted with the fringe white spotted at the end of each vein.

Bionomics. It has multiple generations throughout the year. Eggs are deposited at night on the surface of water, while female adults are standing on it and only after a blood meal. Larvae are encountered in sites that contain fresh or brackish water with floating plants; seepage pools; irrigation ditches; and shallow margins of ponds, streams, and lakes. Breeding occurs in sunlit marshy meadows or closely cropped areas with hoof prints or stock tracks. Adults are strong fliers and feed readily on humans and domestic animals. They are predominantly exophagic and exhibit exophilic resting behavior, although it has shown endophilic behavior in the northern range of its distribution in the Americas. Mating occurs only in twilight, and they are active during the evening and at night but can attack in bright sunlight. It is one of the most important vectors of malaria in the northern South America, Central America, and the Caribbean region (Carpenter and LaCasse, 1955; Sinka et al., 2010).

Distribution. In Texas, it has been found in 14 counties (Fig. 2.13).

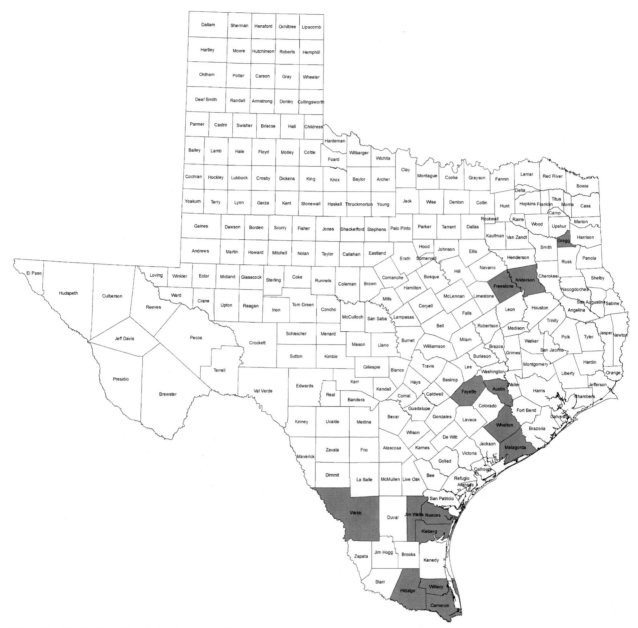

FIG. 2.13 Distribution of *Anopheles albimanus* in Texas.

2.3 *Aedes* Meigen

Due to recent updates in taxonomy, largely on the basis of main characters of the female and male genitalia, *Aedes* has been split into two genera: *Aedes* and *Ochlerotatus* Lynch Arribalzaga (Reinert, 2000). This created confusion and hesitation in conforming to it, because many species needed to change their scientific nomenclature. Traditional nomenclature, however, has continued to be used based on the recommendation of the Journal of Medical Entomology Journal Policy on Names of Aedine Mosquito Genera and Subgenera 2005 and discussion by Savage and Strickman (2004). Therefore such use is in effect in this chapter, wherein the classification proposed by Wilkerson et al. (2015) is available in the Walter Reed Biosystematics Unit (WRBU, 2019). There are 28 species of *Aedes* mosquitoes in Texas and two of them. *Ae. aegypti* and *Ae. albopictus*, are vectors that increase the risk for human vector-borne diseases due to the ability of arboviruses to adopt anthroponotic urban or peridomestic transmission cycles that involve these two mosquito species (Weaver and Reisen, 2010).

2.3.1   Subgenus *Aedimorphus* Theobald

Aedes vexans (Meigen, 1830)

An inland floodwater mosquito that is small to medium with a dark proboscis and short, dark palps where the fourth segment has a few white scales at the base and at the tip. The thorax is brown, and the abdominal segments II to VI have two upside-down, semicircular, or triangular-shaped basal bands of white scales. Their legs are dark scaled with segment 1 of tarsi pale scaled, and hind tarsi have narrow basal white bands. The wings are dark with narrow scales.

Bionomics. Larvae are found in shallow areas subject to inundation, in temporary rain-filled pools, and in floodwater pools. They overwinter in the egg stage and have one generation per year. Larvae may appear periodically after alternate flooding and drying of the eggs during the summer. Eggs are resistant to drying and low temperatures. They may survive more than a year, with hatching occurring in the spring for eggs surviving the previous winter. Larvae are found in freshwater areas free of filamentous algae and decaying vegetation. Adults are abundant in places where they occur and can migrate more than 10 miles from their breeding sites. Females are fierce biters and are active at dusk, after dark, and can feed in shady places during the day. Females feed on any host available including humans, horses, cows, and birds. It is a significant pestiferous species together with *Ae. taeniorhynchus* and *Ae. sollicitans* in Harris County, TX. *Aedes vexans* has been shown to transmit eastern equine encephalitis (EEE), western equine encephalitis (WEE), and Saint Louis encephalitis (SLE) viruses in the laboratory and has also been infected successfully with dog heartworm in the laboratory (WRBU, 2019), West Nile virus (WNV) (CDC, 2016), and Zika virus (ZIKV) in the laboratory (Gendernalik et al., 2017; O'Donnell et al., 2017).

Distribution. It is the most reported *Aedes* species in Texas and found in 118 counties (Fig. 2.14).

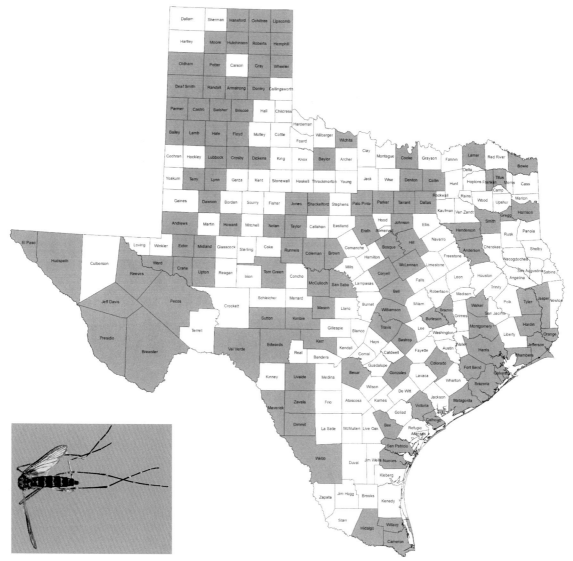

FIG. 2.14   Distribution of *Aedes vexans* in Texas.

2.3.2   Subgenus *Georgecraigius* Reinert, Harbach, and Kitching

Aedes epactius (Dyar and Knab, 1908)

Zavortink (1972) removed *Ae. atropalpus* (Coquillett) from the Group *Finlaya* Theobald and placed it in *Ochlerotatus* Lynch Arribalzaga and raised *Ae. epactius* to full species status leaving two species in the *Atropalpus* group: *Ae. atropalpus* occurring in the Eastern United States and *Ae. epactius* (Rock pool mosquito) in the Southwestern United States and Central America. The latter species differ in the more closely approximated eyes, ornamentation of the scutum, hind femur that is usually dark scaled near the base dorsally, presence of posterior fossa bristles, longer palpus of the male, male genitalia, and the fewer comb scales in larvae.

Bionomics. Much of the ecology of *Ae. epactius* is similar to that of *Ae. atropalpus*. Larvae are found in rock holes along river streams, rock-filled pools, and ground pools. It has also been collected in artificial containers, tree holes, and cavities of Agave leaves. Unlike *Ae. atropalpus*, *Ae. epactius* is not autogenous, that is, requiring a blood meal for oogenesis (egg production and development). They bite humans readily. *Aedes epactius* is a vector of Jamestown Canyon virus (JCV) (Heard et al., 1991).

Distribution. In Texas, it has been found in 55 counties (Fig. 2.15).

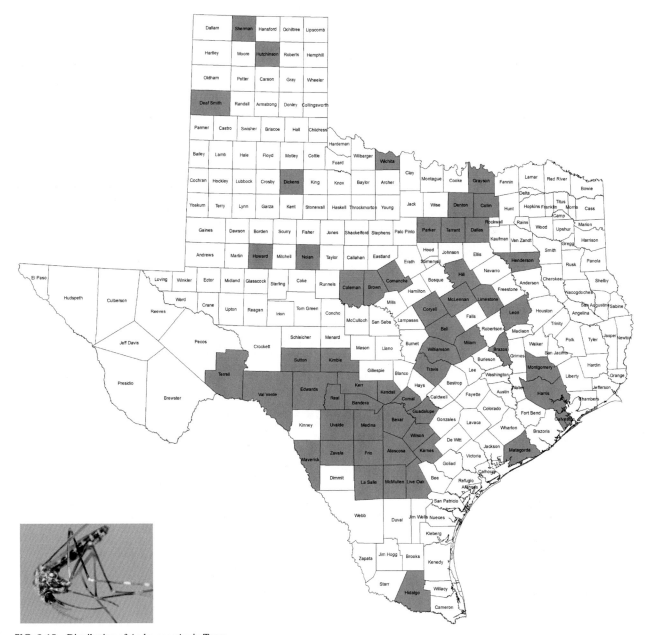

FIG. 2.15   Distribution of *Aedes epactius* in Texas.

2.3.3 Subgenus *Lewnielsenius* Reinert, Harbach, and Kitching

Aedes muelleri (Dyar, 1920)

A medium-sized mosquito with a dark-scaled proboscis, and the palps, legs, and wings have slight metallic blue, purple or violet reflections. The scutum is dark brown to blackish with frosted lines or stripes and a narrow median stripe of white scales followed with laterally brown scales. Abdomen's first tergite has a median patch of white scales, and all other tergites are bronze–brown with narrow basal segmental white bands. Tergites V-VII with basolateral triangular patches.

Bionomics. Larvae are usually found in tree holes and artificial containers in Arizona and in the leaf axils of maguey (Agave) in Mexico. It occurs most commonly in oak-pine forest and cotton-wood forest at higher elevations from 1400 to 2320 m. *Aedes muelleri* has also been found in riparian woodlands and groves of oaks (*Quercus*). In Texas, it has been found at elevations of 1700 m in the Chisos Mountains and is usually associated with *Ae. brelandi*. They are fierce human biters and can feed during the day.

Distribution. In Texas, it has been found in Brewster County (Fig. 2.16).

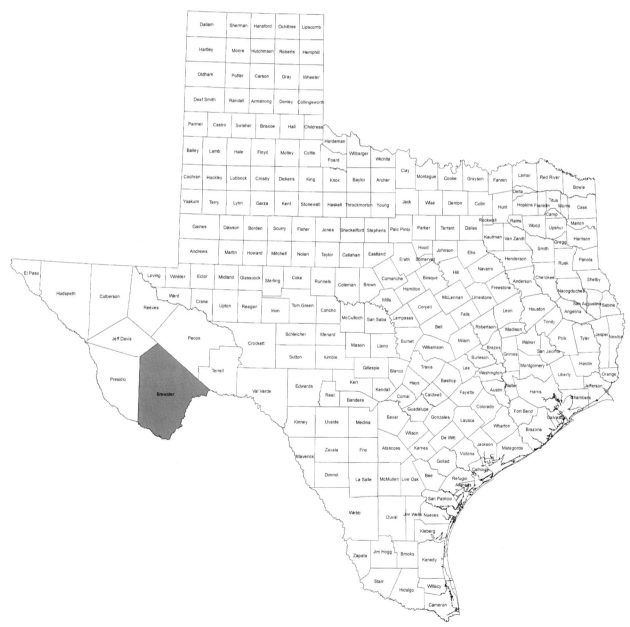

FIG. 2.16 Distribution of *Aedes muelleri* in Texas.

2.3.4 Subgenus *Ochlerotatus* Lynch Arribalzaga

Aedes atlanticus (Dyar and Knab, 1906) and *Aedes tormentor* (Dyar and Knab, 1906)

Aedes atlanticus is a medium-sized, dark brown, black-legged mosquito almost indistinguishable from *Ae. tormentor*. However, morphological features were found and validated for species identification of populations in the area of Houston, TX (Roberts and Scanlon, 1975). The occiput has dark appressed scales laterally and a median longitudinal stripe on the scutum. Distinguishing features between these two species are seen through the examination of male terminalia, their larval pecten teeth and comb scales, or using novel DNA sequence data (Sither et al., 2013).

Bionomics. Both are considered woodland floodwater species as they are found in temporary pools in woodlands and open fields with several types of vegetation after summer rains. Eggs are laid singly and breeding occurs from March to November in the Southern States. In Harris County, their highest numbers have been recorded from June to September. They are usually associated with other woodland species such as *Ae. infirmatus* Dyar and Knab. They are aggressive and persistent biters that can also bite during the day. They feed on small mammals, reptiles, and amphibians and appear to be multivoltine (McGregor and Eads, 1943). The *Ae. atlanticus-Ae. tormentor* complex has been found infected with the California encephalitis virus (CEV) including the Keystone virus (KEYV) in studies conducted in Houston, TX (Kokernot et al., 1974) and has also been found infected with EEE and WNV (CDC, 2016).

Distribution. In Texas, *Ae. atlanticus* has been found in 20 counties (Fig. 2.17) and *Ae. tormentor* in seven counties (Fig. 2.18).

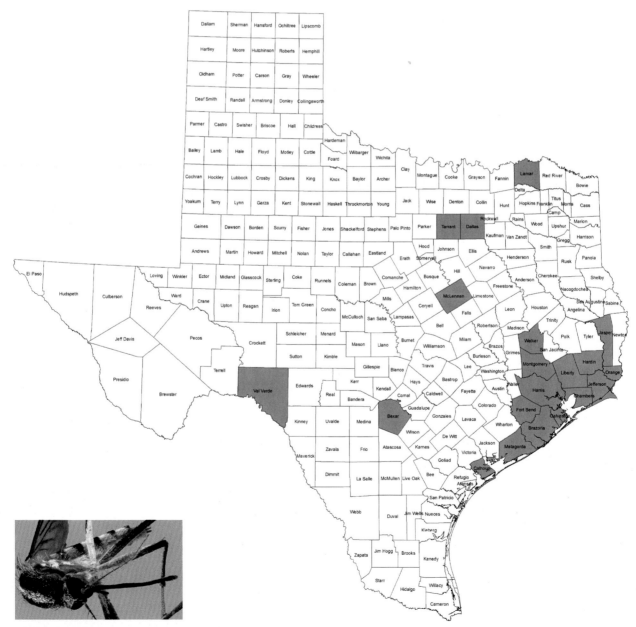

FIG. 2.17 Distribution of *Aedes atlanticus* in Texas.

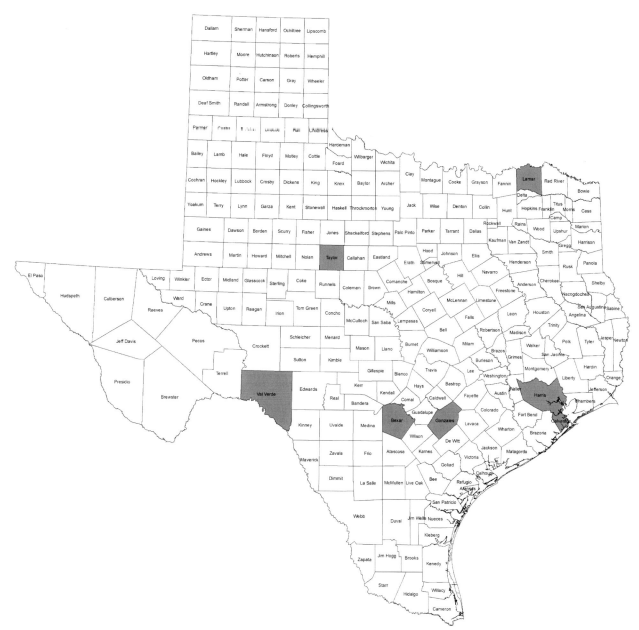

FIG. 2.18 Distribution of *Aedes tormentor* in Texas.

Aedes bimaculatus (Coquillett, 1902)

Aedes bimaculatus is a medium- to large-sized, bright orange rock pool mosquito resembling the coloration of other species such as *Ae. fulvus pallens* Ross and *Ae. campestris* Dyar and Knab. The hind tarsomeres are entirely dark scaled, the scutum is yellow with the exception of a pair of dark brown to black postlateral spots, and the hypostigmal area does not have a dark spot.

Bionomics. It has been found in muddy, temporary ground pools and wells. It has been recorded in the Lower Rio Grande Valley in roadside ditches (Ross, 1943) and woodland pools. Larvae are nearly transparent except for the dark sixth and seventh abdominal segments. Adults can be found resting on the grass near roadside pools.

Distribution. In Texas, it has been found in San Benito, Brownsville, and five other counties (Fig. 2.19).

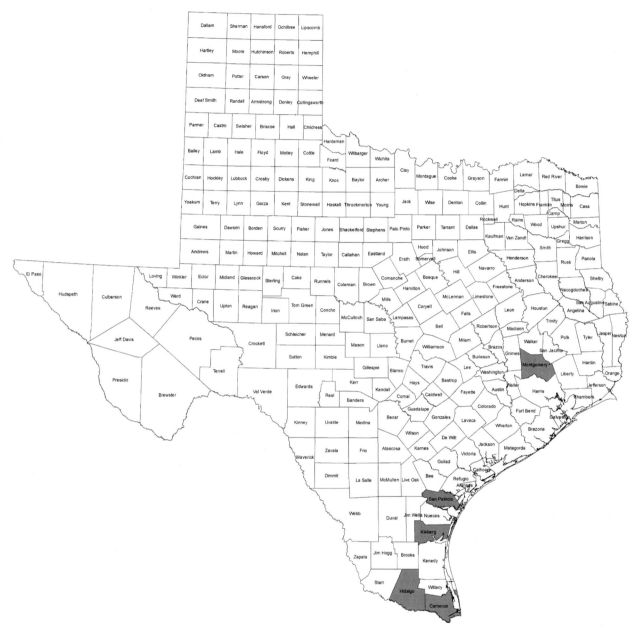

FIG. 2.19 Distribution of *Aedes bimaculatus* in Texas.

Aedes campestris (Dyar and Knab, 1907)

This is a large species with a dark proboscis where the basal two-thirds are speckled with white scales. The thorax is dark brown, the scutum has a broad median stripe of narrow brown scales, and the scutellum has narrow yellowish-white scales. The abdomen's first tergite has a large median patch of cream-white scales, and the remaining tergites are cream white.

Bionomics. It is found in open prairies in the western semiarid plains of the United States. Eggs are deposited singly in the debris around depressions, and after overwintering, they hatch when favorable conditions are present during the following spring and summer. Breeding preference is in pools rich in organic matter. Larvae develop in depressions filled with water from melting snow or rain and open pools in marshy areas. Adults can have one or multiple generations throughout the year and disperse a few miles and are troublesome biters active throughout the day. Studies have shown that WEE has been isolated from *Ae. campestris* along with isolation of other California (CAL) serogroup viruses, and other rhabdoviruses (Clark et al., 1986).

Distribution. It has been found in nine counties in the Trans-Pecos region of Texas (Fig. 2.20).

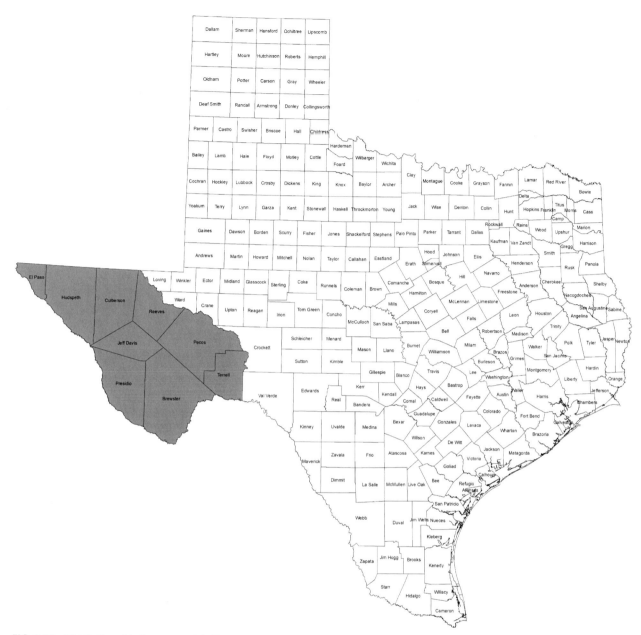

FIG. 2.20   Distribution of *Aedes campestris* in Texas.

Aedes canadensis (Theobald, 1901)

There are two subspecies of *Ae. canadensis*: *Ae. canadensis* (Theobald) and *Ae. canadensis mathesoni* Middlekauff. The one present in Texas is *Ae. C. canadensis* and is a dark brown colored mosquito. The mesonotum is golden brown with patches of white scales on the pleura and without the golden-scaled lyre shape as in *Ae. canadensis mathesoni*. The abdomen has pale lateral spots, and the hind legs have basal and apical pale bands on each tarsi segment with tarsi 5 entirely pale scaled, whereas in mathesoni, tarsi 3 and 4 have pale basal bands only and tarsi 5 is dark.

  Bionomics. It is considered a woodland species found in large numbers near their breeding habitats such as small streams, pools, and ditches that are contiguous to wooded areas. In the south, it overwinters in the egg stage in great numbers after some flooding during the late winter and early spring. Hatching occurs in cold water during the spring with occasional hatching in the fall. Adults emerge from April to early June and can persist in the woods until late summer. Females oviposit once on water or very wet soil. Their biting behavior has been described as seldom troublesome and persistent and is rarely found inside dwellings (Carpenter et al., 1946; Carpenter and LaCasse, 1955). *Aedes canadensis* is a suitable host for *Dirofilaria immitis* in cages and has been implicated as a vector of La Crosse Encephalitis (LAC), EEE, Highland J (HJ), and JC viruses (Andreadis et al., 1994, 1998).

  Distribution. In Texas, it has been found in 20 counties (Fig. 2.21).

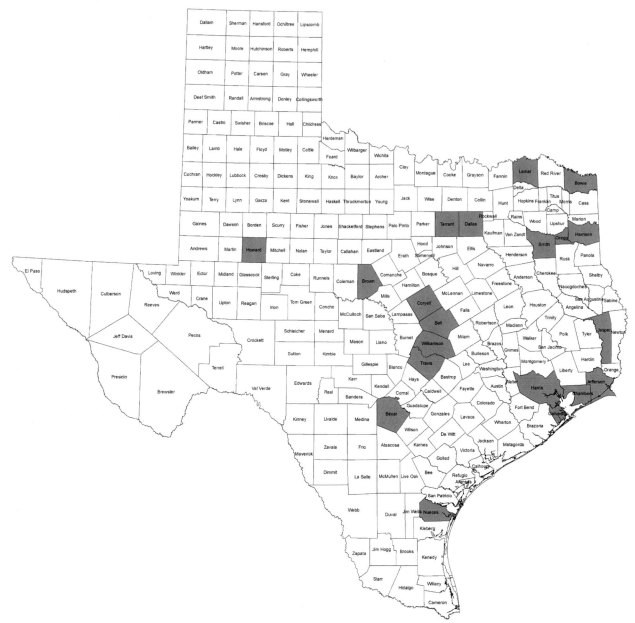

FIG. 2.21 Distribution of *Aedes canadensis* in Texas.

Aedes dorsalis (Meigen, 1830)

A medium-sized, whitish mosquito with white scales on the occiput. The mesonotum has golden-brown median and lateral lines. The pleura has white-scaled patches. The white abdomen has two dorsal black patches on each segment, and the last two segments are often entirely white. The first tarsi are white and black scaled with white bands apically and basally, and the fifth segment is entirely white.

Bionomics. Eggs are placed on soil prone to flooding in the spring after they have passed the winter in that stage. They require a period of partial dryness and might remain viable for several years before hatching. They have multiple generations, and breeding occurs throughout the warm season. Larvae are found in brackish and fresh water and in areas subject to flooding such as tidal flats and mountain snow pools. It is a significant pest for man and animals. Females are fierce biters throughout the day or night but are mostly active during the evening and calm cloudy days. They are found in meadows resting in vegetation after a blood meal. They feed on nectar and blood from humans, livestock, and fowl. They are exophagic but can enter dwellings and feed indoors, too. Western Equine Encephalitis and CEV have been isolated from *Ae. dorsalis* in California (Carpenter and LaCasse, 1955), Cache Valley virus (CVV) (Clark et al., 1986), WNV (CDC, 2016), and Japanese Encephalitis virus (JEV) (Oliveira et al., 2018).

Distribution. In Texas, it has been found in 38 counties (Fig. 2.22).

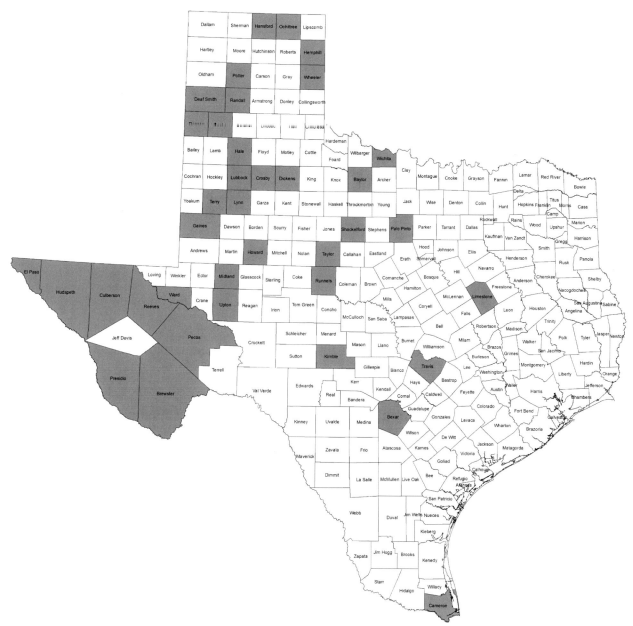

FIG. 2.22 Distribution of *Aedes dorsalis* in Texas.

Aedes dupreei (Coquillett, 1904)

This is a small, dark brown mosquito with a broad longitudinal stripe of silvery-white scales along the scutum and narrow wing scales, and the vertex usually has lateral gray-scale patches (Harrison et al., 2016). It is found with other woodland mosquitoes such as *Ae. canadensis*.

Bionomics. Its larvae's most discernable characteristic is the long anal gills. They are found in woodland pools and temporary summer rain-filled pools. They hide under vegetation and debris at the bottom and rarely come to the surface, making it difficult to detect and collect. Adults are rarely seen in the field but can be collected in light traps. They have multiple generations throughout the year with their greatest numbers from August to September and are not readily attracted to humans. West Nile virus has been detected in this species (CDC, 2016).

Distribution. In Texas, it has been found in four counties (Fig. 2.23).

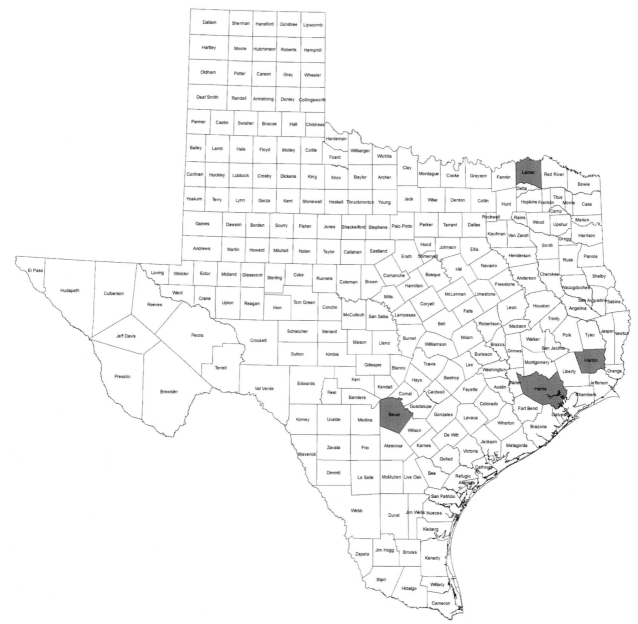

FIG. 2.23   Distribution of *Aedes dupreei* in Texas.

Aedes fulvus pallens (Ross, 1943)

A medium- to large-sized, bright orange-yellow mosquito. The scutum is yellow with two posterolateral dark spots. Wing scales are yellow, broad on costa, subcosta, and vein 1. The pointed abdomen is yellow-scaled with dark apical bands that taper laterally, and segment 7 is almost entirely yellow. Hindtarsus 1 is entirely yellow, and hindtarsal segments 2–5 are primarily dark with intermixed yellow scales.

Bionomics. It overwinters in the egg stage and has multiple broods per year. Larvae are found in temporary woodland pools following rains during the summer months. Females are vicious biters and are active at night, dusk and dawn, and some during the day. They feed on mammals including humans. They are encountered somewhat more frequently than its larvae and have been collected from biting, resting on foliage, and in light trap collections. Although little is known about its role in vector-borne disease transmission, WNV, EEE, and Wyeomyia (WYOV), viruses have been associated with it (Wilkerson et al., 2015).

Distribution. In Texas, it has been found in 11 counties (Fig. 2.24).

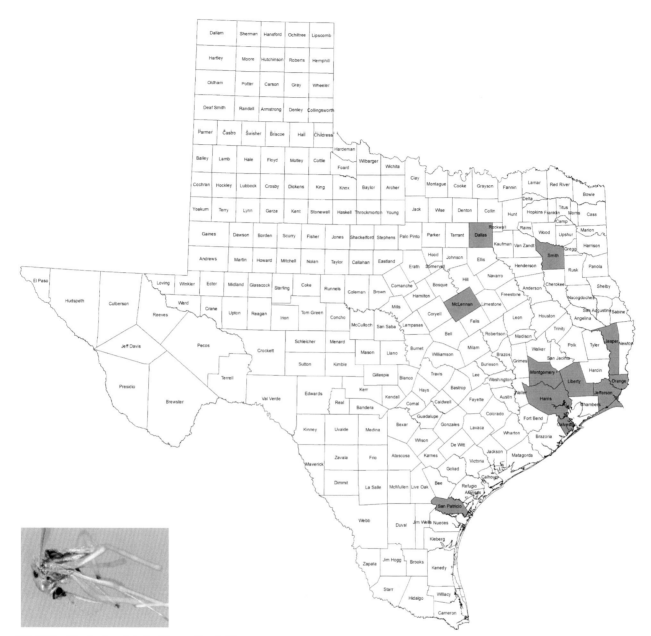

FIG. 2.24   Distribution of *Aedes fulvus pallens* in Texas.

Aedes grossbecki (Dyar and Knab, 1906)

It is a medium- to large-sized mosquito. The proboscis is dark with scattered pale scales. The scutum is dark brown or black with a broad median dark brown stripe and golden-brown anteriorly. The pointed abdomen has pale basal bands, and the hind tarsomeres have broad, pale basal bands.

Bionomics. It has one brood per year. Larvae are found in early spring pools and woodland pools when flooded with cold water. They have also been found in marshes with sedges and grass with water near neutral pH. Females feed on mammals including humans and are fierce, persistent biters. It is short lived and does not venture far from its breeding site. Little is known about its role in pathogen transmission; however, WNV has been detected in it (CDC, 2016).

Distribution. It has been found only in Harris County, TX (Keith, 1979) (Fig. 2.25).

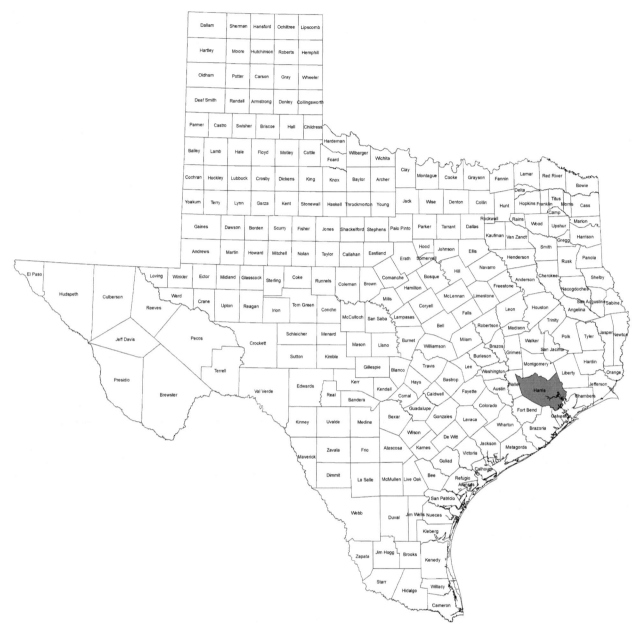

FIG. 2.25   Distribution of *Aedes grossbecki* in Texas.

Aedes infirmatus (Dyar and Knab, 1906)

It is a medium-sized mosquito. The abdomen is dark scaled with pale basal lateral patches. Abdominal tergites VI–VIII are dark scaled medially. The legs are dark scaled, and the hindtibiae are dark from base to apex. Postspiracular setae are present, and the scutum has a broad median longitudinal pale area that ends at or before the level of the wings.

Bionomics. Larvae occur from early spring to late fall in temporary woodland pools following rains and in grassy pools and freshwater swamps. Females are persistent biters feeding at dusk and dawn and during the day in or near wooded areas but seldom enter dwellings. They feed on mammals including humans. Their flight range is near their breeding sites up to a mile and has multiple broods per year. It has been found infected with EEE and Trivittatus (TVT) (Burkett-Cadena, 2013), WNV (CDC, 2016), KEYV, TENV, and WEE (Wellings et al., 1972).

Distribution. In Texas, it has been found in the Lower Rio Grande Valley and in 16 counties (Fig. 2.26).

FIG. 2.26 Distribution of *Aedes infirmatus* in Texas.

Aedes mitchellae (Dyar, 1905)

A medium-sized mosquito with a dark-scaled proboscis that has a white-scaled ring at the middle. The scutum is black, frequently with a pair of narrow longitudinal submedian lines of golden-yellow scales originating at the outer margins of the prescutellar space, and extending forward nearly the full length of the scutum. Abdomen with a pale median longitudinal stripe extending apically on each segment. Tarsi have basal pale bands, and hind tarsomere 5 is entirely pale. Its appearance resembles that of *Ae. sollicitans* (Walker) found in Coastal plains.

Bionomics. Larvae are found in temporary rain-filled pools throughout the year in the south. Its distribution is limited to the Coastal plains. They are fierce biters and are usually found around their breeding sites feeding on mammals including humans at dusk, dawn, and during the day. They have multiple generations per year and overwinter in the egg stage. It has been found naturally infected with EEE posing a low risk to humans because it is uncommon (Varnado et al., 2012). It has also been associated with TENV (Chamberlain et al., 1969; Wilkerson et al., 2015).

Distribution. In Texas, it has been found in 16 counties (Fig. 2.27).

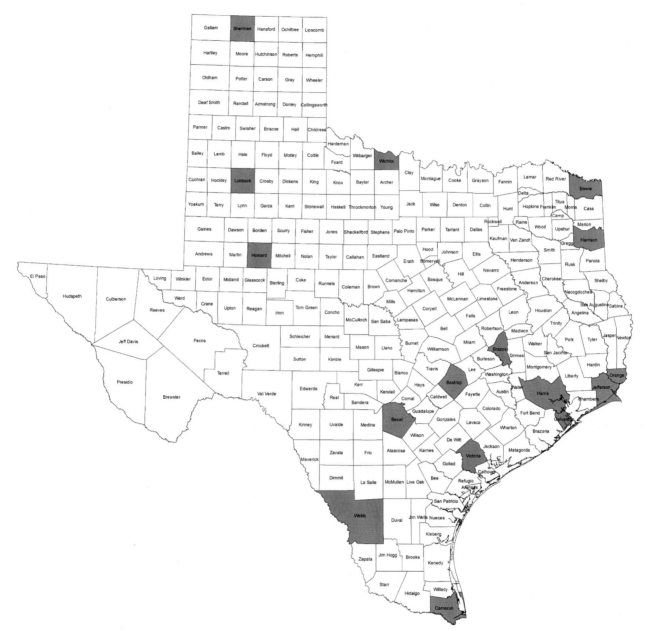

FIG. 2.27 Distribution of *Aedes mitchellae* in Texas.

Aedes nigromaculis (Ludlow, 1907)

A medium-sized mosquito with a dark proboscis that is white ringed near the middle. The scutum is blackish with a median broad stripe of golden-brown scales. Abdomen's first tergite is white scaled and the remaining tergites are white basally, laterally, and medially.

Bionomics. It has several generations per year and overwinters in the egg stage. Breeding season extends from May until late September. Larvae occur mostly in alkaline waters in rain-filled depressions and irrigation ditches. Adults are strong fliers and have been tracked several miles from their breeding locations. They are most active during the evenings but can bite during the day. They feed on all available hosts including humans, horses, cows, and birds. *Aedes nigromaculis* is considered a pest wherever it thrives and has been shown to transmit JE (Oliveira et al., 2018) and was found to be infected with WEE in the laboratory (Blackmore and Winn, 1954).

Distribution. In Texas, it has been found in 71 counties (Fig. 2.28).

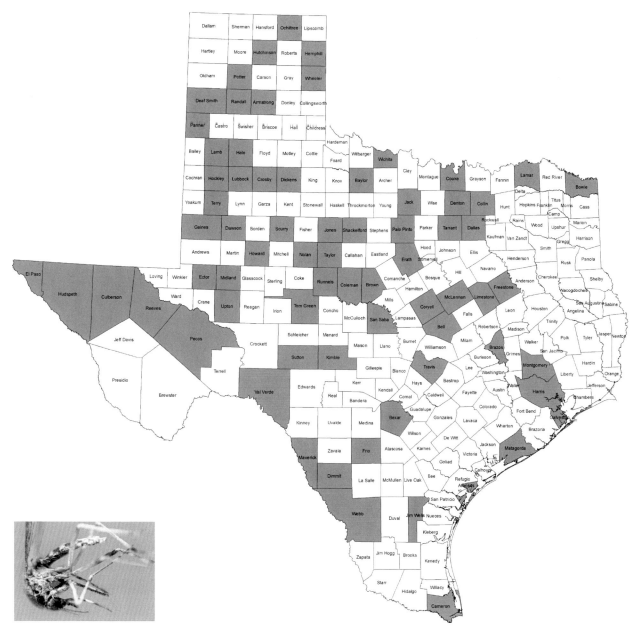

FIG. 2.28 Distribution of *Aedes nigromaculis* in Texas.

Aedes scapularis (Rondani, 1848)

This is a medium-sized, dark-legged *Aedes* mosquito. The anterior half of the mesonotum is brown with a large patch of silvery scales broader than the lateral dark-scaled areas. Abdomen's tergites are bronze-brown scaled with a dorsal median longitudinal white line and narrow basal white bands on each segment. The hind tibia has dark-scaled basal and apical bands.

Bionomics. Larvae are found in temporary rain pools, rock holes, crab holes, marshes, and pools of muddy waters in stream beds in either sun or partial shade. Adults are crepuscular and can disperse a few miles from their breeding sites. It has been collected from the Lower Rio Grande Valley where it occasionally reaches large numbers. At least 15 viruses have been isolated from *Ae. scapularis* including yellow fever virus (YFV) and Venezuelan equine encephalitis (VEE), and it also appears to be a vector of Bancroftian filariasis (Arnell, 1976).

Distribution. In Texas, it has been found in seven counties (Fig. 2.29).

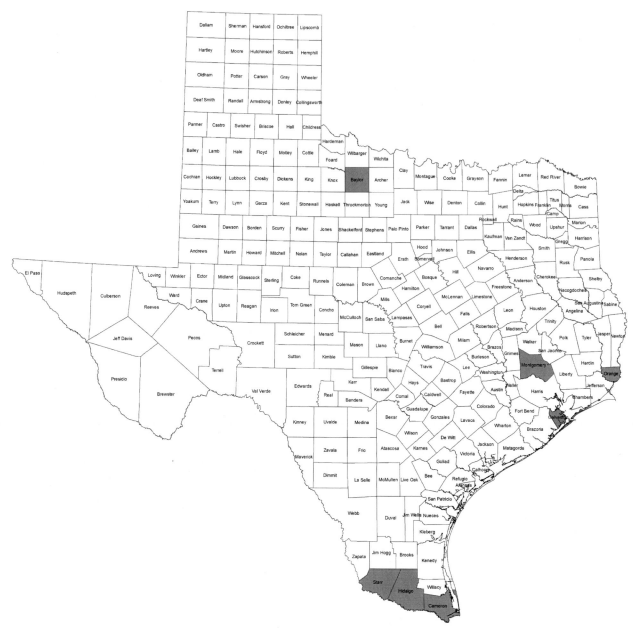

FIG. 2.29 Distribution of *Aedes scapularis* in Texas.

Aedes sollicitans (Walker, 1856)

A medium- to large-sized mosquito known as the eastern or tan saltmarsh mosquito. The proboscis is dark with a white band at the middle. The thorax is golden scaled on sides, and the pointed abdomen has segments with basal pale bands and a distinct median longitudinal pale stripe. Hind tarsi have broad basal pale bands, hind tarsomere 1 has a median pale band, and hind tarsomere 5 is entirely pale. It resembles *Ae. nigromaculis* and *Ae. mitchellae*.

Bionomics. It has several generations per year and overwinters in the egg stage. Larvae occur mostly in salt marshes in coastal areas. It has been found in brackish water inlands, particularly in saltwater waste from oil and gas fields. Tidal aspect of breeding is very important in addition to rainfall. In Harris County, it is present all year long. Its abundance occurs usually from May to October. Adults are strong fliers and migrate long distances of more than 25 miles from their breeding grounds in large numbers. Females are persistent biters that will feed on any available host anytime during the day or night. They rest in vegetation and attack, if disturbed. It is a major pest near the beach and may be encountered in great numbers during the daytime. In 1971 *Ae. sollicitans* was one of the vectors involved in the outbreak of VEE in Texas (Sudia and Newhouse, 1971). It was also found to be infected with WEE and San Angelo virus (SAV) (Sudia et al., 1975). It can also transmit EEE virus, and dog heartworm disease (Varnado et al., 2012) is a potential vector of Rift Valley fever (RVF) (Gargan et al., 1988) and can vector WNV (CDC, 2016).

Distribution. In Texas, it has been found in 76 counties (Fig. 2.30).

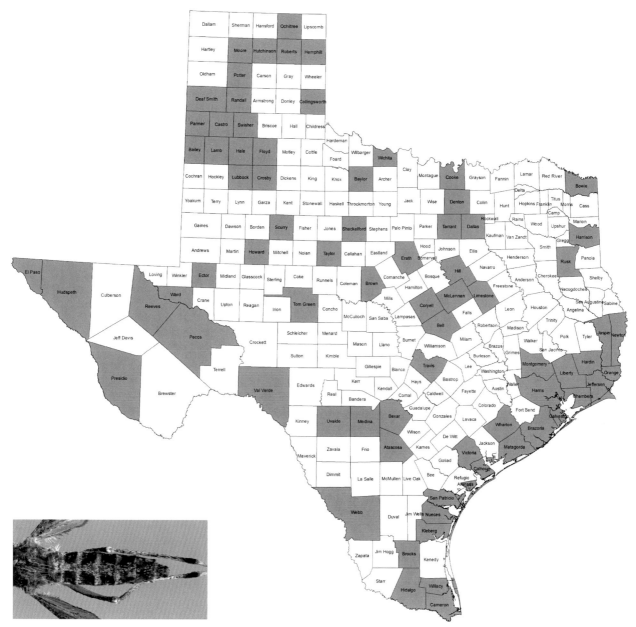

FIG. 2.30 Distribution of *Aedes sollicitans* in Texas.

Aedes sticticus (Meigen, 1838)

It is a floodwater species that is dark brown and white and small- to medium-sized, black-legged *Aedes* mosquito. The proboscis is dark and with a dark median stripe on the scutum, with patches of white scales on either side. The abdomen is dark brown with pale basal bands. Tarsi are dark scaled, and fore femur is speckled with white scales.

Bionomics. Eggs are resistant to desiccation and remain viable for more than three seasons in the absence of flooding and hatch in early spring in woodland pools in April. There is one brood per year but may have two or more broods. Hatching occurs at any time eggs are inundated. Larvae develop from overwintering eggs and are found in floodwater pools in woodlands and rain-filled pools containing leaves or other vegetable matter. Adults can disperse a few miles from their breeding site. Females are vicious biters that can feed during the daytime and early evenings in wooded areas near their breeding sites. They feed on a wide variety of mammals including humans. It is associated with WNV (CDC, 2016), EEE, JCV, and Tahyna virus (TAHV) (Wilkerson et al., 2015).

Distribution. In Texas, it has been found in eight counties (Fig. 2.31).

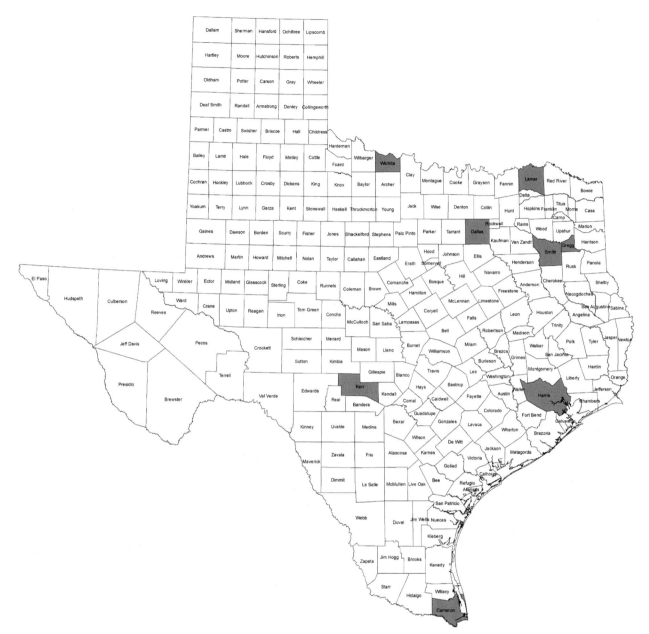

FIG. 2.31 Distribution of *Aedes sticticus* in Texas.

Aedes taeniorhynchus (Wiedemann, 1821)

A dark, medium- to small-sized mosquito commonly known as the Black Saltmarsh mosquito. The proboscis has a white ring near the middle. The mesonotum is golden brown. Palps are short with pale scales at the tips. The abdominal segments have basal pale bands and do not have the median longitudinal stripe as in *Ae. sollicitans*. Last abdominal segments have conspicuous white patches laterally. Postspiracular setae are present, and the hind tarsi have broad basal pale bands.

Bionomics. *Aedes taeniorhynchus* is found associated with *Ae. sollicitans* along the coast, but is more widely distributed. It breeds in saltmarshes flooded by tides or rains, and is occasionally found in fresh water and in inland brackish water swamps, particularly in oil fields. They have multiple continuous generations and overwinter as eggs. Some females are autogenous developing eggs without a blood meal. Larvae and adults are present throughout the year in the south, but their populations thrive following high tides or a combination of high tides and heavy rains during the summer and early fall. Females are fierce biters attacking any time during the day or night and often migrate considerable distances from their breeding sites where they become serious pests. They feed on humans, birds, and other mammals. They rest on vegetation but will attack if their spaces are invaded. It is associated with other species that inhabit and tolerate salt or brackish water, that is, *Ae. sollicitans*, *Cx. salinarius*, *Ae. vexans*, and other saltwater mosquitoes. It is involved in the transmission of dog heartworm disease (Nayar and Rutledge, 2008) and EEE (Varnado et al., 2012). It was also implicated in the VEE epidemic in Texas in 1971 and was found infected with WEE (Sudia et al., 1975). It is also associated with CV, TENV, TVT, and other viruses (Wilkerson et al., 2015) including RVF (Gargan et al., 1988).

Distribution. In Texas, it has been found in 41 counties (Fig. 2.32).

FIG. 2.32 Distribution of *Aedes taeniorhynchus* in Texas.

Aedes thelcter (Dyar, 1918)

A medium-sized, dark-legged mosquito. The proboscis and the palps are brown. The scutum is dark brown with yellowish scales, and the posterior pronotum has narrow curved brown scales. The abdomen is dark brown with triangular basal white bands on each segment. The first abdominal tergite has a median patch of white scales. Lateral white spots are present on the abdominal segments. Legs are brown with the femora, tibia, and first tarsal segment with pale scales. Wing scales are narrow and dark.

Bionomics. It occurs from spring to fall. Larvae may be found in overflow pools from irrigation ditches and temporary rain-filled pools, particularly in the Rio Grande Valley and up the coast as far as Nueces County, TX. The largest collections have been from light traps around Corpus Christi, TX. It was implicated in the VEE epidemic in Texas in 1971 and was also found infected with WEE (Sudia et al., 1975).

Distribution. In addition to the Rio Grande Valley area, it has been found in 44 counties in Texas (Fig. 2.33).

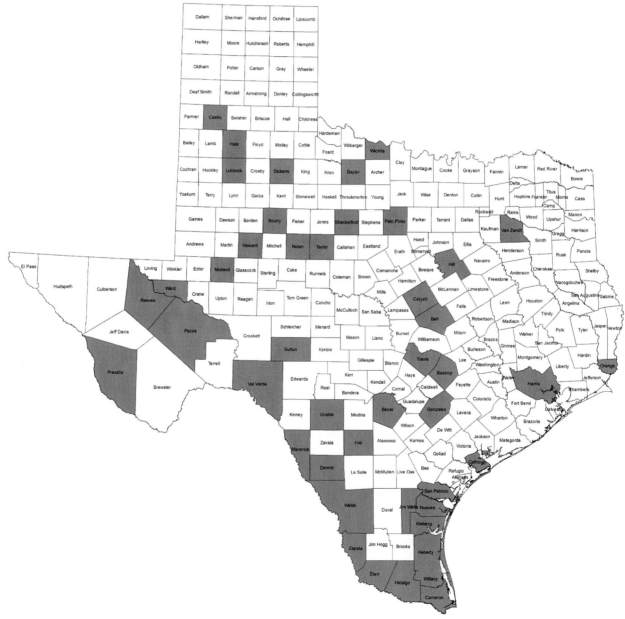

FIG. 2.33 Distribution of *Aedes thelcter* in Texas.

Aedes thibaulti (Dyar and Knab, 1910)

A medium-sized mosquito with an unbanded dark proboscis, short palps, and pointed dark abdomen with pale basolateral patches. The scutum presents a broad median stripe of dark scales and golden scales laterally. Its legs have blue-black scales except for white knee spots and pale inner surfaces of the femora. Wing scales are dark and narrow.

Bionomics. This is a tree hole mosquito usually found in sweet gum, tupelo stump holes, and in tree cavities. It has one generation per year and overwinters in the egg stage. Larvae are found in the hollow bases of gum trees following the flooding of lowland areas. Other habitats include freshwater swamps and dark subterranean voids under uprooted trees of red maples and cypress trees. Adults are found in thickets and woodlands near their breeding sites in the spring. They are active at dusk and dawn and night and day and are fierce biters that feed on mammals including humans and can attack during midday.

Distribution. In Texas, it has been found in three counties (Fig. 2.34).

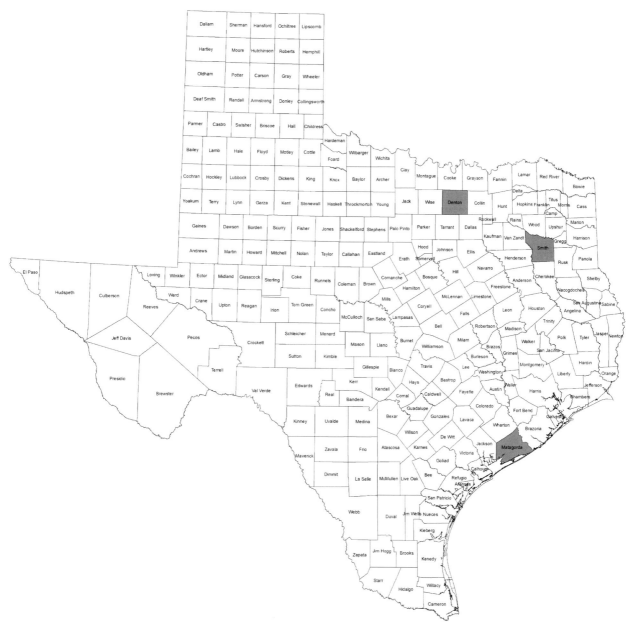

FIG. 2.34 Distribution of *Aedes thibaulti* in Texas.

Aedes trivittatus (Coquillett, 1902)

It is a medium-sized mosquito with a proboscis and palps that are dark. The scutum presents an alternating pattern of black and white stripes on the thorax consisting of a median longitudinal dark stripe edged by two white stripes, which are laterally surrounded by dark-scaled stripes. First abdominal tergite is dark scaled, and the remaining tergites are dark with basal patches of white scales. The legs and the wings are dark.

Bionomics. Eggs are found in shallow depressions subject to rain or overflow of streams where they hatch whenever submerged. Larvae are found anytime during the summer following rains, in floodwater pools, meadows, swamps, and woodlands. It overwinters in the egg stage and has one generation per year. Adult females are persistent biters attacking at dusk and during the daytime, even in bright sunlight when disturbed. They feed on humans, mammals, and reptiles, and their bites are painful. They inhabit open or lightly wooded areas and are found in great numbers in dense woodlands. It is associated with TVT, LAC (Wilkerson et al., 2015), and EEE (Gargan et al., 1988). It has tested positive for WNV (CDC, 2016) and may be involved in the transmission of one or more strains of CE.

Distribution. In Texas, it has been found in 44 counties (Fig. 2.35).

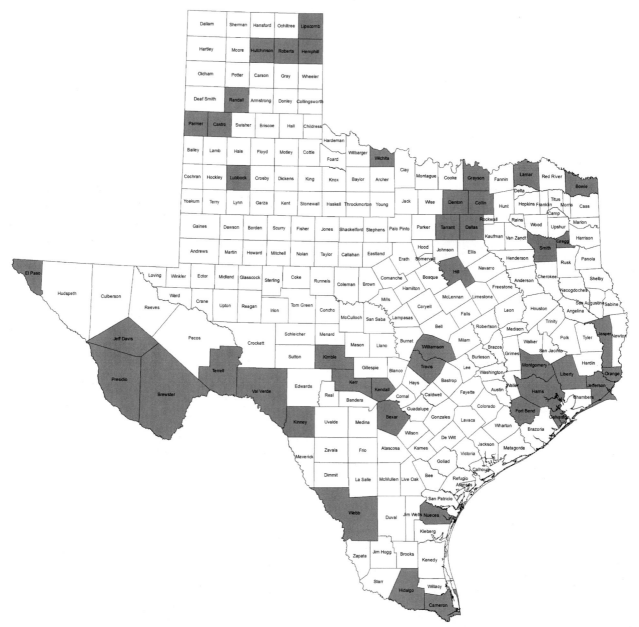

FIG. 2.35 Distribution of *Aedes trivittatus* in Texas.

2.3.5   Subgenus *Protomacleaya* Theobald

Aedes brelandi (Zavortink, 1972)

Aedes brelandi, a tree hole mosquito, is most closely related to *Ae. hendersoni* in the *Triseriatus* group. It has postspiracular setae in a large patch and a median longitudinal stripe of dark scales and silvery-white scales laterally. Adult morphological differences are the frequent presence of acrostichal bristles on the disk of the mesoscutum, darker mesoscutal bristles, whiter light scales of thorax, greater number of postspiracular scales, and the more developed ventral brush in larvae.

Bionomics. Zavortink (1972) collected and reported its larvae from a rot hole in a living evergreen oak tree (*Quercus* spp.) at an elevation of 1170 m in the Chisos Mountains, Big Bend National Park in Brewster County, TX in 1969. The larvae were associated with other tree hole species: *Ae. muelleri* Dyar, *Anopheles judithae* Zavortink, and *Or. kummi*. Females are annoying biters in the vicinity of tree holes. As with all four species of the *Triseriatus* group (*Ae. triseriatus* (Say), *Ae. hendersoni, Ae. brelandi,* and *Ae. zoosophus*), *Ae. brelandi* is susceptible to oral infection of LACV (Paulson et al., 1989).

Distribution. In Texas, it has only been found in Big Bend National Park, Brewster County, TX (Fig. 2.36).

FIG. 2.36   Distribution of *Aedes brelandi* in Texas.

Aedes hendersoni (Cockerell, 1918)

It is one of the mosquitoes in the *Triseriatus* group along with *Ae. triseriatus* and *Ae. brelandi*. *Aedes hendersoni* has been restored to full specific rank and can be separated from *Ae. triseriatus* in the adult and larval stages and male genitalia. The proboscis is dark, unbanded, and with dark short palps. The scutum has a median stripe of dark scales and silvery-white scales laterally. The scutal fossa has numerous well-developed setae, and the base of the coastal vein has a small patch of pale scales (Harrison et al., 2016). Hind tarsomeres are entirely dark scaled.

Bionomics. Larvae are normally found in tree holes with high organic content, in the canopy. It has been found in pitcher plants, and adults have been collected from lights, resting in buildings, and biting humans. It is associated with *Ae. triseriatus* and *Ae. zoosophus*, and sometimes with *Ae. epactius* and *Ae. sierrensis* (Ludlow). Adults have multiple generations per year, frequent areas near their breeding sites and are active at dusk and dawn feeding on sciurids (squirrels and chipmunks) and Procyonids (raccoons) (Nasci, 1982).

Distribution. In Texas, it has been found in nine counties (Fig. 2.37).

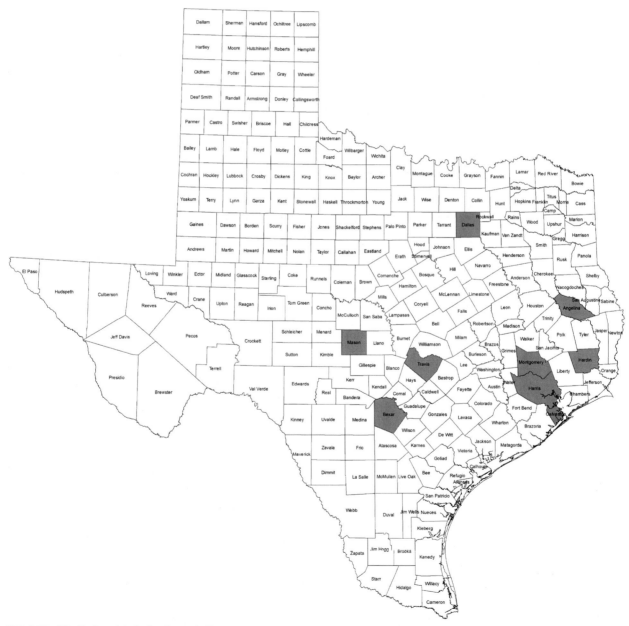

FIG. 2.37 Distribution of *Aedes hendersoni* in Texas.

Aedes triseriatus (Say, 1823)

A medium-sized mosquito commonly known as the Eastern tree hole mosquito with a dark, long proboscis and dark short palps. The scutum is black with a wide longitudinal stripe of narrow dark brown scales margined on each side by white scales. The abdomen, tarsi, and wings are black.

Bionomics. It is a widespread tree hole breeding mosquito. Eggs are laid singly or in groups of 3–5 on sides of cavities just above the water line and hatch when flooded with water. It is a multivoltine species and overwinters in the egg stage. Larvae are usually associated with those of *Ae. hendersoni* and *Ae. zoosophus* and develop in holes of deciduous trees and occasionally in artificial containers such as wooden tubs, barrels, watering troughs, and tires in shaded areas. Females are considered a serious nuisance in areas in or near woodlands and are persistent biters, and their bites are quite painful and lasting. Adults are crepuscular, actively flying at dusk and dawn and feed on mammals (squirrels and other rodents), humans, and birds. *Aedes triseriatus* is an important vector of LACV (Wilkerson et al., 2015), CV, Potosi virus (POTV) (Mitchell et al., 1998), EEE (Armstrong and Andreadis, 2010), and WNV (CDC, 2016). In the laboratory, it was found to be an effective vector of YFV, VEE, and WEE (WRBU, 2019), RVF (Gargan et al., 1988), and JCV was also detected in its eggs (Wilkerson et al., 2015). It is also a vector of dog heartworm disease (Rogers and Newson, 1979).

Distribution. In Texas, it has been found in 68 counties (Fig. 2.38).

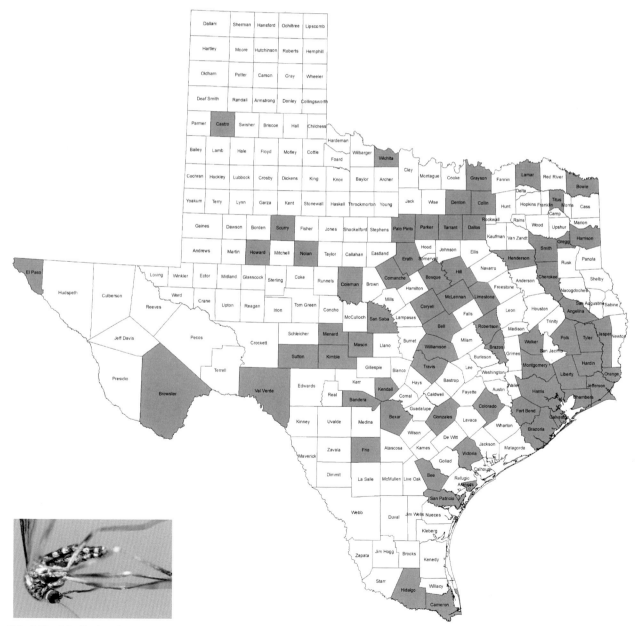

FIG. 2.38 Distribution of *Aedes triseriatus* in Texas.

Aedes zoosophus (Dyar and Knab, 1918)

This is a medium-sized mosquito. The scutum is dark brown with silver scales on the anterior half and with a median golden-brown stripe. The posterior half of the scutum is dark brown with silvery scales bordering the prescutellar space and forming a narrow line on either side of it. The first abdominal tergite has a median patch of dark scales, and the remaining segments are bronze-brown with narrow basal cream-white bands, and silver-white lateral patches. Hind femora have pale scales basally and brownish black on the distal half. Tibiae are dark with basal white markings. The hind tarsi are dark with broad basal white bands. Wing scales are dark with a patch of white scales on the base of costa, subcosta, and vein 1.

Bionomics. Larvae are found in rotten cavities of trees, particularly willows, and occasionally in artificial containers. This species is in the *Zoosophus* Group and is frequently associated with two species of *Protomacleaya* Theobald in the *Triseriatus* group, that is, *Ae. hendersoni* and *Ae. triseriatus*. In Texas, it is usually associated with *Or. signifera* and *Ae. triseriatus* in their breeding habitats. Females feed on humans and are similar to *Ae. triseriatus* and *Ae. varipalpus* (Coquillett) in this respect.

Distribution. In Texas, it has been found in 46 counties (Fig. 2.39).

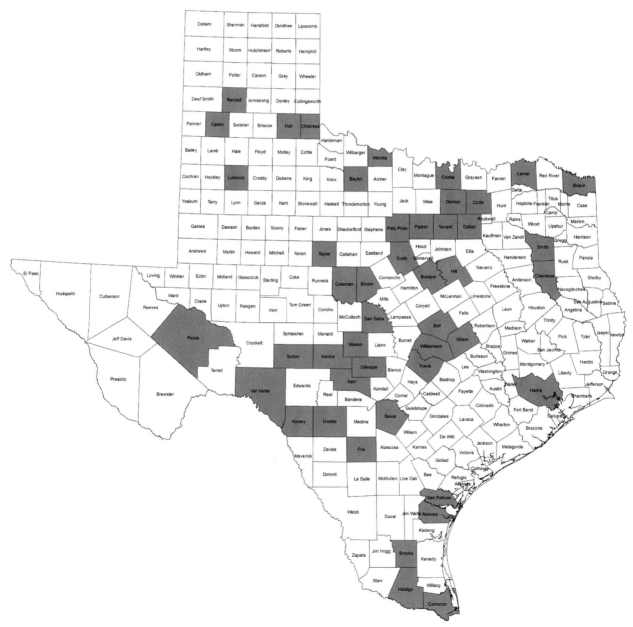

FIG. 2.39 Distribution of *Aedes zoosophus* in Texas.

2.3.6 Subgenus *Stegomyia* Theobald

Aedes aegypti (Linnaeus, 1762)

This is a small- to medium-sized species with a lyre-shaped pattern of pale scales on the scutum as being distinctive of the species. It originated in Africa but, currently, is widespread and found in tropical and subtropical regions of the world. *Aedes aegypti* has been traditionally known as the "yellow fever mosquito."

   Bionomics. *Ae. aegypti* breeds in artificial containers, cans, bottles, buckets, flower pots, bird baths, uncovered barrels, and in any receptacle that can hold water, including tires. It is found near or around human dwellings, especially in urban areas, thus being referred to as peridomestic. It prefers human blood over that of other animals, but it is opportunistic in nature. It not only feeds at dusk and dawn but also can feed at any time of the day. It is multivoltine, that is, having several generations per year, and its eggs can be viable for more than a year in a dry state and can re-emerge after a cold or dry spell. It can survive moderate winters in large containers in coastal cities such as Houston, Galveston, and Texas City (Horsfall, 1955). It is found in a variety of collecting devices throughout the year in Harris County, TX, and its population densities peak in June and July. The diseases it transmits, that is, YFV, dengue (DENV), Mayaro (MAYV), Chikungunya (CHIKV), and ZIKV, are still present in the Americas (CDC, 2017).

   Distribution. In Texas, it has been found in 108 counties. (Fig. 2.40).

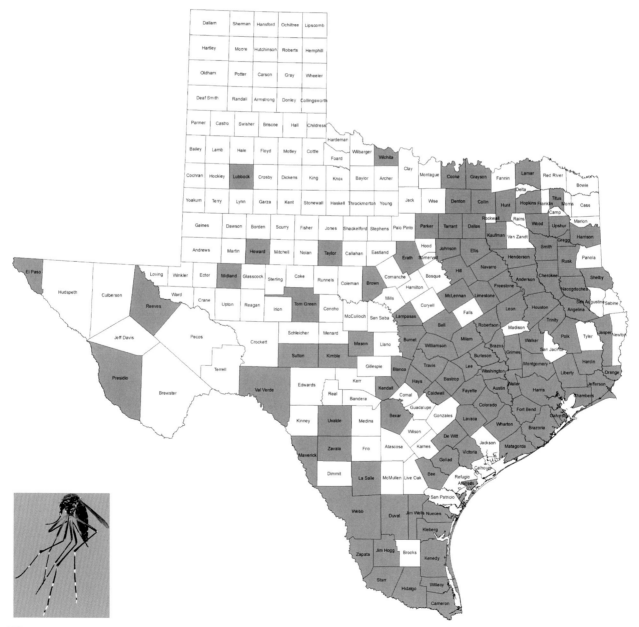

FIG. 2.40 Distribution of *Aedes aegypti* in Texas.

Aedes albopictus (Skuse, 1894)

It is commonly known as the "Asian tiger mosquito," and is native to tropical and subtropical areas of Southeast Asia. It is characterized by the white markings on its legs and a longitudinal white stripe on its thorax. This is a very similar species to *Ae. aegypti* and can be mistaken for the latter when the scales and other morphological markings are rubbed off. To differentiate them apart, the clypeus on *Ae. aegypti* has pale scales, whereas that of *Ae. albopictus* is unscaled and entirely black. The anterior surface of the midfemur in *Ae. aegypti* presents a stripe of white scales, whereas in *Ae. albopictus* is entirely dark scaled. The abdominal sterna III-V in *Ae. aegypti* is white scaled, while in *Ae. albopictus,* they alternate as the pale scales are restricted to basal and median areas and the apicolateral portions are covered in dark scales.

Bionomics. *Aedes albopictus* is a forest species that has adapted to the urban environment. It prefers unpaved dirt and ground vegetation, and unlike *Ae. aegypti*, it is not restricted to the domestic and peridomestic environment. It is an artificial container species, breeding in water-holding containers, and can also breed in tree holes, plant axils such as those of lucky bamboo and bromeliads, and in any other cryptic site that can hold water around dwellings. Its flight range is usually short within a few feet of its breeding site and close to the ground. It lays its eggs a few at a time in several containers contributing to their fast spread. Its dispersal and establishment is facilitated by transport of infected containers. It is an aggressive daytime biter that feeds on humans but can feed on many other mammals. It is more tolerable to the cold than *Ae. aegypti* due to its ability to go into diapause preventing egg hatching during the winter, which facilitates its distribution farther North in the United States. The susceptibility of *Ae. albopictus* to oral ingestion and ability to transmit virus by bite and occurrence in nature make it a competent vector of the following arboviruses: EEE, WEE, VEE, MAYV, CHIKV, RVF, SLE, WNV, YFV, DENV, JEV, ZIKV, LAC, TVT, and JC viruses (Estrada-Franco and Craig, 1995). *Ae. albopictus* may be infected with at least 32 viruses including 13 present in the United States (Vanlandingham et al., 2016).

Distribution. In August 1985, *A. albopictus* was first collected from tire dumps in Harris County and the City of Houston, TX (Sprenger and Whuithiranyagool, 1986). In Texas, it has been found in 92 counties (Fig. 2.41).

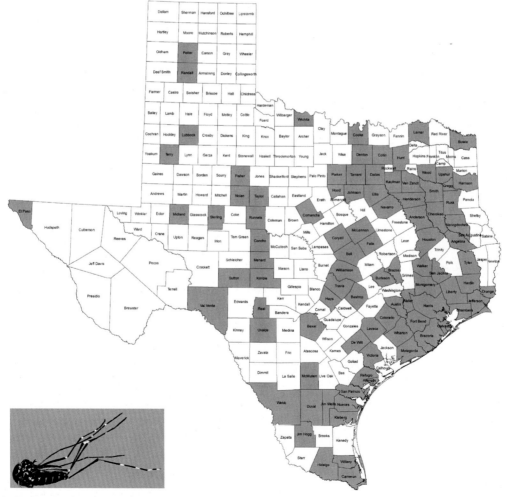

FIG. 2.41 Distribution of *Aedes albopictus* in Texas.

2.4 *Haemagogus* Williston

2.4.1 Subgenus *Haemagogus* Williston

Haemagogus equinus (Theobald, 1903)

A mosquito with antepronotal lobes enlarged and closed together, middorsally. The scutum has broad flat metallic scales and a trilobed scutellum with setae in three distinct groups. The mesopostnotum does not have setae. The hind coxa is in line with the mesomeron, and cell R_2 is as long as vein R_{2+3}. Setae are absent on postspiracular and prespiracular area and subcoastal vein.

Bionomics. It lives in the forest canopy where it lays its eggs in cavities, tree holes, and bamboo. The eggs adhere to the substrate until they are flooded with rain water. They can also be found in leaf axils of bromeliads. In urban environments, they are found in ground pools, rock holes, and artificial containers. Adults are active during the day and are found in tropical and deciduous forests and mangroves. They feed in the forest canopy but can feed on humans in open areas, on coastal and mangrove areas. Several species in this genus including *Hg. equinus* are vectors of sylvatic YFV (Harbach, 2013a,b).

Distribution. In Texas, it has been found in Cameron County (Fig. 2.42).

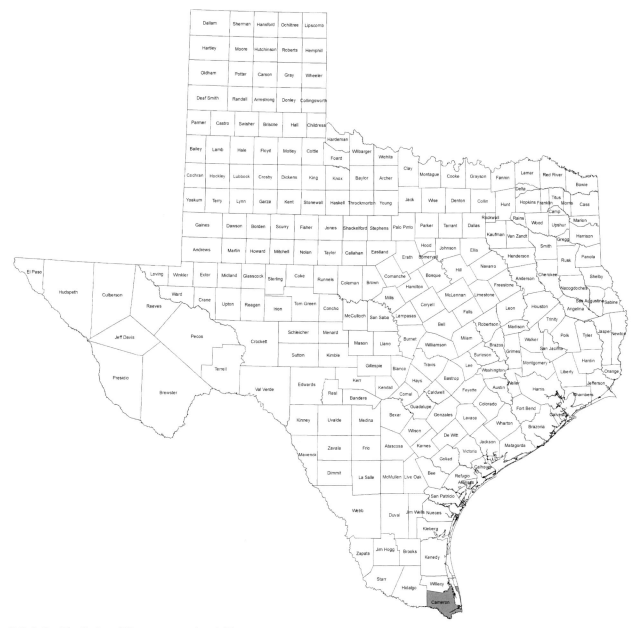

FIG. 2.42 Distribution of *Haemagogus equinus* in Texas.

2.5 *Psorophora* Robineau-Desvoidy

There are 11 species of *Psorophora* mosquitoes found in Texas. *Psorophora columbiae* (Dyar and Knab) tops the list of abundance as it is documented in 43% of the counties followed by *Ps. cyanescens* (Coquillett) (41%), *Ps. signipennis* (Coquillett) (38%), *Ps. ciliata* (Fabricius) (35%), *Ps. discolor* (Coquillett) (30%), and *Ps. ferox* (von Humboldt) (15%). The remaining five species are documented in less than 10% of the counties in Texas, where *Ps. mexicana* (Bellardi) has only been documented in Cameron County, TX (Joyce, 1945; Knight and Stone, 1977; Harrison et al., 2008). These mosquitoes are considered floodwater mosquitoes similar to *Aedes* mosquitoes and are abundant in the Gulf Coast Region laying their eggs singly on nonaquatic surfaces, at the edge of standing water in marshes, wetlands, woodland pools, and roadside ditches where the eggs survive for several years until flooded following heavy rains or high tide. They can be a nuisance after large climatic events such as tropical storms or hurricanes. Females are most active around sunset or in shaded areas during the day. They are strong flyers and fierce biters and can make areas inhabitable due to their large populations. They feed on humans, livestock, and pets. Certain species are considered potential vectors of VEE, SLE, and WNV.

2.5.1 Subgenus *Grabhamia* Theobald

Psorophora columbiae (Dyar and Knab, 1906)

A medium- to large-sized mosquito with a salt and pepper appearance also known as the Dark rice field mosquito. The proboscis has a broad pale band, and the abdomen has apical pale bands that do not interconnect medially in the last abdominal segments. The hind tarsomeres have basal pale bands. Hind tarsomere 1 has a median pale band, and the hind femur presents a preapical pale band. Wing scales are a mix of white and black without a discernable pattern.

Bionomics. It is the most common *Psorophora* found in large numbers along the coast and in southwest Texas. Eggs are deposited on the surface of soil subject to flooding by rainfall or overflow, especially by irrigation canals in the rice fields. Eggs are resistant to drying and to a certain degree of cold and overwinter in the egg stage. Breeding occurs throughout the year as they are multivoltine, and the number of generations is limited or dependent upon the frequency of the inundation of the eggs at summer temperatures. Larvae can be found in shallow ground pools subject to a prior period of drying, in temporary bodies of water, rice fields, roadside ditches, grassy swales, savannahs, and woodland pools. Adults inhabit tall grass, reeds, and herbaceous vegetation and are found on the stems near the ground or under the leaves. They can disperse relative far distances from their breeding sites and are fierce biters attacking at any time of the day or night. They feed on mammals including humans. They can often kill livestock and make it unbearable for people to remain outdoors at night or in shaded areas during the day and also feed on any hosts that enter their resting habitats during the day. It has been found naturally infected with EEE, WNV (Varnado et al., 2012; CDC, 2016), and POTV (Wozniak et al., 2001).

Distribution. In Texas, it has been found in 110 counties (Fig. 2.43).

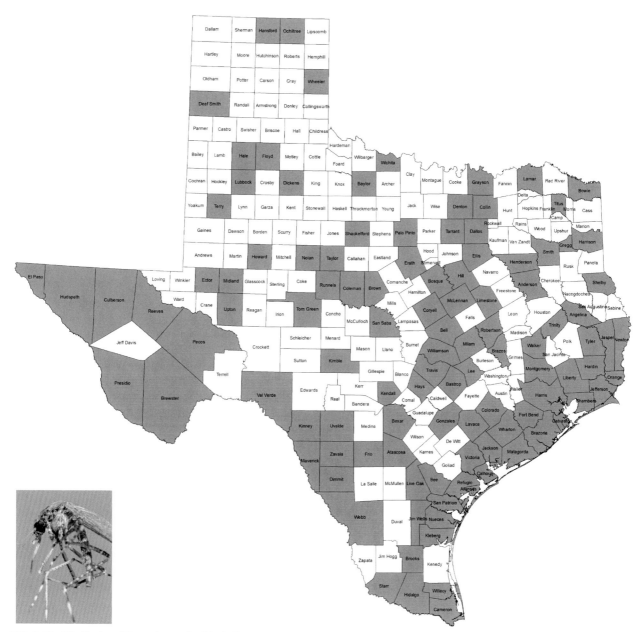

FIG. 2.43   Distribution of *Psorophora columbiae* in Texas.

Psorophora discolor (Coquillett, 1903)

A medium-sized species similar to *Ps. columbiae* in appearance but with brown and white scales. The proboscis has a broad pale band, and the thorax is dark brown covered with narrow pale to golden brown scales. The first abdominal ter-gite has a median patch of grayish-white scales, and the rest of the segments are covered by pale or white scales speckled with dark scales, which are more evident basolaterally. The hind tarsi have pale basal bands, and the hind femur has a faint preapical band. The wings have pale scale areas on some veins and pale and brown scales in other areas of the wings.

Bionomics. A widespread mosquito occurring throughout the year in low numbers in temporary rain pools and semi-permanent water over most of Texas. Its eggs may be abundant on old, dry cow dung and are resistant to drying for up to 2 years. Larvae are found in ground pools that are flooded during the summer. They inhabit shallow collections of water in barnyards and pastures, especially if water is fouled with cattle manure. Farm ponds and rice fields are also suitable habitats for their development. This is a multivoltine species as *Ps. columbiae* and overwinters in the egg stage. Adults are active during the daytime and at dusk. In areas where they proliferate, they are troublesome biters, especially at night. Females feed on mammals including humans and are a pest for both humans and livestock. Cattle provide the main source of blood meals, while horses, mules, and hogs are intermediate sources. It is associated with VEE (Wilkerson et al., 2015).

Distribution. In Texas, it has been found in 75 counties (Fig. 2.44).

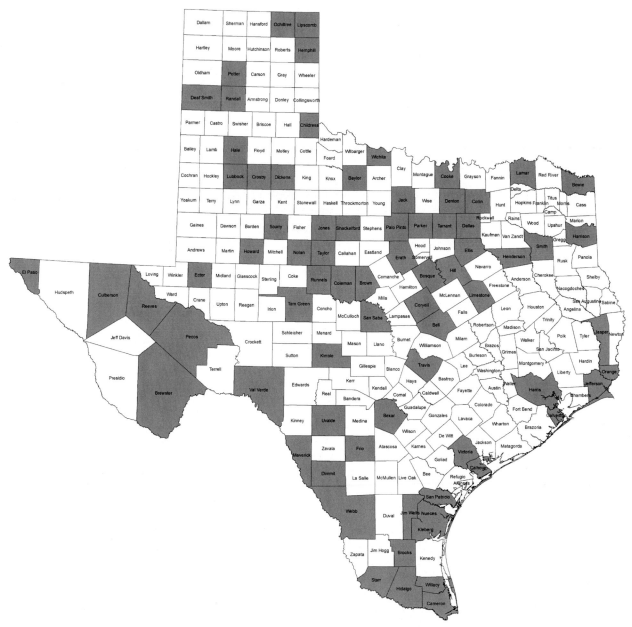

FIG. 2.44   Distribution of *Psorophora discolor* in Texas.

Psorophora signipennis (Coquillett, 1904)

A medium-sized mosquito with a dark proboscis that has a wide median whitish-yellow band in the middle. The palps are short and dark and speckled with few pale scales. The thorax is dark brown with fine narrow golden-brown scales becoming yellowish on the sides and the prescutellar space. The first abdominal segment has a median patch of white scales, and the remaining tergites on the abdomen are mainly white-scaled speckled with dark scales. The femora are dark and speckled with pale scales, and knee spots are present. The hind tarsus with segment 1 is white ringed basally, and dark ringed subbasally and apically. Segments 2 and 5 are pale scaled on basal third or basal half with the rest being dark.

Bionomics. It is often found in open areas rather than woodlands. Larvae are found in grassy pools, temporary ground pools exposed to full sun, flooded meadows after summer rains, and irrigated areas where they can have several generations per year. It overwinters in the egg stage. Adult females can be aggressive biters that attack in sunlight or shaded areas from early morning to late at night, feeding on humans and livestock.

Distribution. In Texas, it has been found in 96 counties (Fig. 2.45).

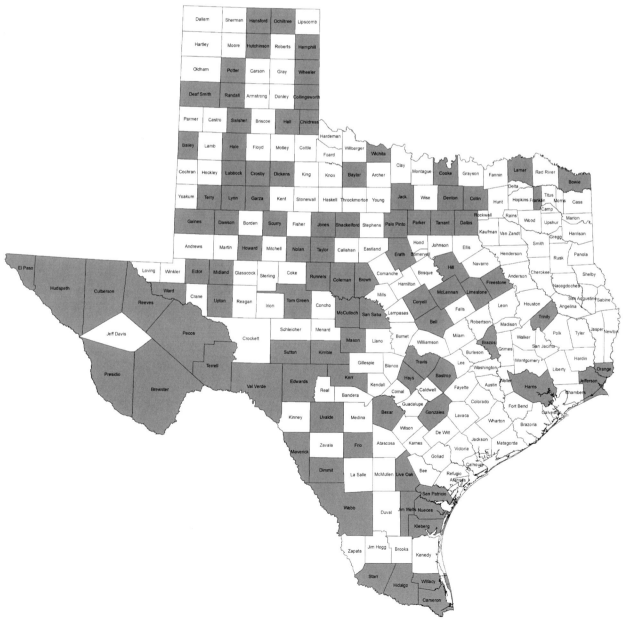

FIG. 2.45 Distribution of *Psorophora signipennis* in Texas.

2.5.2 Subgenus *Janthinosoma* Lynch Arribalzaga

Psorophora cyanescens (Coquillett, 1902)

A medium- to large-sized mosquito. The proboscis, palps, wings, and tarsi are dark purple. The thorax has yellow or golden-green scales. The abdomen presents apical median triangular patches of pale scales on segments 2–6 and do not interconnect medially on last segments as they do in segments 1 and 2. Tarsi are entirely purple as in *Ps. ferox*, *Ps. mathesoni*, and *Ps. horrida*.

Bionomics. Its eggs are laid in moist soil and require a period of dryness before they hatch once they are flooded after rainfall or overflowing. Hatching does not occur the same year; they are laid and overwinter in the egg stage. Larvae develop in temporary pools in open or partially shaded areas. They have multiple generations, and abundance is greater after rains in June and August in the south central United States. Adults are vicious biters and are active at dusk, dawn, night, and at any time during the day. They proliferate in areas following summer rains and are a burden to livestock and interfere with farm work. They are found in woodlands and thickets but will come out to open fields seeking for a blood meal. They feed on large mammals including humans and on any warm-blooded animal that invade their habitat any time of the day. They feed for long periods until they are unable to fly. They can disperse distances of 2 miles or more. They are not known to transmit any human pathogens, although VEE has been documented (Wilkerson et al., 2015).

Distribution. In Texas, it has been found in 103 counties (Fig. 2.46).

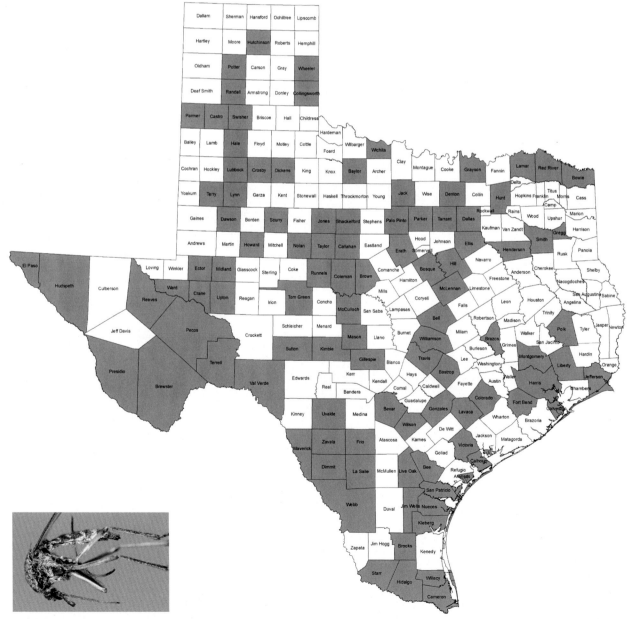

FIG. 2.46 Distribution of *Psorophora cyanescens* in Texas.

Psorophora ferox (von Humboldt, 1819)

A medium-sized woodland mosquito with a dark-scaled proboscis and short dark palps. The prespiracular and postspiracular setae are present on the side of thorax, and the first abdominal segment has a median patch of purple scales. The remaining tergites are dark scaled with a purplish reflection on the dorsum and with apicolateral triangular patches of whitish-yellow to golden-yellow scales. The middle tibia and tarsi are dark scaled, and segments 4 and 5 of the hind tarsi are white scaled, giving it the common name of "white sox mosquito."

Bionomics. Eggs can be found on soil under surface of plant debris in depressions and shaded floodplains of rivers and smaller streams. Eggs can withstand seasonal dryness, cold, and prolonged flooding. It overwinters in the egg stage and has multiple generations per year. Larvae occur in transient woodland pools, temporary rain-filled pools, pot holes, and overflow pools along streams. Adults are active at dusk, dawn, and at night and can disperse in the vicinity of their breeding sites. They are abundant in woods and thickets. Females are vicious biters and venture into the open field seeking a host to feed upon. They feed on any animal that ventures into their habitats including mammals, humans, birds, and reptiles, at any time of the day or night. Several viruses have been detected including EEE (Armstrong and Andreadis, 2010; Oliver et al., 2018), WNV (CDC, 2016), VEE (Chamberlain et al., 1956), SLE, WYOV (Wilkerson et al., 2015), Ilheus virus, Una virus (Turell et al., 2005), and CV (Andreadis et al., 2014), and it is also known to carry eggs of *Dermatobia* (Arnett, 1949; Carpenter and LaCasse, 1955).

Distribution. In Texas, it has been found in 37 counties (Fig. 2.47).

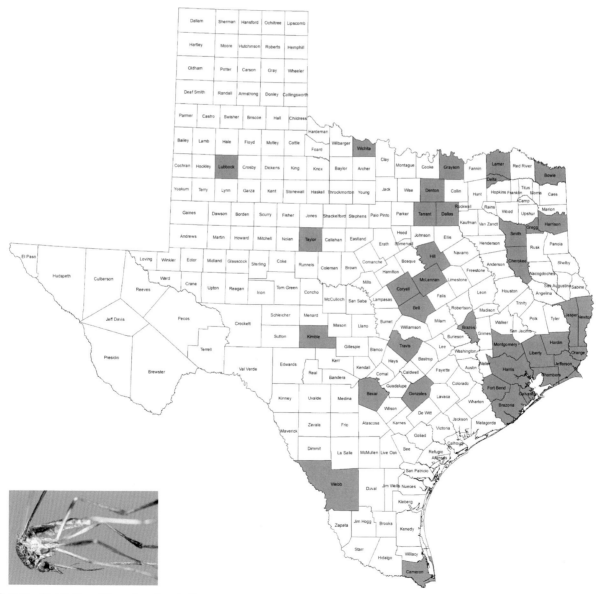

FIG. 2.47   Distribution of *Psorophora ferox* in Texas.

Psorophora horrida (Dyar and Knab, 1908)

A medium-sized, woodland mosquito with a long dark-scaled proboscis and short palps that are less than one-third the length of the proboscis where the fourth segment is curved and about equal in length to the other three combined. Both, the prespiracular and postspiracular setae, are present, and the scutum has a broad median longitudinal dark-scaled stripe bordered laterally by white scales. The front and middle femora are dark purplish scaled, and the half basal part of the hind tarsi are pale scaled.

Bionomics. It is multivoltine and overwinters in the egg stage as other floodwater mosquitoes. Adults are active at dusk, dawn, and at night, feeding on mammals and rest on vegetation near their breeding sites.

Distribution. In Texas, it has been found in 22 counties (Fig. 2.48).

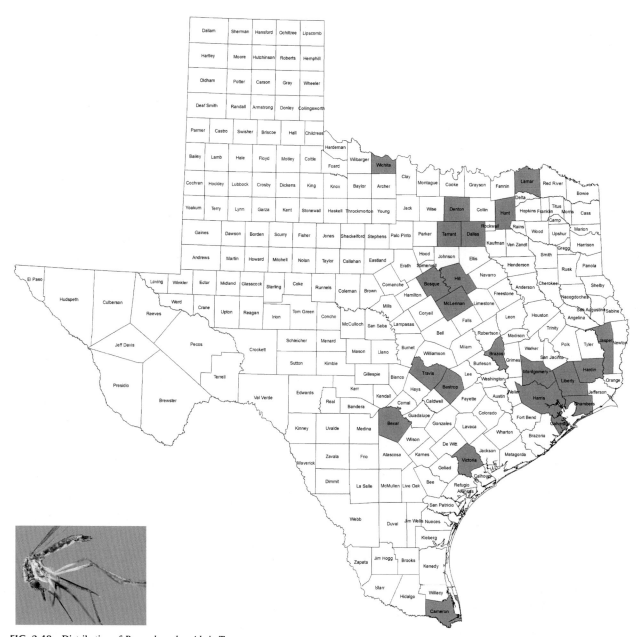

FIG. 2.48   Distribution of *Psorophora horrida* in Texas.

Psorophora longipalpus (Randolph and O'Neill, 1944)

A medium-sized mosquito that resembles *Ps. horrida* but lacks the knee spots and the patch of white scales at the base of the hind coxa. The proboscis is long, dark scaled with a violet reflection; the palps are more than one-third as long as the proboscis and have the same coloration as the proboscis. The scutum has a broad median dark stripe delimited laterally by white to yellowish scales. The first abdominal segment has a patch of yellowish-white scales, and the remaining segments are dark with apical patches of white scales laterally on segments 4–6. The front and middle femora are dark, and the posterior surface is pale. Basal two-thirds of hind femur are pale. The front and middle tibiae and tarsi are entirely dark with a violet reflection.

Bionomics. Larvae develop in heavily shaded temporary pools after summer rains. In the southcentral United States, it overwinters in the egg stage and may have several generations in 1 year. It is usually associated with other *Psorophora* and *Aedes* species. Males can be easily recognized while resting by their bulbous terminalia. They rest in shaded vegetated areas near their breeding sites.

Distribution. In Texas, it has been found in 16 counties (Fig. 2.49).

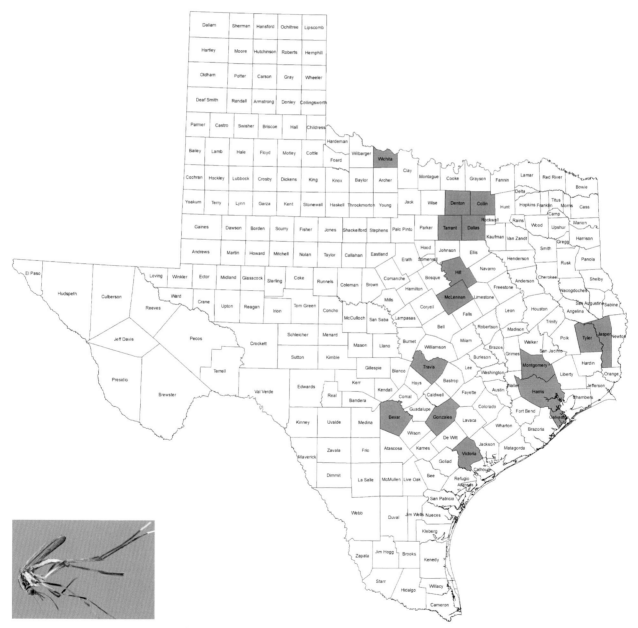

FIG. 2.49 Distribution of *Psorophora longipalpus* in Texas.

Psorophora mathesoni (Belkin and Heinemann, 1975)

A small- to medium-sized mosquito with a dark proboscis and dark, short palps. The thorax has a broad longitudinal median dark brown stripe bordered laterally by yellowish-white scales. The first abdominal tergite has white scales, and the remaining segments are dark scaled with a purplish reflection and apicolateral pale patches. The legs are mostly purple and resemble *Ps. ferox* with the exception of the fourth segment of the hind tarsus, which is white scaled.

Bionomics. Eggs are deposited similarly as other *Psorophora* species; hatching any time, sites are flooded with water at summer temperatures. They overwinter in the egg stage and are resistant to drying, freezing, and submersion. They can have more than one generation per year, especially during the summer months. Larvae develop in temporary pools, in woodland pools floating in a mat of dense plant debris, and in association with other *Psorophora* species such as *Ps. ferox*. They disperse near their breeding sites and are active at dusk, dawn, and night. Females are vicious biters attacking during the day in woodlands and feed on birds, reptiles, and mammals, including humans.

Distribution. In Texas, it has been found in 10 counties (Fig. 2.50).

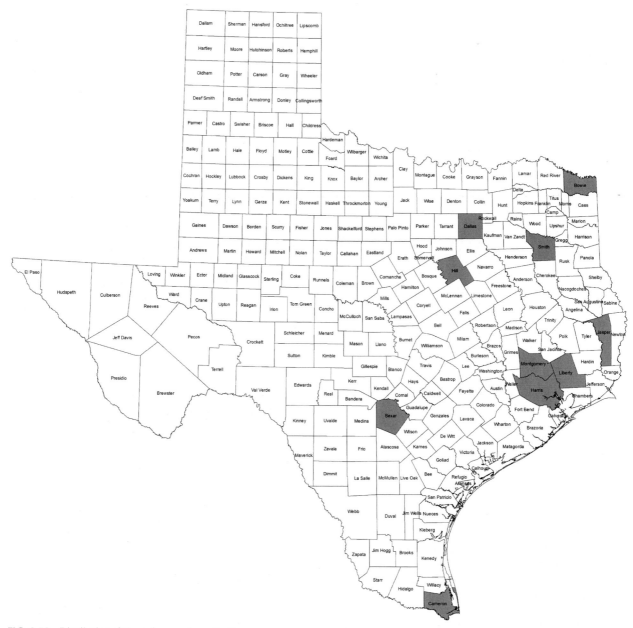

FIG. 2.50 Distribution of *Psorophora mathesoni* in Texas.

Psorophora mexicana (Bellardi, 1859)

It is a small-sized mosquito characterized by having the last hind tarsal segment white scaled. It is a distinct species and not a southern variant of one of the other US species. Thus under new characteristic diagnosis, it has the antennal pedicel without scales; the subspiracular area and proepimeron are bare and with either pale scales on basal 0.33 of hind tarsomere or with base of hind tarsomere 5 with dark scales.

Bionomics. Unlike other floodwater species, it was found biting on horses and humans close to salt water suggesting that it does not breed in temporary pools of fresh water. It is believed that cryptic habitats such as mangrove areas and brackish water are the reason for the lack of larval stages collected coupled with their rapid development.

Distribution. It has only been collected from Brownsville in Cameron County, TX (Joyce, 1945; Knight and Stone, 1977) (Fig. 2.51).

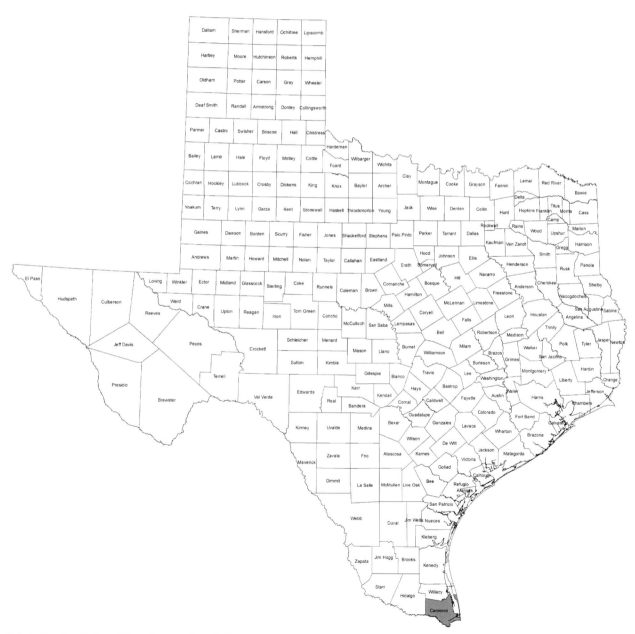

FIG. 2.51 Distribution of *Psorophora mexicana* in Texas.

2.5.3 Subgenus *Psorophora* Robineau-Desvoidy

Psorophora ciliata (Fabricius, 1794)

This is a large dark brown mosquito with a proboscis that is yellowish and dark apically and with long scales basally. The palps are short, about one-third the length of the proboscis. It has a median golden stripe on the thorax. The abdominal scales are pale yellow to brown. Their legs have long, dark, erect scales on the femur and tibiae, and the hind tarsi have broad pale bands or rings basally.

Bionomics. Eggs are deposited in moist soil under plant debris in areas subject to inundation such as in woodlands and meadows. It has multiple generations per year and is abundant from March to October in the South and from May to September farther north. Eggs hatch any time in dried sites flood with water at summer temperatures, whereas those eggs that have been submerged during the winter do not hatch the following summer until they have been dried. They overwinter in the egg stage and are resistant to submergence, freezing, and drying. Larvae occur in temporary rain-filled pools and are predacious, commonly feeding on other larvae of some associated species such as *Ps. columbiae*. Larvae can also be found in transient ground pools, rice fields, ditches, and woodland pools. Adults can disperse moderate distances from their breeding sites and are active at dusk, dawn, and night but attack any time during the day. They are an outdoor species inhabiting herbaceous vegetation and can enter buildings to feed. They are aggressive biters that feed on mammals including humans and are considered troublesome in open fields or near woods. It has been found naturally infected with EEE (Varnado et al., 2012), associated with WNV (CDC, 2016), TENV (Wozniak et al., 2001), and VEE (Wilkerson et al., 2015).

Distribution. In Texas, it has been found in 89 counties (Fig. 2.52).

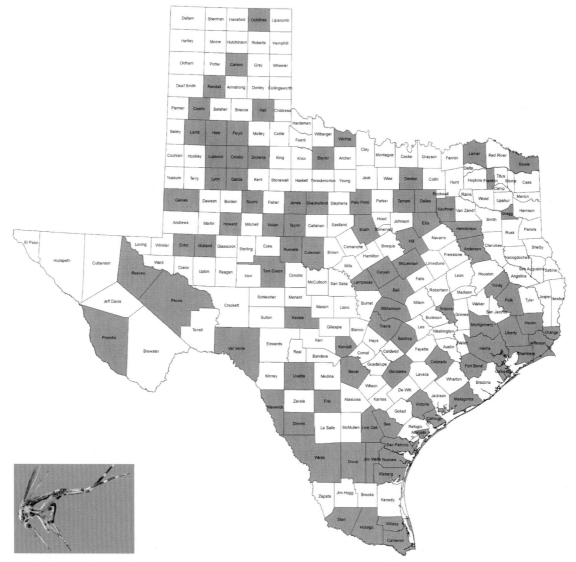

FIG. 2.52 Distribution of *Psorophora ciliata* in Texas.

Psorophora howardii (Coquillett, 1901)

A large mosquito with a long proboscis that is brown and darker apically, and the palps are about half the length of the proboscis. The occiput has white scales and black setae. The mesonotum has a median longitudinal black stripe with dark setae and patches of white scales laterally. The first abdominal segment has a broad patch of white scales, and the rest of the segments are blue-black scaled dorsally and with white scales laterally and apically. As in *Ps. ciliata*, it has tufts of dark scales on the hind tibia and femur and pale bands at the base of each hind tarsi segment.

Bionomics. Two or more broods may appear from one site when flooding occurs. Eggs lie on soil covered by leaves under complete shade by forest canopy and are resistant to drying, flooding, and prolonged freezing as those of *Ps. ciliata*. *Psorophora howardii* overwinters in the egg stage. Larvae are found mostly in unshaded or partly shaded temporary rain-filled pools, and drainage ditches, and the last three larval instars are predacious feeding on larvae of other associated species and can feed on other organisms that may cross their path in the aquatic environment such as water fleas, tadpoles, and midges. When overcrowding occurs, they can become cannibalistic. Larvae are found from March to October in the south and May to September in the northern part of its range. Adults are active anytime during the day or night, at dusk and at dawn. They are persistent biters and are found near their larval habitats dispersing up to 2 miles. They feed primarily on mammals and will attack any time their habitats are invaded during the day. It is associated with WNV (CDC, 2016).

Distribution. In Texas, it has been found in 20 counties (Fig. 2.53).

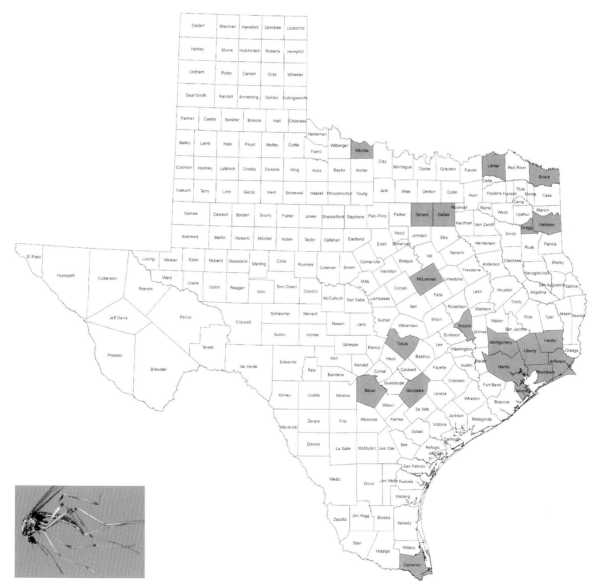

FIG. 2.53 Distribution of *Psorophora howardii* in Texas.

2.6 *Culex* Linnaeus

There are 19 species of *Culex* mosquitoes in Texas. The most predominant species is the main vector of SLE and WNV, that is, *Culex quinquefasciatus*, the "Southern house mosquito" documented in 59% of Texas, followed by *Cx. tarsalis* Coquillett (55%), *Cx salinarius* Coquillett (40%) and *Cx erraticus* Dyar, and Knab (39%), whereas the least documented are *Cx. pilosus* (Dyar and Knab) (Harrison County) and *Cx. arizonensis* Bohart (Culberson County). Some species are vectors of viruses including WNV, SLE, WEE, EEE, and Filariasis.

2.6.1 Subgenus *Culex* Linnaeus

Culex chidesteri (Dyar, 1921)

A brown medium-sized mosquito. The proboscis is dark scaled with a broad median area of pale scales. The scutum is brown with middorsal acrostichal seta present, and the pleura has several patches of pale scales. The abdomen's first tergite has a patch of dark bronze-brown scales, and the remaining tergites are dark with a narrow basal pale band extending to basolateral white patches.

Bionomics. Larvae have been collected in grassy roadside pools, ponds and ditches, and at the edge of ponds overgrown with hyacinths and cattails. It breeds throughout the year, and some of the associated species are *An. albimanus*, *Cx. peccator*, *Cx. erraticus*, *Cx. salinarius*, and *Mansonia titillans* Walker.

Distribution. In Texas, it has been found in Cameron, Hidalgo, and Val Verde counties (Fig. 2.54).

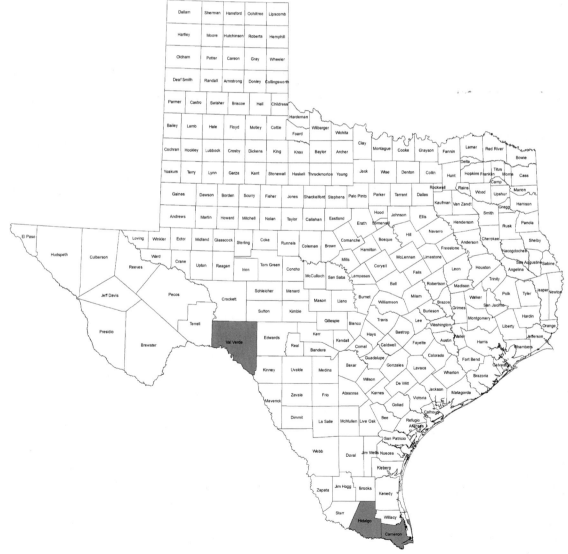

FIG. 2.54 Distribution of *Culex chidesteri* in Texas.

Culex coronator (Dyar and Knab, 1906)

A small- to medium-size mosquito with a dark-scaled proboscis that is pale ventrally near the middle. The first abdomen's tergite has a median patch of dark bronze–brown scales, and the remaining tergites have a white basal band joining basolateral patches. Abdominal sterna are mostly pale scaled. The tarsi are white ringed, basally and apically.

Bionomics. *Culex coronator* occurs in rain-filled pools and artificial containers such as livestock water troughs, and a variety of different habitats, such as stagnant or slow moving ground pools and seeps, ditches and in sunny or shaded habitats. They have multiple broods per year and overwinter in the adult stage. Adults are active at dusk, night, and dawn where females feed on mammals and occasionally on humans and possibly on birds. It has tested positive for WNV in Texas and Louisiana (Varnado et al., 2012). Other viruses such as VEE and SLE have been isolated from field collected females (Burkett-Cadena, 2013).

Distribution. In Texas, it was found in the southwestern counties, particularly in the Lower Rio Grande Valley, and recently in 127 counties (Sames et al., 2019), although their map depicts 125 counties. However, the missing 2 counties

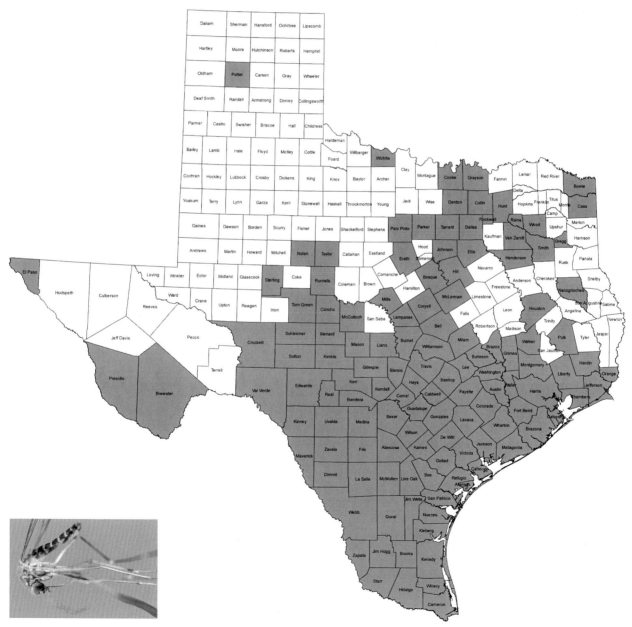

FIG. 2.55 Distribution of *Culex coronator* in Texas.

(Aransas and Calhoun) are included in Table 1 of their article. Its distribution now extends mainly to all the Texas Gulf Coastal plains (Gulf Prairies and Marshes, Pineywoods, Post Oak Savannah, Blackland Prairies and Cross Timbers), Edward's Plateau (Hill country) and the South Texas Plains (Fig. 2.55).

Culex declarator (Dyar and Knab, 1906)

In the United States, *Cx. declarator* has been found in Texas (Fisk and LeVan, 1940) and recently in Florida (Darsie and Shroyer, 2004). However, the specimens from Texas were recorded as *Culex virgultus* Theobald (Breland, 1953, 1954). This species though became unrecognized, and *declarator* was resurrected for *virgultus* (Stone, 1956). *Culex declarator* adults resemble those of *Cx. nigripalpus*. Moreover, there are seven listed synonyms for *Cx. declarator* that indicate it is a very variable species and the characteristics that separates it from other Nearctic *Culex* may not be the best one to distinguish it (Darsie and Ward, 1981).

The following characters are described for *Cx. virgultus* by Carpenter and LaCasse (1955): a medium- to small-size mosquito. The proboscis is dark with a median pale area, ventrally. The thorax is brown, and the scutum has fine narrow bronze-brown curved scales and of lighter color on the anterior and lateral margins of the prescutellar space. The first tergite has a median patch of dark scales with bronze to metallic blue-green reflection with narrow basal bands of white scales widening at the sides to form lateral spots. The legs are dark with some bronze reflection where the femora and tibiae are pale scaled. The morphological characters key out similar to *Cx. coronator*, but hind tarsomere 5 has a basal band of pale scales (Darsie and Ward, 2005).

Bionomics. The larvae are found in rock pools, swamps, cement drains, rot cavities in trees and coconut husks. Adults have been collected by several trapping devices including CDC light traps baited with dry ice and by sweeping vegetation in diurnal resting habitats with mechanical aspirators. Isolations of SLE have been made from pools of *Cx. declarator* in Brazil (Monath et al., 1980) and in Trinidad along with the Turlock virus (Aitken et al., 1964, 1969). Although experimental

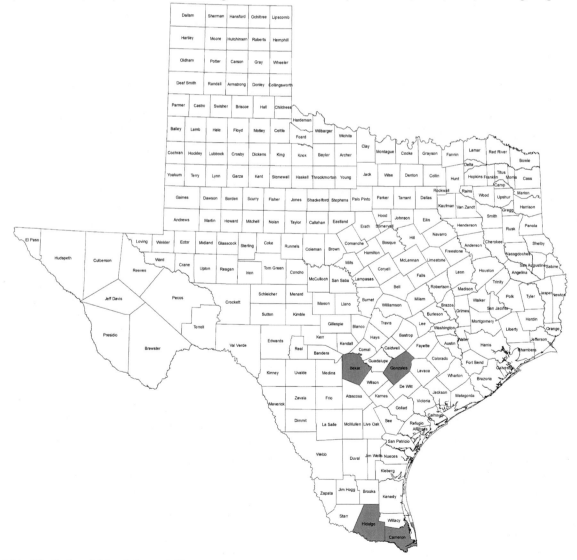

FIG. 2.56 Distribution of *Culex declarator* in Texas.

infection with VEE has been conducted unsuccessfully, the species was able to get infected in the midgut tissues (Turell, 1999). It may also be a potential vector of dog heartworm disease (Labarthe et al., 1998).

Distribution. In Texas, it has been found in four counties (Fig. 2.56).

Culex erythrothorax (Dyar, 1907)

A medium-sized, dark-legged *Culex* mosquito. The proboscis is dark scaled with some pale scales ventrally. The scutum is reddish brown with fine hairlike golden scales. The first tergite on the abdomen has a median patch of reddish-brown scales. Tergites II–VII have brownish-black scales, and the eight tergite is mostly pale scaled. Hind tarsomeres are dark or with narrow basal bands of pale scales.

Bionomics. This is a species of the marshes bearing tule, cattail, or bulrush (*Scirpus*), *Typha*, and other grasses. Larvae are found in shallow ponds with heavy vegetation. It attacks viciously and causes a painful bite and feeds on humans and birds. Its abundance occurs late in the season.

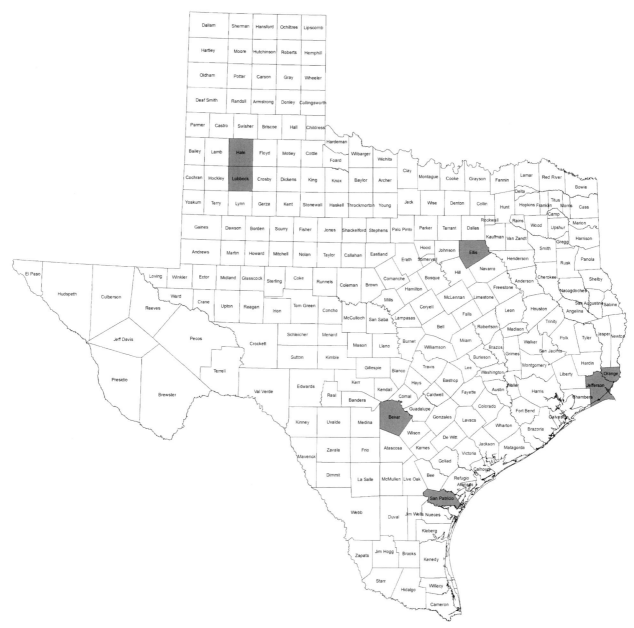

FIG. 2.57 Distribution of *Culex erythrothorax* in Texas.

Distribution. In Texas, it has been found in seven counties (Fig. 2.57).

Culex interrogator (Dyar and Knab, 1906)

A small mosquito with a dark-scaled proboscis with a median pale area ventrally. Middorsal acrostichal setae are present in scutum, which is brown with golden-brown scales, and without a pair of submedian pale spots. The first abdominal tergite has a patch of dark scales, and remaining tergites are dark scaled with basal bands of white scales widening at the sides. Wing cell R_2 3.0–4.0 times the length of vein R_{2+3}.

Bionomics. Larvae are found in foully waters in ground pools, wheel ruts, tree holes, rain barrels, and other similar aquatic habitats. They have also been found in irrigations seepage areas in Texas. It has been recently recorded in Florida from storm water drains containing considerable amounts of grass cuttings and from a roadside swale. Associated species found in the storm drains were *Cx. nigripalpus* Theobald, *Cx. quinquefasciatus*, and *Cx. salinarius*.

Distribution. In Texas, it has been found in five counties (Fig. 2.58).

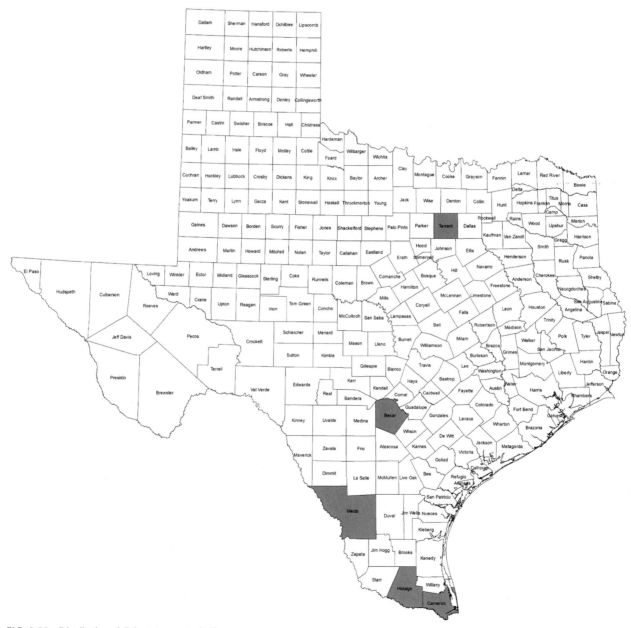

FIG. 2.58   Distribution of *Culex interrogator* in Texas.

Culex nigripalpus (Theobald, 1901)

A medium-sized mosquito with dark brown palps and proboscis. The latter is usually pale underneath on the basal half. The thorax is dark brown. The scale patches on thoracic pleura are in groups of fewer than six scales. The abdomen is brown scaled with basolateral white spots and sometimes with basal white bands on some segments.

Bionomics. Larvae are found in ditches, ground pools, grassy pools, and permanent or semipermanent marshes. They are occasionally found in artificial containers and in leaf axils of plants. Larvae and adults are present throughout the year in the extreme south. It is multivoltine and overwinters in the adult stage. They are most abundant in coastal regions in mild winter temperatures. They have a tendency to move northward during warm periods but die off when exposed to extended periods of below-freezing temperatures. They are active at dusk and dawn and feed on large mammals, including humans. It is considered a vector of EEE, SLE, and WNV (WRBU, 2019).

Distribution. In Texas, it has been found in 23 counties (Fig. 2.59).

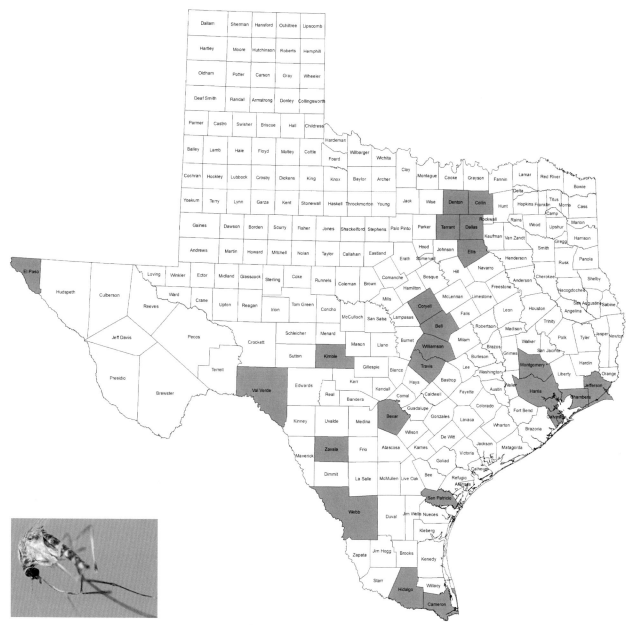

FIG. 2.59 Distribution of *Culex nigripalpus* in Texas.

Culex quinquefasciatus (Say, 1823)

The "southern house mosquito" is a small to medium mosquito and is very similar to *Cx. pipiens* Linnaeus, the "northern house mosquito." They can crossbreed where they overlap (36°–39° north latitude) (Barr, 1957) and intermediate forms can arise. The thorax is light brown, and the abdomen is dark brown. It has pale, convex, or crescent moon–shaped basal bands narrowing laterally where it may or occasionally may not join the lateral patches of pale scales on each segment.

Bionomics. Larvae are found in highly organic, polluted waters such as catch basins, storm drains, ditches, cesspools, ground pools, rain barrels, tubs, and other artificial containers. Eggs are laid in rafts. They have multiple generations throughout the year and overwinter in the adult stage. They are active at dusk, dawn, and at night where they feed on birds and mammals, including humans. They enter houses readily and are found in dark, moist, shaded areas such as those underneath pier and beam houses. In Harris County, TX, they are most abundant in May-June and reach a second smaller peak sometime in the fall in October-November. It is a vector of avian malaria, lymphatic filariasis, SLE, WEE, WNV, and dog heartworm disease (Carpenter and LaCasse, 1955; Sirivanakarn, 1976).

Distribution. In Texas, it is widely distributed and has been found in 150 counties (Fig. 2.60).

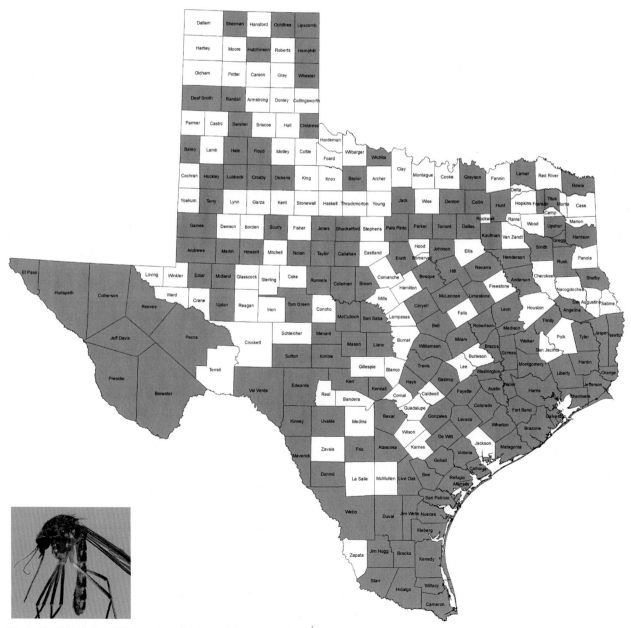

FIG. 2.60 Distribution of *Culex quinquefasciatus* in Texas.

Culex restuans (Theobald, 1901)

A medium-sized mosquito with dark short palps and a dark proboscis that has some pale scales on the underside. The thorax is brown to reddish brown and has two small pale spots on the scutum; a characteristic that differentiates it from other *Culex* species. The first abdominal segment has a median patch of dark bronze-brown scales. The rest of the segments have a white band similar to those in *Cx. quinquefasciatus*, but it is less prominent or more evenly rounded in the middle and usually joining the basolateral patches of pale scales. Wing cell R_2 is 4.5 or more the length of vein R_{2+3}.

Bionomics. Eggs are deposited in rafts in the meniscus of water during the early morning hours of the day. Larvae are found in a variety of aquatic habitats such as ditches, pools in streams, woodland pools, and artificial containers. They can inhabit bodies of water that are clean or with high organic material. They have been associated with decaying grass and leaves and fouly water with dead animals. They have several generations per year and overwinter in the adult stage. They are most abundant in the cooler months from spring and early summer but can be present throughout the year in the southern part of its range. They are active during the night, dusk and dawn, feeding on birds, and mammals, including humans. They are regarded as troublesome biters, can enter dwellings, and disperse up to 2 miles. It is a vector of WNV (Sardelis et al., 2001; Ebel et al., 2005), SLE (Chamberlain et al., 1959; Mitchell et al., 1980), EEE (Morris et al., 1973, 1975), and WEE (Norris, 1946).

Distribution. In Texas, it has been found in 76 counties (Fig. 2.61).

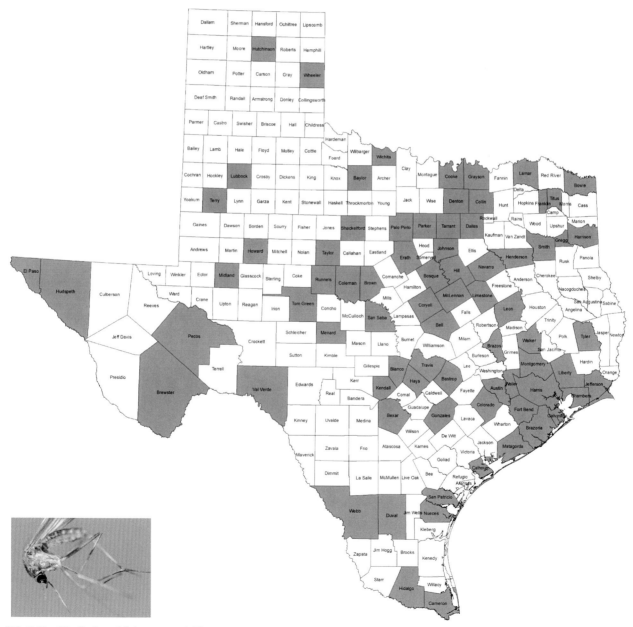

FIG. 2.61 Distribution of *Culex restuans* in Texas.

Culex salinarius (Coquillett, 1904)

A medium-size mosquito with a dark proboscis and short palps known as the saltwater or brackish water mosquito. The thorax is light to dark brown with golden-brown scales, and the thoracic pleura have patches of six or more white scales. The abdominal segments have bronze or copper basal bands, and segments 7 and 8 are almost entirely covered with pale bronze scales commonly referred to as "dingy yellow" scales.

Bionomics. It is associated with brackish waters. Eggs are deposited in rafts. Larvae can be found in grassy pools, fresh or brackish water, ditches, and ponds and occasionally in rain barrels, bilge water in boats, and cattle tracks and sometimes in stump holes. It has also been found in permanent and semipermanent ponds, marshes, and freshwater swamps. It has multiple generations per year, and females overwinter in their hibernating adult stage farther north. Adults can be collected in dwellings especially during hibernation in outbuildings and other resting places. They can disperse moderate distances of up to 5 miles during migration times. Adults are active at dusk, night, and dawn feeding on birds, mammals, and humans. They are most abundant in the Atlantic and Gulf coastal regions. It is considered a vector of WNV (Sardelis et al., 2001) and SLE and a secondary vector of EEE in other southern states (WRBU, 2019).

Distribution. In Texas, it has been found in 102 counties (Fig. 2.62).

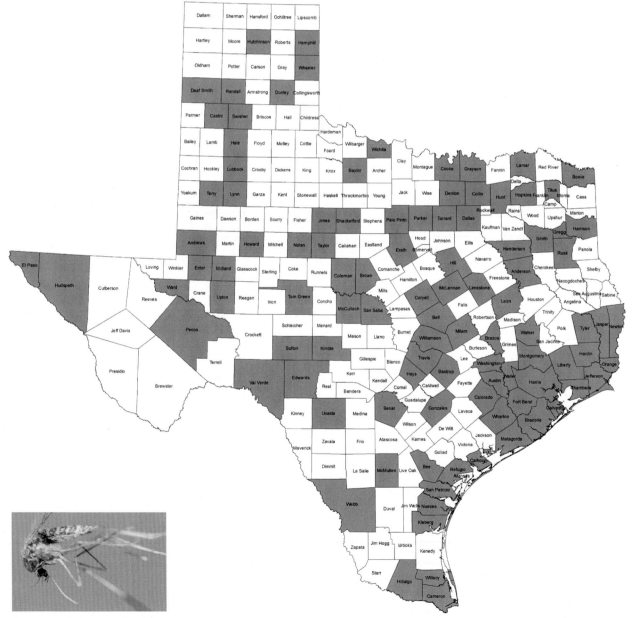

FIG. 2.62 Distribution of *Culex salinarius* in Texas.

Culex stigmatosoma (Dyar, 1907)

A medium-size mosquito known as the banded foul water mosquito with a dark-scaled proboscis that has a median white band similar to that on *Cx. tarsalis*. These two species have white bands in their hind legs similar to *Cx. coronator.* The palps are short, dark, and tipped with white. The wing scales are narrow and dark with a few white scales on the costa.

Bionomics. Larvae are found in foully water such as that from sewage plants and street drains and in polluted water around dairies. They are occasionally found in clean water and sometimes in artificial containers. Females feed mainly on birds but can feed on mammals, including humans, sheep, dogs, pigs, and reptiles. Western equine encephalitis has been isolated from wild-caught *Cx. stigmatosoma* in California (Hammon and Reeves, 1943b). It has also been identified as a WNV carrier (Goddard et al., 2002) and a potential SLE vector (Reisen et al., 1992).

Distribution. In Texas, it has been found in 27 counties (Fig. 2.63).

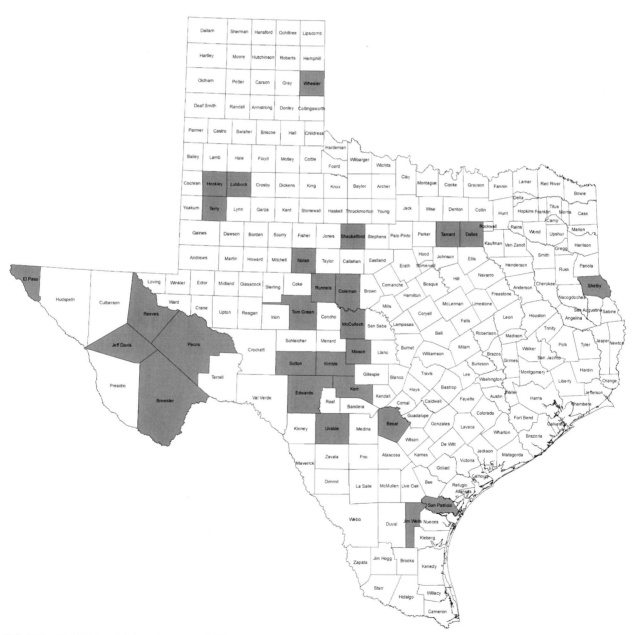

FIG. 2.63 Distribution of *Culex stigmatosoma* in Texas.

Culex tarsalis (Coquillett, 1896)

A mosquito of similar size and coloration as *Cx. stigmatosoma* and *Cx. coronator.* Its scutum has two narrow submedian pale stripes extending about half the length from the posterior end and ending in two pale spots. The abdominal segments have basal bands with abdominal segment II having a triangular patch of pale scales. The legs are dark scaled, and the posterior surface of the femora and tibiae are white scaled. A narrow line of white scales or row of spots on the anterior surface of femora and tibiae are present and often on the first tarsi.

Bionomics. Eggs are laid in rafts on water surfaces and are able to withstand drying. Larvae can be found in a variety of aquatic habitats with emergent vegetation; in grassy marshes, seeps, irrigation wastes, ditches, hoof prints, ornamental pools, woodland pools, ponds, rain barrels and fouly water; and in clear, unpolluted waters in mountainous areas. They are also found at high elevations in clear or fouly water, fresh or alkaline, and shaded or sunlit areas. Larval development ranges from late spring to early fall with several generations throughout the year as they occur in Texas, and peak abundance is usually during August or September. They can fly considerable distances from their breeding sites, and the females are persistent and painful biters active at dusk, night, and dawn. They feed on birds and mammals, including humans, horses, and cows as incidental hosts. Adults readily enter dwellings for blood meals and hide in shaded areas during the day. They overwinter in the adult stage hibernating during the winter in basements, caves, cellars, and outbuildings protected from the cold. It is a vector of WEE, SLE (Hammon and Reeves, 1943a, b), CEV (Carpenter and LaCasse, 1955), and WNV (CDC, 2016).

Distribution. In Texas, it has been found in 140 counties (Fig. 2.64).

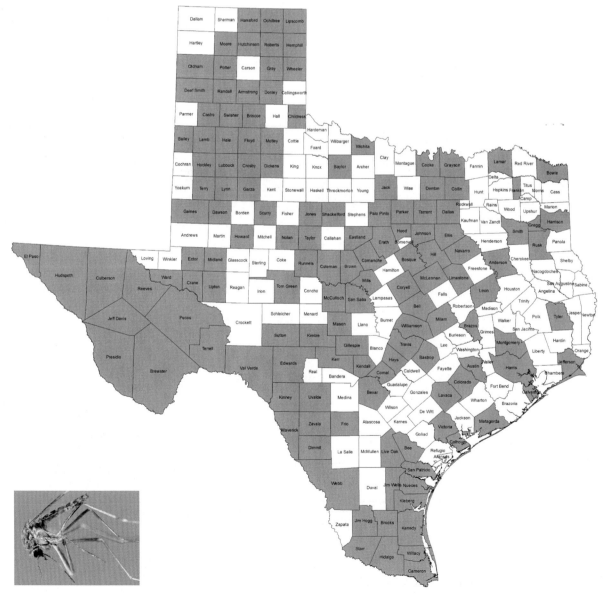

FIG. 2.64 Distribution of *Culex tarsalis* in Texas.

Culex thriambus (Dyar, 1921)

It is a medium-size mosquito that resembles *Cx. coronator* as it has a broad median white area on the ventral side of a dark proboscis but is not ringed completely. The palps are short and dark. The hind tarsomeres have distinct basal and apical white scales, and the ventral side of the abdomen has median triangular patches of black scales on the abdominal sterna. This characteristic is very close in appearance to other two *Culex* species that occur in Texas: *Cx. tarsalis* and *Cx. stigmatosoma*.

Bionomics. *Culex thriambus* larvae are found in riparian systems, marshes, and in small dirty pools by the river in Kerrville, TX. They have also been found in leaf-filled rock pools besides streams. Peak seasonality occurs during September through November, and females enter diapause during winter. It is a potential vector of WNV (CDC, 2016).

Distribution. In Texas, it has been found in 41 counties (Fig. 2.65).

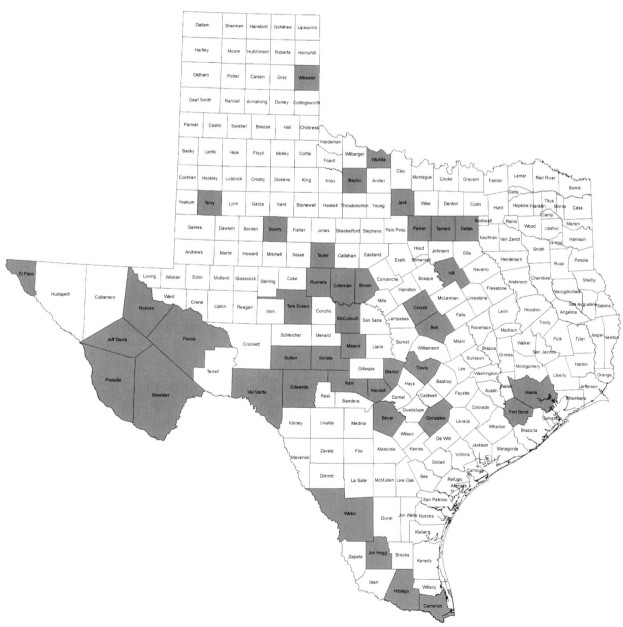

FIG. 2.65   Distribution of *Culex thriambus* in Texas.

2.6.2 Subgenus *Melanoconion* Theobald

There are four species present in Texas from the 160 belonging to the *Melanoconion* subgenus: *Cx. abominator* Dyar and Knab, *Cx. erraticus* (Dyar and Knab), *Cx. peccator* Dyar and Knab, *and Cx. pilosus* (Dyar and Knab).

Culex abominator (Dyar and Knab, 1909)

A small dark brown mosquito with a long, dark proboscis that is slightly swollen at the tip. The occiput has light-golden scales and brown erect forked scales on the posterodorsal region where the anterodorsal and lateral regions have broad appressed dingy-white scales. The scutum does not have middorsal acrostichal setae. The mesokatepisternum has a patch of six or more pale scales, and the mesepimeron has a light integumentary area. Hind tarsomeres are entirely dark scaled.

Bionomics. Very little is known about this species, but it has been collected in the summer and fall in grass overhanging a permanent pool near New Braunfels, TX. Eggs are laid in floating rafts on the water surface or on emerging vegetation just above the waterline. Larvae and pupae are found in ground pools including temporary ponds, pools from drying streams or overflows, or from rain water collected on the ground, at the edges of slow moving streams and in swamps in tropical forests. The importance of this species is its ability to attack man although some species of *Melanoconion* have shown preference to avian hosts. It can be a troublesome pest.

Distribution. In Texas, it has been found in 23 counties (Fig. 2.66).

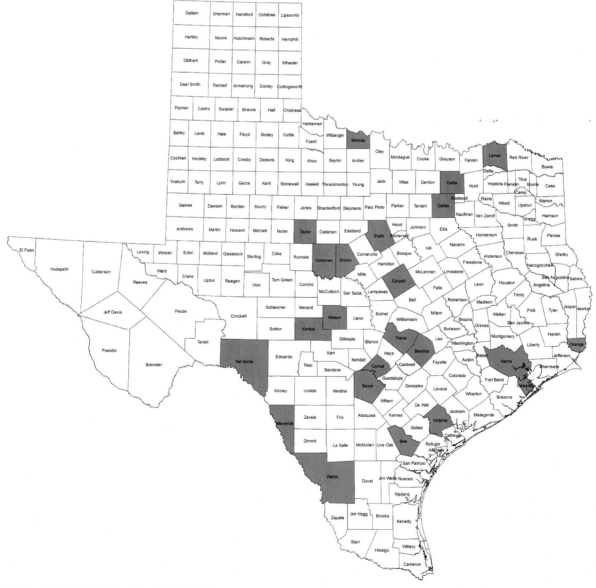

FIG. 2.66 Distribution of *Culex abominator* in Texas.

Culex erraticus (Dyar and Knab, 1906)

A small dark brown mosquito from the subgenus *Melanoconion* with patches of white scales in the scutum and abdomen with narrow white basal bands. It has a thorax with patches of white scales ventrally and laterally and a patch of broad white scales on the mesepimeron.

Bionomics. Eggs are laid in masses glued to the surfaces of floating leaves. Larvae can be found in grassy areas in ponds, lakes, marshes, and streams associated with *An. quadrimaculatus* in the Southern United States. Breeding occurs throughout the year in the south as the species has multiple generations per year but are more frequent from May to October farther north. They overwinter in the adult stage. They prefer to feed on avian hosts but can attack lizards, mammals, and humans at night outdoors. They are not troublesome biters but can become annoying to humans in the woods in late summer. It is known to transmit EEE and has tested positive for WNV in the Southeastern United States (Varnado et al., 2012).

Distribution. In Texas, it has been found in 100 counties (Fig. 2.67).

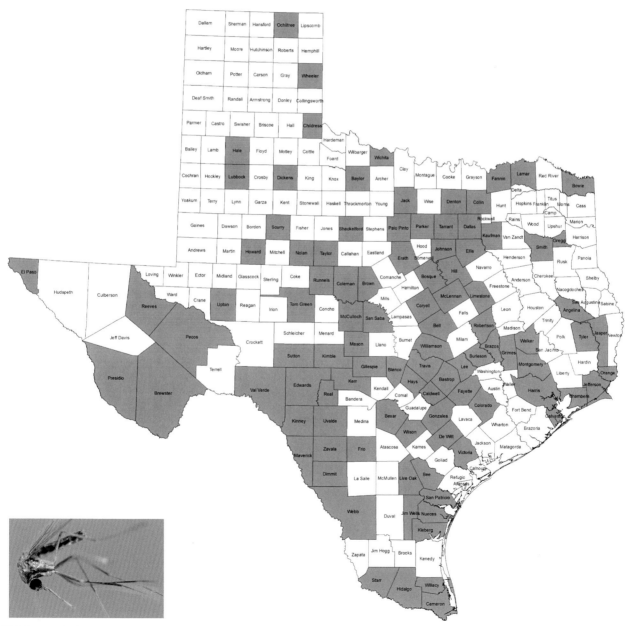

FIG. 2.67   Distribution of *Culex erraticus* in Texas.

Culex peccator (Dyar and Knab, 1909)

This is a small, dark brown mosquito with diminutive patches of white scales on the thorax and abdomen resembling those in the subgenus *Melanoconion* that are present in Texas. The proboscis is long, narrow, and swollen at the tip, and the occiput has broad dark brown scales anteromedially. Their hind tarsomeres are entirely dark.

Bionomics. Larvae have been found in ground pools, grassy pools in swampy areas, and pools along streams and marshy areas. They have multiple broods per year and overwinter in the adult stage. Adults do not fly far from their breeding sites and feed on amphibians, reptiles, and occasionally birds. They very seldom feed on humans or other mammals. Although it has no significant role in the transmission of pathogens, EEE has been detected in wild-collected mosquitoes (Cupp et al., 2004; Burkett-Cadena et al., 2008).

Distribution. In Texas, it has been found in 11 counties (Fig. 2.68).

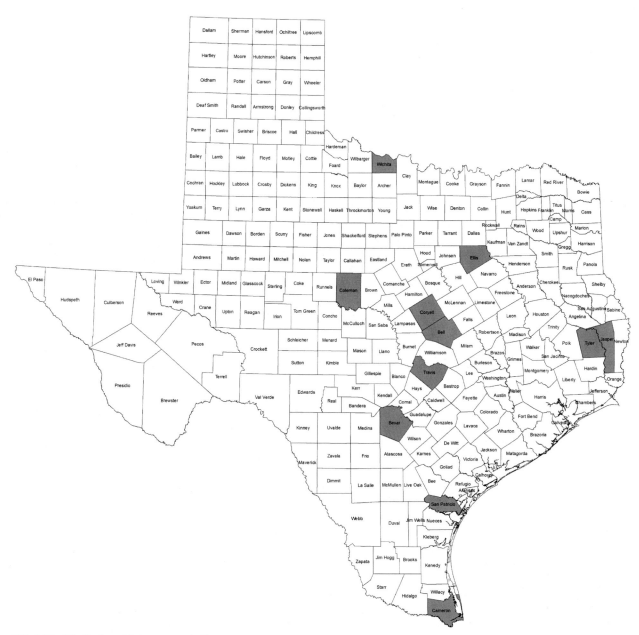

FIG. 2.68 Distribution of *Culex peccator* in Texas.

Culex pilosus (Dyar and Knab, 1906)

A small, dark brown mosquito from the subgenus *Melanoconion* that looks similar to other mosquitoes belonging to this subgenus. It can be distinguished by the size and location of the white patches on its thorax.

Bionomics. Its eggs are able to withstand drying. Larvae can be found in temporary and semipermanent pools including ditches, streams, grassy pools, and floodwater areas and have been collected from April to December in the extreme south. It is multivoltine and overwinters in the adult stage.

Distribution. In Texas, it has been found only in Harrison County (Fig. 2.69).

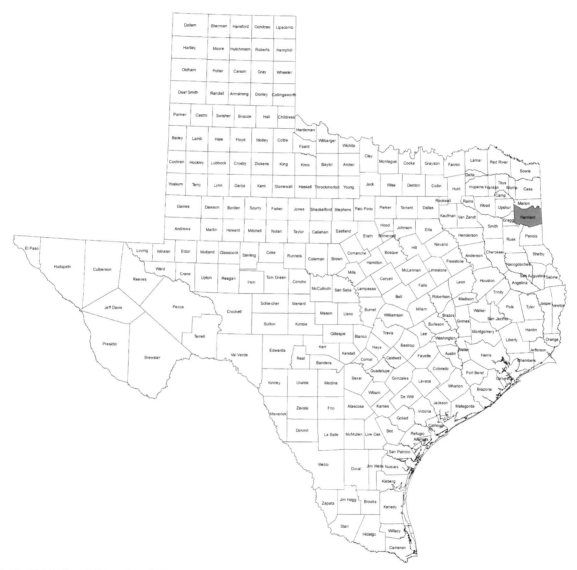

FIG. 2.69 Distribution of *Culex pilosus* in Texas.

2.6.3 Subgenus *Neoculex* Dyar

Culex apicalis (Adams, 1903)

A small brown mosquito with dark brown proboscis and palps that are short and usually with a few pale scales on the subterminal segment and at the base of the terminal segment. The scutum has middorsal acrostichal setae, and the mesonotum is golden with two faint brown median lines and long dark brown hairs. Its abdomen has narrow apical white bands of scales widening laterally on each segment.

Bionomics. Larvae are found in permanent and semipermanent pools, in streams, woodland pools, and swamps. They occur throughout the year in the southern part of its range. Adults are found in vegetation and in shelters near their breeding sites. They are rarely found inside dwellings. Adults feed on cold-blooded animals but are not known to feed on humans.

Distribution. In Texas, it has been found in Brewster and Jeff Davis Counties (Fig. 2.70).

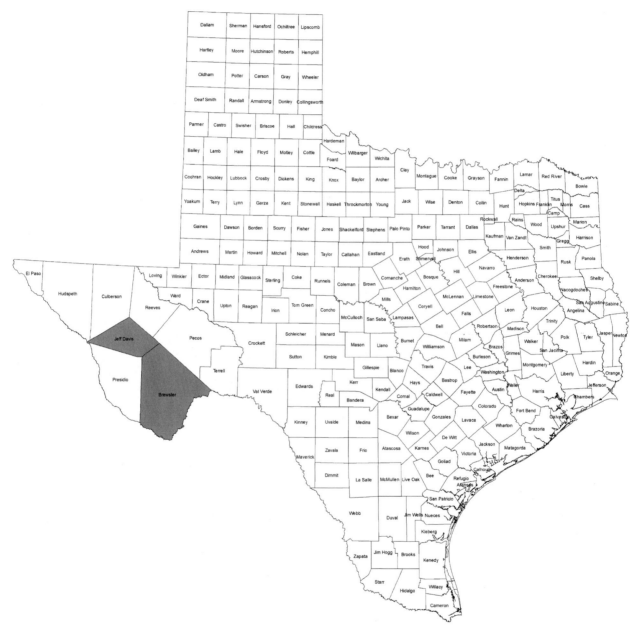

FIG. 2.70 Distribution of *Culex apicalis* in Texas.

Culex arizonensis (Bohart, 1948)

A small mosquito with a dark, long proboscis similar to *Cx. apicalis*, but its palps are completely dark and short. The thorax is dark brown, and the scutum has narrow brown scales with middorsal acrostical setae. The scutal angle has whitish scales extending posteriorly in two submedian lines, one on either side of the prescutellar space. The pleura contains patches of broad white scales. The first abdominal tergite has a median patch of dark and pale scales, and the remaining segments are dark scaled with a narrow apical band on each segment.

Bionomics. It is found in shaded creek bed pools associated with riparian forests. Bohart (1948) observed larvae in large quantities in all stages and were usually associated with *Cx. apicalis,* and females were numerous, but did not attempt to bite. Breeding occurred continuously throughout the season. In Texas, it was discovered in Dog Spring, a wellspring in the Guadalupe Mountains National Park at an elevation of over 2000 m (Reeves and Darsie, 2003).

Distribution. In 2002, *Cx. arizonensis* was found for the first time on the western limestone wall of dry wash Upper Dog Canyon on the northern border of the Guadalupe Mountains National Park, Culberson County, TX (Fig. 2.71).

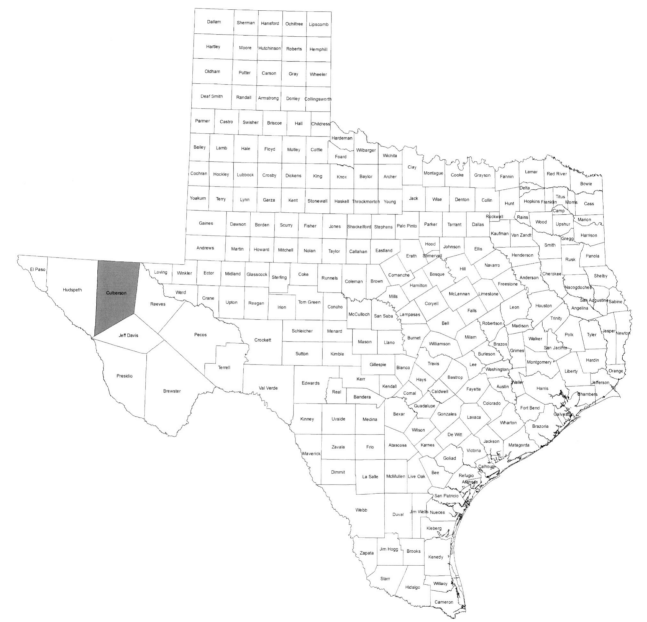

FIG. 2.71   Distribution of *Culex arizonensis* in Texas.

Culex territans (Walker, 1856)

A small mosquito with dark proboscis, palps, legs, and wings. The thorax is light brown, and the abdomen is dark brown with white scales apically or with lateral patches. The ventral side of the abdomen is entirely white scaled.

Bionomics. Egg rafts are laid directly on water. Larvae are found in semipermanent or permanent pools in streams, freshwater swamps, ponds, tree holes, and drainage ditches, but do not favor foully water. They have multiple generations per year and overwinter in the adult stage as inseminated females in colder areas. They are present throughout the year in the Southern United States. Dispersion is restricted to small distances from their breeding sites. Biting activity is at dusk and after dark. They are rarely encountered inside dwellings but are found in vegetation and shelter places near their larval habitats during the day. Females are not known to bite humans, but feed mainly on cold-blooded animals, particularly frogs and water rats and birds. It is a species in which WNV has been detected (CDC, 2016).

Distribution. In Texas, it has been found in 48 counties (Fig. 2.72).

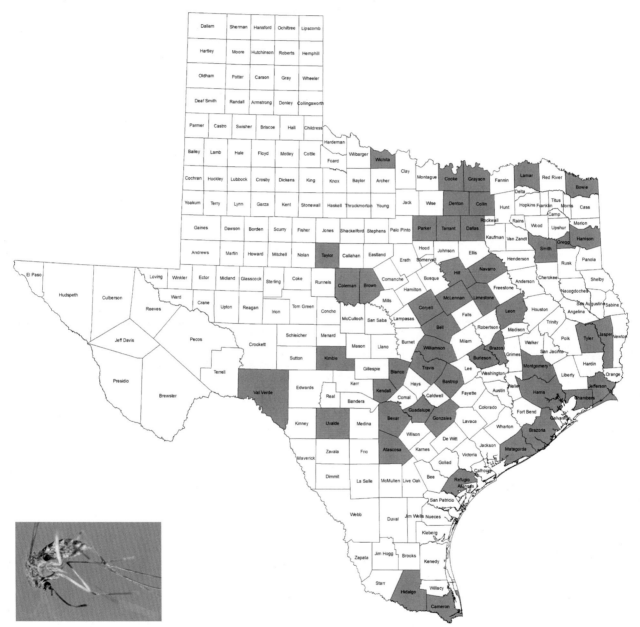

FIG. 2.72 Distribution of *Culex territans* in Texas.

2.7 *Deinocerites* Theobald

The distinguishing characteristics in the adults of *Deinocerites* are the antennae, which are longer than the proboscis, and the first flagellar segment, which is longer than the clypeus. However, the adults of *Deinocerites mathesoni* have the shortest antennae in the *Deinocerites* genus and lack the patch of pale scales on the mesepimeron, and their cerci lack the specialized setae as in *De. cancer*. *Deinocerites pseudes*, on the contrary, have a patch of translucent scales on the mesepimeron as do those in the *Epitedeus* group.

Bionomics. *Deinocerites* larvae and adults are mostly found in crab holes. However, larvae can be found in post holes, rock holes, tree holes, artificial containers, and ground pools formed by flooding of depressions containing crab holes. Adults have a short dispersal range and are crepuscular or nocturnal. During the day, they rest in the upper portion of crab burrows. *Deinocerites mathesoni* have been collected in light traps miles away from crab holes, but they seldom fly far from their crab hole habitats. Although there is no specificity of association of any mosquito species with a particular species of crab, *De. mathesoni* has been found associated with *Uca pugilator* (Bosc) and *Uca subcylindrica* (Stimpson), whereas *De. pseudes* has been found associated with *Cardisoma crassum* Smith (Mouthless crab) or *C. guanhumi* Latreille (Great Land Crab). It is not unusual to find more than one species of *Deinocerites* in the same crab hole and other species of *Culex*, *Aedes*, *Anopheles*, and others depending on the habitat (Adames, 1971). Feeding preference is not well known, but species in this genus feed on mammals including humans, birds, reptiles, and amphibians. Although VEE and SLE have been isolated in wild-caught mosquitoes from this genus (Adames 1971), specifically *De. pseudes*, they are not important pests of humans and probably do not play a role in pathogen transmission.

Distribution. In Texas, *Deinocerites mathesoni* has been found in four counties (Fig. 2.73), whereas *De. pseudes* has been found in three counties (Fig. 2.74).

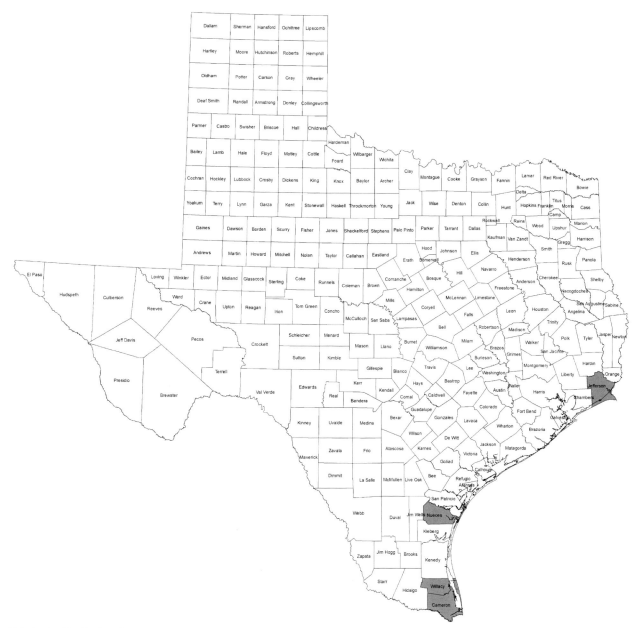

FIG. 2.73   Distribution of *Deinocerites mathesoni* in Texas.

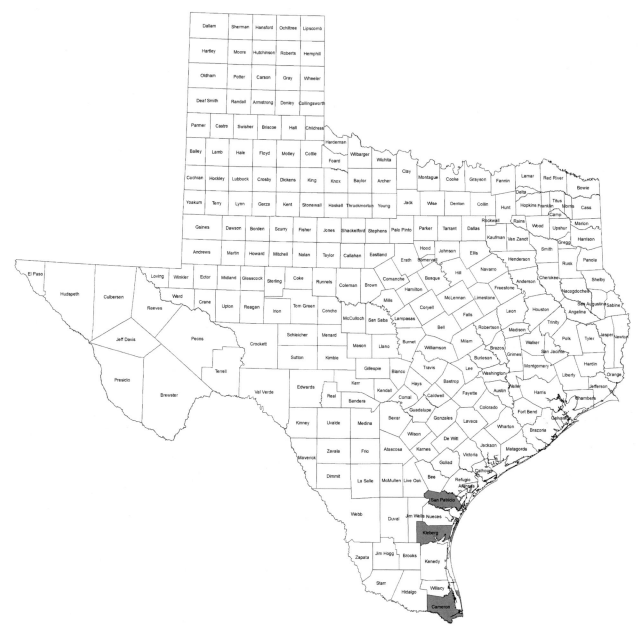

FIG. 2.74   Distribution of *Deinocerites pseudes* in Texas.

2.8   *Culiseta* Felt

2.8.1   Subgenus *Climacura* Howard, Dyar and Knab

Culiseta melanura (Coquillett, 1902)

A dark brown medium-size mosquito known as the "black-tailed mosquito." Occasionally, basal pale bands on the abdominal terga or lateral patches of pale scales may also be found on the ventral side of the abdomen. The palps are dark and short, whereas the dark proboscis is long, slender, and projects downwards. A few dark spiracular bristles are present and no postspiracular bristles. A tuft of setae is also present on the subcosta at the base of the wing, underneath.

Bionomics. Eggs are laid singly and float separately on the surface of developmental sites. Larvae are encountered in permanent bodies of water, particularly in swamps. It is a bog mosquito and found usually in pools over muck. It occurs in sites where the water is cold and springlike and in woodland pools. It has also been found in permanent water in winter where the sites were frozen over the surface. It has multiple generations per year and overwinters in the larval stage under the ice farther north. It has been found associated with tree hole and transient water species. They fly considerable distances from their breeding sites and are active at night, dusk, and dawn. They not only feed primarily on birds but also can feed on humans. It is considered a vector of EEE with the virus found in wild-caught mosquitoes (Chamberlain et al., 1951; Armstrong and Andreadis, 2010).

Distribution. In Texas, it has been found in four counties (Fig. 2.75).

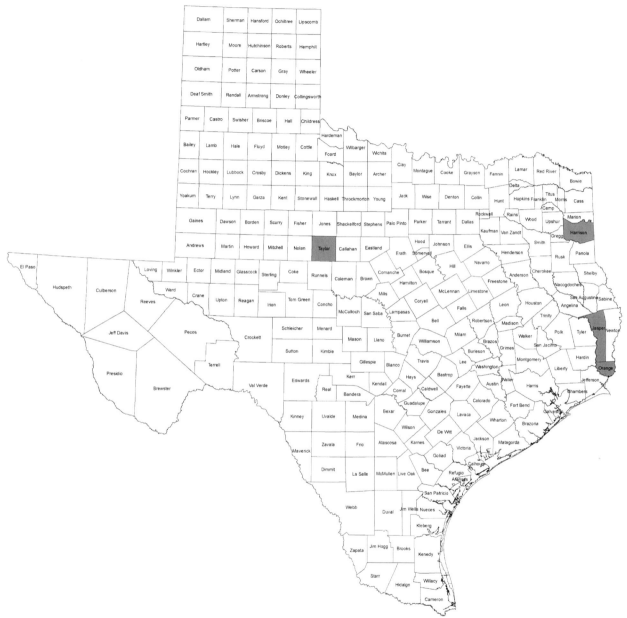

FIG. 2.75 Distribution of *Culiseta melanura* in Texas.

2.8.2 Subgenus *Culiseta* Felt

Culiseta incidens (Thomson, 1869)

A large mosquito known as the "cool weather mosquito" that has a dark scaled, long proboscis with a few pale scales intermixed and short dark palps speckled with pale scales. The thorax is dark brown with the prespiracular setae present and the postspiracular setae absent. The abdomen has basal yellowish-white bands; the legs are dark brown with some pale or white scales, and the tarsi are dark with basal pale bands that are more distinct on the hind tarsi. The wings are dark with dense patches of dark scales, and a tuft of setae is present on the underside of the wing at the base of the subcosta. A small area of white scales is usually present on the basal part of the costa.

Bionomics. Eggs are laid in rafts. Larvae can be found in a wide variety of habitats including brackish pools along the coast, ditches, ponds, clear woodland pools, creeks, stagnant and polluted waters, transient pools, and artificial containers including discarded tires. It has multiple generations, and its dispersal is in close proximity to its breeding location. It is a pest of large animals and is annoying to humans in certain areas. Females hibernate in the colder regions and can be found during the winter in sheltered places. Due to its successful experimental transmission, it may be a potential vector of SLE (Hammon and Reeves, 1943a), WEE (Hammon and Reeves, 1943b), Japanese B encephalitis virus (JEV) (Reeves and Hammon, 1946), and WNV (CDC, 2016).

Distribution. In Texas, it has been found in seven counties (Fig. 2.76).

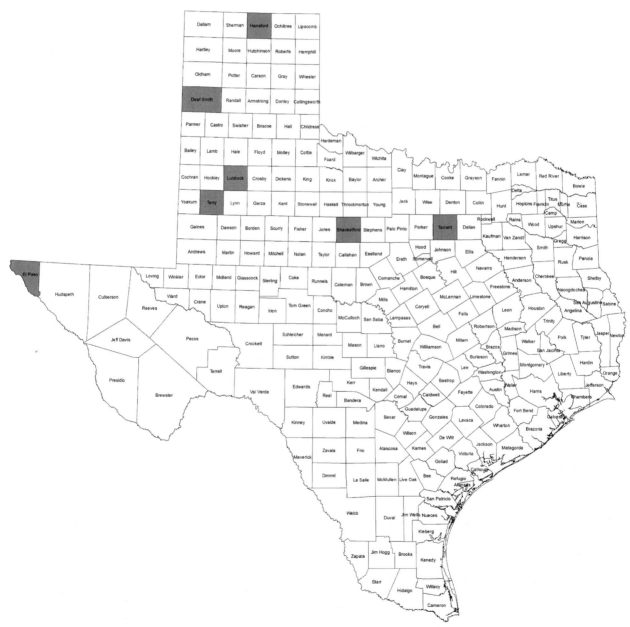

FIG. 2.76 Distribution of *Culiseta incidens* in Texas.

Culiseta inornata (Williston, 1893)

A similar mosquito to *Cs. incidens* in terms of its size and coloration and a tuft of setae at the base of the costa on the base of the wing. The abdominal terga has pale basal bands, and the wings have intermixed dark and pale scales on anterior veins (costa, subcosta, and vein 1), and the hind tarsomeres 1 and 2 have dark and pale scales.

Bionomics. Larvae occur in semipermanent pools, ditches, and occasionally in artificial containers. They can also be encountered in brackish waters in coastal marshes and edges of ponds with emergent vegetation. In the South, breeding occurs from autumn to spring months and survives the summer in the egg stage, whereas in the Northern part, adults hibernate during the winter and the larvae are found from spring to fall. They have multiple generations and can disperse considerable distances, although they generally fly close to their breeding site. They are persistent biters and feed on large mammals including humans. It has been found naturally infected with WEE (Hammon and Reeves, 1943b), a proven experimental vector of JEV (Reeves and Hammon, 1946) and a vector of WNV (CDC, 2016).

Distribution. In Texas, it has been found in 96 counties (Fig. 2.77).

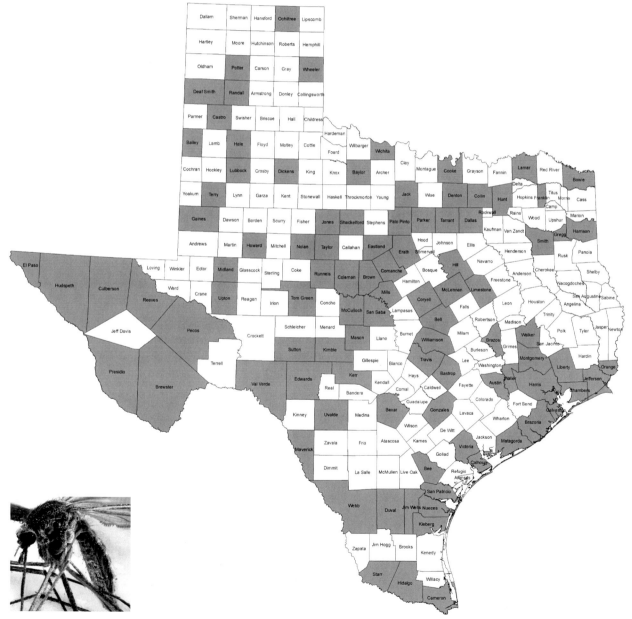

FIG. 2.77   Distribution of *Culiseta inornata* in Texas.

2.9 *Coquillettidia* Dyar

2.9.1 Subgenus *Coquillettidia* Dyar

Coquillettidia perturbans (Walker, 1856)

A moderately large mosquito known as the salt and pepper or cattail mosquito with a black proboscis with scattered white scales and a white ring at the middle. The scutum is dark brown intermixed with pale golden scales, and the postspiracular bristles are absent. Its legs are dark brown with broad basal white bands on each segment. The hind tarsi have broad basal pale bands, hind tarsomere 1 has a median pale band, and hind tibiae have a broad preapical pale band. Wing scales are broad.

Bionomics. Eggs are laid in rafts in waters where heavy vegetation is present. After hatching, both larvae and pupae attach to roots or submerged stems of plants with their modified siphon and respiratory trumpets. They overwinter as larvae, and adults emerge in spring and summer. Adults are strong fliers and bite mainly at night and can occasionally feed during the day in shady places when their habitat is invaded. They are considered a troublesome nuisance. They have one or more broods per year and can disperse long distances. They feed on humans, mammals, and birds, and enter houses to feed on humans. It transmits EEE (WRBU, 2019).

Distribution. In Texas, it has been found in 36 counties (Fig. 2.78).

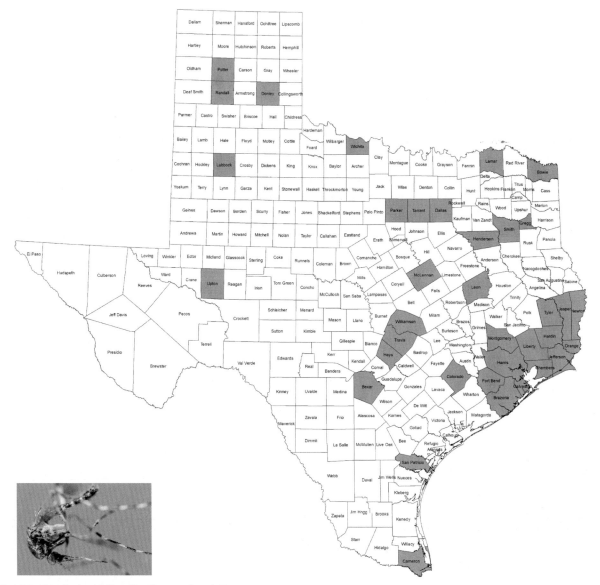

FIG. 2.78 Distribution of *Coquillettidia perturbans* in Texas.

2.10 *Mansonia* Blanchard

2.10.1 Subgenus *Mansonia* Blanchard

Mansonia titillans (Walker, 1848)

A medium-sized, dark brown with golden scale mosquito with a dark proboscis speckled with pales scales and having a narrow white ring on the apical third of its length. The palps are of similar coloration as the proboscis and the tips are white. The first abdominal segment has a median area of pale scales, and the remaining tergites are dark scaled with scattered yellow scales laterally and a few yellow and white scales apically. The hind tarsi present basal pale bands, and the wing scales are broad and blended with brown and white scales.

Bionomics. They inhabit permanent waters in association with aquatic plants such as freshwater swamps, lake margins, and drainage ditches. Eggs are deposited in masses glued to the underside of vegetation such as water lettuce (*Pistia*) and water hyacinth (*Eichhornia crassipes*). They oviposit on these plants by landing on leaves and curving their abdomen underneath them to deposit the eggs. Larvae and pupae attach to the submerged roots of plants with their siphon to obtain their oxygen supply. It is usually associated with the closely related *Cq. perturbans*. Adults are aggressive biters active at dusk and at dawn but will feed any time after sundown. They feed on all available mammals, including humans. They can disperse moderate to long distances from their breeding sites. They are most common in the fall, whereas its sister species *Cq. perturbans* is most common in late spring. They are known to transmit VEE (Carpenter and LaCasse, 1955).

Distribution. In Texas, it has been found in the Rio Grande Valley in association with water hyacinths and in 19 counties (Fig. 2.79).

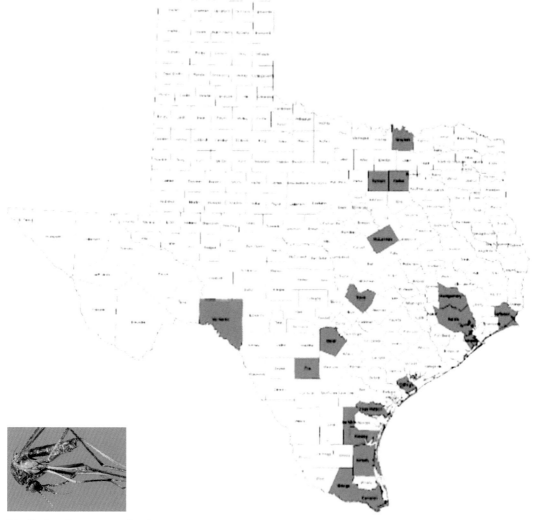

FIG. 2.79   Distribution of *Mansonia titillans* in Texas.

2.11 *Orthopodomyia* Theobald

A medium-size mosquitoes, distinctively ornamented and with six distinguishing longitudinal lines of white scales on the dark thorax. They have a dark proboscis, and their legs have black and white scales. Both species are very similar to each other. Larvae are the only stages where they can be distinguished from each other. The only morphological difference in adult females is found in the pale scales extending to the apex of the second abdominal tergite in *Or. alba*. The bands of *Or. alba* on the hind tarsal segments 1–2 and 2–3 are as broad as in *Or. signifera* but are more evenly placed on the joints.

Bionomics. Eggs and larvae are found mainly in rotten cavities of trees and on occasions in artificial containers. Breeding occurs throughout the year in the south as they have multiple generations and overwinter in the larval stage. They may overwinter in the adult stage in Texas. Dispersal is short near their breeding sites. In the south, cold tolerance has been observed in the larval stage as long as they do not reach a freezing point. They inhabit forests and are active at night. They feed primarily on birds.

Distribution. In Texas, *Or. alba* has been found in 14 counties (Fig. 2.80) and *Or. signifera* in 22 counties (Fig. 2.81).

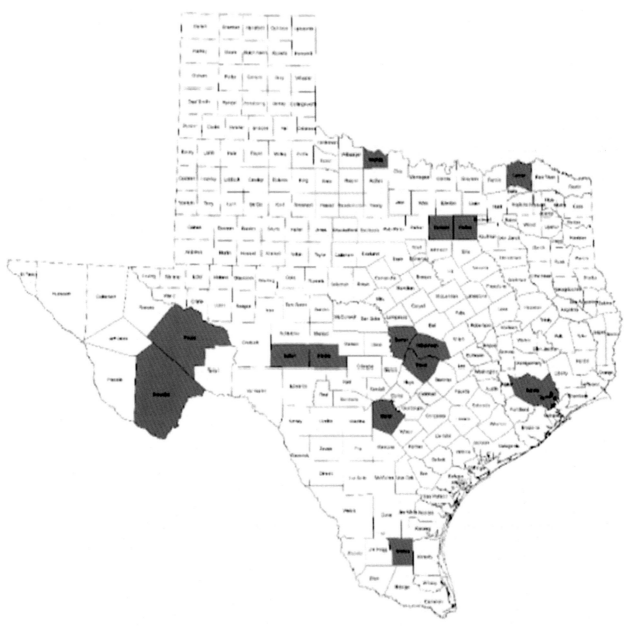

FIG. 2.80 Distribution of *Orthopodomyia alba* in Texas.

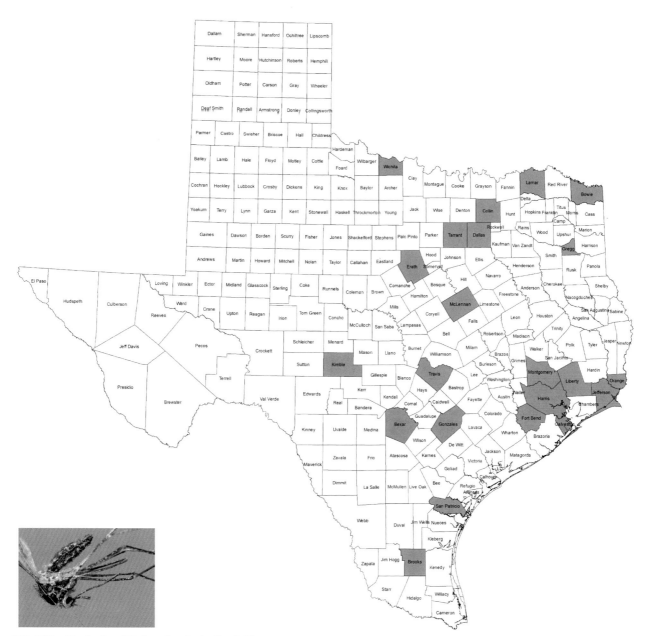

FIG. 2.81   Distribution of *Orthopodomyia signifera* in Texas.

Orthopodomyia kummi (Edwards, 1939)

A medium-sized reddish-brown to black mosquito. The palps are dark scaled with some scattered white scales. The thorax is reddish brown to black, and their legs have the same coloration on forecoxa; the mid- and hind coxa are a bit lighter in color. Adults may be identified by the presence of a transverse line of pale scales on the anterior face of the proepisternum, a line of white scales extending from the base of wing vein R to the separation of vein R_S and the dark scaled base of vein 1A. Abdominal tergite I is tan to brown, and sternites II–VII are whitish to brown. Sternite VIII is brown to dark brown. The abdomen has dark scales brown to black, and tergite II has white scales in a large middorsal patch, which may extend to the apical margin and in small to large basolateral patch or narrow to broad basal band.

 Bionomics. Its larvae have been found in tree holes, open-ended bamboo internodes, artificial containers, and in abandoned cesspools. They are found in association with *Or. signifera* in Southern Arizona (Zavortink, 1968) and with *Ae. brelandi* in Texas (Zavortink, 1972). Adults are active during darkness and feed on birds. They could possibly be important in the maintenance of arboviruses in sylvan environments or in the transmission of them to domestic fowl. *Orthopodomyia kummi* and *Or. signifera* are basically allopatric and are known to occupy the same tree holes in only one area, and It is possible that there are hibrids due to introgression (Zavortink, 1968).

 Distribution. In Texas, it has only been found in Chisos Mountains Basin, Big Bend National Park in Brewster County (Fig. 2.82).

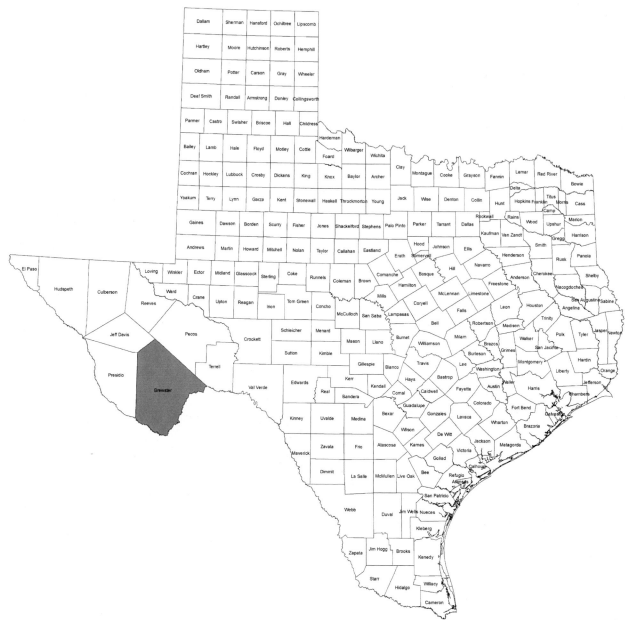

FIG. 2.82   Distribution of *Orthopodomyia kummi* in Texas.

2.12   *Toxorhynchites* Theobald

2.12.1   Subgenus *Lynchiella* Lahille

Toxorhynchites moctezuma (Dyar and Knab, 1906)

Hind tarsomere 4 of adult female is completcly white scaled, and the hind tarsomere 5 is completely dark scaled. The pleuron has yellowish scales on a light brown integument, and the top of the head has purple or copper scales. The upper postpronotum and sides of the scutum have blue and greenish blue scales.

Bionomics. Its larvae have been collected from a variety of aquatic habitats such as bamboo internodes, tree holes, and artificial containers such as discarded tires, flower vases, and plastic buckets, always in total or partial shade and with abundant leaves at bottom. Adults usually rest in shaded vegetation and are more active during dusk.

Distribution. In Texas, it has been found in Brownsville in Cameron County (Fig. 2.83).

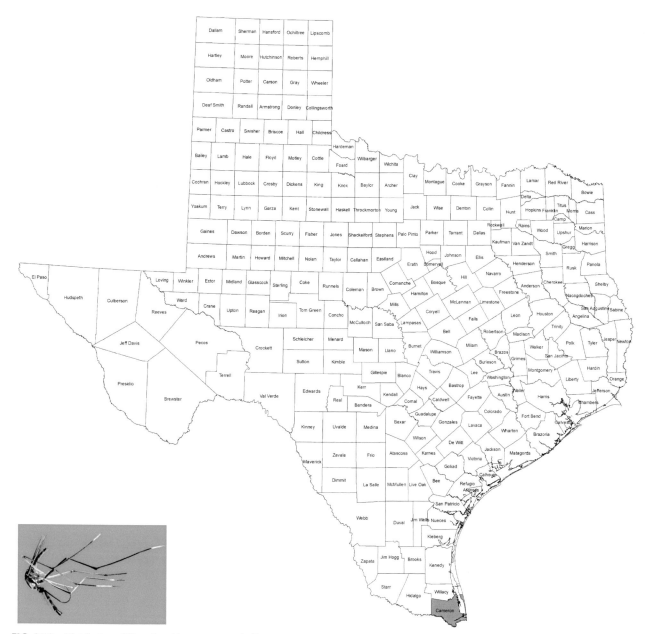

FIG. 2.83 Distribution of *Toxorhynchites moctezuma* in Texas.

Toxorhynchites rutilus septentrionalis (Dyar and Knab, 1906)

A very large ornamented mosquito indistinguishable from other subspecies *Tx. rutilus* with a long recurved downwards proboscis and blue-violet reflecting palps that are about two-thirds as long as the proboscis. The mesonotum is dark in the center and lighter around the edges with an anterior median pale stripe. The abdomen appears with glistening metallic green with purplish reflections and lateral golden spots on each segment. The males of *Tx. rutilus* have a pale band on the tarsus of the fore legs, whereas those of *Tx. rutilus septentrionalis* are entirely dark.

Bionomics. Eggs and larvae are found in tree cavities and in artificial containers including rain barrels, tubs, buckets, cans, jars, rock holes, and tires, especially those of large vehicles such as tractors. They have multiple broods per year, and their eggs hatch by splitting lengthwise into two halves. Larvae are predacious, feeding on other associated mosquito larvae such as those of *Anopheles*, *Aedes*, and *Orthopodomyia*. Development may be long requiring months to reach adulthood. It overwinters in the larval stage, especially mature instars, but it may overwinter as an adult in Texas (Breland, 1948a). Adults are active during the day feeding on nectar of flowers and other plant juices near their breeding sites. They do not transmit any human pathogens since they do not bite or feed on blood.

Distribution. In Texas, it has been found in 22 counties (Fig. 2.84).

2.13 *Uranotaenia* Lynch Arribalzaga

2.13.1 Subgenus *Pseudoficalbia* Theobald

Uranotaenia anhydor syntheta (Dyar and Shannon, 1924)

A very small mosquito similar to *Ur. anhydor*. The proboscis is long, dark, and swollen at the tip. The palps are short and dark. Unlike *Ur. anhydor*, the thorax of *Ur.* anhydor *syntheta* has a line of iridescent blue scales extending anteriorly from the base of the wing and broken into two segments at the scutal angle. The abdomen is black with lateral pale spots, and the hind tarsi are dark scaled.

Bionomics. Larvae have been found in still water along the margins of shallow grassy ditches exposed to sunshine and in small depressions along streams containing water hyacinths in Texas (Breland, 1948b). In Mexico, it is a winter species most abundant in November and with scanty generations in the spring and summer.

Distribution. In Texas, it has been found in 39 counties (Fig. 2.85).

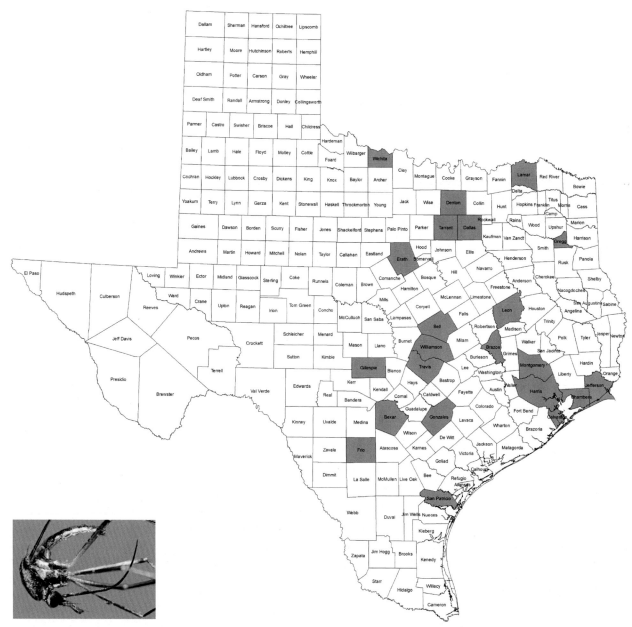

FIG. 2.84 Distribution of *Toxorhynchites rutilus septentrionalis* in Texas.

2.13.2 Subgenus *Uranotaenia* Lynch Arribalzaga

Uranotaenia lowii (Theobald, 1901)

A very small mosquito with a dark thorax and abdomen and iridescent blue scales on the sides of the head, thorax, abdomen, and wings. The proboscis is long, dark, and swollen at the tip. The palps are short and dark scaled. The scutum has a median dark stripe without the median iridescent blue-scaled stripe present on *Ur. sapphirina* (Osten-Sacken). The abdominal segments are dark where the third, fifth, and sixth segment have patches of iridescent blue scales apicolaterally, and the eighth segment is retracted into the seventh segment. The second marginal wing cell is shorter than vein R$_{2+3}$.

Bionomics. In certain areas, larvae occur in ground pools such as rocky pools, borrow pits, and hoof prints, while in others, they have been found in sunny pools with vegetation. Larvae have been found under sunlight in grassy shallow margins of ponds and lakes, and in permanent fresh water in coastal plains. They are commonly associated with species of *Culex*, specifically with *Cx. interrogator* in Texas. They have multiple generations per year and overwinter in the adult stage. Females are nocturnal, feed on amphibians and reptiles, and rarely bite humans.

Distribution. In Texas, it has been found in 36 counties (Fig. 2.86).

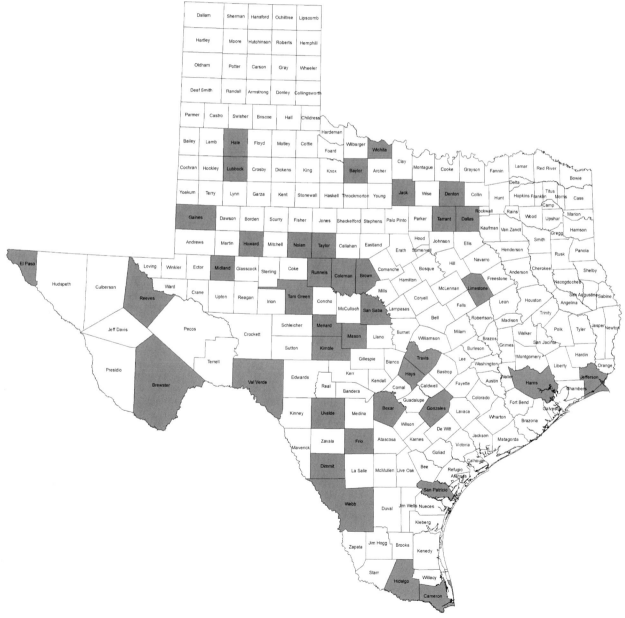

FIG. 2.85   Distribution of *Uranotaenia anhydor syntheta* in Texas.

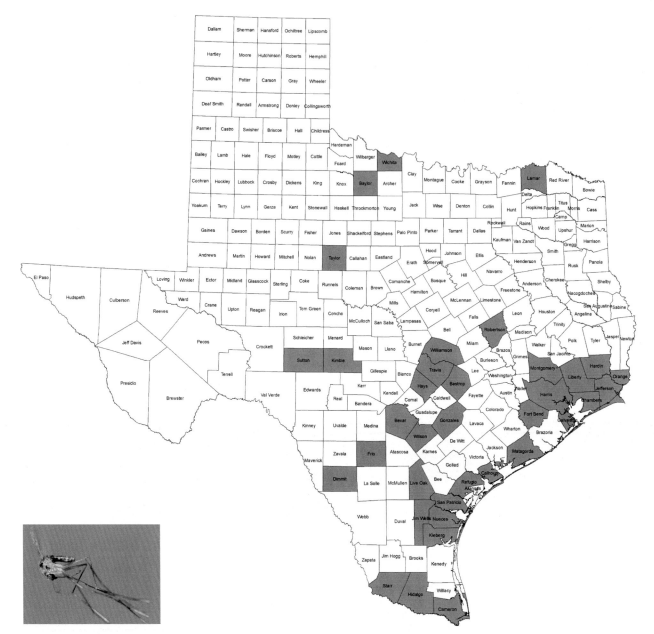

FIG. 2.86   Distribution of *Uranotaenia lowii* in Texas.

Uranotaenia sapphirina (Osten-Sacken, 1868)

A very small mosquito like *Ur. lowii*. The proboscis is long, dark brown and often with iridescent blue scales basally and swollen at the tip. The palps are very short and dark. Unlike *Ur. lowii*, it has a longitudinal median line of blue scales on the thorax and on the sides of the scutum and also lacks the blue scales on the abdomen. Its hind tarsi are dark scaled, whereas in *Ur. lowii,* they are pale scaled.

Bionomics. Larvae are found in similar habitats as *Ur. lowii* such as in permanent pools, ponds, and lakes with floating or emergent vegetation and swamps and rice growing areas where rice plants grow over fast producing shade preventing high afternoon temperatures. They are multivoltine and overwinter as adults. They can disperse considerable distances despite their tiny size and rarely bite humans as they feed mainly on amphibians and reptiles. Adults are found resting on damp places such as culverts, hollow trees, and vegetation near their larval habitats and are in great numbers during the winter months in the south.

Distribution. In Texas, it has been found in 39 counties (Fig. 2.87).

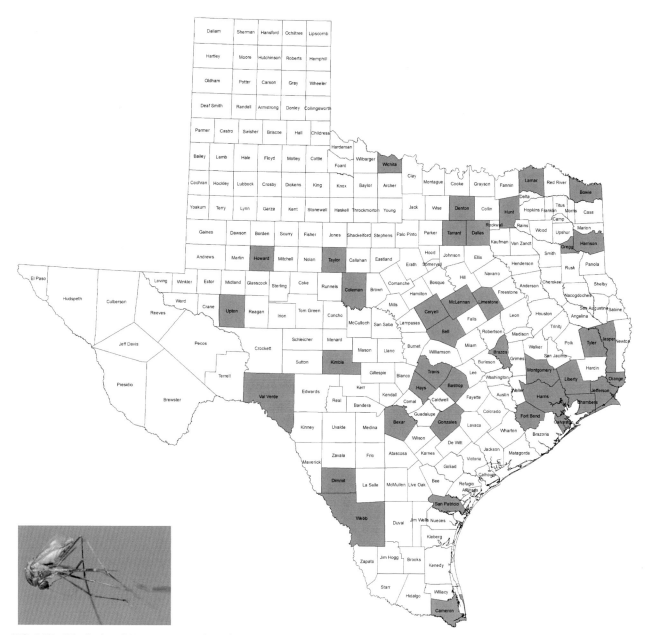

FIG. 2.87 Distribution of *Uranotaenia sapphirina* in Texas.

Appendix 2.1

Number of mosquito species reported by county

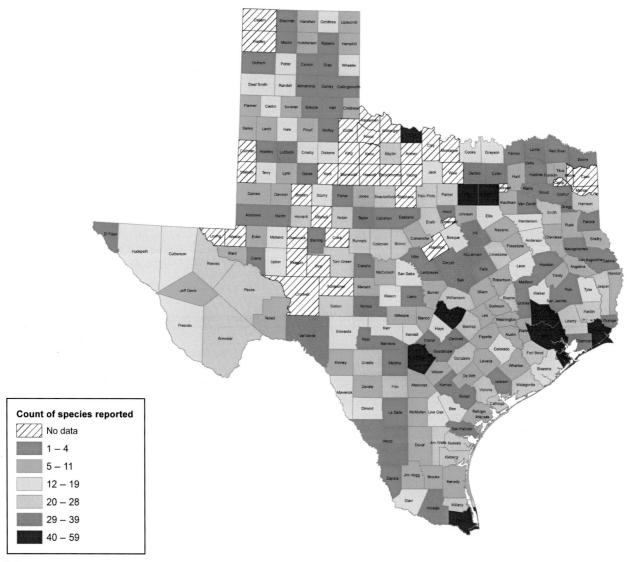

FIG. 2.88   Number of mosquito species reported by county.

TABLE 2.2 Mosquito species by county.

Species/county	Anderson	Andrews	Angelina	Aransas	Archer	Armstrong	Atascosa	Austin	Bailey	Bandera	Bastrop	Baylor	Bee	Bell	Bexar	Blanco	Borden	Bosque
Aedes aegypti	X		X	X				X			X		X	X	X	X		
Aedes albopictus	X		X	X				X			X			X	X			
Aedes atlanticus															X			
Aedes bimaculatus																		
Aedes brelandi																		
Aedes campestris																		
Aedes canadensis												X		X	X			
Aedes dorsalis															X			
Aedes dupreei															X			
Aedes epactius							X			X				X	X			
Aedes fulvus pallens																		
Aedes grossbecki																		
Aedes hendersoni			X												X			
Aedes infirmatus															X			
Aedes mitchellae						X					X				X			
Aedes muelleri																		
Aedes nigromaculis				X								X		X	X			
Aedes scapularis												X						
Aedes sollicitans				X			X		X			X		X	X			
Aedes sticticus																		
Aedes taeniorhynchus				X								X			X			
Aedes thelcter											X	X		X	X			
Aedes thibaulti																		
Aedes tormentor										X								
Aedes triseriatus			X	X									X	X	X			X
Aedes trivittatus															X			

Continued

TABLE 2.2 Mosquito species by county.—cont'd

Species/county	Anderson	Andrews	Angelina	Aransas	Archer	Armstrong	Atascosa	Austin	Bailey	Bandera	Bastrop	Baylor	Bee	Bell	Bexar	Blanco	Borden	Bosque
Aedes vexans	X	X				X			X		X	X	X	X	X			X
Aedes zoosophus												X		X	X			X
Anopheles albimanus	X							X										
Anopheles atropos															X			
Anopheles barberi															X			
Anopheles bradleyi																		
Anopheles crucians	X		X					X			X		X	X	X			
Anopheles franciscanus																		
Anopheles freeborni																		
Anopheles judithae																		
Anopheles pseudopunctipennis		X					X			X	X	X		X	X	X		X
Anopheles punctipennis	X		X					X		X	X	X	X	X	X	X		X
Anopheles quadrimaculatus	X		X				X	X			X		X	X	X			X
Anopheles smaragdinus																		
Anopheles walkeri																		
Coquillettidia perturbans															X			
Culex abominator											X		X		X			
Culex apicalis																		
Culex arizonensis																		
Culex chidesteri																		
Culex coronator				X			X	X		X	X	X	X	X	X	X		X
Culex declarator															X			
Culex erraticus			X								X	X	X	X	X	X		X
Culex erythrothorax															X			
Culex interrogator															X			
Culex nigripalpus														X	X			

Species	1	2	3	4	5	6	7	8	9	10	11	12	13	14	15
Culex peccator				X	X										
Culex pilosus															
Culex quinquefasciatus	X			X	X	X	X	X	X	X	X			X	X
Culex restuans	X			X	X		X	X		X					
Culex salinarius			X	X	X	X	X	X		X					X
Culex stigmatosoma				X									X		
Culex tarsalis	X			X	X	X	X	X	X	X					X
Culex territans			X	X	X			X			X				
Culex thriambus			X	X	X										
Culiseta incidens															
Culiseta inornata				X	X	X	X	X	X	X					
Culiseta melanura															
Deinocerites mathesoni															
Deinocerites pseudes															
Haemagogus equinus															
Mansonia titillans				X											
Orthopodomyia alba				X											
Orthopodomyia kummi															
Orthopodomyia signifera				X											
Psorophora ciliata				X	X	X	X	X						X	X
Psorophora columbiae	X			X	X	X	X	X			X			X	X
Psorophora cyanescens	X			X	X	X	X	X						X	
Psorophora discolor	X			X	X		X								
Psorophora ferox				X	X										
Psorophora horrida	X			X				X							
Psorophora howardii				X											
Psorophora longipalpus				X											
Psorophora mathesoni				X											
Psorophora mexicana															
Psorophora signipennis				X	X		X	X	X						

Continued

TABLE 2.2 Mosquito species by county.—cont'd

Species/county	Anderson	Andrews	Angelina	Aransas	Archer	Armstrong	Atascosa	Austin	Bailey	Bandera	Bastrop	Baylor	Bee	Bell	Bexar	Blanco	Borden	Bosque
Toxorhynchites rutilus septentrionalis														X	X			
Uranotaenia syntheta												X			X			
Uranotaenia lowii											X	X			X			
Uranotaenia sapphirina											X			X	X			
	12	4	10	11	0	3	8	12	6	5	25	24	16	33	59	8	0	15

Species/county	Bowie	Brazoria	Brazos	Brewster	Briscoe	Brooks	Brown	Burleson	Burnet	Caldwell	Calhoun	Callahan	Camp	Cameron	Carson	Cass	Castro	Chambers
Aedes aegypti		X	X				X	X	X	X				X				X
Aedes albopictus	X	X	X					X						X				X
Aedes atlanticus		X									X							X
Aedes bimaculatus														X				
Aedes brelandi				X														
Aedes campestris				X														
Aedes canadensis	X			X			X											X
Aedes dorsalis														X			X	
Aedes dupreei																		
Aedes epactius			X				X											
Aedes fulvus pallens																		
Aedes grossbecki																		
Aedes hendersoni																		
Aedes infirmatus														X				
Aedes mitchellae	X		X											X				
Aedes muelleri				X										X				
Aedes nigromaculis	X		X				X							X				
Aedes scapularis						X								X				
Aedes sollicitans	X	X					X				X			X			X	X
Aedes sticticus														X				
Aedes taeniorhynchus	X	X									X			X			X	X
Aedes thelcter											X			X			X	
Aedes thibaulti																		
Aedes tormentor																		
Aedes triseriatus	X	X	X	X										X			X	X
Aedes trivittatus	X			X										X			X	

Continued

TABLE 2.2 Mosquito species by county.—cont'd

Species/county	Bowie	Brazoria	Brazos	Brewster	Briscoe	Brooks	Brown	Burleson	Burnet	Caldwell	Calhoun	Callahan	Cameron	Camp	Carson	Cass	Castro	Chambers
Aedes vexans	X	X	X	X	X		X	X			X		X				X	X
Aedes zoosophus	X					X	X						X				X	
Anopheles albimanus													X					
Anopheles atropos																		
Anopheles barberi											X							
Anopheles bradleyi											X		X					
Anopheles crucians	X	X	X				X	X			X		X					X
Anopheles franciscanus																		
Anopheles freeborni																		X
Anopheles judithae				X														
Anopheles pseudopunctipennis		X	X	X			X		X	X			X					X
Anopheles punctipennis	X	X	X	X			X	X	X		X		X					X
Anopheles quadrimaculatus	X	X	X	X			X	X	X		X		X					X
Anopheles smaragdinus			X															
Anopheles walkeri																		
Coquillettidia perturbans	X	X											X					X
Culex abominator							X											
Culex apicalis				X														

Species																
Culex arizonensis																
Culex chidesteri											X					
Culex coronator	X	X	X		X		X	X	X	X		X				X
Culex declarator												X		X		
Culex erraticus	X	X	X			X	X		X			X				X
Culex erythrothorax																
Culex interrogator												X				
Culex nigripalpus												X				X
Culex peccator												X				
Culex pilosus																
Culex quinquefasciatus	X	X	X		X	X				X	X	X				X
Culex restuans	X	X	X			X				X		X				X
Culex salinarius	X	X	X			X				X		X			X	X
Culex stigmatosoma			X													
Culex tarsalis	X	X	X	X	X	X				X		X			X	
Culex territans	X	X				X	X					X				X
Culex thriambus			X			X						X				
Culiseta incidens																
Culiseta inornata	X	X	X		X	X				X		X			X	X
Culiseta melanura																
Deinocerites mathesoni												X				
Deinocerites pseudes												X				
Haemagogus equinus												X				
Mansonia titillans										X		X				
Orthopodomyia alba	X	X	X					X								
Orthopodomyia kummi	X		X													

Continued

TABLE 2.2 Mosquito species by county.—cont'd

Species/county	Bowie	Brazoria	Brazos	Brewster	Briscoe	Brooks	Brown	Burleson	Burnet	Caldwell	Calhoun	Callahan	Cameron	Camp	Carson	Cass	Castro	Chambers
Orthopodomyia signifera	X					X												
Psorophora ciliata	X		X								X		X		X		X	X
Psorophora columbiae	X	X	X	X		X	X				X	X	X					X
Psorophora cyanescens	X		X	X		X	X				X		X				X	
Psorophora discolor	X			X		X	X				X		X					
Psorophora ferox	X	X	X										X					X
Psorophora horrida			X										X					X
Psorophora howardii	X		X										X					X
Psorophora longipalpus																		
Psorophora mathesoni	X												X					
Psorophora mexicana													X					
Psorophora signipennis	X		X	X			X						X					
Toxorhynchites rutilus septentrionalis			X	X			X						X					X
Uranotaenia syntheta											X		X					
Uranotaenia lowii	X		X										X					X
Uranotaenia sapphirina													X					X
	32	20	29	27	2	10	25	9	6	4	22	2	54	0	1	1	13	30

Species/county	Cherokee	Childress	Clay	Cochran	Coke	Coleman	Collin	Collingsworth	Colorado	Comal	Comanche	Concho	Cooke	Coryell	Cottle	Crane	Crockett
Aedes aegypti	X						X		X				X				
Aedes albopictus	X						X		X		X	X	X	X			
Aedes atlanticus																	
Aedes bimaculatus																	
Aedes brelandi																	
Aedes campestris																	
Aedes canadensis														X			
Aedes dorsalis																	
Aedes dupreei																	
Aedes epactius						X	X			X	X			X			
Aedes fulvus pallens																	
Aedes grossbecki																	
Aedes hendersoni																	
Aedes infirmatus							X										
Aedes mitchellae																	
Aedes muelleri																	
Aedes nigromaculis						X	X						X	X			
Aedes scapularis																	
Aedes sollicitans								X					X	X			
Aedes sticticus																	
Aedes taeniorhynchus							X		X								
Aedes thelcter														X			
Aedes thibaulti																	
Aedes tormentor																	
Aedes triseriatus	X					X	X		X		X			X			

Continued

TABLE 2.2 Mosquito species by county.—cont'd

Species/county	Cherokee	Childress	Clay	Cochran	Coke	Coleman	Collin	Collingsworth	Colorado	Comal	Comanche	Concho	Cooke	Coryell	Cottle	Crane	Crockett
Aedes trivittatus							X										
Aedes vexans						X	X		X				X	X		X	
Aedes zoosophus	X	X				X	X						X				
Anopheles albimanus																	
Anopheles atropos																	
Anopheles barberi																	
Anopheles bradleyi																	
Anopheles crucians	X						X		X					X			
Anopheles franciscanus																	
Anopheles freeborni																	
Anopheles judithae																	
Anopheles pseudopunctipennis						X	X		X					X			
Anopheles punctipennis		X				X	X		X				X	X			
Anopheles quadrimaculatus	X	X				X	X		X	X				X			
Anopheles smaragdinus																	
Anopheles walkeri																	
Coquillettidia perturbans									X								
Culex abominator						X	X			X				X			
Culex apicalis																	
Culex arizonensis																	
Culex chidesteri																	

Species												
Culex coronator	X			X	X	X		X	X	X	X	
Culex declarator												
Culex erraticus				X				X		X	X	X
Culex erythrothorax												
Culex interrogator												
Culex nigripalpus				X				X		X		
Culex peccator				X				X			X	
Culex pilosus												
Culex quinquefasciatus				X				X		X	X	X
Culex restuans				X	X			X	X	X	X	
Culex salinarius				X	X			X	X	X	X	
Culex stigmatosoma											X	
Culex tarsalis		X		X	X	X	X	X	X	X	X	X
Culex territans				X				X	X	X	X	
Culex thriambus				X				X			X	
Culiseta incidens												
Culiseta inornata				X		X	X	X	X	X	X	
Culiseta melanura												
Deinocerites mathesoni												
Deinocerites pseudes												
Haemagogus equinus												
Mansonia titillans												
Orthopodomyia alba												
Orthopodomyia kummi												
Orthopodomyia signifera									X	X		
Psorophora ciliata				X	X			X			X	
Psorophora columbiae				X	X			X		X	X	

Continued

TABLE 2.2 Mosquito species by county.—cont'd

Species/county	Cherokee	Childress	Clay	Cochran	Coke	Coleman	Collin	Collingsworth	Colorado	Comal	Comanche	Concho	Cooke	Coryell	Cottle	Crane	Crockett
Psorophora cyanescens						X		X	X							X	
Psorophora discolor		X				X	X						X	X			
Psorophora ferox	X													X			
Psorophora horrida																	
Psorophora howardii																	
Psorophora longipalpus							X										
Psorophora mathesoni																	
Psorophora mexicana																	
Psorophora signipennis		X				X	X	X					X	X			
Toxorhynchites rutilus septentrionalis																	
Uranotaenia syntheta						X											
Uranotaenia lowii																	
Uranotaenia sapphirina						X								X			
	7	8	0	0	0	26	29	3	19	5	5	2	15	30	0	3	1

Species/county	Crosby	Culberson	Dallam	Dallas	Dawson	Deaf Smith	Delta	Denton	DeWitt	Dickens	Dimmit	Donley	Duval	Eastland	Ector	Edwards	El Paso	Ellis
Aedes aegypti				X				X	X				X				X	X
Aedes albopictus				X				X	X				X				X	X
Aedes atlanticus				X														
Aedes bimaculatus																		
Aedes brelandi																		
Aedes campestris		X															X	
Aedes canadensis				X														
Aedes dorsalis	X	X				X				X							X	
Aedes dupreei																		
Aedes epactius				X		X		X		X						X		
Aedes fulvus pallens				X														
Aedes grossbecki																		
Aedes hendersoni				X														
Aedes infirmatus				X				X										
Aedes mitchellae																		
Aedes muelleri																		
Aedes nigromaculis	X	X		X	X	X		X		X	X				X		X	
Aedes scapularis																		
Aedes sollicitans	X			X		X		X							X		X	
Aedes sticticus				X													X	
Aedes taeniorhynchus				X														
Aedes thelcter										X	X							
Aedes thibaulti								X										
Aedes tormentor																		

Continued

TABLE 2.2 Mosquito species by county.—cont'd

Species/county	Crosby	Culberson	Dallam	Dallas	Dawson	Deaf Smith	Delta	Denton	DeWitt	Dickens	Dimmit	Donley	Duval	Eastland	Ector	Edwards	El Paso	Ellis
Aedes triseriatus				X				X									X	
Aedes trivittatus				X				X									X	
Aedes vexans	X			X	X	X		X		X	X	X			X	X	X	
Aedes zoosophus				X				X										
Anopheles albimanus																		
Anopheles atropos																		
Anopheles barberi				X														
Anopheles bradleyi		X																
Anopheles crucians				X				X	X					X				
Anopheles franciscanus																	X	
Anopheles freeborni																	X	
Anopheles judithae																		
Anopheles pseudopunctipennis	X			X	X	X	X	X	X	X	X			X		X	X	
Anopheles punctipennis	X	X		X		X		X			X					X	X	
Anopheles quadrimaculatus				X				X	X		X						X	X
Anopheles smaragdinus																		
Anopheles walkeri																		
Coquillettidia perturbans				X								X						
Culex abominator				X														
Culex apicalis																		

Culex arizonensis	X														
Culex chidesteri															
Culex coronator		X			X	X		X		X			X	X	X
Culex declarator															
Culex erraticus		X			X	X	X	X					X	X	X
Culex erythrothorax															X
Culex interrogator															
Culex nigripalpus		X			X									X	X
Culex peccator															X
Culex pilosus															
Culex quinquefasciatus	X	X		X	X	X	X	X		X		X	X	X	
Culex restuans		X			X					X				X	
Culex salinarius		X		X	X				X			X	X	X	
Culex stigmatosoma		X											X	X	
Culex tarsalis	X	X	X	X	X		X	X	X		X	X	X	X	X
Culex territans		X			X										
Culex thriambus		X											X	X	
Culiseta incidens				X										X	
Culiseta inornata	X	X		X	X		X			X	X		X	X	
Culiseta melanura															
Deinocerites mathesoni															
Deinocerites pseudes															
Haemagogus equinus															
Mansonia titillans		X													
Orthopodomyia alba		X													
Orthopodomyia kummi															

Continued

TABLE 2.2 Mosquito species by county.—cont'd

Species/county	Crosby	Culberson	Dallam	Dallas	Dawson	Deaf Smith	Delta	Denton	DeWitt	Dickens	Dimmit	Donley	Duval	Eastland	Ector	Edwards	El Paso	Ellis
Orthopodomyia signifera				X														
Psorophora ciliata	X			X				X		X	X		X		X			X
Psorophora columbiae		X		X		X		X		X	X				X		X	X
Psorophora cyanescens	X			X	X			X		X	X				X		X	X
Psorophora discolor	X	X		X		X		X		X	X				X		X	X
Psorophora ferox				X			X	X										
Psorophora horrida				X				X										
Psorophora howardii				X				X										
Psorophora longipalpus				X														
Psorophora mathesoni				X														
Psorophora mexicana																		
Psorophora signipennis	X	X		X	X	X		X		X	X				X	X	X	
Toxorhynchites rutilus septentrionalis				X				X										
Uranotaenia syntheta				X				X			X						X	
Uranotaenia lowii											X							
Uranotaenia sapphirina				X				X			X							
	12	12	0	50	6	15	2	35	8	15	18	4	7	4	11	13	31	13

Species/county	Erath	Falls	Fannin	Fayette	Fisher	Floyd	Foard	Fort Bend	Franklin	Freestone	Frio	Gaines	Galveston	Garza	Gillespie	Glasscock	Goliad	Gonzales
Aedes aegypti	X			X				X		X			X				X	
Aedes albopictus		X			X			X					X					
Aedes atlanticus								X					X					
Aedes bimaculatus																		
Aedes brelandi																		
Aedes campestris																		
Aedes canadensis													X					
Aedes dorsalis												X						
Aedes dupreei																		
Aedes epactius											X		X					
Aedes fulvus pallens													X					
Aedes grossbecki																		
Aedes hendersoni													X					
Aedes infirmatus													X					X
Aedes mitchellae													X					
Aedes muelleri																		
Aedes nigromaculis	X									X	X	X	X					X
Aedes scapularis													X					
Aedes sollicitans	X					X							X					
Aedes sticticus						X		X										
Aedes taeniorhynchus								X					X					
Aedes thelcter											X							X
Aedes thibaulti																		
Aedes tormentor													X					X
Aedes triseriatus	X							X			X		X					X

Continued

TABLE 2.2 Mosquito species by county.—cont'd

Species/county	Erath	Falls	Fannin	Fayette	Fisher	Floyd	Foard	Fort Bend	Franklin	Freestone	Frio	Gaines	Galveston	Garza	Gillespie	Glasscock	Goliad	Gonzales
Aedes trivittatus								X					X					
Aedes vexans	X					X		X	X				X					X
Aedes zoosophus	X										X				X			
Anopheles albimanus				X						X								
Anopheles atropos													X					
Anopheles barberi													X					
Anopheles bradleyi													X					
Anopheles crucians	X			X				X		X	X		X					
Anopheles franciscanus																		
Anopheles freeborni												X						
Anopheles judithae																		
Anopheles pseudopunctipennis	X		X	X							X	X	X		X			X
Anopheles punctipennis	X		X	X				X			X	X	X	X	X			
Anopheles quadrimaculatus	X		X	X							X		X				X	X
Anopheles smaragdinus								X										
Anopheles walkeri																		
Coquillettidia perturbans								X										
Culex abominator	X												X					
Culex apicalis													X					
Culex arizonensis																		

Species	1	2	3	4	5	6	7	8	9	10	11	12
Culex chidesteri												
Culex coronator	X		X		X		X		X		X	X
Culex declarator												X
Culex erraticus	X				X		X		X			X
Culex erythrothorax												
Culex interrogator												
Culex nigripalpus							X					
Culex peccator												
Culex pilosus												
Culex quinquefasciatus	X	X	X	X	X	X	X				X	X
Culex restuans	X		X	X			X					X
Culex salinarius	X		X				X					X
Culex stigmatosoma												
Culex tarsalis	X	X		X	X	X	X	X	X			
Culex territans												X
Culex thriambus			X									X
Culiseta incidens												
Culiseta inornata	X			X		X	X					X
Culiseta melanura												
Deinocerites mathesoni												
Deinocerites pseudes												
Haemagogus equinus												
Mansonia titillans					X		X					
Orthopodomyia alba												
Orthopodomyia kummi												
Orthopodomyia signifera	X		X				X					X

TABLE 2.2 Mosquito species by county.—cont'd

Species/county	Erath	Falls	Fannin	Fayette	Fisher	Floyd	Foard	Fort Bend	Franklin	Freestone	Frio	Gaines	Galveston	Garza	Gillespie	Glasscock	Goliad	Gonzales
Psorophora ciliata	X					X		X	X		X	X	X	X				X
Psorophora columbiae	X					X		X			X		X					X
Psorophora cyanescens	X							X			X		X		X			X
Psorophora discolor	X										X		X					
Psorophora ferox								X					X					X
Psorophora horrida													X					
Psorophora howardii													X					X
Psorophora longipalpus													X					X
Psorophora mathesoni																		
Psorophora mexicana																		
Psorophora signipennis	X								X	X	X	X		X				X
Toxorhynchites rutilus septentrionalis	X										X		X		X			X
Uranotaenia syntheta											X	X						
Uranotaenia lowii								X			X		X					X
Uranotaenia sapphirina								X					X					X
	25	1	4	9	1	7	0	23	7	5	22	11	47	4	8	0	4	28

Species/county	Gray	Grayson	Gregg	Grimes	Guadalupe	Hale	Hall	Hamilton	Hansford	Hardeman	Hardin	Harris	Harrison	Hartley	Haskell	Hays	Hemphill	Henderson
Aedes aegypti		X	X	X							X	X	X			X		X
Aedes albopictus			X								X	X	X			X		X
Aedes atlanticus											X	X						
Aedes bimaculatus																		
Aedes brelandi																		
Aedes campestris																		
Aedes canadensis			X						X			X	X					
Aedes dorsalis						X											X	
Aedes dupreei											X	X						
Aedes epactius		X			X							X						X
Aedes fulvus pallens												X						
Aedes grossbecki												X						
Aedes hendersoni											X	X						
Aedes infirmatus			X									X						
Aedes mitchellae												X	X					
Aedes muelleri																		
Aedes nigromaculis						X						X					X	
Aedes scapularis																		
Aedes sollicitans						X					X	X	X				X	
Aedes sticticus			X									X						
Aedes taeniorhynchus											X	X						
Aedes thelcter						X						X						
Aedes thibaulti																		
Aedes tormentor												X						
Aedes triseriatus		X	X								X	X	X					X

Continued

TABLE 2.2 Mosquito species by county.—cont'd

Species/county	Gray	Grayson	Gregg	Grimes	Guadalupe	Hale	Hall	Hamilton	Hansford	Hardeman	Hardin	Harris	Harrison	Hartley	Haskell	Hays	Hemphill	Henderson
Aedes trivittatus		X	X									X					X	
Aedes vexans	X		X			X			X		X	X	X				X	X
Aedes zoosophus							X					X						
Anopheles albimanus			X															
Anopheles atropos												X	X					
Anopheles barberi												X						
Anopheles bradleyi												X						
Anopheles crucians			X								X	X	X			X		X
Anopheles franciscanus																		
Anopheles freeborni																		
Anopheles judithae																		
Anopheles pseudopunctipennis					X	X					X	X				X		
Anopheles punctipennis		X	X			X						X				X		X
Anopheles quadrimaculatus		X	X	X							X	X				X		X
Anopheles smaragdinus																		X
Anopheles walkeri																		
Coquillettidia perturbans			X								X	X				X		X
Culex abominator												X						
Culex apicalis																		
Culex arizonensis																		
Culex chidesteri																		
Culex coronator		X	X	X	X						X	X				X		X
Culex declarator																		

Species	1	2	3	4	5	6	7	8	9	10	11	12
Culex erraticus			X		X			X		X	X	
Culex erythrothorax								X		X		
Culex interrogator												
Culex nigripalpus					X							
Culex peccator												
Culex pilosus				X								
Culex quinquefasciatus	X	X	X	X	X	X		X		X	X	
Culex restuans	X		X	X	X							
Culex salinarius	X	X	X	X	X	X		X			X	
Culex stigmatosoma												
Culex tarsalis		X	X	X	X		X	X			X	X
Culex territans				X	X				X			
Culex thriambus					X							
Culiseta incidens							X					
Culiseta inornata				X	X			X		X	X	
Culiseta melanura				X	X							
Deinocerites mathesoni												
Deinocerites pseudes												
Haemagogus equinus												
Mansonia titillans					X							
Orthopodomyia alba											X	
Orthopodomyia kummi												
Orthopodomyia signifera					X				X			
Psorophora ciliata	X				X	X		X		X	X	
Psorophora columbiae	X		X	X	X	X	X	X		X	X	
Psorophora cyanescens	X				X			X		X	X	
Psorophora discolor	X				X			X				
Psorophora ferox		X		X	X	X				X	X	
Psorophora horrida				X	X	X						

Continued

TABLE 2.2 Mosquito species by county.—cont'd

Species/county	Gray	Grayson	Gregg	Grimes	Guadalupe	Hale	Hall	Hamilton	Hansford	Hardeman	Hardin	Harris	Harrison	Hartley	Haskell	Hays	Hemphill	Henderson
Psorophora howardii			X								X	X	X					
Psorophora longipalpus												X						
Psorophora mathesoni												X						
Psorophora mexicana																		
Psorophora signipennis			X			X	X		X			X				X	X	
Toxorhynchites rutilus septentrionalis			X									X						
Uranotaenia syntheta						X										X		
Uranotaenia lowii											X	X				X		
Uranotaenia sapphirina			X									X	X			X		
	2	16	29	5	4	19	3	0	6	0	22	57	22	0	0	18	10	18

Species/county	Hidalgo	Hill	Hockley	Hood	Hopkins	Houston	Howard	Hudspeth	Hunt	Hutchinson	Irion	Jack	Jackson	Jasper	Jeff Davis	Jefferson	Jim Hogg
Aedes aegypti	X	X			X	X	X		X					X		X	X
Aedes albopictus	X			X		X			X							X	X
Aedes atlanticus														X		X	
Aedes bimaculatus	X																
Aedes brelandi																	
Aedes campestris								X							X		
Aedes canadensis							X							X		X	
Aedes dorsalis							X	X									
Aedes dupreei																	
Aedes epactius	X	X					X			X							
Aedes fulvus pallens																	
Aedes grossbecki																	
Aedes hendersoni																	
Aedes infirmatus	X	X												X		X	
Aedes mitchellae							X									X	
Aedes muelleri																	
Aedes nigromaculis			X				X	X		X		X					
Aedes scapularis	X									X							
Aedes sollicitans	X	X					X	X						X		X	
Aedes sticticus																	
Aedes taeniorhynchus	X						X									X	
Aedes thelcter		X					X										
Aedes thibaulti																	
Aedes tormentor							X										
Aedes triseriatus	X	X								X				X		X	
Aedes trivittatus	X	X												X	X	X	

Continued

TABLE 2.2 Mosquito species by county.—cont'd

Species/county	Hidalgo	Hill	Hockley	Hood	Hopkins	Houston	Howard	Hudspeth	Hunt	Hutchinson	Irion	Jack	Jackson	Jasper	Jeff Davis	Jefferson	Jim Hogg
Aedes vexans	X	X					X	X		X				X	X	X	
Aedes zoosophus	X	X															
Anopheles albimanus	X																
Anopheles atropos																X	
Anopheles barberi		X												X			
Anopheles bradleyi																X	
Anopheles crucians	X	X										X		X		X	
Anopheles franciscanus															X		
Anopheles freeborni																	
Anopheles judithae																	
Anopheles pseudopunctipennis	X	X					X	X				X			X	X	
Anopheles punctipennis	X	X					X	X				X		X		X	
Anopheles quadrimaculatus	X	X							X				X	X		X	X
Anopheles smaragdinus																X	
Anopheles walkeri																X	
Coquillettidia perturbans														X		X	
Culex abominator																	
Culex apicalis															X		
Culex arizonensis																	
Culex chidesteri	X																
Culex coronator	X	X				X			X				X			X	X
Culex declarator	X																

Species											
Culex erraticus	X		X				X		X	X	
Culex erythrothorax	X									X	
Culex interrogator	X										
Culex nigripalpus	X		X							X	
Culex peccator				X						X	
Culex pilosus											
Culex quinquefasciatus	X	X	X		X	X	X		X	X	X
Culex restuans	X	X	X		X	X				X	X
Culex salinarius	X	X	X	X	X	X	X			X	X
Culex stigmatosoma		X					X	X			
Culex tarsalis	X	X	X		X	X	X		X	X	X
Culex territans	X									X	
Culex thriambus	X		X		X		X		X	X	X
Culiseta incidens											
Culiseta inornata	X		X		X	X	X		X	X	
Culiseta melanura				X							
Deinocerites mathesoni										X	
Deinocerites pseudes											
Haemagogus equinus											
Mansonia titillans	X									X	
Orthopodomyia alba											
Orthopodomyia kummi											
Orthopodomyia signifera										X	
Psorophora ciliata	X		X		X		X		X	X	
Psorophora columbiae	X		X		X	X	X	X		X	
Psorophora cyanescens	X		X	X	X	X			X	X	X
Psorophora discolor	X		X		X				X	X	
Psorophora ferox	X									X	
Psorophora horrida	X			X					X	X	

Continued

TABLE 2.2 Mosquito species by county.—cont'd

Species/county	Hidalgo	Hill	Hockley	Hood	Hopkins	Houston	Howard	Hudspeth	Hunt	Hutchinson	Irion	Jack	Jackson	Jasper	Jeff Davis	Jefferson	Jim Hogg
Psorophora howardii																X	
Psorophora longipalpus		X												X			
Psorophora mathesoni		X												X			
Psorophora mexicana																	
Psorophora signipennis	X	X					X	X		X		X				X	
Toxorhynchites rutilus septentrionalis																X	
Uranotaenia syntheta	X						X					X				X	
Uranotaenia lowii	X								X							X	
Uranotaenia sapphirina							X							X		X	
	39	32	4	2	2	3	26	15	10	11	0	13	3	26	10	45	7

Species/county	Jim Wells	Johnson	Jones	Karnes	Kaufman	Kendall	Kenedy	Kent	Kerr	Kimble	King	Kinney	Kleberg	Knox	La Salle	Lamar	Lamb	Lampasas	Lavaca
Aedes aegypti	X	X			X	X	X			X			X		X	X		X	X
Aedes albopictus	X	X			X					X			X			X			X
Aedes atlanticus																X			
Aedes bimaculatus													X						
Aedes brelandi																			
Aedes campestris																			
Aedes canadensis																X			
Aedes dorsalis										X									
Aedes dupreei																X			
Aedes epactius				X		X			X	X					X				
Aedes fulvus pallens																			
Aedes grossbecki																			
Aedes hendersoni																			
Aedes infirmatus																			
Aedes mitchellae																			
Aedes muelleri																			
Aedes nigromaculis	X		X							X						X	X		
Aedes scapularis																			
Aedes sollicitans													X			X	X		
Aedes sticticus									X										
Aedes taeniorhynchus	X												X						X
Aedes thelcter	X						X						X						
Aedes thibaulti																			
Aedes tormentor																X			
Aedes triseriatus						X				X						X			
Aedes trivittatus						X			X	X		X				X			

Continued

TABLE 2.2 Mosquito species by county.—cont'd

Species/county	Jim Wells	Johnson	Jones	Karnes	Kaufman	Kendall	Kenedy	Kent	Kerr	Kimble	King	Kinney	Kleberg	Knox	La Salle	Lamar	Lamb	Lampasas	Lavaca
Aedes vexans		X	X						X	X						X	X		
Aedes zoosophus									X	X		X				X			
Anopheles albimanus	X												X						
Anopheles atropos																			
Anopheles barberi																X			
Anopheles bradleyi																			
Anopheles crucians	X				X								X			X			
Anopheles franciscanus																			
Anopheles freeborni																			
Anopheles judithae																			
Anopheles pseudopunctipennis	X					X	X		X	X		X	X			X			X
Anopheles punctipennis		X	X			X	X		X	X						X			
Anopheles quadrimaculatus	X				X		X						X			X		X	X
Anopheles smaragdinus																			
Anopheles walkeri																X			
Coquillettidia perturbans										X									
Culex abominator																X			
Culex apicalis																			
Culex arizonensis																			
Culex chidesteri																			
Culex coronator	X	X		X		X	X		X	X		X	X		X			X	X
Culex declarator																			
Culex erraticus	X	X			X				X	X		X	X			X			
Culex erythrothorax																			

Species													
Culex interrogator													
Culex nigripalpus								X					
Culex peccator													
Culex pilosus													
Culex quinquefasciatus	X	X		X	X	X	X	X			X		
Culex restuans		X		X		X	X	X					
Culex salinarius	X			X		X	X	X					
Culex stigmatosoma	X				X	X							
Culex tarsalis	X	X		X	X	X	X	X			X	X	
Culex territans		X				X	X	X					
Culex thriambus						X		X					
Culiseta incidens													
Culiseta inornata	X			X		X	X	X					
Culiseta melanura													
Deinocerites mathesoni													
Deinocerites pseudes							X						
Haemagogus equinus													
Mansonia titillans	X				X		X	X					
Orthopodomyia alba								X			X		
Orthopodomyia kummi													
Orthopodomyia signifera								X			X		
Psorophora ciliata	X			X		X	X	X			X		
Psorophora columbiae	X			X		X	X	X			X	X	
Psorophora cyanescens	X				X	X	X	X	X		X		X
Psorophora discolor	X	X				X	X	X			X		
Psorophora ferox						X		X			X		
Psorophora horrida								X					
Psorophora howardii								X			X		
Psorophora longipalpus								X					

Continued

TABLE 2.2 Mosquito species by county.—cont'd

Species/county	Jim Wells	Johnson	Jones	Karnes	Kaufman	Kendall	Kenedy	Kent	Kerr	Kimble	King	Kinney	Kleberg	Knox	La Salle	Lamar	Lamb	Lampasas	Lavaca
Psorophora mathesoni																			
Psorophora mexicana																			
Psorophora signipennis	X		X						X	X			X			X			
Toxorhynchites rutilus septentrionalis																X			
Uranotaenia syntheta										X									
Uranotaenia lowii	X									X			X			X			
Uranotaenia sapphirina										X						X			
	22	9	11	2	7	14	9	0	15	33	0	8	24	0	4	39	5	4	10

Species/county	Lee	Leon	Liberty	Limestone	Lipscomb	Live Oak	Llano	Loving	Lubbock	Lynn	Madison	Marion	Martin	Mason	Matagorda	Maverick	McCulloch
Aedes aegypti	X	X	X	X					X					X	X	X	
Aedes albopictus			X	X					X						X		
Aedes atlanticus			X												X		
Aedes bimaculatus																	
Aedes brelandi																	
Aedes campestris																	
Aedes canadensis																	
Aedes dorsalis				X						X							
Aedes dupreei																	
Aedes epactius		X		X		X			X						X	X	
Aedes fulvus pallens			X														
Aedes grossbecki																	
Aedes hendersoni														X			
Aedes infirmatus																	
Aedes mitchellae									X								
Aedes muelleri																	
Aedes nigromaculis				X					X						X	X	
Aedes scapularis																	
Aedes sollicitans			X	X					X						X		
Aedes sticticus																	
Aedes taeniorhynchus			X														
Aedes thelcter									X						X	X	
Aedes thibaulti															X		
Aedes tormentor																	
Aedes triseriatus			X	X										X			

Continued

TABLE 2.2 Mosquito species by county.—cont'd

Species/county	Lee	Leon	Liberty	Limestone	Lipscomb	Live Oak	Llano	Loving	Lubbock	Lynn	Madison	Marion	Martin	Mason	Matagorda	Maverick	McCulloch
Aedes trivittatus			X		X				X								
Aedes vexans					X				X	X				X	X	X	X
Aedes zoosophus									X					X			
Anopheles albimanus															X		
Anopheles atropos																	
Anopheles barberi																	
Anopheles bradleyi																	
Anopheles crucians	X	X	X	X		X	X								X		
Anopheles franciscanus									X								
Anopheles freeborni																	
Anopheles judithae																	
Anopheles pseudopunctipennis			X	X		X			X					X		X	
Anopheles punctipennis	X	X	X	X		X	X		X					X		X	X
Anopheles quadrimaculatus	X	X	X	X		X	X		X							X	
Anopheles smaragdinus																	
Anopheles walkeri																	
Coquillettidia perturbans		X	X						X								
Culex abominator														X		X	
Culex apicalis																	
Culex arizonensis																	
Culex chidesteri																	
Culex coronator	X		X			X	X							X	X	X	X
Culex declarator																	
Culex erraticus	X			X		X			X					X		X	X

Species																
Culex erythrothorax									X							
Culex interrogator																
Culex nigripalpus																
Culex peccator																
Culex pilosus																
Culex quinquefasciatus	X	X	X	X	X	X	X		X	X	X		X	X	X	X
Culex restuans	X	X				X				X				X	X	
Culex salinarius	X	X	X		X	X			X	X	X		X	X	X	
Culex stigmatosoma			X										X	X		X
Culex tarsalis	X		X	X	X	X	X		X	X	X		X	X	X	X
Culex territans	X		X				X				X				X	
Culex thriambus											X			X		X
Culiseta incidens										X						
Culiseta inornata		X	X	X		X	X		X	X	X		X	X	X	X
Culiseta melanura																
Deinocerites mathesoni																
Deinocerites pseudes																
Haemagogus equinus																
Mansonia titillans																
Orthopodomyia alba			X													
Orthopodomyia kummi																
Orthopodomyia signifera																
Psorophora ciliata	X	X	X	X	X	X		X		X			X	X	X	X
Psorophora columbiae	X	X	X	X	X	X	X		X	X			X	X	X	X
Psorophora cyanescens	X	X	X	X	X	X				X			X		X	X
Psorophora discolor		X	X		X	X				X					X	X
Psorophora ferox		X	X			X				X						
Psorophora horrida		X	X			X										

Continued

TABLE 2.2 Mosquito species by county.—cont'd

Species/county	Lee	Leon	Liberty	Limestone	Lipscomb	Live Oak	Llano	Loving	Lubbock	Lynn	Madison	Marion	Martin	Mason	Matagorda	Maverick	McCulloch
Psorophora howardii			X														
Psorophora longipalpus																	
Psorophora mathesoni			X														
Psorophora mexicana																	
Psorophora signipennis				X		X			X	X				X		X	X
Toxorhynchites rutilus septentrionalis		X															
Uranotaenia syntheta				X		X			X					X			
Uranotaenia lowii			X	X											X		
Uranotaenia sapphirina			X	X													
	6	12	28	22	5	14	5	0	31	7	1	0	1	18	22	19	12

Species/county	McLennan	McMullen	Medina	Menard	Midland	Milam	Mills	Mitchell	Montague	Montgomery	Moore	Morris	Motley	Nacogdoches	Navarro
Aedes aegypti	X				X	X				X				X	X
Aedes albopictus	X	X		X	X					X				X	X
Aedes atlanticus	X									X					
Aedes bimaculatus										X					
Aedes brelandi															
Aedes campestris															
Aedes canadensis															
Aedes dorsalis					X										
Aedes dupreei															
Aedes epactius	X	X	X			X				X					
Aedes fulvus pallens	X									X					
Aedes grossbecki															
Aedes hendersoni										X					
Aedes infirmatus										X					
Aedes mitchellae															
Aedes muelleri															
Aedes nigromaculis	X				X					X					
Aedes scapularis										X					
Aedes sollicitans	X		X							X	X				
Aedes sticticus															
Aedes taeniorhynchus			X							X					
Aedes thelcter					X										
Aedes thibaulti															
Aedes tormentor															
Aedes triseriatus	X			X						X					

TABLE 2.2 Mosquito species by county.—cont'd

Species/county	McLennan	McMullen	Medina	Menard	Midland	Milam	Mills	Mitchell	Montague	Montgomery	Moore	Morris	Motley	Nacogdoches	Navarro
Aedes trivittatus										X					
Aedes vexans	X				X					X	X				
Aedes zoosophus						X									
Anopheles albimanus															
Anopheles atropos										X					
Anopheles barberi										X					
Anopheles bradleyi															
Anopheles crucians	X			X						X				X	
Anopheles franciscanus															
Anopheles freeborni															
Anopheles judithae															
Anopheles pseudopunctipennis	X	X			X	X				X					
Anopheles punctipennis	X			X			X			X				X	X
Anopheles quadrimaculatus	X	X			X	X				X					
Anopheles smaragdinus															
Anopheles walkeri															
Coquillettidia perturbans	X									X					
Culex abominator															
Culex apicalis															
Culex arizonensis															
Culex chidesteri														X	
Culex coronator	X	X	X	X		X	X			X					
Culex declarator															
Culex erraticus	X									X					

Species								
Culex erythrothorax								
Culex interrogator								
Culex nigripalpus					X			
Culex peccator								
Culex pilosus								
Culex quinquefasciatus	X	X	X		X			X
Culex restuans	X	X			X			X
Culex salinarius		X	X		X			
Culex stigmatosoma								
Culex tarsalis		X	X	X	X	X	X	X
Culex territans					X			X
Culex thriambus								
Culiseta incidens								
Culiseta inornata		X		X	X			
Culiseta melanura								
Deinocerites mathesoni								
Deinocerites pseudes								
Haemagogus equinus								
Mansonia titillans					X			
Orthopodomyia alba								
Orthopodomyia kummi								
Orthopodomyia signifera		X			X			
Psorophora ciliata		X	X		X			
Psorophora columbiae		X			X			
Psorophora cyanescens					X			
Psorophora discolor								
Psorophora ferox					X			
Psorophora horrida					X			
Psorophora howardii					X			

Continued

TABLE 2.2 Mosquito species by county.—cont'd

Species/county	McLennan	McMullen	Medina	Menard	Midland	Milam	Mills	Mitchell	Montague	Montgomery	Moore	Morris	Motley	Nacogdoches	Navarro
Psorophora longipalpus	X									X					
Psorophora mathesoni										X					
Psorophora mexicana					X										
Psorophora signipennis	X														
Toxorhynchites rutilus septentrionalis										X					
Uranotaenia syntheta				X	X										
Uranotaenia lowii	X									X					
Uranotaenia sapphirina										X					
	33	6	4	8	18	10	4	0	0	44	3	0	1	5	7

Species/county	Newton	Nolan	Nueces	Ochiltree	Oldham	Orange	Palo Pinto	Panola	Parker	Parmer	Pecos	Polk	Potter	Presidio	Rains	Randall	Reagan	Real
Aedes aegypti			X			X			X			X		X				
Aedes albopictus		X	X			X			X				X			X		X
Aedes atlanticus						X												
Aedes bimaculatus																		
Aedes brelandi																		
Aedes campestris											X			X				
Aedes canadensis			X															
Aedes dorsalis				X			X			X	X		X	X		X		
Aedes dupreei																		
Aedes epactius		X							X									X
Aedes fulvus pallens						X												
Aedes grossbecki																		
Aedes hendersoni																		
Aedes infirmatus																		
Aedes mitchellae						X												
Aedes muelleri																		
Aedes nigromaculis		X		X			X			X	X		X			X		
Aedes scapularis						X												
Aedes sollicitans	X		X	X		X				X	X		X	X		X		
Aedes sticticus																		
Aedes taeniorhynchus	X	X	X			X					X							
Aedes thelcter		X	X			X	X							X				
Aedes thibaulti																		
Aedes tormentor																		
Aedes triseriatus		X				X	X		X			X		X				
Aedes trivittatus			X			X				X				X		X		

Continued

TABLE 2.2 Mosquito species by county.—cont'd

Species/county	Newton	Nolan	Nueces	Ochiltree	Oldham	Orange	Palo Pinto	Panola	Parker	Parmer	Pecos	Polk	Potter	Presidio	Rains	Randall	Reagan	Real
Aedes vexans		X	X	X	X	X	X		X	X	X		X	X		X		
Aedes zoosophus			X				X		X		X					X		
Anopheles albimanus			X															
Anopheles atropos			X			X												
Anopheles barberi						X	X		X									
Anopheles bradleyi			X			X	X		X		X							
Anopheles crucians		X	X			X	X	X	X		X			X				
Anopheles franciscanus											X							
Anopheles freeborni																		
Anopheles judithae																		
Anopheles pseudopunctipennis		X	X			X	X		X		X		X	X				X
Anopheles punctipennis		X	X			X	X	X	X		X							
Anopheles quadrimaculatus	X	X	X			X	X		X									
Anopheles smaragdinus						X									X			
Anopheles walkeri						X												
Coquillettidia perturbans	X					X			X				X			X		
Culex abominator						X												
Culex apicalis																		
Culex arizonensis																		
Culex chidesteri																		
Culex coronator		X	X			X	X					X	X	X	X			X
Culex declarator												X						
Culex erraticus		X	X	X		X	X		X		X		X	X				X
Culex erythrothorax						X												

Species	1	2	3	4	5	6	7	8	9	10	11	12	13
Culex interrogator													
Culex nigripalpus													
Culex peccator													
Culex pilosus													
Culex quinquefasciatus	X	X			X		X	X	X		X	X	X
Culex restuans					X		X	X				X	X
Culex salinarius	X				X		X	X	X			X	X
Culex stigmatosoma					X								
Culex tarsalis	X	X	X		X		X	X		X	X	X	X
Culex territans							X						
Culex thriambus		X			X		X						
Culiseta incidens													
Culiseta inornata	X	X	X		X		X	X	X		X	X	X
Culiseta melanura									X				
Deinocerites mathesoni												X	
Deinocerites pseudes													
Haemagogus equinus													
Mansonia titillans													
Orthopodomyia alba					X								
Orthopodomyia kummi									X				
Orthopodomyia signifera													
Psorophora ciliata	X	X		X	X			X	X		X	X	X
Psorophora columbiae		X			X		X	X	X		X	X	X
Psorophora cyanescens	X	X	X	X	X	X		X	X			X	X
Psorophora discolor	X		X		X		X	X	X		X	X	X
Psorophora ferox									X				X
Psorophora horrida													
Psorophora howardii													
Psorophora longipalpus													

Continued

TABLE 2.2 Mosquito species by county.—cont'd

Species/county	Newton	Nolan	Nueces	Ochiltree	Oldham	Orange	Palo Pinto	Panola	Parker	Parmer	Pecos	Polk	Potter	Presidio	Rains	Randall	Reagan	Real
Psorophora mathesoni																		
Psorophora mexicana																		
Psorophora signipennis		X	X	X		X	X		X		X		X	X		X		
Toxorhynchites rutilus septentrionalis																		
Uranotaenia syntheta		X																
Uranotaenia lowii			X			X												
Uranotaenia sapphirina						X												
	9	24	29	12	2	37	23	2	24	6	25	5	13	19	2	16	0	5

Species/county	Red River	Reeves	Refugio	Roberts	Robertson	Rockwall	Runnels	Rusk	Sabine	San Augustine	San Jacinto	San Patricio	San Saba	Schleicher	Scurry
Aedes aegypti		X			X			X							
Aedes albopictus			X				X	X			X	X			
Aedes atlanticus															
Aedes bimaculatus												X			
Aedes brelandi															
Aedes campestris		X													
Aedes canadensis															
Aedes dorsalis		X					X								
Aedes dupreei															
Aedes epactius															
Aedes fulvus pallens												X			
Aedes grossbecki															
Aedes hendersoni															
Aedes infirmatus															
Aedes mitchellae															
Aedes muelleri															
Aedes nigromaculis		X					X						X		X
Aedes scapularis															
Aedes sollicitans		X		X				X				X			X
Aedes sticticus															
Aedes taeniorhynchus												X			
Aedes thelcter		X										X			X
Aedes thibaulti															
Aedes tormentor															
Aedes triseriatus					X							X	X		X

Continued

TABLE 2.2 Mosquito species by county.—cont'd

Species/county	Red River	Reeves	Refugio	Roberts	Robertson	Rockwall	Runnels	Rusk	Sabine	San Augustine	San Jacinto	San Patricio	San Saba	Schleicher	Scurry
Aedes trivittatus				X											
Aedes vexans		X		X			X					X	X		
Aedes zoosophus												X	X		
Anopheles albimanus															
Anopheles atropos															
Anopheles barberi															
Anopheles bradleyi															
Anopheles crucians		X							X	X	X	X			
Anopheles franciscanus															
Anopheles freeborni															
Anopheles judithae															
Anopheles pseudopunctipennis		X					X					X	X		X
Anopheles punctipennis							X			X		X	X		X
Anopheles quadrimaculatus					X		X	X	X	X	X	X			
Anopheles smaragdinus															
Anopheles walkeri															
Coquillettidia perturbans												X			
Culex abominator															
Culex apicalis															
Culex arizonensis															
Culex chidesteri						X									
Culex coronator			X				X					X		X	

Species									
Culex declarator	X								
Culex erraticus	X			X	X		X	X	X
Culex erythrothorax							X		
Culex interrogator									
Culex nigripalpus							X		
Culex peccator							X		
Culex pilosus									
Culex quinquefasciatus	X	X		X	X	X	X	X	X
Culex restuans					X		X		
Culex salinarius		X				X	X	X	
Culex stigmatosoma	X				X		X		
Culex tarsalis	X		X		X	X	X	X	X
Culex territans		X							
Culex thriambus	X				X				X
Culiseta incidens									
Culiseta inornata	X				X		X	X	
Culiseta melanura									
Deinocerites mathesoni									
Deinocerites pseudes							X		
Haemagogus equinus									
Mansonia titillans							X		
Orthopodomyia alba									
Orthopodomyia kummi									
Orthopodomyia signifera							X		
Psorophora ciliata	X	X			X		X		X
Psorophora columbiae	X	X		X	X		X	X	
Psorophora cyanescens	X				X		X		X

Continued

TABLE 2.2 Mosquito species by county.—cont'd

Species/county	Red River	Reeves	Refugio	Roberts	Robertson	Rockwall	Runnels	Rusk	Sabine	San Augustine	San Jacinto	San Patricio	San Saba	Schleicher	Scurry
Psorophora discolor		X					X					X	X		X
Psorophora ferox															
Psorophora horrida															
Psorophora howardii															
Psorophora longipalpus															
Psorophora mathesoni															
Psorophora mexicana															
Psorophora signipennis		X					X					X	X		X
Toxorhynchites rutilus septentrionalis												X			
Uranotaenia syntheta		X					X					X	X		
Uranotaenia lowii			X		X							X			
Uranotaenia sapphirina												X			
	1	21	7	4	7	1	21	7	2	3	3	37	16	1	14

Species/county	Shackelford	Shelby	Sherman	Smith	Somervell	Starr	Stephens	Sterling	Stonewall	Sutton	Swisher	Tarrant	Taylor	Terrell	Terry	Throckmorton	Titus
Aedes aegypti		X		X		X				X		X	X				X
Aedes albopictus				X				X		X		X	X		X		
Aedes atlanticus												X					
Aedes bimaculatus																	
Aedes brelandi																	
Aedes campestris														X			
Aedes canadensis				X								X					
Aedes dorsalis	X												X		X		
Aedes dupreei																	
Aedes epactius			X	X						X		X		X			
Aedes fulvus pallens																	
Aedes grossbecki																	
Aedes hendersoni																	
Aedes infirmatus				X								X	X				
Aedes mitchellae			X														
Aedes muelleri																	
Aedes nigromaculis	X									X		X	X		X		
Aedes scapularis						X											
Aedes sollicitans	X			X							X	X	X				
Aedes sticticus												X					
Aedes taeniorhynchus																	
Aedes thelcter	X					X							X				
Aedes thibaulti				X						X							
Aedes tormentor													X				
Aedes triseriatus				X						X		X					X

Continued

TABLE 2.2 Mosquito species by county.—cont'd

Species/county	Shackelford	Shelby	Sherman	Smith	Somervell	Starr	Stephens	Sterling	Stonewall	Sutton	Swisher	Tarrant	Taylor	Terrell	Terry	Throckmorton	Titus
Aedes trivittatus				X								X		X			
Aedes vexans	X			X						X	X	X	X		X		X
Aedes zoosophus				X						X		X	X				
Anopheles albimanus																	
Anopheles atropos																	
Anopheles barberi												X					
Anopheles bradleyi																	
Anopheles crucians		X		X		X						X					
Anopheles franciscanus																	
Anopheles freeborni																	
Anopheles judithae																	
Anopheles pseudopunctipennis	X					X				X		X	X	X	X		
Anopheles punctipennis	X	X		X						X		X	X	X			
Anopheles quadrimaculatus		X		X		X						X	X				
Anopheles smaragdinus																	
Anopheles walkeri																	
Coquillettidia perturbans				X								X					
Culex abominator													X				
Culex apicalis																	
Culex arizonensis																	
Culex chidesteri								X									
Culex coronator				X		X				X		X	X				
Culex declarator																	

	1	2	3	4	5	6	7	8	9	10	11	12	13	14
Culex erraticus	X		X	X			X		X	X				
Culex erythrothorax														
Culex interrogator									X					
Culex nigripalpus									X					
Culex peccator														
Culex pilosus														
Culex quinquefasciatus	X	X	X	X			X	X	X	X		X		X
Culex restuans	X		X						X	X		X		X
Culex salinarius	X		X					X	X	X		X		X
Culex stigmatosoma		X					X		X			X		
Culex tarsalis	X		X	X			X		X	X	X	X		
Culex territans			X						X	X				
Culex thriambus									X	X		X		
Culiseta incidens									X			X		
Culiseta inornata			X	X			X		X	X		X		
Culiseta melanura									X	X				
Deinocerites mathesoni														
Deinocerites pseudes														
Haemagogus equinus														
Mansonia titillans				X					X					
Orthopodomyia alba				X			X		X					
Orthopodomyia kummi														
Orthopodomyia signifera									X					
Psorophora ciliata	X		X	X			X		X	X				
Psorophora columbiae	X	X	X	X			X		X	X		X		X
Psorophora cyanescens	X		X	X			X		X	X	X	X		
Psorophora discolor	X		X	X			X		X	X				
Psorophora ferox	X								X	X				
Psorophora horrida									X					

Continued

TABLE 2.2 Mosquito species by county.—cont'd

Species/county	Shackelford	Shelby	Sherman	Smith	Somervell	Starr	Stephens	Sterling	Stonewall	Sutton	Swisher	Tarrant	Taylor	Terrell	Terry	Throckmorton	Titus
Psorophora howardii												X					
Psorophora longipalpus												X					
Psorophora mathesoni				X													
Psorophora mexicana																	
Psorophora signipennis	X					X				X	X	X	X	X	X		
Toxorhynchites rutilus septentrionalis												X					
Uranotaenia syntheta						X				X		X	X				
Uranotaenia lowii													X				
Uranotaenia sapphirina												X	X				
	20	7	3	28	0	17	0	2	0	22	7	47	33	8	16	0	7

Species/county	Tom Green	Travis	Trinity	Tyler	Upshur	Upton	Uvalde	Val Verde	Van Zandt	Victoria	Walker	Waller	Ward	Washington	Webb	Wharton	Wheeler
Aedes aegypti	X	X	X		X		X	X		X	X	X		X	X	X	
Aedes albopictus		X		X	X		X	X	X	X	X	X			X	X	
Aedes atlanticus								X			X						
Aedes bimaculatus																	
Aedes brelandi																	
Aedes campestris																	
Aedes canadensis		X															
Aedes dorsalis		X				X							X				X
Aedes dupreei																	
Aedes epactius		X					X	X									
Aedes fulvus pallens																	
Aedes grossbecki																	
Aedes hendersoni		X															
Aedes infirmatus																	
Aedes mitchellae										X					X		
Aedes muelleri																	
Aedes nigromaculis	X	X				X		X							X		X
Aedes scapularis																	
Aedes sollicitans	X	X					X	X		X			X		X	X	X
Aedes sticticus		X								X							
Aedes taeniorhynchus															X	X	
Aedes thelcter		X					X	X	X				X		X		
Aedes thibaulti																	
Aedes tormentor								X									
Aedes triseriatus		X		X				X		X	X						

Continued

TABLE 2.2 Mosquito species by county.—cont'd

Species/county	Tom Green	Travis	Trinity	Tyler	Upshur	Upton	Uvalde	Val Verde	Van Zandt	Victoria	Walker	Waller	Ward	Washington	Webb	Wharton	Wheeler
Aedes trivittatus		X						X							X		
Aedes vexans	X	X		X		X	X	X		X	X		X		X		X
Aedes zoosophus		X					X	X									
Anopheles albimanus															X	X	
Anopheles atropos		X															
Anopheles barberi																	
Anopheles bradleyi																	
Anopheles crucians		X		X		X		X		X	X	X	X		X	X	
Anopheles franciscanus													X				
Anopheles freeborni								X									
Anopheles judithae																	
Anopheles pseudopunctipennis	X	X		X		X	X	X			X		X		X		X
Anopheles punctipennis	X	X		X			X	X			X		X	X	X	X	X
Anopheles quadrimaculatus		X		X		X	X	X		X	X	X		X	X	X	
Anopheles smaragdinus																	
Anopheles walkeri																	
Coquillettidia perturbans		X		X		X											
Culex abominator		X						X		X					X		
Culex apicalis																	
Culex arizonensis																	
Culex chidesteri								X									
Culex coronator	X	X					X	X	X	X	X	X		X	X	X	

Species
Culex declarator
Culex erraticus
Culex erythrothorax
Culex interrogator
Culex nigripalpus
Culex peccator
Culex pilosus
Culex quinquefasciatus
Culex restuans
Culex salinarius
Culex stigmatosoma
Culex tarsalis
Culex territans
Culex thriambus
Culiseta incidens
Culiseta inornata
Culiseta melanura
Deinocerites mathesoni
Deinocerites pseudes
Haemagogus equinus
Mansonia titillans
Orthopodomyia alba
Orthopodomyia kummi
Orthopodomyia signifera
Psorophora ciliata
Psorophora columbiae

Continued

TABLE 2.2 Mosquito species by county.—cont'd

Species/county	Tom Green	Travis	Trinity	Tyler	Upshur	Upton	Uvalde	Val Verde	Van Zandt	Victoria	Walker	Waller	Ward	Washington	Webb	Wharton	Wheeler
Psorophora cyanescens	X	X				X	X	X		X			X		X		X
Psorophora discolor	X	X					X	X		X					X		
Psorophora ferox		X													X		
Psorophora horrida		X								X							
Psorophora howardii		X															
Psorophora longipalpus		X		X						X							
Psorophora mathesoni																	
Psorophora mexicana																	
Psorophora signipennis	X	X	X			X	X	X					X		X		X
Toxorhynchites rutilus septentrionalis		X															
Uranotaenia syntheta	X	X					X	X							X		
Uranotaenia lowii		X															
Uranotaenia sapphirina		X		X		X		X							X		
	21	48	5	18	3	16	24	37	3	21	16	9	11	6	33	12	17

Species/county	Wichita	Wilbarger	Willacy	Williamson	Wilson	Winkler	Wise	Wood	Yoakum	Young	Zapata	Zavala	Total counties
Aedes aegypti	X		X	X				X			X	X	108
Aedes albopictus	X		X	X				X					92
Aedes atlanticus													20
Aedes bimaculatus													5
Aedes brelandi													1
Aedes campestris													9
Aedes canadensis				X									20
Aedes dorsalis	X												38
Aedes dupreei													4
Aedes epactius	X			X	X							X	55
Aedes fulvus pallens													11
Aedes grossbecki													1
Aedes hendersoni													9
Aedes infirmatus													16
Aedes mitchellae	X												16
Aedes muelleri													1
Aedes nigromaculis	X												71
Aedes scapularis													7
Aedes sollicitans	X		X										76
Aedes sticticus	X												8
Aedes taeniorhynchus	X		X										41
Aedes thelcter	X		X								X		45
Aedes thibaulti													3
Aedes tormentor													7

Continued

TABLE 2.2 Mosquito species by county.—cont'd

Species/county	Wichita	Wilbarger	Willacy	Williamson	Wilson	Winkler	Wise	Wood	Yoakum	Young	Zapata	Zavala	Total counties
Aedes triseriatus	X			X									68
Aedes trivittatus	X			X									44
Aedes vexans	X		X	X								X	118
Aedes zoosophus	X			X									46
Anopheles albimanus			X										14
Anopheles atropos													10
Anopheles barberi													11
Anopheles bradleyi					X								6
Anopheles crucians	X		X	X	X								84
Anopheles franciscanus													7
Anopheles freeborni													4
Anopheles judithae													1
Anopheles pseudopunctipennis	X		X	X	X						X	X	110
Anopheles punctipennis	X		X	X									106
Anopheles quadrimaculatus	X		X	X								X	113
Anopheles smaragdinus								X					7
Anopheles walkeri													2
Coquillettidia perturbans	X			X									36
Culex abominator	X												23
Culex apicalis													1
Culex arizonensis													2
Culex chidesteri													3
Culex coronator	X		X	X	X			X			X	X	127

Species						
Culex declarator						4
Culex erraticus	X	X		X	X	100
Culex erythrothorax						7
Culex interrogator						5
Culex nigripalpus		X		X		23
Culex peccator	X					11
Culex pilosus						1
Culex quinquefasciatus	X	X				150
Culex restuans	X					76
Culex salinarius	X	X				102
Culex stigmatosoma						27
Culex tarsalis	X	X		X		140
Culex territans	X	X				48
Culex thriambus	X					41
Culiseta incidens						7
Culiseta inornata	X	X				96
Culiseta melanura						4
Deinocerites mathesoni		X				4
Deinocerites pseudes						3
Haemagogus equinus						1
Mansonia titillans		X				19
Orthopodomyia alba	X					14
Orthopodomyia kummi						1
Orthopodomyia signifera	X					22
Psorophora ciliata	X	X				89
Psorophora columbiae	X	X		X		110
Psorophora cyanescens	X	X	X	X		103
Psorophora discolor	X	X				75
Psorophora ferox	X					37

Continued

TABLE 2.2 Mosquito species by county.—cont'd

Species/county	Wichita	Wilbarger	Willacy	Williamson	Wilson	Winkler	Wise	Wood	Yoakum	Young	Zapata	Zavala	Total counties
Psorophora horrida	X												22
Psorophora howardii	X												20
Psorophora longipalpus	X												16
Psorophora mathesoni													10
Psorophora mexicana													1
Psorophora signipennis	X		X										96
Toxorhynchites rutilus septentrionalis	X			X									22
Uranotaenia syntheta	X			X									36
Uranotaenia lowii	X			X	X								39
Uranotaenia sapphirina	X											11	36
	45	0	21	27	7	0	0	4	0	0	4		

Acknowledgments

We thank Vence Salvato for his assistance in preparing the distribution maps, Gene White for providing the mosquito pictures, Susan Real for some of the pictures while at Harris County, and Aldo I. Ortega-Morales for his valuable comments and assistance in connecting Chapters 2 and 9 of this book. We are grateful to the many individuals from Texas counties, cities, and other municipalities for their contribution on the occurrence of mosquito species in their localities, and to Leopoldo Rueda for his assistance in this chapter.

References and further reading

Adames, A.J., 1971. Mosquito studies (Diptera: Culicidae). XXIV. A revision of the crabhole mosquitoes of the genus *Deinocerites*. Contr. Am. Ent. Inst. 7 (2), 1–155.

Aitken, T.H.G., Downs, W.G., Spence, L., Jonkers, A.H., 1964. St. Louis encephalitis virus isolations in Trinidad, West Indies, 1953–1962. Am. J. Trop. Med. Hyg. 13, 450–451.

Aitken, T.H.G., Spence, L., Jonkers, A.H., Downs, W.G., 1969. A 10-year survey of Trinidadian arthropods for natural virus infections (1953–1963). J. Med. Entomol. 6, 207–215.

AMCA (American Mosquito Control Association), 2019a. Mosquito Info-American Mosquito Control Association. Available from: https://www.mosquito.org/default.aspx. (Accessed March 26, 2019).

AMCA (American Mosquito Control Association), 2019b. Mosquito Info-American Mosquito Control Association. Available from: https://www.mosquito.org/page/mosquitoinfo. (Accessed March 26, 2019).

Andreadis, T., Capotosto, P., Shope, R., Tirrell, S., 1994. Mosquito and arbovirus surveillance in Connecticut, 1991–1992. J. Am. Mosq. Control Assoc. 10, 556–564.

Andreadis, T., Anderson, J., Tirrell-Peck, S., 1998. Multiple isolations of eastern equine encephalitis and highlands J viruses from mosquitoes (Diptera: Culicidae) during a 1996 epizootic in southeastern Connecticut. J. Med. Entomol. 35, 296–302.

Andreadis, T.G., Armstrong, P.M., Anderson, J.F., Main, A.J., 2014. Spatial-temporal analysis of Cache Valley virus (Bunyaviridae: Orthobunyavirus) infection in anopheline and culicine mosquitoes (Diptera: Culicidae) in the Northeastern United States, 1997–2012. Vector Borne Zoonotic Dis. 14 (10), 713–773.

Armstrong, P.M., Andreadis, T.G., 2010. Eastern equine encephalitis virus in mosquitoes and their role as bridge vectors. Emerg. Inf. Dis. 16, 1869–1874. PMID: 21122215. PMCID: PMC3294553, https://doi.org/10.3201/eid1612.100640.

Arnell, J.H., 1973. Mosquito studies (Diptera, Culicidae) XXXII. A revision of the genus *Haemagogus*. Contr. Am. Ent. Inst. 10 (2), 1–174.

Arnell, J.H., 1976. Mosquito studies (Diptera, Culicidae) XXXIII. A revision of the *Scapularis* Group of *Aedes (Ochlerotatus)*. Contr. Am. Ent. Inst. 13 (3), 1–144.

Arnett, R., 1949. Notes on the distribution, habits, and habitats of some Panama Culicines (Diptera: Culicidae). J. N. Y. Entomol. Soc. 57 (4), 233–251.

Bang, F.B., Quinby, G.E., Simpson, T.W., 1943. Studies on *Anopheles walkeri* Theobald conducted at Reelfoot Lake, Tennessee, 1935–1941. J. Am. Trop. Med. Hyg. 23 (2), 247–273.

Barr, A.R., 1957. The distribution of *Culex p. pipiens* and *Cx. p. quinquefasciatus* in North America. Am. J. Trop. Med. Hyg. 6 (1), 153–165.

Belkin, J.N., Heinemann, S.J., 1975. *Psorophora (Janthinosoma) mathesoni*, sp. nov. for "*varipes*" of the Southeastern U.S.A. Mosq. Syst. 7 (4), 363–366.

Belkin, J.N., Hogue, C.L., 1959. A review of the crabhole mosquitoes of the genus *Deinocerites* (Diptera: Culicidae). 14(6) University of California Press, Berkeley and Los Angeles. 411–458.

Belkin, J.N., McDonald, W.A., 1956. A population of *Uranotaenia anhydor* from Death Valley, with descriptions of all stages and discussion of the complex (Diptera: Culicidae). Ann. Ent. Soc. Am. 49 (2), 105–132.

Belkin, J.N., Schick, R.X., Heinemann, S.J., 1966. Mosquito studies (Diptera: Culicidae). VI. Mosquitoes originally described from North America. Contr. Am. Ent. Inst. 1 (6), 1–39.

Blackmore, J.S., Winn, J.F., 1954. *Aedes nigromaculis* (Ludlow), mosquito naturally infected with western equine encephalomyelitis virus. Proc. Soc. Exp. Biol. Med. 87 (2), 328–329.

Blosser, E.M., Burkett-Cadena, N.D., 2017. *Culex (Melanoconion) panocossa* from peninsular Florida, USA. Acta Trop. 167, 59–63. https://doi.org/10.1016/j.actatropica.2016.12.024.

Bohart, R.M., 1948. The subgenus *Neoculex* in America North of Mexico (Diptera, Culicidae). Ann. Ent. Soc. Am. 41, 330–345.

Breland, O.P., 1948a. Notes on the mosquito, *Uranotaenia syntheta* Dyar and Shannon (Diptera: Culicidae). Mosq. News 8, 108–109.

Breland, O.P., 1948b. The biology and the immature stages of the mosquito, *Megarhinus septentrionalis* Dyar and Knab. Ann. Ent. Soc. Am. 42, 38–47.

Breland, O.P., 1953. Keys to the larvae of Texas mosquitoes with notes on recent synonymy IN The genus *Culex* Linnaeus. Tx. J. Sci. 5, 114–119.

Breland, O.P., 1954. Notes on the *Culex virgultus* complex (Diptera: Culicidae). Mosq. News 14 (2), 68–71.

Breland, O.P., 1956. Some remarks on Texas mosquitoes. Mosq. News 16 (2), 94–97.

Breland, O.P., 1960. Restoration of the name, *Aedes hendersoni* Cockerell, and its elevation to full specific rank (Diptera: Culicidae). Ann. Ent. Soc. Am. 53 (5), 600–606.

Burket-Cadena, N.D., Blosser, E.M., 2017. *Aedeomyia squamipennis* (Diptera: Culicidae) in Florida, USA, a new state and country record. J. Med. Entomol. 54 (3), 788–792.

Burkett-Cadena, N., 2013. Mosquitoes of the Southern United States. University of Alabama Press, Tuscaloosa, AL. 208 p.

Burkett-Cadena, N.D., Graham, S.P., Hassan, K.H., Guyer, C., Eubanks, M.D., Katholi, C.R., Unnasch, T.R., 2008. Blood feeding patterns of potential arbovirus vectors of the genus *Culex* targeting ectothermic hosts. J. Am. Trop. Med. Hyg. 79 (5), 809–815.

Byrd, B.D., Wesson, D.M., Harrison, B.A., 2009. Regional problems identifying the fourth instar larvae of *Orthopodomyia signifera* (Coquillett) and *Orthopodomyia kummi* Edwards (Diptera: Culicidae). Proc. Ent. Soc. Wash. 111 (3), 752–754.

Byrd, B.D., Harrison, B.A., Zavortink, T.J., Wesson, D.M., 2012. Sequence, secondary structure, and phylogenetic analyses of the ribosomal internal transcribed spacer 2 (ITS2) in members of the North American *signifera* group of *Orthopodomyia* (Diptera: Culicidae). J. Med. Entomol. 49 (6), 1189–1197.

Carpenter, S.J., LaCasse, W.J., 1955. Mosquitoes of North America (North of Mexico). University of California Press, Berkeley, CA.

Carpenter, S.J., Middlekauff, W.W., Chamberlain, R.W., 1946. The Mosquitoes of the Southern United States East of Oklahoma and Texas. Am. Midl Nat. Monogr. 3:1–292, The University Press, Notre Dame, IN.

CDC (Centers for Disease Control and Prevention), 1986. Epidemiologic Notes and Reports Update: *Aedes albopictus* Infestation—United States. Available from: https://www.cdc.gov/mmwr/preview/mmwrhtml/00001699.htm. (Accessed September 7, 2018).

CDC (Centers for Disease Control and Prevention), 2016. Mosquito species in which West Nile virus has been detected, United States, 1999–2016. Available from: https://www.cdc.gov/westnile/resources/pdfs/MosquitoSpecies1999-2016.pdf. (Accessed September 7, 2018).

CDC (Centers for Disease Control and Prevention), 2017. Estimated potential range of *Aedes aegypti* and *Aedes albopictus* in the United States, 2017. Available from: https://www.cdc.gov/zika/vector/range.html. (Accessed September 7, 2018).

Chamberlain, R., 1980. In: Monath, T.P. (Ed.), History of St. Louis Encephalitis. American Public Health Association, Washington, DC, St. Louis Encephalitis, pp. 3–61.

Chamberlain, R.W., Rubin, H., Kissling, R.E., Edison, M.E., 1951. Recovery of virus of eastern equine encephalomyelitis from a mosquito, *Culiseta melanura* (Coquillett). Proc. Soc. Exp. Biol. Med. 77, 396–397.

Chamberlain, R.W., Sikes, R.K., Nelson, D.B., 1956. Infection of *Mansonia perturbans* and *Psorophora ferox* mosquitoes with Venezuelan equine encephalomyelitis virus. Proc. Soc. Exp. Biol. Med. 91, 215–216.

Chamberlain, R.W., Sudia, W.D., Gillett, J.D., 1959. St. Louis encephalitis virus in mosquitoes. Am. J. Hyg. 70 (3), 221–236.

Chamberlain, R.W., Sudia, W.D., Coleman, P.H., 1969. Isolations of an arbovirus of the Bunyamwera group (Tensaw virus) from mosquitoes in the Southeastern United State, 1960–1963. Am. J. Trop. Med. Hyg. 18 (1), 92–97.

Clark, G.C., Crabbs, C.L., Bailey, C.L., Calisher, C.H., Craig Jr., G.B., 1986. Identification of *Aedes campestris* from New Mexico: with notes on the isolation of Western equine encephalitis and other arboviruses. J. Am. Mosq. Control Assoc. 2 (4), 529–534.

Cupp, E.W., Zhang, D., Yue, X., Cupp, M.S., Guyer, C., Sprenger, T.R., Unnasch, T.R., 2004. Identification of reptilian and amphibian blood meals from mosquitoes in an eastern equine encephalomyelitis virus focus in Central Alabama. J. Am. Trop. Med. Hyg. 71 (3), 272–276.

Daily Breeze, 2018. Meet the new Aedes mosquito spreading misery around a wide swath of Southern California [Press release]. Retrieved 10 September 2018 from: www.dailybreeze.com.

Darsie Jr., R.F., 1973. A record of changes in mosquito taxonomy in the Unites States of America 1955–1972. Mosq. Syst. 5 (2), 187–193.

Darsie Jr., R.F., Shroyer, D.A., 2004. *Culex (Culex) declarator*, a mosquito species new to Florida. J. Am. Mosq. Control Assoc. 20 (3), 224–227.

Darsie Jr., R.F., Ward, R.A., 1981. Identification and geographical distribution of the mosquitoes of North America, North of Mexico. American Mosquito Control Association, Fresno, CA.

Darsie Jr., R.F., Ward, R.A., 2005. Identification and Geographical Distribution of the Mosquitoes of North America, North of Mexico, second ed. University Press of Florida, Gainesville.

Darsie Jr., R.F., Vlack, J.J., Fussell, E.M., 2002. New addition to the mosquito fauna of the United States, *Anopheles grabhamii* (Diptera: Culicidae). J. Med. Entomol. 39 (3), 430–431.

De Buen, A.M., 1952. *Orthopodomyia kummi* Edwards, 1939. Mosquito nuevo para Mexico. Descripcion de la larva y de la pupa (Diptera, Culicidae). Anales del Instituto de Biologia de la Universidad Nacional de Mexico 13, 242–252.

Duhrkopf, R.E., 1994. A survey of container-breeding mosquitoes in McLennan County, Texas. Tex. J. Sci. 46 (2), 127–132.

Dyar, H.G., 1923. The mosquitoes of the Yellowstone National Park (Diptera, Culicidae). Ins. Ins. Mens. 11 (1–3), 36–46.

Dyar, H.G., Knab, F., 1917. New American mosquitoes. Insecutor Inscitiae Menstrus 5 (10–12), 165–169.

Eads, R.B., 1946. A new record of *Anopheles albimanus* in Texas. J. Econ. Entomol. 39 (3), 420.

Eads, R.B., Menzies, G.C., Ogden, L.J., 1951. Distribution records of West Texas mosquitoes. Mosq. News 11 (1), 41–47.

Eads, R.B., Foyle, J.G., Peel, R.E., 1960. Mosquito densities in Orange County. Texas. Mosq. News 20 (1), 49–52.

Easton, E.R., Price, M.A., Graham, O.H., 1968. The collection of biting flies in West Texas with malaise and animal-baited traps. Mosq. News 28 (3), 465–469.

Ebel, G.D., Rochlin, I., Longaker, J., Kramer, L.D., 2005. *Culex restuans* (Diptera: Culicidae) relative abundance and vector competence for West Nile virus. J. Med. Entomol. 42 (5), 838–843.

Eldridge, B.F., Harbach, R.E., 1989. *Culex stigmatosoma* Dyar, 1907 and *C. thriambus* Dyar 1921 (Insecta: Diptera): proposed conservation of the specific names by the suppression of *C. peus* Speiser, 1904. Bull. Zool. Nomencl. 46 (4), 247–249.

Estrada-Franco, J.G., Craig Jr., G.B., 1995. Biology, Disease Relationships, and Control of *Aedes albopictus*. Pan-American Health Organization (PAHO), Washington, DC. 49 p. (Technical paper 42).

Fisk, F.W., 1941. *Deinocerites spanius* at Brownsville, Texas, with notes on its biology and a description of the larva. Ann. Ent. Soc. Am. 34 (3), 543–550.

Fisk, F.W., LeVan, J.H., 1940. Mosquito collections at Brownsville, Texas. J. Econ. Entomol. 33, 944–945.

Floore, T.G., Harrison, B.A., Eldridge, B.F., 1976. The Anopheles (Anopheles) crucians subgroup in the United States (Diptera: Culicidae). Mosq. Syst. 8 (1), 1–109.

Fournier, P.V., Teltow, G.J., Snyder, J.L., 1989. Medical Entomology Section Training Manual. Mosquito-Borne Encephalitis Surveillance and Distribution Records of Texas Mosquitoes. Texas Department of Health Bureau of Laboratories, Austin. 115 p.

Gaffigan, T.V., Wilkerson, R.C., Pecor, J.E., Stoffer, J.A., Anderson, T., 2011. Systematic Catalog of Culicidae. Walter Reed Biosystematic Unit (WRBU), Division of Entomology. Walter Reed Army Institute of Research (WRAIR), Silver Spring, Maryland. http://www.mosquitocatalog.org.

Gargan, T.P., Clark, G.G., Dohm, D.J., Turell, M.J., Bailey, C.L., 1988. Vector potential of selected North American mosquito species for Rift Valley fever virus. Am. J. Trop. Med. Hyg. 38, 440–446.

Gendernalik, K.L., Weger-Lucarelli, J., Garcia Luna, S.M., Fauver, J.R., Ruckert, C., Murrieta, R.A., Bergren, N., Samaras, D., Nguyen, C., Kading, R.C., Ebel, G.D., 2017. American *Aedes vexans* mosquitoes are competent vectors of Zika virus. J. Med. Entomol. 96 (6), 1338–1340.

Goddard, L.B., Roth, A.E., Reisen, W.K., Scott, T.W., 2002. Vector competence of California mosquitoes for West Nile virus. Emerg. Inf. Dis. 8 (12), 1383–1391. https://doi.org/10.3201/eid0812.020536.

Hammon, W.M., Reeves, W.C., 1943a. Laboratory transmission of St. Louis encephalitis virus by three genera of mosquitoes. J. Exp. Med. 78 (4), 241–253.

Hammon, W.M., Reeves, W.C., 1943b. Laboratory transmission of Western equine encephalitis virus by mosquitoes of the genera *Culex* and *Culiseta*. J. Exp. Med. 78 (6), 425–434.

Hammon, W.M., Reeves, W.C., Galindo, P., 1945. Epidemiology studies of encephalitis in the San Joaquin Valley of California, 1943, with the isolation of viruses from mosquitoes. Am. J. Hyg. 42 (3), 299–306.

Harbach, R.E., 2013a. Mosquito Taxonomic Inventory. http://mosquito-taxonomic-inventory.info/. (Accessed February 11, 2019).

Harbach, R.E., 2013b. Genus *Haemagogus*. Mosquito Taxonomic Inventory. http://mosquito-taxonomic-inventory.info/simpletaxonomy/term/6099/. (Accessed August 26, 2017).

Harbach, R.E., 2018. Culicipedia: Species-Group, Genus-Group and Family-Group Names in Culicidae (Diptera). CABI, Wallingford, Oxfordshire, UK. 378 pp.

Harbach, R.E., Kitching, I.J., 1998. Phylogeny and classification of the Culicidae (Diptera). Syst. Entomol. 23 (4), 327–370.

Harbach, R.E., Knight, K.L., 1980. Taxonomists' Glossary of Mosquito Anatomy. Plexus, Marlton, NJ. 413 pp.

Harrison, B.A., Varnado, W., Whitt, P.B., Goddard, J., 2008. New diagnostic characters for females of *Psorophora (Janthinosoma)* species in the United States, with notes on *Psorophora mexicana* (Bellardi) (Dipter: Culicidae). J. Vector Ecol. 33 (2), 232–237.

Harrison, B.A., Byrd, B.D., Sither, C.B., Whitt, P.B., 2016. The Mosquitoes of the Mid-Atlantic Region: An Identification Guide. Mosquito and Vector-Borne Infectious Diseases Laboratory Publication 2016-1, Western Carolina University, Cullowhee, NC. 201 p.

Heard, P.B., Zhang, M.B., Grimstad, P.R., 1991. Laboratory transmission of Jamestown Canyon and snowshoe hare viruses (Bunyaviridae: California serogroup) by several species of mosquitoes. J. Am. Mosq. Control Assoc. 7, 94–102.

Hill, S.O., Smittle, B.J., Philips, F., 1958. Distribution of Mosquitoes in the Fourth US Army Area, Entomol. Div., Fourth US Army Med. Lab., Fort Sam Houston, Texas, 155 pp.

Horsfall, W.R., 1955. Mosquitoes: Their Bionomics and Relation to Disease. Ronald Press Co, New York. 723 p.

Journal Policy on Names of Aedine Mosquito Genera and Subgenera, 2005. J. Med. Entomol. 42 (4), 511. https://doi.org/10.1093/jmedent/42.4.511.

Joyce, C.R., 1945. The occurrence of *Psorophora mexicana* (Bellardi) in the United States. Mosq. News 5 (3), 86.

Keith, R.D., 1979. The occurrence of *Aedes grossbecki* in Texas. Mosq. News 39 (4), 797.

King, W.V., Bradley, G.H., 1937. Notes on *Culex erraticus* and related species in the United States. Ann. Ent. Soc. Am. 30 (2), 345–357.

Knight, K.L., 1978. Supplement to a Catalog of the Mosquitoes of the World (Diptera: Culicidae). Thomas Say Foundation. Suppl. to Volume VI. 107 p.

Knight, K.L., Stone, A., 1977. A Catalog of the Mosquitoes of the World (Diptera: Culicidae), second ed. vol. VI. Thomas Say Foundation, College Park, MD. 611 p.

Kokernot, R.H., Hayes, J., Boyd, K.R., Sullivan, P.S., 1974. Arbovirus studies in Houston, Texas, 1968–1970. J. Med. Entomol. 11 (4), 419–425.

Labarthe, N., Serrao, M.L., Melo, Y.E., de Oliveira, S.J., Lourenco-de-Oliveira, R., 1998. Mosquito frequency and feeding habits in an exotic canine dirofilariasis area of Niteroi, State of Rio de Janeiro, Brazil. Mem. Inst. Oswaldo Cruz 93, 145–154.

McDonald, W.A., Belkin, J.N., 1960. *Orthopodomyia kummi* new to the United States (Diptera: Culicidae). Proc. Ent. Soc. Wash. 62 (4), 249–250.

McGregor, T., Eads, R.B., 1943. Mosquitoes of Texas. J. Econ. Entomol. 36 (6), 938–940.

McPhatter, L.P., Mahmood, F., Debboun, M., 2012. Survey of mosquito Fauna in San Antonio, Texas. J. Am. Mosq. Control Assoc. 28 (3), 240–247.

Menzies, G.C., Eads, R.B., Harmston, F.C., 1955. The discovery of *Culex erythrothorax* Dyar in Texas. Mosq. News 15 (4), 235–236.

Mitchell, C.J., Francy, D.B., Monath, T.P., 1980. Arthropod vectors. In: Monath, T.P. (Ed.), St. Louis Encephalitis. American Public Health Association, Washington, DC, pp. 313–379.

Mitchell, C.J., Haramis, L.D., Karabatsos, N., Smith, G.C., Starwalt, V.J., 1998. Isolation of La Crosse, Cache Valley, and Potosi viruses from *Aedes* mosquitoes (Diptera: Culicidae) collected at used-tire sites in Illinois during 1994–1995. J. Med. Entomol. 35, 573–577. https://doi.org/10.1093/jmedent/35.4.573. 9701947.

Monath, T.P., Cropp, C.B., Bowen, G.S., Kemp, G.E., Mitchell, C.J., Gardner, J.J., 1980. Variation in virulence for mice and rhesus monkeys among St. Louis encephalitis virus strains of different origin. Am. J. Trop. Med. Hyg. 29, 948–962.

Moore, C.G., Francy, B., Eliason, D.A., Bailey, R.E., Campos, E.G., 1990. *Aedes albopictus* and other container-inhabiting mosquitoes in the United States: results of an eight-city survey. J. Am. Mosq. Control Assoc. 6 (2), 173–178.

Morris, C.D., Whitney, E., Bast, T.F., Deibel, R., 1973. An outbreak of eastern equine encephalomyelitis in upstate New York during 1971. Am. J. Trop. Med. Hyg. 22 (4), 561–566.

Morris, C.D., Caines, A.R., Woodall, J.P., Bast, T.F., 1975. Eastern equine encephalomyelitis in upstate New York, 1972–1974. Am. J. Trop. Med. Hyg. 24 (6), 986–991.

Nasci, R.S., 1982. Differences in host choice between the sibling species of treehole mosquitoes *Aedes triseriatus* and *Aedes hendersoni*. Am. J. Med. Trop. Hyg. 31 (2), 411–415.

Nava, M.R., Debboun, M., 2016. A taxonomic checklist of the mosquitoes of Harris County, Texas. J. Vector Ecol. 41, 190–194. https://doi.org/10.1111/jvec.12212.

Nayar, J.K., Rutledge, C.R., 2008. Mosquito-Borne Dog Heartworm Disease. EDIS. ENY-648. (1 August 2018).

Nelson, M.J., 1986. *Aedes aegypti*: Biology and Ecology. Pan-American Health Organization (PAHO), Washington, DC. 50 pp.

Nielsen, E.T., Nielsen, A.T., 1953. Field observations on the habits of *Aedes taeniorhynchus*. Ecology 34, 141–156.

Nielsen, L.T., Linam, J.H., Arnell, J.H., Zavortink, T.J., 1968. Distribution al and biological notes on the tree hole mosquitoes of the Western United States. Mosq. News 28 (3), 361–365.

Norris, M., 1946. Recovery of a strain of Western equine encephalitis virus from *Culex restuans* (Theo.) (Diptera: Culicidae). Can J. Res. E 24, 63–70.

O'Donnell, K.L., Bixby, M.A., Morin, K.J., Bradley, D.S., Vaughan, J.A., 2017. Potential of a northern population of *Aedes vexans* (Diptera: Culicidae) to transmit Zika virus. J. Med. Entomol. 54 (5), 1354–1359.

Oliveira, A.R.S., Cohnstaedt, L.W., Cernicchiaro, N., 2018. Japanese encephalitis virus: placing disease vectors in the epidemiological triad. Ann. Ent. Soc. Am. https://doi.org/10.1093/aesa/say025.

Oliver, J., Lukacik, G., Kokas, J., Campbell, S.R., Kramer, L.D., Sherwood, J.A., Howard, J.J., 2018. Twenty years of surveillance for eastern equine encephalitis virus in mosquitoes in New York State from 1993 to 2012. Parasit. Vectors 11, 362.

O'Meara, G.F., Craig Jr., J.B., 1970a. Geographical variation in *Aedes atropalpus* (Diptera: Culicidae). Ann. Ent. Soc. Am. 63 (5), 1392–1400.

O'Meara, G.F., Craig Jr., J.B., 1970b. A new subspecies of *Aedes atropalpus* (Coquillett) from southwestern United States (Diptera: Culicidae). Proc. Ent. Soc. Wash. 72 (4), 1392–1400.

O'Neill, K., Ogden, L.G., Eyles, D.E., 1944. Additional species of mosquitoes found in Texas. J. Econ. Entomol. 37 (4), 555–556.

Parsons, R., 2003. Mosquito Control—Texas Style. Wing Beats 14 (1), 4–38.

Paulson, S.L., Grimstad, P.R., Craig, G.B., 1989. Midgut and salivary barriers to La Crosse virus dissemination in mosquitoes of the *Aedes triseriatus* group. Med. Vet. Entomol. 3 (2), 113–123.

Porter, J.E., 1946. The larvae of *Uranotaenia syntheta* (Diptera, Culicidae). Am. Mid. Nat. 35, 535–537.

Reeves, W.K., Darsie Jr., R.F., 2003. Discovery of *Culex (Neoculex) arizonensis* in Texas (Diptera: Culicidae). J. Am. Mosq. Control Assoc. 19 (1), 87–88.

Reeves, W.K., Hammon, W.M., 1946. Laboratory transmission of Japanese B encephalitis virus by seven species (three genera) of north American mosquitoes. J. Exp. Med. 83 (3), 185–194.

Reinert, J.F., 2000. New classification for the composite genus *Aedes* (Diptera: Culicidae), elevation of subgenus *Ochlerotatus* to generic rank, reclassification of the other subgenera, and notes on certain subgenera and species. J. Am. Mosq. Control Assoc. 16 (3), 175–188.

Reinert, J.F., Kaiser, P.E., Seawright, J.A., 1997. Analysis of the *Anopheles (Anopheles) quadrimaculatus* complex of sibling species (Diptera: Culicidae) using morphological, cytological, molecular, genetic, biochemical, and ecological techniques in an integrated approach. J. Am. Mosq. Control Assoc. 13 (Suppl), 1–102.

Reisen, W.K., Milby, M.M., Presser, S.B., Hardy, J.L., 1992. Ecology of mosquitoes and St. Louis encephalitis virus in the Los Angeles Basin of California 1987–1990. J. Med. Entomol. 29 (4), 582–598.

Roberts, D.R., Scanlon, J.E., 1975. The ecology and behavior of *Aedes atlanticus* D. & K. and other species with reference to keystone virus in the Houston area, Texas. J. Med. Entomol. 12 (5), 537–546.

Roberts, D.R., Scanlon, J.E., 1979. Field studies on the population biology of immature stages of six woodland mosquito species in the Houston, Texas area. Mosq. News 39 (1), 26–34.

Rogers, J.S., Newson, H.D., 1979. Comparisons of *Aedes hendersoni* and *Ae. triseriatus* as potential vectors of *Dirofilaria immitis*. Mosq. News 39, 463–466.

Ross, E.S., 1943. The identity of *Aedes bimaculatus* (Coquillett) and a new subspecies of *Aedes fulvus* (Wiedemann) from the United States (Diptera, Culicidae). Proc. Ent. Soc. Wash. 45 (6), 143–151.

Rueger, M.E., Druce, S., 1950. New mosquito distribution records for Texas. Mosq. News 10 (2), 60–63.

Sames, W.J., Dacko, N.M., Bolling, B.G., Bosworth, A.B., Swiger, S.L., Duhrkopf, R.E., Burton, R.G., 2019. Distribution of *Culex coronator* in Texas. J. Am. Mosq. Control Assoc. 35 (1), 55–64.

Sardelis, M.R., Turell, M.J., Dohm, D.J., O'Guinn, M.L., 2001. Vector competence of selected north American *Culex* and *Coquillettidia* mosquitoes for West Nile virus. Emerg. Inf. Dis. 7 (6), 1018–1022.

Savage, H.M., 2005. Classification of mosquitoes in tribe Aedini (Diptera: Culicidae): paraphyliphobia, and classification versus Cladystic analysis. J. Med. Entomol. 42 (6), 923–927.

Savage, H.M., Strickman, D., 2004. The genus and subgenus categories within Culicidae and placement of *Ochlerotatus* as a subgenus *of Aedes*. J. Am. Mosq. Control Assoc. 20, 208–214.

Shepard, J.J., Andreadis, T.G., Vossbrinck, C.R., 2006. Molecular phylogeny and evolutionary relationship among mosquitoes (Diptera: Culicidae) from the Northeastern United States based on small subunit ribosomal DNA (18S rDNA) sequences. J. Med. Entomol. 43 (3), 443–454.

Shroyer, D.A., Harrison, B.A., Bintz, B.J., Wilson, M.R., Sither, C.B., Byrd, B.D., 2015. *Aedes pertinax*, a newly recognized mosquito species in the United States. J. Am. Mosq. Control Assoc. 31 (1), 97–100. https://doi.org/10.2987/14-6447R.1.

Sinka, M.E., Rubio-Palis, Y., Manguin, S., Patil, A.P., Temperley, W.H., Gething, P.W., Van Boeckel, T., Kabaria, C.W., Harbach, R.E., Hay, S.I., 2010. The dominant *Anopheles* vectors of human malaria in the Americas: occurrence data, distribution maps and bionomic précis. Parasit. Vectors 3, 72. https://doi.org/10.1186/1756-3305-3-72.

Sirivanakarn, S., 1976. Medical entomology studies III. A revision of the subgenus *Culex* in the oriental region (Diptera: Culicidae). Contr. Am. Ent. Inst. 12 (2), 1–272.

Sither, C.B., Hopkins, V.E., Harrison, B.A., Bintz, B.J., Hickman, E.Y., Brown, J.S., Wilson, M.R., Byrd, B.D., 2013. Differentiation of *Aedes atlanticus* and *Aedes tormentor* by restriction fragment length polymorphisms of the second internal transcribed spacer. J. Am. Mosq. Control Assoc. 29 (4), 376–379.

Siverly, R.E., 1972. Mosquitoes of Indiana. Indiana State Board of Health, Indianapolis, IN. 126 pp.

Sprenger, D., Whuithiranyagool, T., 1986. The discovery and distribution of *Aedes albopictus* in Harris County, Texas. J. Am. Control Assoc. 2 (2), 217–219.

Stone, A., 1956. Correction in taxonomy and nomenclature of mosquitoes. Proc. Entomol. Soc. Wash. 58, 333–343.

Strickman, D., 1988a. *Culex stigmatosoma* and *Cx. peus*: Identification of female adults in the United States. J. Am. Mosq. Control Assoc. 4 (4), 555–556.

Strickman, D., 1988b. Redescription of the holotype of *Culex (Culex) peus* Speiser and taxonomy of *Culex (Culex) stigmatosoma* Dyar and *thriambus* Dyar (Diptera: Culicidae). Proc. Entomol. Soc. Wash. 90 (4), 483–494.

Sublette, M.S., Sublette, J.E., 1970. Distributional records of mosquitoes on the Southern High Plains with a checklist of species from New Mexico and Texas. Mosq. News 30 (4), 533–538.

Sudia, W.D., Newhouse, V.F., 1971. Venezuelan equine encephalitis in Texas, 1971: informational report. Mosq. News 31 (3), 350–351.

Sudia, W.D., Newhouse, V.F., Beadle, L.D., Miller, D.L., Johnston Jr., J.G., Young, R., Calisher, C.H., Maness, K., 1975. Epidemic Venezuelan equine encephalitis in North America in 1971: vector studies. Am. J. Epidemiol. 101 (1), 17–35.

Texas A&M, 2013. Mosquitoes of Texas. Retrieved from: http://agrilife.org/aes/public-health-vector-and-mosquito-control/mosquitoes-of-texas/.

Texas State Health Department, 1944. The Mosquitoes of Texas. 100 p.

TMCA (Texas Mosquito Control Association), 2001a. Organized Mosquito Control Districts in Texas. Available from: http://texasmosquito.org/Districts.html. (Accessed September 7, 2018).

TMCA (Texas Mosquito Control Association). 2001b. Mosquitoes found in the state of Texas. In: Fournier, P.V., Teltow G, Snyder J, eds. Medical Entomology Section Training Manual [Internet]. Austin, TX: Texas Department of Health (Accessed 15 April 2015) Available from: http://www.texasmosquito.org/checklist.html

Turell, M.J., 1999. Vector competence of three Venezuelan mosquitoes (Diptera: Culicidae) for an epizootic IC strain of Venezuelan equine encephalitis virus. J. Med. Entom. 36 (4), 407–409.

Turell, M.J., O'Guinn, M.L., Jones, J.W., Sardelis, M.R., Dohm, D.J., Watts, D.M., Fernandez, R., Travassos Da Rosa, A., Guzman, H., Tesh, R., Rossi, C.A., Ludwig, G.V., Mangiafico, J.A., Kondig, J., Wasieloski Jr., L., Pecor, J., Zyzak, M., Schoeler, C., Mores, C.N., Calampa, C., Lee, J.S., Klein, T.A., 2005. Isolation of viruses from mosquitoes (Diptera: Culicidae) collected in the Amazon Basin region of Peru. J. Med. Entom. 42 (5), 891–898.

Turner, R.L., 1924. A new mosquito from Texas (Diptera: Culicidae). Ins. Ins. Mens. 12, 84.

Uejio, C.K., Hayden, M.H., Zielinski-Gutierrez, E., Lopez, J.L.R., Barrera, R., Amador, M., Thompson, G., Waterman, S.H., 2014. Biological control of mosquitoes in scrap tires in Brownsville, Texas USA and Matamoros, Tamaulipas, Mexico. J. Am. Mosq. Control Assoc. 30, 130–135.

Vanlandingham, D.L., Higgs, S., Yan-Jang, S.H., 2016. *Aedes albopictus* (Diptera: Culicidae) and mosquito-borne viruses in the United States. J. Med. Entom. 53 (5), 1024–1028.

Varnado, W.C., Goddard, J., Harrison, B., 2012. Identification guide to adult mosquitoes in Mississippi. Mississippi State University Extension Service117.

Weaver, S.C., Reisen, W.K., 2010. Present and future Arboviral threats. Antivir. Res. 85 (2), 328–345. https://doi.org/10.1016/j.antiviral.2009.10.008.

Wellings, F.M., Lewis, A.L., Pierce, L.V., 1972. Agents encountered during arboviral ecological studies: Tampa Bay area, Florida, 1963–1970. Am. Soc. Trop. Med. Hyg. 21, 201–213. (PMID: 4400812).

Wikipedia, 2018a. United States Physiographic Region. Available from: https://en.wikipedia.org/wiki/United_States_physiographic_region. (Accessed September 7, 2018).

Wikipedia, 2018b. List of Geographical Regions in Texas. Available from: https://en.wikipedia.org/wiki/List_of_geographical_regions_in_Texas. (Accessed September 7, 2018).

Wikipedia, 2018c. County United States. Available from: https://en.wikipedia.org/wiki/County_ (United_States). (Accessed September 7, 2018).

Wilkerson, R.C., Reinert, J.F., Li, C., 2004. Ribosomal DNA ITS2 sequences differenciate six species in the *Anopheles crucians* complex (Diptera: Culicidae). J. Med. Entomol. 41 (3), 392–401.

Wilkerson, R.C., Linton, Y.-M., Fonseca, D.M., Schultz, T.R., Price, D.C., Strickman, D.A., 2015. Making mosquito taxonomy useful: a stable classification of tribe Aedini that balances utility with current knowledge of evolutionary relationships. PLoS One 10 (7), e0133602. https://doi.org/10.1371/journal.pone.0133602.

Wiseman, J.S., 1965. A list of mosquito species reported from. Texas. Mosq. News 25 (1), 58–59.

Wolff, T.A., Nielsen, L.T., 2007. The Mosquitoes of New Mexico. University of New Mexico Press, Albuquerque, NM. 114 pp.

Wolff, T.A., Nielsen, L.T., Hayes, R.O., 1975. A current list and bibliography of the mosquitoes of New Mexico. Mosq. Syst. 7 (1), 13–18.

Wozniak, A., Dowda, H., Tolson, M., Karabatsos, N., Vaughan, D., Turner, P., Ortiz, D., Wills, W., 2001. Arbovirus surveillance in South Carolina, 1996–98. J. Am. Mosq. Control Assoc. 17, 73–78.

WRBU (Walter Reed Biosystematics Unit), 2019. Systematic Catalogue of Culicidae. Available from: http://www.mosquitocatalog.org/default.aspx?pgID=2. (Accessed 26 March 2019).

Zavortink, T.J., 1968. Mosquito studies (Diptera: Culicidae). VIII. A prodrome of the genus *Orthopodomyia*. Contr. Am. Ent. Inst. 3 (2), 1–221.

Zavortink, T.J., 1972. Mosquito studies (Diptera: Culicidae). XXVIII. The new world species formerly placed in *Aedes (Finlaya)*. Contr. Am. Ent. Inst. 8 (3), 1–206.

Chapter 3

Key to the Genera of Adult Female Mosquitoes of Texas

Leopoldo M. Rueda

Walter Reed Biosystematics Unit, Department of Entomology, Smithsonian Institution, Suitland, MD, United States

Chapter Outline

3.1 Adult morphology

The external morphology of adult female mosquito is so complicated and sometimes simply ignored. The most important approach to identify mosquitoes is locating the correct morphological structure on the specimens. Repeated examinations of mosquitoes under a high-powered microscope are essential processes of learning mosquito identification. Any relevant identification keys will not be helpful unless the user is well prepared, with better knowledge of mosquito morphology, and well equipped with appropriate microscopes and related equipment.

The adult exoskeleton is hardened and characterized with the presence or absence of setae, scales, pigmentations, and structural shapes (lengths and widths) that are important diagnostic features of certain genera and species of mosquitoes. Many characters in the keys are provided on the morphology photographs or illustrations, with arrows pinpointing the structures that are necessary to identify the genus or species. Many individual photographs or illustrations used in the key couplets to separate genera and species have been simplified by removing extraneous structures such as setae, scales, and other body parts that are not important for the identification of the genus or species in that couplet.

Adult mosquitoes, like other arthropods including insects, have bilateral symmetry. The wings, legs, sides of head, thorax, and abdomen are paired; thus if one side is rubbed or damaged and a character is missing, that character should be examined on the other side. Some illustrations or photos in the keys only show one-half of the structures, since the other side is just a mirror image of the first side. Dissecting microscopes, with higher magnifications (at least 80×) must be used since numerous small characters are included in these adult keys.

Dichotomous keys are prepared here, with selected photographs on most key couplets as shown on various figures. These keys are a series of paired or contrasting statements (or couplets) that allow the user to accurately identify an unknown mosquito specimen at genus or species level. Most couplets may contain a few valid characters to assist in identifying rubbed or damaged structures or even missing parts.

Major morphological characters are used to identify adult female mosquitoes. They include the scales (colors, sizes, shapes, and statures), setae (single or branched, sizes, rows or clusters, and colors), color patterns (bands, stripes, spots, bands, and speckled) and length ratios (legs, proboscis, antennae, and palpi).

Before using these keys to the adult females, the sex of the adults must be determined, using the following features: (a) female, abdomen with rounded or pointed tip, without distinct external structures except cerci; antennae with sparse short whorl setae. (b) Male, abdomen ending with obvious male genitalia; antennae look "bushy," with numerous, long, whorl setae.

To help the reader the adult morphology (or habitus) is shown in Fig. 3.1, together with other parts such as the head and thorax (Fig. 3.2), part of abdomen (Fig. 3.3), and hindleg (Fig. 3.4). Specific diagnostic characters of some body parts

Mosquitoes, Communities, and Public Health in Texas. https://doi.org/10.1016/B978-0-12-814545-6.00003-1

not included in Figs. 3.1–3.4 are shown in other figures of selected key couplets. The terminology used in the keys follows Harbach and Knight (1980, 1982), except wing venation (Belkin, 1962).

    The illustrations presented in the keys were developed mainly by making freehand computer images based on multiple printed references, photographs, and preserved specimens as guides. The primary reference materials used in the developments of the key illustrations include Carpenter and LaCasse (1955), Belkin (1962), Rueda et al. (1998), Rueda (2004), Darsie and Ward (2005), Harrison et al. (2016), and WRBU (2018). Colored photographs of mosquitoes were used, principally from Walter Reed Biosystematics Units (WRBU) and additional collections from previous mosquito biosystematics projects of L.M. Rueda.

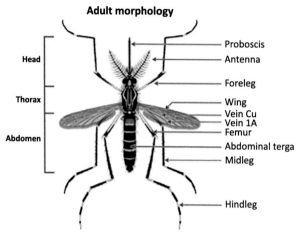

FIG. 3.1   Dorsal view of adult female mosquito—*Aedes aegypti.*

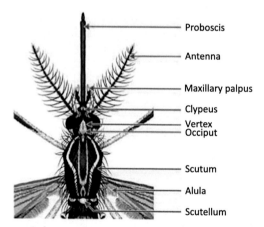

FIG. 3.2   Dorsal view of adult head and thorax—*Aedes aegypti.*

FIG. 3.3   Lateral view of adult head, thorax, and abdomen (part)—*Aedes aegypti.*

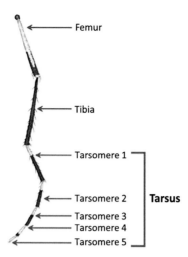

FIG. 3.4 Anterior view of hindleg—*Aedes aegypti.*

3.2 Key to the genera of adult females

1 Head. Proboscis strongly bent forward and attenuated in distal half (Fig. 3.5A). Wing. Distinct emargination present just beyond tip of vein CuA (Fig. 3.5C) .... *Toxorhynchites*
Head. Proboscis nearly straight and not attenuated in distal half (Fig. 3.5B). Wing. Emargination absent just beyond tip of vein CuA (Fig. 3.5D) 2

2(1) Thorax. Scutellum evenly rounded (Fig. 3.5E). Head. Maxillary palpus approximately equal to length of proboscis (Fig. 3.6A) ....*Anopheles*
Thorax. Scutellum with three distinct lobes (Fig. 3.5F). Head. Maxillary palpus shorter, less than 0.8 length of proboscis (Fig. 3.6B) 3

3(2) Wing. Cell R_2 shorter than vein R_{2+3} (Fig. 3.6C). Vein 1A reaching wing margin before a perpendicular line drawn from front margin through base of Cu_1 (Fig. 3.6C). Thorax. Lines of metallic blue scales usually present (Fig. 3.6D) .... *Uranotaenia*
Wing. Cell R_2 at least as long as vein R_{2+3}. Vein 1A reaching wing margin beyond a perpendicular line drawn from front margin through base of Cu_1. Thorax. Lines of metallic blue scales absent 4

4(3) Thorax. Postspiracular setae present (Fig. 3.6E) 5
Thorax. Postspiracular setae absent (Fig. 3.6F) 7

5(4) Abdomen. Apex bluntly rounded (Fig. 3.7A). Wing. Dorsal scales very broad (Fig. 3.7C) .... *Mansonia (Ma. titillans)*
Abdomen. Apex bluntly tapered (Fig. 3.7B). Wing. Dorsal scales long and slender (Fig. 3.7D) 6

6(5) Thorax. Prespiracular setae present (Fig. 3.7E). Abdomen. Terga usually with apical pale bands or lateral pale patches .... *Psorophora*
Thorax. Prespiracular setae absent (Fig. 3.7F). Abdomen. Terga usually with basal pale and/or lateral pale patches (Fig. 3.7G) .... *Aedes* (in part)

7(4) Wing. Base of vein Sc on underside of wing with patch of short setae (Fig. 3.8A). Thorax. Prespiracular setae present .... *Culiseta*
Wing. Base of vein Sc on underside of wing without patch of setae. Thorax. Prespiracular setae absent 8

8(7) Thorax. Scutum covered with broad flat metallic scales (Fig. 3.8B). Thorax. Antepronotum large, approaching mid-dorsally (Fig. 3.8C) .... *Haemagogus (Hg. equinus)*
Thorax. Scutum not covered with broad flat metallic scales. Thorax. Antepronotum small, not approaching mid-dorsally 9

9(8) Thorax. Scutum with narrow lines of white scales (Fig. 3.8D). Legs. Tarsomere 1 of fore- and midlegs longer than other four tarsomeres combined (Fig. 3.8E) .... *Orthopodomyia*

Thorax. Scutum without narrow lines of white scales. Legs. Tarsomere 1 of fore- and midlegs shorter than other four tarsomeres combined (Fig. 3.8F) 10

10(9) Wing. Dorsal scales very broad Fig. 3.9A). Leg. Hind tibia with broad preapical pale band ....
.... *Coquillettidia (Cq. perturbans)*

Wing. Dorsal scales narrow and long (Fig. 3.9B). Leg. Hind tibia without broad preapical pale band 11

11(10) Head. Antenna with flagellomere 1 longer than flagellomere 2 (Fig. 3.9E) .... *Deinocerites*

Head. Antenna with flagellomere 1 about as long as flagellomere 2 (Fig. 3.9F) 12

12(11) Abdomen. Apex usually bluntly rounded (Fig. 3.9C); tergum with pale, yellowish or non-silvery basolateral patches of scales (Fig. 3.9C). Leg. Pulvillus present (Fig. 3.9H) .... *Culex*

Abdomen. Apex usually tapered (Fig. 3.9D); tergum with silvery basolateral patches of scales (Fig. 3.9D). Leg. Pulvillus absent (Fig. 3.9G) .... *Aedes*

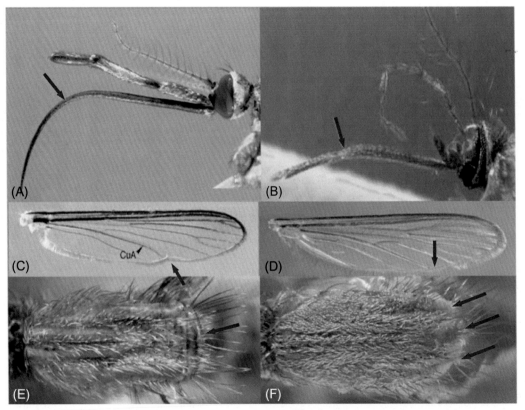

FIG. 3.5 (A) Head showing *(arrow)* recurved proboscis—*Toxorhynchites* sp. (B) Head showing *(arrow)* nonrecurved proboscis—*Culex quinquefasciatus.* (C) Wing showing *(arrow)* distinct emargination—*Toxorhynchites* sp. (D) Wing showing *(arrow)* without emargination—*Culex quinquefasciatus.* (E) Thorax showing *(arrow)* rounded scutellum—*Anopheles* sp. (F) Thorax showing *(arrows)* trilobed scutellum—*Culex quinquefasciatus.*

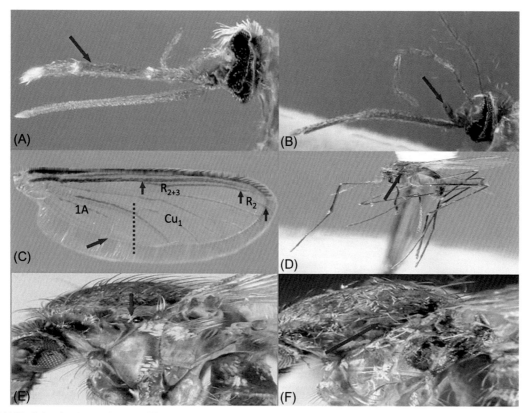

FIG. 3.6 (A) Head showing *(arrow)* long maxillary palpus—*Anopheles* sp. (B) Head showing *(arrow)* short maxillary palpus—*Culex quinquefasciatus.* (C) Wing showing *(arrows)* cell R_2 shorter than vein R_{2+3} and other veins—1A, Cu_1—*Uranotaenia* sp. (D) Thorax showing *(arrow)* line of iridescent scales—*Uranotaenia* sp. (E) Thorax showing *(arrow)* the presence of postspiracular setae—*Mansonia* sp. (F) Thorax showing *(arrow)* absence of post-spiracular setae—*Culex quinquefasciatus.*

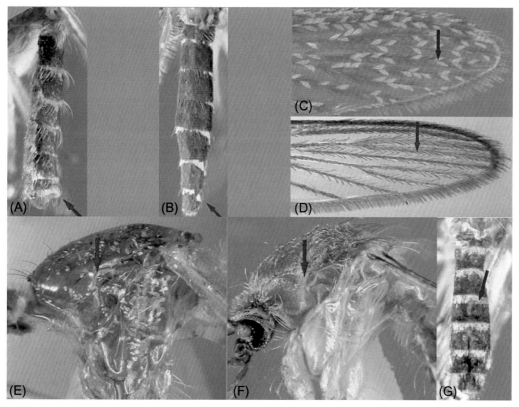

FIG. 3.7 (A) Abdomen showing *(arrow)* rounded apex—*Mansonia* sp. (B) Abdomen showing *(arrow)* tapered apex—*Aedes albopictus.* (C) Wing showing *(arrow)* very broad dorsal scales—*Mansonia* sp. (D) Wing showing *(arrow)* long and slender dorsal scales—*Aedes vexans.* (E) Thorax showing *(arrow)* the presence of prespiracular setae—*Psorophora ferox.* (F) Thorax showing *(arrow)* the absence of prespiracular setae—*Culex sp.* (G) Abdomen showing *(arrow)* basal transverse pale bands—*Aedes vexans.*

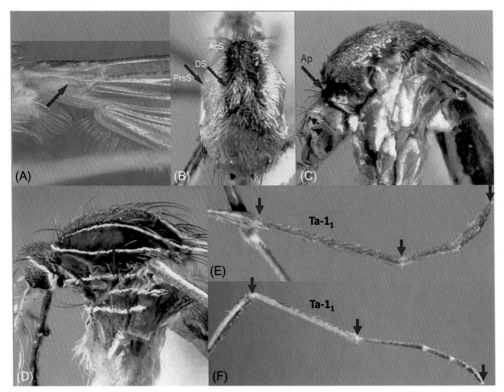

FIG. 3.8 (A) Wing showing *(arrow)* base of vein Sc with row of setae ventrally—*Culiseta* sp. (B) Thorax showing *(arrows)* scutum covered with broad flat metallic scales (*Acs*, arostichal seta; *DS*, dorsocentral seta; *PrsS*, prescutal suture)—*Haemagogus* sp. (C) Thorax showing *(arrow)* antepronotum (Ap) large, approaching middorsally—*Haemagogus* sp. (D) Thorax showing *(arrow)* pleuron and scutum with narrow lines of pale scales—*Orthopodomyia* sp. (E) Leg showing tarsomere 1 (Ta-1$_1$) longer than other four tarsomeres combined—*Orthopodomyia* sp. (F) Leg showing tarsomere 1 (Ta-1$_1$) shorter than other four tarsomeres combined—*Culex quinquefasciatus.*

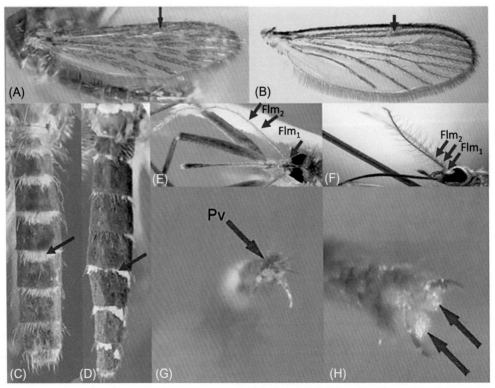

FIG. 3.9 (A) Wing showing *(arrow)* very broad dorsal scales—*Coquillettidia perturbans.* (B) Wing showing *(arrow)* long and slender dorsal scales—*Culex quinquefasciatus.* (C) Abdomen showing *(arrow)* tergum with pale, yellowish, or nonsilvery basolateral patches of scales—*Culex quinquefasciatus.* (D) Abdomen showing *(arrow)* tergum with silvery basolateral patches of scales—*Aedes albopictus.* (E) Head showing *(arrows)* antenna with flagellomere 1 (Flm$_1$) longer than flagellomere 2 (Flm$_2$)—*Deinocerites mathesoni.* (F) Head showing *(arrows)* antenna with flagellomere 1 (Flm$_1$) about as long as flagellomere 2 (Flm$_2$)—*Aedes atlanticus.* (G) Leg tarsomere 5 showing *(arrow)* absence of pulvillus (Pv)—*Mansonia* sp. (H) Leg tarsomere 5 showing *(arrow)* presence of pulvillus (Pv)—*Culex quinquefasciatus.*

Acknowledgments

Great appreciations go to J. Stoffer for taking photographs when she was under contract at Walter Reed Biosystematics Unit and T. Litwak for *Aedes* illustrations. Special thanks go to G. White for assisting with sorting and retrieving specimens for initial photographs and B.P. Rueda for the enormous help in finalizing the photographs used in the keys. Thanks also to B.P. Rueda and C.R. Summers for the valuable comments and suggestions on this chapter.

References and further reading

Belkin, J.N., 1962. The Mosquitoes of the South Pacific (Diptera, Culicidae). vols. I & II. University of California Press, Berkely, CA. 608 pp.

Carpenter, S.J., LaCasse, W.J., 1955. Mosquitoes of North America (North of Mexico). University of California Press, Berkeley and Los Angeles, CA. 360 pp. +127 pl.

Darsie Jr., R.F., Ward, R.A., 2005. Identification and Geographical Distribution of the Mosquitoes of North America, North of Mexico. University Press of Florida, Gainesville, FL. 383 pp.

Harbach, R.E., Knight, K.L., 1980. Taxonomists' Glossary of Mosquito Anatomy. Plexus Publishing, Marlton, NJ. 415 pp.

Harbach, R.E., Knight, K.L., 1982. Corrections and additions to taxonomists' glossary of mosquito anatomy. Mosq. Syst. 13, 201–217.

Harrison, B.A., Byrd, B.D., Sither, C.B., Whitt, P.B., 2016. The Mosquitoes of the Mid-Atlantic Region: An Identification Guide. Western Carolina University, Cullowhee, NC. 201 pp.

Rueda, L.M., 2004. Pictorial keys for the identification of mosquitoes (Diptera: Culicidae) associated with dengue virus transmission. Zootaxa 589, 1–60.

Rueda, L.M., Stockwell, S.A., Pecor, J.E., Gaffigan, T., 1998. Key to the Mosquito Genera of the World (INTKEY Module). Walter Reed Biosystematics Unit, Smithsonian Institution, Washington, DC. [CD]. Also, in The Diptera Dissemination Disk – vol. 1, North American Dipterists Society, Washington, DC.

Walter Reed Biosystematics Unit (WRBU), 2018. Mosquito Identification Keys. Walter Reed Biosystematics Unit, Smithsonian Institution, Suitland, MD. Available from: http://www.wrbu.org/aors/northcom_Keys.html. (Accessed December 15, 2018).

Chapter 4

Key to the Species of Adult Female Mosquitoes of Texas

Leopoldo M. Rueda

Walter Reed Biosystematics Unit, Department of Entomology, Smithsonian Institution, Suitland, MD, United States

Chapter Outline

4.1 Key to species of adult females of genus *Aedes* (= *Ae.*)

1 Thorax. Scutum with two large posterolateral black spots 2
Thorax. Scutum without large posterolateral black spots 3

2(1) Thorax. Hypostigmal area with dark spot. Abdomen. Terga II–VI with basal yellow scales and apical dark scales .... *Ae. fulvus pallens*
Thorax. Hypostigmal area without dark spot. Abdomen. Terga II–VI entirely with yellow scales .... *Ae. bimaculatus*

3(1) Leg. Hindtarsomeres with pale bands (Fig. 4.1A and B) 4
Leg. Hindtarsomeres entirely dark (Fig. 4.1C) 17

4(3) Leg. Hindtarsomeres with basal pale bands only (Fig. 4.1A) 5
Leg. Hindtarsomeres with basal and apical pale bands (Fig. 4.1B) 14

5(4) Head. Proboscis with distinct median pale band (Fig. 4.1D) 6
Head. Proboscis either dark scaled, with scattered pale scales, or with ventral pale area, but without complete median pale band (Fig. 4.1E) 9

6(5) Abdomen. Terga with median longitudinal stripe or row of spots (Fig. 4.1F) 7
Abdomen. Terga without median longitudinal stripe or row of spots (Fig. 4.1G) .... *Ae. taeniorhynchus*

7(6) Wing. All dark scales (Fig. 4.1H). Thorax. Postprocoxal membrane without scales (Fig. 4.4B) ....*Ae. mitchellae*
Wing. Speckled with dark and pale scales (Fig. 4.1I). Thorax. Postprocoxal membrane with white scales (Fig. 4.4A) 8

8(7) Leg. Hindtarsomere 1 with distinct median yellow band. Abdomen. Terga with white basolateral patches (Fig. 4.3D) .... *Ae. sollicitans*
Leg. Hindtarsomere 1 usually without distinct median yellow band; if present, then whitish band. Abdomen. Terga with yellow basolateral patches .... *Ae. nigromaculis* (in part)

9(5) Thorax. Scutum with white lyre-shaped scale markings or with narrow median white stripe (Fig. 4.3A and B) 10

Thorax. Scutum with other scale markings or markings absent (Fig. 4.3C) 11

10(9) Thorax. Scutum black or brown with a pair of submedian longitudinal white stripes, but without median longitudinal white stripe or with white lyre-shape markings (Fig. 4.3A); mesepimeron with two well-separated white scale patches (Fig. 4.2A). Head. Clypeus with white scales (Fig. 4.2C). Leg. Midfemur with anterior longitudinal white stripe (Fig. 4.2E) .... *Ae. aegypti*

Thorax. Scutum with narrow median longitudinal white stripe (Fig. 4.3B); mesepimeron with white scale patches not separated, forming V-shaped white patch (Fig. 4.2B). Head. Clypeus without white scales (Fig. 4.2D). Leg. Midfemur without anterior longitudinal white stripe (Fig. 4.2F) .... *Ae. albopictus*

11(9) Leg. Hindfemur with basal half entirely pale-scaled .... *Ae. zoosophus*

Leg. Hindfemur with basal half anteriorly dark-scaled or with mixed dark and pale scales 12

12(11) Abdomen. Terga II–VI with basal pale bands with median notch (Fig. 4.3E). Leg. Hindtarsomeres with narrow basal pale bands, band length rarely more than two times diameter of tarsomere (Fig. 4.3F) .... *Ae. vexans*

Abdomen. Terga II–VI with basal pale bands without median notch. Leg. Hindtarsomeres with broad basal pale bands, band length at least three times diameter of tarsomere Fig. 4.3G). Abdomen. Terga II–VI with basal pale bands without median notch 13

13(12) Thorax. Scutum with prominent lateral white scales. Wing. Scales very broad, truncate, with evenly mixed dark and pale scales .... *Ae. grossbecki*

Thorax. Scutum with mixed lateral brown and pale scales. Wing. Scales narrow, slender with unevenly mixed dark and pale scales usually with rounded tips .... *Ae. nigromaculis* (in part)

14(4) Wing. With mixed dark and white scales, usually with distinct white areas and speckled with black scales on other areas. Thorax. Postprocoxal scale patch present (Fig. 4.4A) 15

Wing. With all dark scales or with some pale scales on anterior veins. Thorax. Postprocoxal scale patch absent (Fig. 4.4B) 16

15(14) Wing. Vein R_{4+5} with more dark scales than veins R_2 and R_3. Leg. Foreclaw almost straight at middle or base of tooth .... *Ae. dorsalis*

Wing. Vein R_{4+5} with as many dark scales as veins R_2 and R_3. Leg. Foreclaw slightly curved near or base of tooth .. .. *Ae. campestris*

16(14) Wing. Base of vein C with small white patch. Thorax. Scutum with median area of dark brown or golden scales and usually large anterior lateral patches of pale scales. Head. Palpus almost entirely dark-scaled .... *Ae. epactius*

Wing. Base of vein C entirely dark. Thorax. Scutum entirely covered with brown scales. Head. Palpus with scattered pale scales and white-scaled apex .... *Ae. canadensis*

17(3) Thorax. Scutum with pale cream-colored scales laterally 18

Thorax. Scutum with silver white, yellow, or tan bronze scales laterally 27

18(17) Thorax. Scutum with broad dark brown median longitudinal stripe, much broader posteriorly 19

Thorax. Scutum with two narrow dark submedian longitudinal stripes separated by yellow median stripe; if the two dark stripes are joined, then the dark area is not broader posteriorly 21

19(18) Thorax. Scutum with broad median dark stripe on anterior half, reaching laterally to dorsocentral setal rows (Fig. 4.4D); with lateral silvery white scales; scutal fossa with 1–4 weakly developed setae. Wing. Base of vein C with black scales .... *Ae. triseriatus*

Thorax. Scutum with median silvery white, pale yellow or dark scales; scutal fossa with numerous and well-developed setae. Wing. Extreme base of vein C with pale scales 20

20(19) Thorax. Postspiracular scale patch small or absent; scutal fossa with lightly pigmented setae .... *Ae. hendersoni*

Thorax. Postspiracular scale patch large; scutal fossa with darkly pigmented setae .... *Ae. brelandi*

21(18) Thorax. Scutum with two submedian longitudinal white stripes, separated by narrow median longitudinal brown stripe (Fig. 4.4E) .... *Ae. trivittatus*

Thorax. Scutum with median longitudinal pale stripe or having wide anterior area 22

22(21) Thorax. Scutum with wide median silvery white or yellow patch on anterior half (Fig. 4.4F) 23

Thorax. Scutum with median longitudinal white stripe from head to scutellum 24

23(22) Leg. Hind tibia with basal and apical dark-scaled bands and median pale band. Abdomen. Terga VI–VIII with median light-colored scales .... *Ae. scapularis*

Leg. Hind tibia entirely with dark scales or without median pale band. Abdomen. Terga VI–VIII with median dark scales .... *Ae. infirmatus*

24(22) Abdomen. Terga II–VIII with basal pale bands. Thorax. Scutum with submedian longitudinal stripes of dark scales .... *Ae. muelleri*

Abdomen. Terga II–VIII with basolateral pale patches. Thorax. Scutum without submedian longitudinal stripes of dark scales 25

25(24) Thorax. Scutum with median longitudinal white stripe usually wider than brown lateral areas. Wing. Length about 2.5 mm (small-sized species) .... *Ae. dupreei*

Thorax. Scutum with median longitudinal white stripe usually narrower than brown lateral areas. Wing. Length equal to or greater than 3.0 mm (medium-sized species) 26

26(25) Thorax. Scutum with median longitudinal pale stripe of equal width on anterior and posterior ends (Fig. 4.4C). Head. Lateral black scale patches extend forward to reach the edge of eye .... *Ae. atlanticus*

Thorax. Scutum with median longitudinal pale stripes having wider anterior end. Head. Lateral black scale patches do not reach the eyes, with 2–3 rows of narrow white scales bordering the eyes .... *Ae. tormentor*

27(17) Thorax. Scutum with broad median longitudinal stripe of dark scales. Abdomen. Terga II–IV without complete basal pale bands .... *Ae. thibaulti*

Thorax. Scutum with two narrower longitudinal stripes of brown scales. Abdomen. Terga II–IV with basal pale bands 28

28(27) Abdomen. Terga with median basal triangular patches of pale scales .... *Ae. thelcter*

Abdomen. Terga with basal pale scales having other pattern (not triangular) .... *Ae. sticticus*

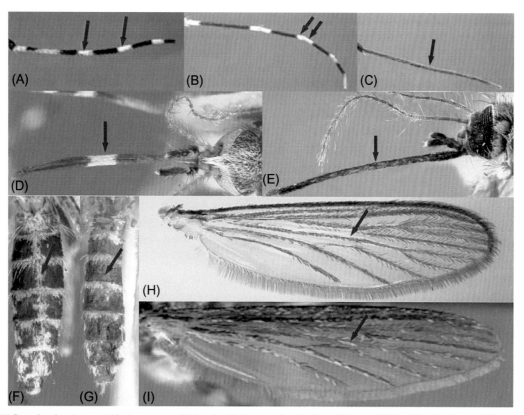

FIG. 4.1 (A) Leg showing *(arrows)* hindtarsomere with basal pale bands only—*Aedes sollicitans.* (B) Leg showing *(arrows)* hindtarsomeres with basal and apical pale bands—*Aedes canadensis.* (C) Leg showing *(arrow)* hindtarsomere entirely dark—*Aedes triseriatus.* (D) Head showing *(arrow)* proboscis with distinct median pale band—*Aedes sollicitans.* (E) Head showing *(arrow)* proboscis without median pale band—*Aedes vexans.* (F) Abdomen showing *(arrow)* terga with median longitudinal stripe—*Aedes sollicitans.* (G) Abdomen showing *(arrow)* terga without median longitudinal stripe—*Aedes taeniorhynchus.* (H) Wing showing *(arrow)* all dark scales—*Aedes vexans.* (I) Wing showing *(arrow)* intermixed dark and pale scales—*Aedes sollicitans.*

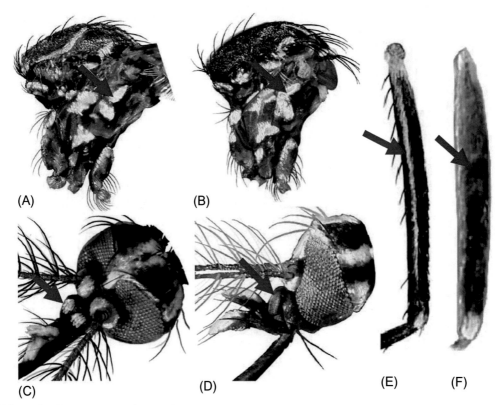

FIG. 4.2 (A) Thorax showing *(arrow)* mesepimeron with two well-separated white scale patches—*Aedes aegypti.* (B) Thorax showing *(arrow)* mese-pimeron with V-shaped white scale patches—*Aedes albopictus.* (C) Head showing *(arrow)* clypeus with white scales—*Aedes aegypti.* (D) Head showing *(arrow)* clypeus without white scales—*Aedes albopictus.* (E) Leg showing *(arrow)* midfemur with anterior longitudinal white stripe—*Aedes aegypti.* (F) Leg showing *(arrow)* midfemur without anterior longitudinal white stripe—*Aedes albopictus.*

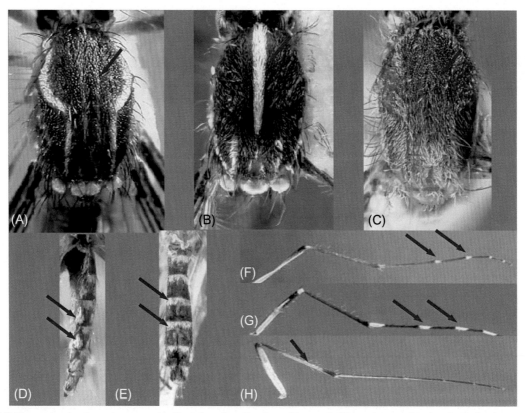

FIG. 4.3 (A) Thorax showing *(arrow)* scutum with silvery lyre-shaped scale markings and without narrow yellow longitudinal stripe—*Aedes aegypti.* (B) Thorax showing *(arrow)* scutum with narrow white median stripe—*Aedes albopictus.* (C) Thorax showing *(arrow)* scutum without narrow white median narrow stripe—*Aedes vexans.* (D) Abdomen showing *(arrows)* terga with white basolateral patches—*Aedes sollicitans.* (E) Abdomen showing *(arrows)* terga with basal pale bands with median notch—*Aedes vexans.* (F) Leg showing *(arrows)* hindtarsomeres with narrow basal pale bands—*Aedes vexans.* (G) Leg showing *(arrows)* hindtarsomeres with broad basal pale bands—*Aedes albopictus.* (H) Leg showing *(arrow)* hind tibia with median pale band in between basal and apical dark-scaled bands—*Aedes scapularis.*

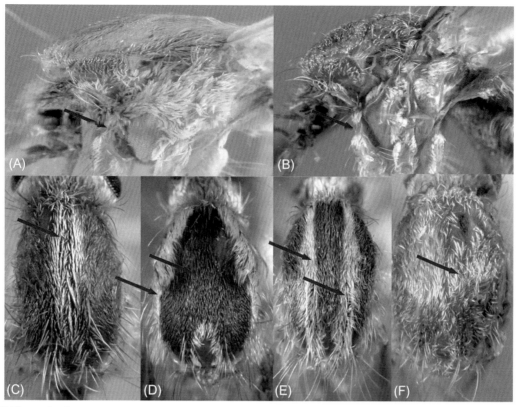

FIG. 4.4 (A) Thorax showing *(arrow)* the presence of postprocoxal scale patch—*Aedes dorsalis.* (B) Thorax showing *(arrow)* the absence of post-procoxal scale patch—*Aedes canadensis.* (C) Thorax showing *(arrow)* scutum with median pale stripe—*Aedes atlanticus.* (D) Thorax. Scutum with median dark stripe and lateral silvery white scales—*Aedes triseriatus.* (E) Thorax showing *(arrow)* scutum with pair of submedian pale-scaled stripes, separated by dark stripe of about same width—*Aedes trivittatus.* (F) Thorax showing *(arrow)* scutum with anteromedian patch of silvery scales—*Aedes scapularis.*

4.2 Key to species of adult females of genus *Anopheles* (= *An.*)

1 Wing. Pale-scaled spots present (Fig. 4.5A) 2

Wing. Pale-scaled spots absent, entirely dark-scaled or with silvery apical fringe spot (Fig. 4.5B) 5

2(1) Leg. Hindtarsomeres 3–5 all pale-scaled and hindtarsomeres 1 and 5 with basal 0.5 dark-scaled (Fig. 4.6A) ....
.... *An. albimanus*

Leg. Hindtarsomeres 1–5 all dark-scaled (Fig. 4.6B) 3

3(2) Wing. Vein C with apical pale spot or dark-scaled; vein A with three dark-scaled spots (Fig. 4.5D) ....
.... *An. crucians,* *An. bradleyi,* and *An. georgianus*

Wing. Vein C with apical and subcostal pale spots; vein A with less than three dark-scaled spots 4

4(3) Wing. Vein 1A with two dark spots (basal and apical) (Fig. 4.5E). Head. Palpi all dark-scaled (Fig. 4.6C) ....
.... *An. punctipennis*

Wing. Vein 1A with one dark spot (Fig. 4.5C). Head. Palpi with pale-scaled rings (Fig. 4.6D) ....
.... *An. pseudopunctipennis*

Wing. Mostly dark-scaled. Head. Palpomere 5 all dark-scaled .... *An. franciscanus*

5(1) Wing. Dark-scaled spots absent; wing length about 3.0 mm; small species. Thorax. Scutal setae about 0.5 width of
scutum 6

Wing. Dark-scaled spots present; wing length 4.0 mm or more; large species. Thorax. Scutal setae less than 0.5 width
of scutum (Fig. 4.7B) 7

6(5) Thorax. Proepisternum with 6–11 setae; anterior acrostichal setae dark. Leg. Forecoxa with 19 or more setae .... *An. barberi*

Thorax. Proepisternum with 2–5 setae; anterior acrostichal setae amber. Leg. Forecoxa with 18 or fewer setae .... *An. judithae*

7(5) Head. Frontal tuft with some pale setae (Fig. 4.6E); palpi all dark-scaled. Wing. With four distinct dark-scaled spots (Fig. 4.5F) 8

Head. Frontal tuft all with dark setae; palpi usually with apical pale rings. Wing. Without distinct dark-scaled spots 10

8(7) Wing. Basal part of vein Cu with scales having truncate apices (Fig. 4.5G) .... *An. freeborni*

Wing. Basal part of vein Cu with scales having rounded apices (Fig. 4.5H) 9

9(8) Thorax. Scutal fossa usually with 21–45 setae (Fig. 4.7C); prealar area with 6–12 setae. Head. Interocular area usually with 7–12 setae (Fig. 4.6F) .... *An. quadrimaculatus*

Thorax. Scutal fossa usually with 8–20 setae; prealar area with 1–5 setae. Head. Interocular area usually with 4–6 setae .... *An. smaragdinus*

10(7) Thorax. Halter usually with pale-scaled capitellum. Head. Occiput medioanteriorly with pale-scaled patch (Fig. 4.7A). Leg. Femora with apical pale-scaled spots .... *An. walkeri*

Thorax. Halter with dark-scaled capitellum. Head. Occiput all dark-scaled. Leg. Femora without or few apical pale scales .... *An. atropos*

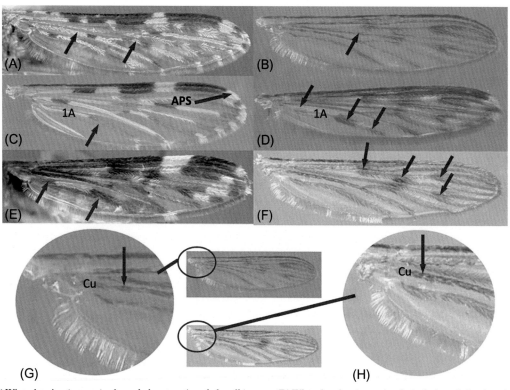

FIG. 4.5 (A) Wing showing *(arrows)* pale-scaled spots—*Anopheles albimanus*. (B) Wing showing *(arrow)* entirely dark-scaled—*Anopheles freeborni*. (C) Wing showing *(arrows)* vein C with apical pale spot (APS) and vein 1A with one dark spot—*Anopheles pseudopunctipennis* (D) Wing showing *(arrows)* vein 1A with three dark-scaled spots—*Anopheles crucians* (E) Wing showing *(arrows)* vein 1A with two dark spots—*Anopheles punctipennis*. (F) Wing showing *(arrows)* four distinct dark-scaled spots—*Anopheles quadrimaculatus*. (G) Wing showing *(arrow)* basal part of vein Cu with scales having truncate apices—*Anopheles freeborni*. (H) Wing showing *(arrow)* basal part of vein Cu with scales having rounded apices—*Anopheles quadrimaculatus*.

FIG. 4.6   (A) Leg showing *(arrow)* hindtarsomeres 3–5 all pale-scaled and hindtarsomeres 1 and 5 with basal 0.5 dark-scaled—*Anopheles albimanus*. (B) Leg showing hindtarsomeres 1–5 all dark-scaled—*Anopheles pseudopunctipennis*. (C) Head. Palpi all dark-scaled—*Anopheles punctipennis*. (D) Head. Palpi with pale-scaled rings—*Anopheles pseudopunctipennis*. (E) Head showing *(arrow)* frontal tuft with some pale setae—*Anopheles quadrimaculatus*. (F) Head showing *(arrow)* interocular area usually with 7–12 setae—*Anopheles quadrimaculatus*.

FIG. 4.7   (A) Head showing *(arrow)* occiput medioanteriorly with pale-scaled patch—*Anopheles walkeri*. (B) Thorax showing *(arrow)* scutal setae (SS) less than 0.5 width of scutum—*Anopheles freeborni*. (C) Thorax showing *(arrow)* scutal fossa (SF) usually with 21–45 setae—*Anopheles quadrimaculatus*.

4.3 Key to species of adult females of genus *Culex* (= *Cx.*)

1 Thorax. Scutum with median acrostichal setae (Fig. 4.8A). Head. Occiput with narrow appressed scales (Fig. 4.8E) 2

Thorax. Scutum without median acrostichal setae. Head. Occiput with broad appressed scales (Fig. 4.8F) (subgenus *Melanoconion*) 16

2(1) Abdomen. Terga with basal pale bands or patches (Fig. 4.9A) (subgenus *Culex*) 3

Abdomen. Terga with apical pale bands or patches or entirely dark scaled (subgenus *Neoculex*) 14

3(2) Leg. Hindtarsomeres with distinct basal and apical pale bands (Fig. 4.10C) 4

Leg. Hindtarsomeres entirely dark (Fig. 4.10D), or with narrow basal pale bands 8

4(3) Head. Proboscis with complete pale-scaled ring (Fig. 4.10A) 5

Head. Proboscis without complete pale-scaled ring (Fig. 4.10B) 6

5(4) Leg. Forefemur and tibia with pale stripe or line of spots on anterior surface. Abdomen. Sterna with V-shaped dark-scaled marks .... *Cx. tarsalis*

Leg. Forefemur and tibia without pale stripe or line of spots. Abdomen. Sterna with oval-shaped dark-scaled marks .... *Cx. stigmatosoma*

6(4) Abdomen. Sterna with median triangular dark-scaled areas .... *Cx. thriambus*

Abdomen. Sterna without median triangular dark-scaled areas, mostly pale-scaled 7

7(6) Leg. Hindtarsomere 5 with basal and apical pale bands .... *Cx. coronator*

Leg. Hindtarsomere 5 with basal pale bands .... *Cx. declarator*

8(3) Thorax. Scutum and thoracic pleura reddish brown. Leg. Coxae reddish brown .... *Cx. erythrothorax*

Thorax. Scutum and thoracic pleura brown (Fig. 4.8B). Leg. Coxae brown (not reddish brown) 9

9(8) Abdomen. Terga entirely dark or with very narrow basal pale bands on segments II–III 10

Abdomen. Terga with distinct basal pale bands on segments II–VII 12

10(9) Thorax. Pleura with patches in groups of less than six scales, mostly in lower mesokatepisternum (Fig. 4.11B). Abdomen. Tergum VII mostly dark-scaled, with large patches of white scales, and usually with narrow basal white band (Fig. 4.9B) .... *Cx. nigripalpus*

Thorax. Pleura with patches in groups of six or more scales (Fig. 4.11C). Abdomen. Terga usually with narrow basal white or coppery band, VII mostly dark or coppery-scaled 11

11(10) Abdomen. Terga VII–VIII nearly covered with copper-colored scales (Fig. 4.9C) .... *Cx. salinarius*

Abdomen. Tergum VII mostly with dark scales .... *Cx. chidestri*

12(9) Abdomen. Terga with basal pale bands rounded posteriorly (Fig. 4.9A). Thorax. Scutum without white-scaled spots (Fig. 4.8A and C) .... *Cx. quinquefasciatus*

Abdomen. Terga with basal pale bands almost straight posteriorly (Fig. 4.9D). Thorax. Scutum with or without white-scaled spots 13

13(12) Wing. Cell R_2 usually about 1.3 or more length of vein R_{2+3} (Fig. 4.11A); wing length 4.0 mm or greater. Thorax. Scutum usually with two small round white-scaled spots (Fig. 4.8D) .... *Cx. restuans*

Wing. Cell R_2 usually about 1.2 or less length of vein R_{2+3}; wing length 3.0 or less. Thorax. Scutum without white-scaled spots .... *Cx. interrogator*

14(2) Abdomen. Terga IV–VI with apicolateral pale patches extending basally at least 0.5 of terga. Head. Palpi about 2.0 length of antennal flagellomere 4 .... *Cx. territans*

Abdomen. Terga IV–VI with dorsolateral pale patches extending basally more than 0.3 of terga. Head. Palpi 2.5 or more length of antennal flagellomere 4 15

15(14) Head. Palpi with pale scales .... *Cx. apicalis*

Head. Palpi entirely dark-scaled .... *Cx. arizonensis*

16(1) Thorax. Mesepimeron with large patch of broad scales .... *Cx. erraticus*

Thorax. Mesepimeron without scales or with few narrow scales 17

17(16) Thorax. Upper mesokatepisternum without scales or with five or less pale scales .... *Cx. pilosus*

Thorax. Upper mesokatepisternum with scales or with six or more pale scales 18

18(17) Head. Occiput with broad white scales anteromedially .... *Cx. abominator*

Head. Occiput with broad dark brown scales anteromedially .... *Cx. peccator*

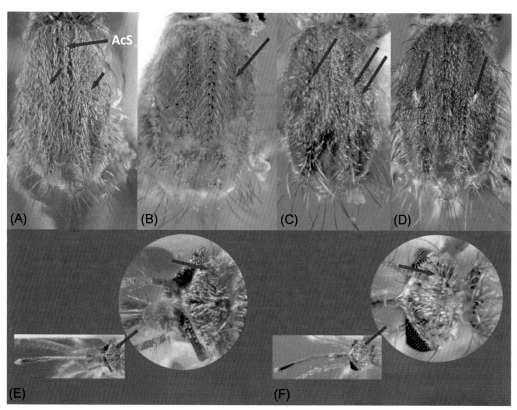

FIG. 4.8 (A) Thorax showing *(arrow)* scutum with median acrostichal setae (AcS) and without white spots (two *short arrows*), *Culex quinquefasciatus*. (B) Thorax showing *(arrow)* brown scutum (not reddish brown), *Culex nigripalpus* (C) Thorax showing *(arrows)* scutum without white-scaled spots, *Culex erraticus*. (D) Thorax showing *(arrows)* two white-scaled spots, *Culex restuans* (E) Head showing *(arrow)* occiput with narrow appressed scales, *Culex quinquefasciatus*. (F) Head showing *(arrow)* occiput with broad appressed scales, *Culex erraticus*.

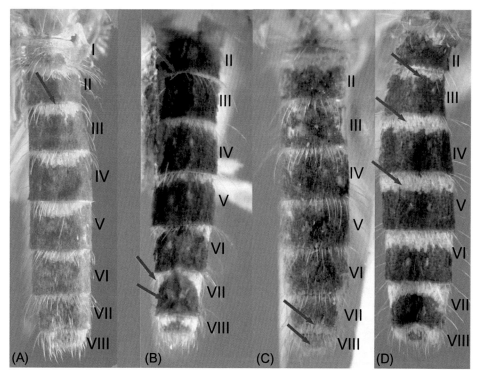

FIG. 4.9 (A) Abdomen showing *(arrows)* terga with basal pale bands, rounded posteriorly, *Culex quinquefasciatus*. (B) Abdomen showing *(arrows)* dark tergum III with narrow basal pale bands and tergum VII mostly dark scaled, with large patches of white scales, *Culex nigripalpus*. (C) Abdomen showing *(arrows)* terga VII–VIII nearly covered with copper-colored scales, *Culex salinarius*. (D) Abdomen showing *(arrows)* terga III–V with basal pale bands almost straight posteriorly, *Culex restuans*.

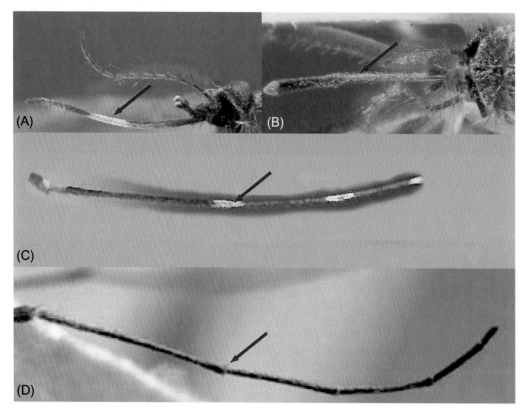

FIG. 4.10 (A) Head showing *(arrow)* proboscis with complete pale-scaled ring, *Culex tarsalis.* (B) Head showing *(arrow)* proboscis without pale-scaled ring, *Culex quinquefasciatus.* (C) Leg showing *(arrow)* hindtarsomeres with distinct basal and apical pale bands, *Culex tarsalis.* (D) Leg showing *(arrow)* hindtarsomeres entirely dark, *Culex nigripalpus.*

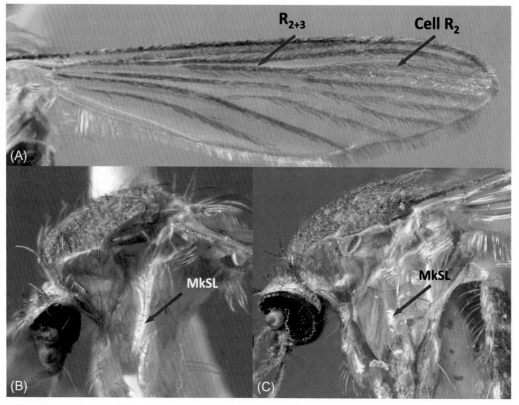

FIG. 4.11 (A) Wing showing *(arrow)* cell R$_2$ 1.3 or more length of vein R$_{2+3}$; *Culex restuans.* (B) Thorax showing *(arrow)* pleura with patches in groups of less than six scales, mostly in lower mesokatepisternum (MkSL), *Culex nigripalpus.* (C) Thorax showing *(arrow)* pleura with patches in groups of six or more scales, *Culex salinarius.*

4.4 Key to species of adult females of genus *Culiseta* (= *Cs.*)

1 Abdomen. Terga without basal pale bands (Fig. 4.12A) .... *Cs. melanura*
  Abdomen. Terga with basal pale bands 2
2(1) Leg. Hindtarsomeres with pale-scaled bands on all or some segments. Wing. Dense patch of dark scales present ....
  .... *Cs. incidens*
  Leg. Hindtarsomeres without pale-scaled bands (Fig. 4.12B). Wing. Dense patch of dark scales absent (Fig. 4.12C) .....
  .... *Cs. inornata*

FIG. 4.12 (A) Abdomen showing *(arrows)* terga without basal pale bands, *Culiseta melanura.* (B) Leg showing *(arrow)* hindtarsomeres without pale-scaled bands, *Culiseta inornata.* (C) Wing showing *(arrow)* the absence of dense patch of dark scales, *Culiseta inornata.*

4.5 Key to species of adult females of genus *Deinocerites* (= *De.*)

1 Thorax. Mesepimeron with a patch of translucent scales (Fig. 4.13) .... *De. pseudes*
2 Thorax. Mesepimeron without scales .... *De. mathesoni*

4.6 Key to species of adult females of genus *Orthopodomyia* (= *Or.*)

1 Thorax. Proepisternum with pale-scaled line on anterior part; mesokatepisternum with very narrow scaled line. Wing.
  Base of vein 1A dark-scaled .... *Or. kummi*
  Thorax. Proepisternum without pale-scaled line on anterior part (Fig. 4.14A); mesokatepisternum with broad scaled
  line (Fig. 4.14A). Wing. Base of vein 1A pale-scaled 2
2(1) Thorax. Lower mesokatepisternal setae 0–3 (Fig. 4.14A); scutellum usually without white scales on lateral lobes
  (Fig. 4.14B). Wing. Base of vein R_{4+5} usually dark-scaled .... *Or. alba*
  Thorax. Lower mesokatepisternal setae four or more; scutellum usually with white scales on lateral lobes. Wing. Base
  of vein R_{4+5} usually with patch of pale scales .... *Or. signifera*

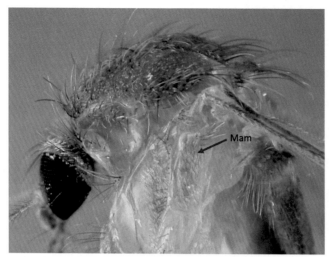

FIG. 4.13   Thoracic mesepimeron showing (*arrow*, Mam) translucent scales—*Deinocerites pseudes.*

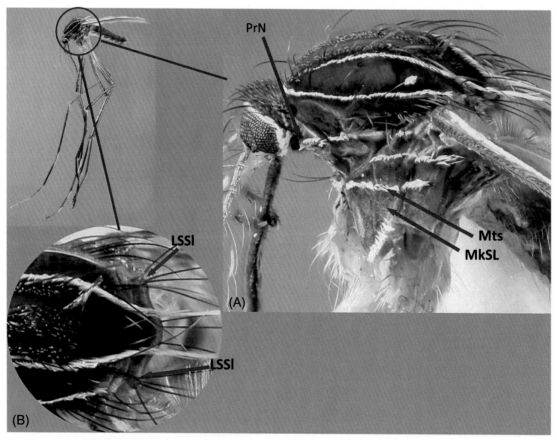

FIG. 4.14   (A) Thorax showing (*arrow*) proepisternum (PrN) without pale-scaled line on anterior part; mesokatepisternum (*arrow*, Mts) with broad scale line and with 0–3 lower mesokatepisternal setae (MkSL)—*Orthopodomyia* sp. (B) Lateral scutellar lobes (*arrows*, LSSl) without white scales–*Orthopodomyia* sp.

4.7 Key to species of adult females of genus *Psorophora* (= *Ps.*)

1 Wing. Dorsal surface of wing with black (or brown) and white scales on most veins (Fig. 4.15A). Leg. Hindfemur with narrow submedian white-scaled band (Fig. 4.15C); tarsal claws long, simple, without basal tooth (subgenus *Grabhamia*) 2

Wing. Dorsal surface of wing with all veins having dark scales, except veins C and Sc with a few pale scales (Fig. 4.15B). Leg. Hindfemur without submedian pale-scaled band (Fig. 4.15D); tarsal claws with distinct basal tooth 5

2(1) Leg. Hindtarsomere 1 dark-scaled, except narrow basal pale-scaled band .... *Ps. pygmaea*

Leg. Hindtarsomere 1 largely pale-scaled or with median pale-scaled band 3

3(2) Leg. Hindtarsomere 1 with basal and median pale-scaled bands (Fig. 4.15E). Wing. Veins with black and white scales randomly scattered .... *Ps. columbiae*

Leg. Hindtarsomere 1 largely pale-scaled bands. Wing. Some veins with brown and white scales not randomly scattered 4

4(3) Wing. Fringe with alternating dark and pale-scaled spots; vein A pale-scaled apically .... *Ps. signipennis*

Wing. Fringe with uniform dark scales; vein 1A dark-scaled apically .... *Ps. discolor*

5(1) Leg. Hindfemur and tibia apices with long, erect, shaggy scales (Fig. 4.16A); hindtarsomere 5 not entirely pale-scaled (subgenus *Psorophora*) 6

Leg. Hindfemur and tibia apices usually without erect scales; if shaggy, then hindtarsomere 5 entirely pale-scaled (subgenus *Janthinosoma*) 7

6(5) Thorax. Scutum with narrow median stripe of golden scales. Head. Proboscis yellow scaled in distal 0.5 except for labella (Fig. 4.16B) .... *Ps. ciliata*

Thorax. Scutum with narrow median stripe of dark scales. Head. Proboscis dark skilled .... *Ps. howardii*

7(5) Leg. Hindtarsomeres all dark-scaled. Abdomen. Terga with dorsal patches of golden scales .... *Ps. cyanescens*

Leg. Hindtarsomeres not all dark-scaled. Abdomen. Terga with apicolateral patches of white to yellow scales 8

8(7) Leg. Hindtarsomeres 1, 2, 3, and 5 dark-scaled; hindtarsomere 4 pale-scaled on at least one side .... *Ps. mathesoni*

Leg. Hindtarsomeres 3 (part), 4, and 5, or only hindtarsomere 5 pale-scaled (Fig. 4.15F) 9

9(8) Leg. Hindtarsomeres 1, 2, 3, and 4 dark-scaled; hindtarsomere 5 pale-scaled .... *Ps. mexicana*

Leg. Hindtarsomeres 1, 2, 3 (part) dark-scaled; hindtarsomeres 3 (part), 4, and 5 pale-scaled 10

10(9) Thorax. Scutum with dark brown and golden-yellow scales (Fig. 4.16C). Abdomen. Tergum 1 with median purple scales (Fig. 4.16D) .... *Ps. ferox*

Thorax. Scutum with broad longitudinal median stripe of dark scales laterally. Abdomen. Tergum 1 with median pale scales 11

11(10) Leg. Femora with apical pale spots. Head. Palpus less than 0.25 length of proboscis; palpomere 4 less than length of palpomeres 1–3 .... *Ps. horrida*

Leg. Femora without apical pale spots. Head. Palpus more 0.3 length of proboscis; palpomere 4 about 1.5 length palpomeres 1–3 .... *Ps. longipalpus*

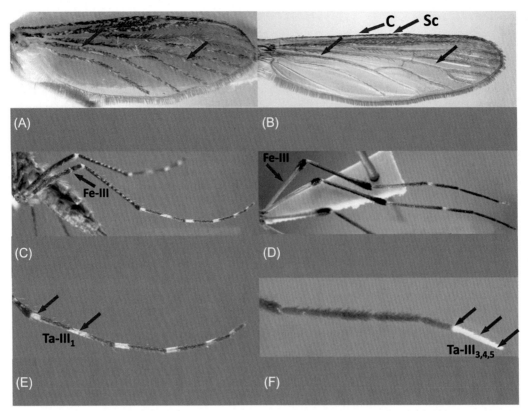

FIG. 4.15 (A) Wing showing *(arrows)* all veins with dark and pale scales—*Psorophora columbiae*. (B) Wing showing all veins with dark scales, except veins C and Sc with a few pale scales—*Psorophora ciliata*. (C) Leg showing *(arrow)* hindfemur (Fe-III) with subapical pale-scaled band—*Psorophora columbiae*. (D) Leg showing *(arrow)* hindfemur (Fe-III) without subapical pale-scaled band—*Psorophora ciliata*. (E) Leg showing hindtarsomere 1 (Ta-III$_1$) with basal and median pale-scaled bands— *Psorophora columbiae*. (F) Leg showing hindtarsomeres 3 (part) (Ta-III$_3$), 4 (Ta-III$_4$), and 5 (Ta-III$_5$), pale-scaled—*Psorophora ferox*.

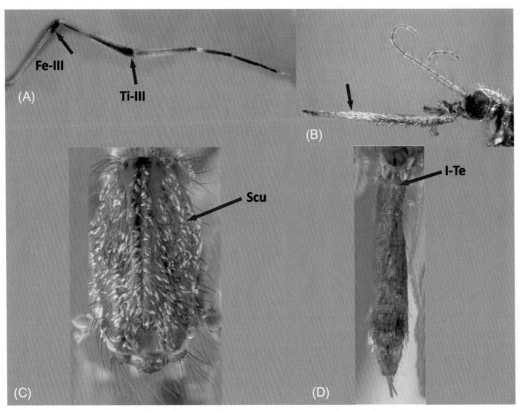

FIG. 4.16 (A) Leg showing *(arrows)* hindfemur (Fe-III) and tibia (TI-III) apices with long, erect, shaggy scales—*Psorophora ciliata*. (B) Head showing *(arrow)* proboscis with yellow scales—*Psorophora ciliata*. (C) Thorax showing *(arrow)* scutum (Scu) with dark brown and golden-yellow scales—*Psorophora ferox*. (D) Abdomen showing *(arrow)* tergum 1 (I-Te) with median purple scales—*Psorophora ferox*.

Key to the Species of Adult Female Mosquitoes of Texas **Chapter | 4 191**

4.8 Key to species of adult females of genus *Toxorhynchites* (= *Tx.*)

1 Leg. Hindtarsomere 5 usually completely white-scaled (Fig. 4.17A). Abdomen. Lateral parts of abdominal terga II–VII with golden-yellow scales (Fig. 4.17B). Wing. Veins C and R completely dark purple-scaled (Fig. 4.17C) ....
.... *Tx. rutilus septentrionalis*
Leg. Hindtarsomere 5 usually completely dark-scaled. Abdomen. Lateral parts of abdominal terga II–VII with pale violet or lilac scales. Wing. Veins C and R dark purple-scaled on basal half .... *Tx. moctezuma*

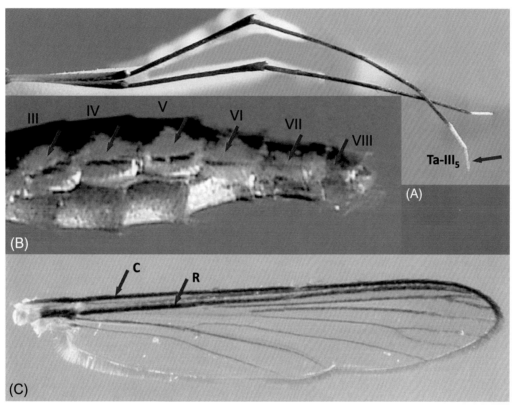

FIG. 4.17 (A) Leg showing *(arrow)* hindtarsomere 5 (Ta-III$_5$) completely white-scaled—*Toxorhynchites rutilus septentrionalis*. (B) Abdomen showing *(arrow)* lateral parts of abdominal terga II–VII (II–VIII-Te) with golden-yellow scales, *Toxorhynchites rutilus septentrionalis*. (C) Wing showing *(arrows)* veins C and R dark purple- scaled on basal half, *Toxorhynchites moctezuma*.

4.9 Key to species of adult females of genus *Uranotaenia* (= *Ur.*)

1 Leg. Hindtarsomeres 3 (part), 4, and 5 pale-scaled .... *Ur. lowii*
Leg. Hindtarsomeres all dark-scaled (Fig. 4.18A) 2
2(1) Thorax. Scutum with median longitudinal row of shiny iridescent blue scales .... *Ur. sapphirina*
Thorax. Scutum without median longitudinal row of iridescent blue scales (Fig. 4.18B) .... *Ur. anhydor syntheta*

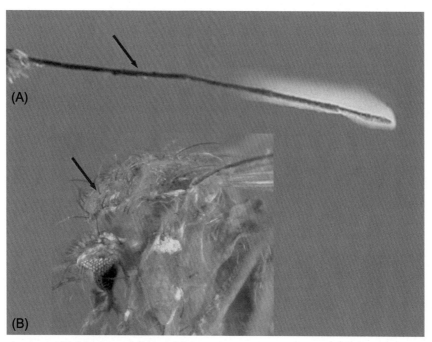

FIG. 4.18 (A) Leg showing *(arrow)* hindtarsomeres all dark scaled, *Uranotaenia* sp. (B) Thorax showing *(arrow)* scutum without median longitudinal row of iridescent blue scales, *Uranotaenia anhydor syntheta*.

Acknowledgments

Great appreciations go to J. Stoffer for taking photographs when she was under contract at Walter Reed Biosystematics Unit, G. White for assisting with sorting and retrieving specimens for initial photographs, and B.P. Rueda for the enormous help in finalizing the photographs used in the keys. Thanks also to B.P. Rueda and C.R. Summers for valuable comments and suggestions on this chapter.

References and further reading

Belkin, J.N., 1962. The Mosquitoes of the South Pacific (Diptera, Culicidae). vols. I & II. University of California Press, Berkely, CA. 608 pp.

Carpenter, S.J., LaCasse, J.A., 1955. Mosquitoes of North America (North of Mexico). University of California Press, Berkeley and Los Angeles, CA. 360 pp. +127 pl.

Darsie Jr., R.F., Ward, R.A., 2005. Identification and Geographical Distribution of the Mosquitoes of North America, North of Mexico. University Press of Florida, Gainesville, FL. 383 pp.

Harbach, R.E., Knight, K.L., 1980. Taxonomists' Glossary of Mosquito Anatomy. Plexus Publishing, Marlton, NJ. 415 pp.

Harbach, R.E., Knight, K.L., 1982. Corrections and additions to taxonomists' glossary of mosquito anatomy. Mosq. Syst. 13, 201–217.

Harrison, B.A., Byrd, B.D., Sither, C.B., Whitt, P.B., 2016. The Mosquitoes of the Mid-Atlantic Region: An Identification Guide. Western Carolina University, Cullowhee, NC. 201 pp.

Rueda, L.M., 2004. Pictorial keys for the identification of mosquitoes (Diptera: Culicidae) associated with dengue virus transmission. Zootaxa 589, 1–60.

Rueda, L.M., Stockwell, S.A., Pecor, J.E., Gaffigan, T., 1998. Key to the Mosquito Genera of the World (INTKEY Module). Walter Reed Biosystematics Unit, Smithsonian Institution, Washington, DC. [CD]. Also, in The Diptera Dissemination Disk – vol. 1, North American Dipterists Society, Washington, DC.

Walter Reed Biosystematics Unit (WRBU), 2018. Mosquito Identification Keys. Walter Reed Biosystematics Unit, Smithsonian Institution, Suitland, MD. Available from: http://www.wrbu.org/aors/northcom_Keys.html. (Accessed December 15, 2018).

Chapter 5

Key to Genera of Fourth Instar Mosquito Larvae of Texas

Leopoldo M. Rueda

Walter Reed Biosystematics Unit, Department of Entomology, Smithsonian Institution, Suitland, MD, United States

Chapter Outline

5.1 Larval morphology

The external morphology of the fourth instar larvae is not simple and often too difficult to learn it. Like adults the most important approach to identify the larvae is locating the correct morphological structures on the specimens. Repeated examinations of mosquitoes under a high-powered microscope (particularly compound microscope for slide-mounted larvae) are essential processes of learning mosquito identification. Any available identification larval keys will not be helpful unless the user is well prepared, with better knowledge of mosquito larval morphology, and well equipped with appropriate microscopes and related equipment.

Many characters in the larval keys are provided on the morphology photographs or illustrations, with arrows pinpointing the structures that are necessary to identify the genus or species. Many individual illustrations or photographs used in the key couplets to separate genera and species have been simplified by removing extraneous structures such as setae and other body parts that are not important for the identification of the genus or species in that couplet.

Larvae of mosquitoes, like their adults, have bilateral symmetry. The head, thorax, and abdomen are paired; thus if one side is damaged and a character is missing, that character should be examined on the other side. Some illustrations or photographs in the keys only show one-half of the structures, since the other side is just a mirror image of the first side. Compound microscopes, with higher magnifications (at least 100×), must be used for slide-mounted specimens since numerous small characters are included and illustrated in these larval keys.

Dichotomous keys are prepared here, with selected illustrations and photographs on most key couplets as shown on various figures. These keys are a series of paired or contrasting statements (or couplets) that allow the user to accurately identify an unknown mosquito specimen at genus or species level. Most couplets may contain a few valid characters to assist in identifying damaged or missing structures.

Major morphological characters are used to identify the fourth instar larvae. They include the pecten spines (sizes, shapes, and denticles), setae (single or branched, shapes—blunt, pointed, and palmate—sizes, rows, colors, and alveolus location), anal papillae (sizes, shapes, and visible trachea), siphon (shapes, spines, acus, pigments, or bands), and length ratios (anal papillae, antennae, and siphon).

Before using these keys the specific features of the fourth instar larvae must be determined, using the following characters: siphon and saddle with complete sclerotization and larger size compared with the first, second, or third instar larvae.

To help the reader, the larval morphology (or habitus) is shown in Fig. 5.1, together with other parts such as head and thorax (Fig. 5.2) and part of abdomen (Fig. 5.3). Specific diagnostic characters of some body parts not included in Figs. 5.1–5.3 are shown in other figures of selected key couplets. The terminology used in the keys follows Harbach and Knight (1980, 1982).

Mosquitoes, Communities, and Public Health in Texas. https://doi.org/10.1016/B978-0-12-814545-6.00005-5

The illustrations presented in the keys were developed mainly by making freehand computer images based on multiple printed references, photographs, and preserved specimens as guides. The primary reference materials used in the developments of the key illustrations include Carpenter and LaCasse (1955), Belkin (1962), Rueda et al. (1998), Rueda (2004), Darsie and Ward (2005), Harrison et al. (2016), and WRBU (2018). Colored photographs of mosquitoes were used, principally from Walter Reed Biosytematics Unit (WRBU) and additional collections from previous mosquito biosystematics projects of L.M. Rueda.

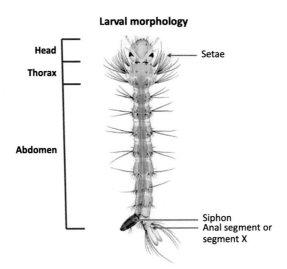

FIG. 5.1 Dorsal view of fourth instar mosquito larva (segments VIII and X, lateral view)—*Aedes aegypti.*

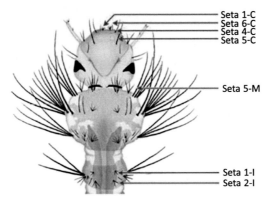

FIG. 5.2 Dorsal view of larval head, thorax, and abdomen (part)—*Aedes aegypti.*

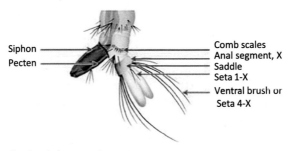

FIG. 5.3 Lateral view of larval abdomen (part)—*Aedes aegypti.*

5.2 Key to genera of fourth instar mosquito larvae

1 Abdomen. Respiratory siphon absent (Fig. 5.4A); seta 1 palmate, usually with well-developed leaflets in some segments (Fig. 5.4C) .... *Anopheles*
Abdomen. Respiratory siphon present (Fig. 5.4B); seta 1 never palmate 2

2(1) Abdomen. Siphon with sclerotized saw-toothed process apically, modified for piercing plant tissues (Fig. 5.4D) 3
Abdomen. Siphon without sclerotized saw-toothed process apically (Fig. 5.4E) 4

3(2) Head. Antenna with distal part to setae 2, 3-A less than 0.5 length of proximal part; antenna about as long as width of head (Fig. 5.4F). Abdomen. Anal segment with multiple sharply pointed setae .... *Mansonia (Ma. titillans)*
Head. Antenna with distal part to setae 2, 3-A as long as or longer than proximal part; antenna longer than width of head, articulated, with distal portion elongate, about as long as basal portion. Abdomen. Anal segment with single or multibranched setae or none .... *Coquillettidia (Cq. perturbans)*

4(2) Abdomen. Siphon without pecten spines 5
Abdomen. Siphon with pecten spines 6

5(4) Abdomen. Segment VIII with comb scales (Fig. 5.4G). Head. Lateral palatal brush with numerous fine simple filaments .... *Orthopodomyia*
Abdomen. Segment VIII without comb scales (Fig. 5.4H). Head. Lateral palatal brush with some thick filaments .... *Toxorhynchites (Tx. moctezuma* and *Tx. rutilus septentrionalis)*

6(4) Head. Longer than wide (Fig. 5.5A). Abdomen. Segment VIII with large lateral comb plate having comb scales (Fig. 5.5C) .... *Uranotaenia*
Head. Wider than long (Fig. 5.5B). Abdomen. Segment VIII without comb plate or with small comb plate having comb scales 7

7(6) Head. Widest near level of bases of antennae. Abdomen. Segment X with dorsal and ventral sclerotized plates (Fig. 5.5D) .... *Deinocerites*
Head. Widest in posterior 0.5. Abdomen. Segment X with single sclerotized saddle (Fig. 5.5E) 8

8(7) Abdomen. Siphon with at least a basal pair of ventral setae (Fig. 5.5F) .... *Culiseta*
Abdomen. Siphon without ventral setae near base and setae scattered elsewhere 9

9(8) Abdomen. Siphon with at least three pairs of setae (Fig. 5.6A) 10
Abdomen. Siphon with one pair of setae (Fig. 5.6B) 11

10(9) Abdomen. Segment X completely encircled by saddle .... *Culex*
Abdomen. Segment X not completely encircled by saddle .... *Aedes* (in part)

11(9) Abdomen. Segment X completely encircled by saddle having midventral row of setal tufts of seta 4-X .... *Psorophora*
Abdomen. Segment X usually not completely encircled by saddle, or if so, then setal tufts of seta 4-X at posterior to it 12

12(11) Abdomen. Segment X with saddle having distinct posterior aciculae (Fig. 5.6C); seta 3-VII single, longer than tergum (Fig. 5.6D) .... *Haemagogus (Hg. equinus)*
Abdomen. Segment X with saddle having indistinct or small posterior aciculae; seta 3-VII single or branched, shorter than tergum (Fig. 5.6E) .... *Aedes* (in part)

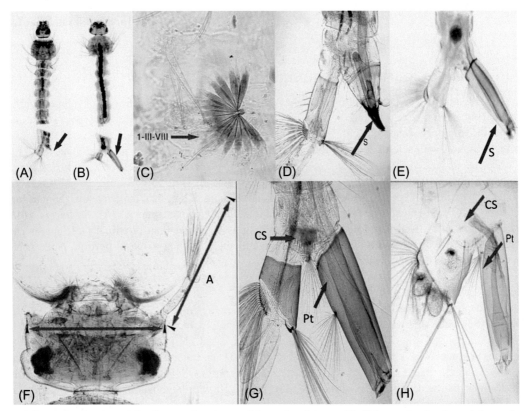

FIG. 5.4 (A) Abdomen showing *(arrow)* the absence of respiratory siphon—*Anopheles* sp. (B) Abdomen showing *(arrow)* the respiratory siphon—*Aedes vexans*. (C) Abdominal seta 1 palmate *(arrow)*, usually present in abdominal segments III–VIII—*Anopheles* sp. (D) Abdomen showing *(arrow)* siphon (S) with sclerotized saw-toothed process apically—*Mansonia* sp. (E) Abdomen showing *(arrow)* siphon (S) without sclerotized saw-toothed process apically—*Culiseta* sp. (F) Head showing *(arrow)* antenna (A) about as long as width of head *(arrow)*—*Mansonia* sp. (G) Abdomen showing *(arrow)* segment VIII with comb scales (CS) and without pecten (Pt)—*Orthopodomyia* sp. (H) Abdomen showing *(arrow)* segment VIII without comb scales (CS) and without pecten (Pt)—*Toxorhynchites* sp.

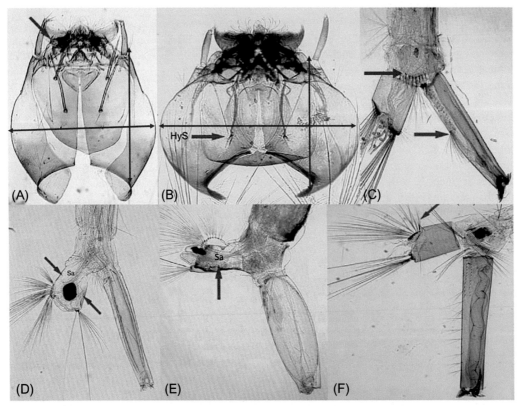

FIG. 5.5 (A) Head showing *(arrow)* longer than wide—*Uranotaenia* sp. (B) Head showing *(arrows)* wider than long (hypostomal suture, HyS)—*Culiseta* sp. (C) Abdomen showing *(arrow)* segment VIII with large lateral comb plate having comb scales—*Uranotaenia* sp. (D) Abdomen showing *(arrows)* segment X (saddle, Sa) with dorsal and ventral sclerotized plates—*Deinocerites* sp. (E) Abdomen showing *(arrow)* segment X with single sclerotized plates or saddle (Sa)—*Psorophora* sp. (F) Abdomen showing *(arrow)* siphon with ventral setae—*Culiseta melanura*.

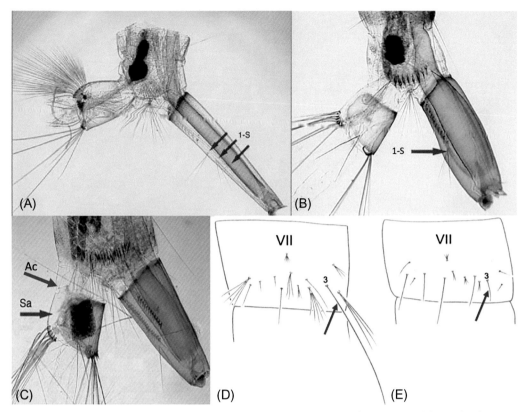

FIG. 5.6 (A) Abdomen showing *(arrows)* siphon with three pairs of setae (1-S)—*Culex quinquefasciatus*. (B) Abdomen showing *(arrow)* siphon with one pair of setae (1-S)—*Aedes* sp. (C) Abdomen showing *(arrow)* segment X with saddle (Sa) having distinct posterior aciculae (Ac)—*Haemagogus* sp. (D) Abdomen showing *(arrow)* seta 3-VII single, longer than tergum—*Haemagogus* sp. (E) Abdomen showing *(arrow)* seta 3-VII branched, shorter than tergum—*Aedes* sp.

Acknowledgments

Great appreciations go to J. Stoffer for taking photographs when she was under contract at Walter Reed Biosystematics Unit and T. Litwak for *Aedes* illustrations. Special thanks go to G. White for assisting with sorting and retrieving specimens for initial photographs and B.P. Rueda for the support and enormous help in finalizing the photographs used in the keys. Thanks also to B.P. Rueda and C.R. Summers for their valuable comments and suggestions on this chapter.

References and further reading

Belkin, J.N., 1962. The Mosquitoes of the South Pacific (Diptera, Culicidae). vols. I & II. University of California Press, Berkely, CA. 608 pp.

Carpenter, S.J., LaCasse, W.J., 1955. Mosquitoes of North America (North of Mexico). University of California Press, Berkeley and Los Angeles, CA. 360 pp. +127 pl.

Darsie Jr., R.F., Ward, R.A., 2005. Identification and Geographical Distribution of the Mosquitoes of North America, North of Mexico. University Press of Florida, Gainesville, FL. 383 pp.

Harbach, R.E., Knight, K.L., 1980. Taxonomists' Glossary of Mosquito Anatomy. Plexus Publishing, Marlton, NJ. 415 pp.

Harbach, R.E., Knight, K.L., 1982. Corrections and additions to taxonomists' glossary of mosquito anatomy. Mosq. Syst. 13, 201–217.

Harrison, B.A., Byrd, B.D., Sither, C.B., Whitt, P.B., 2016. The Mosquitoes of the Mid-Atlantic Region: An Identification Guide. Western Carolina University, Cullowhee, NC. 201 pp.

Rueda, L.M., 2004. Pictorial keys for the identification of mosquitoes (Diptera: Culicidae) associated with dengue virus transmission. Zootaxa 589, 1–60.

Rueda, L.M., Stockwell, S.A., Pecor, J.E., Gaffigan, T., 1998. Key to the Mosquito Genera of the World (INTKEY Module). Walter Reed Biosystematics Unit, Smithsonian Institution, Washington, DC. [CD]. Also, in The Diptera Dissemination Disk – vol. 1, North American Dipterists Society, Washington, DC.

Walter Reed Biosystematics Unit (WRBU), 2018. Mosquito Identification Keys. Walter Reed Biosystematics Unit, Smithsonian Institution, Suitland, MD. Available from: http://www.wrbu.org/aors/northcom_Keys.html. (Accessed December 15, 2018).

Chapter 6

Key to the Species of Fourth Stage Mosquito Larvae of Texas

Leopoldo M. Rueda

Walter Reed Biosystematics Unit, Department of Entomology, Smithsonian Institution, Suitland, MD, United States

Chapter Outline

6.1   Key to species of fourth instar larvae of genus *Aedes* (= *Ae.*)

1   Abdomen. Saddle completely encircling segment X (Fig. 6.1A) 2
     Abdomen. Saddle incomplete, not encircling segment X (Fig. 6.1B) 14
2(1)   Abdomen. Siphon with one or more distal pecten spines widely spaced (Fig. 6.1C) 3
       Abdomen. Siphon with pecten spines almost evenly spaced (Fig. 6.1D) 5
3(2)   Abdomen. Siphon with seta 1-S inserted distally beyond pecten row (Fig. 6.2A) or rarely beside most distal pecten spine; segment VIII with comb scale with median spine at least 4.0 length of minute basal spines .... *Ae. nigromaculis*
       Abdomen. Siphon with seta 1-S inserted within pecten row (Figs. 6.2B and 6.3C); segment VIII with comb scale with median spine not more than 2.0 length of submedian spines or fringed with subequal spines 4
4(3)   Abdomen. Segment VIII with comb scale with median spine longer than submedian spines (Fig. 6.2C). Head. Seta 6-C single (Fig. 6.2E) .... *Ae. thelcter*
       Abdomen. Segment VIII with comb scale with median spine fringed with subequal spines (Fig. 6.2D). Head. Seta 6-C usually with two or three branches .... *Ae. fulvus pallens*
5(2)   Abdomen. Siphon with seta 1-S attached within pecten row (Figs. 6.2B and 6.3C) 6
       Abdomen. Siphon with seta 1-S attached distally beyond pecten row (Fig. 6.2A) 7
6(5)   Abdomen. Segment VIII with 30 or more comb scales; each comb scale fringed with subequal spines (Fig. 6.2D) .... *Ae. bimaculatus*
       Abdomen. Segment VIII with 13 or less comb scales in a single row (Fig. 6.3C); each comb scale with a very long sharp median spine and minute basal spines .... *Ae. tormentor*
7(5)   Abdomen. Segment VIII with comb scale with median spine at least 4.0 length of submedian spines. Thorax. Dorsal integument smooth 8
       Abdomen. Segment VIII with comb scale with median spine not more than 3.0 length of submedian spines or fringed with subequal spines. Thorax. Dorsal integument aculeate 11

8(7) Abdomen. Segment X with anal papillae extremely long, at least eight times length of saddle and having darkly pigmented trachea (Fig. 6.3A); seta 2-X with two or three branches (Fig. 6.3A) .... *Ae. dupreei*

Abdomen. Segment X with anal papillae short, less than six times length of saddle and lacking dark trachea; seta 2-X with at least four branches 9

9(8) Abdomen. Segment VIII with 8 or less (usually 4–6) large comb scales .... *Ae. atlanticus*

Abdomen. Segment VIII with 10 or more small comb scales 10

10(9) Abdomen. Siphon short, length 2.0–2.5 times width at base; dorsal pair of anal papillae shorter than dorsal length of saddle; and pecten spines reaching middle of siphon or more distally (Fig. 6.4A). Head. Seta 1-A long, reaching tip of antenna (Fig. 6.4B) .... *Ae. sollicitans*

Abdomen. Siphon length 3.0–3.5 times width at base, dorsal pair of anal papillae longer than dorsal length of saddle, and pecten spines not reaching middle of siphon. Head. Seta 1-A short, not reaching tip of antenna (Fig. 6.3B) .... *Ae. mitchellae*

11(7) Abdomen. Segment VIII with comb scale with median spine long, not more than 4.0 times length of submedian spines 12

Abdomen. Segment VIII with comb scale with subequal median and submedian spines 13

12(11) Abdomen. Segment VIII with comb scale with median spine 2.0–2.5 times length of adjacent lateral spines; seta 2-VIII with two or three branches, as long as seta 1-VIII. Thorax. Prothorax seta 1-P slightly longer than seta 4-P (Fig. 6.4C) .... *Ae. infirmatus*

Abdomen. Segment VIII with comb scale with median spine less than 2.0 times length of adjacent lateral spines (Fig. 6.5A); seta 2-VIII single much longer than seta 1-VIII (Fig. 6.5A). Thorax. Prothorax seta 1-P more than 2.0 times length of seta 4-P .... *Ae. trivittatus*

13(11) Abdomen. Setae 6-III-V with two to five branches; segment X with anal papillae shorter than length of saddle (Fig. 6.5B) .... *Ae. taeniorhynchus*

Abdomen. Setae 6-III-V single (rarely double); segment X with anal papillae as long or longer than length of saddle .... *Ae. scapularis*

14(1) Abdomen. Siphon with one or more distal pecten spines widely spaced (Fig. 6.1C) 15

Abdomen. Siphon with pecten spines almost evenly spaced (Fig. 6.1D) 17

15(14) Abdomen. Siphon with seta 1-S inserted within pecten spine row (Fig. 6.2B) .... *Ae. epactius*

Abdomen. Siphon with seta 1-S inserted beyond pecten spine row (Fig. 6.2A) 16

16(15) Abdomen. Segment VIII with at least 18 comb scales; each scale with median spine fringed with subequal spines (Fig. 6.2D); saddle distinctly incomplete, encircling about 70% of segment X .... *Ae. campestris*

Abdomen. Segment VIII with not more than 17 comb scales; each scale with median spine longer than submedian spines (Fig. 6.2C); saddle nearly complete, encircling about 90%–95% of segment X .... *Ae. vexans*

17(14) Head. Seta 1-A single or double (Fig. 6.6C); antenna usually smooth or with tiny spines 18

Head. Seta 1-A with more than three branches (Figs. 6.3A and 6.4B); antenna usually with coarse spines 23

18(17) Abdomen. Segment VIII with comb scale having pointed unfringed median spine (Fig. 6.6D and E) 19

Abdomen. Segment VIII with comb scale having blunt apex, evenly fringed with short spines 20

19(18) Abdomen. Ventral brush (4-X) with five pairs of setae; seta 4-a,b X branched (Fig. 6.6A); segment VIII with comb scale with stout, submedian spines (Fig. 6.6D). Head. Seta 7-C single. Thorax. Mesothorax and metathorax with large curved spines on lateral support plates .... *Ae. aegypti*

Abdomen. Ventral brush (4-X) with four pairs of setae; seta 4-a,b X single (Fig. 6.6B); segment VIII with comb scale without stout submedian spines (Fig. 6.6E). Head. Seta 7-C with two or three branches. Thorax. Mesothorax and metathorax with small spines on lateral support plates .... *Ae. albopictus*

20(18) Abdomen. Saddle extending more than 0.6 distance to midventral line; seta 1-X attached more dorsal of ventral edge of saddle (Fig. 6.7A); ventral brush (4-X) with at least seven pairs of setae .... *Ae. zoosophus*

Abdomen. Saddle extending less than 0.6 distance to midventral line; seta 1-X attached near ventral edge of saddle; ventral brush (4-X) with five or six pairs of setae 21

21(20) Abdomen. Ventral brush (4-X) with six pairs of setae; acus usually attached to siphon or detached and too close to its base; segment X with dorsal pair of anal papillae longer than ventral pair (Fig. 6.7B) .... *Ae. triseriatus*

Abdomen. Ventral brush (4-X) with five pairs of setae; acus detached and not close to base of siphon; segment X with dorsal pair of anal papillae as long as ventral pair 22

22(21) Abdomen. Ventral brush (4-X) with two most anterior pairs of setae double .... *Ae. hendersoni*

Abdomen. Ventral brush (4-X) with two most anterior pairs of setae having three or four branches .... *Ae. brelandi*

23(17) Abdomen. Segment VIII with comb scale with median spine 1.5 or more length of submedian spines ....

.... *Ae. sticticus*

Abdomen. Segment VIII with comb scale with median spine less than 1.5 length of submedian spines 24

24(23) Head. Seta 5-C with at least four branches; seta 6-C with at least three branches (Fig. 6.7C) 25

Head. Setae 5-C and 6-C usually single or with two to four branches (Fig. 6.7D) 26

25(24) Abdomen. Siphon length more than 4.0 times width at base; segment VIII with comb scale with median and sub-median spines much stouter than lateral spines. Head. Setae 5-7-C nearly inserted in straight line; antenna length nearly equal head length at midline .... *Ae. thibaulti*

Abdomen. Siphon length less than 4.0 times width at base; segment VIII with comb scale fringed with subequal spines. Head. Setae 5-7-C inserted not in straight line; antenna length about half head length at midline ....

.... *Ae. canadensis*

26(24) Abdomen. Pecten spines extending beyond basal 0.5 of siphon, with two most apical spines much stouter than preceding two .... *Ae. campestris*

Abdomen. Pecten spines not extending beyond basal 0.5 of siphon, with two most apical spines not stouter than preceding two 27

27(26) Abdomen. Segment VIII with comb scales with median spine approximately equal length of lateral adjacent spines; seta 1-X about 0.5 length of saddle (Fig. 6.8A). Head. Setae 5,6-C usually single .... *Ae. dorsalis*

Abdomen. Segment VIII with comb scales with median spine 1.5–2.75 times length of adjacent lateral spines; seta 1-X almost equal to length of saddle (Fig. 6.8B). Head. Setae 5,6-C usually branched .... *Ae. grossbecki*

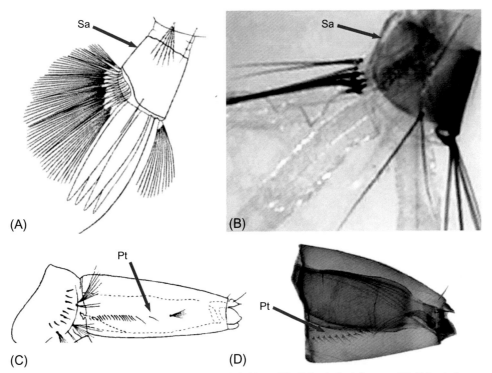

FIG. 6.1 (A) Abdominal segment X showing *(arrow)* completely encircled by saddle (Sa)—*Aedes infirmatus.* (B) Abdominal segment X showing not completely encircled by saddle (Sa, *arrow*)—*Aedes albopictus.* (C) Siphon showing *(arrow)* pecten spines (Pt) detached apically—*Aedes* sp. (D) Siphon showing *(arrow)* pecten spines (Pt) almost evenly spaced—*Aedes aegypti.*

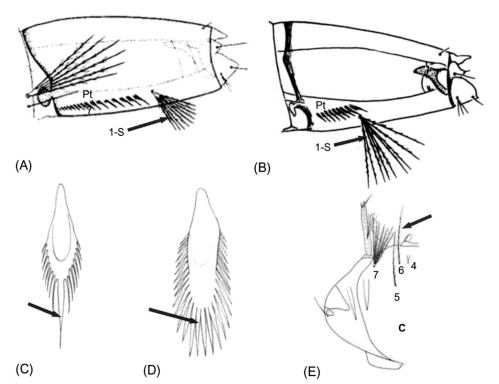

FIG. 6.2 (A) Siphon showing *(arrow)* seta 1-S inserted distally beyond pecten row (Pt)—*Aedes* sp. (B) Siphon showing seta 1-S inserted within pecten row (Pt)—*Aedes scapularis.* (C) Comb scale showing *(arrow)* median spine longer than submedian spines—*Aedes thelcter.* (D) Comb scale showing *(arrow)* median spine fringed with subequal spines—*Aedes fulvus pallens.* (E) Head (C) showing *(arrow)* single seta 6-C—*Aedes thelcter.*

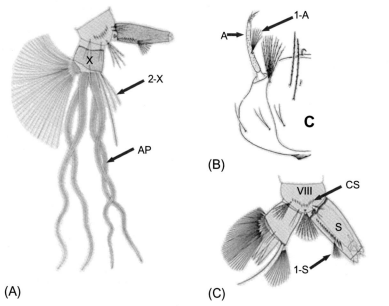

FIG. 6.3 (A) Abdominal segment X showing *(arrows)* long anal papillae (AP) and seta 2-X with two branches—*Aedes dupreei.* (B) Head (C) showing *(arrow)* short antennal seta 1-A not reaching tip of antenna (A)—*Aedes mitchellae.* (C) Abdominal segment VIII showing *(arrow)* comb scales (CS) in a single row; siphon (S) with seta 1-S *(arrow)*—*Aedes tormentor.*

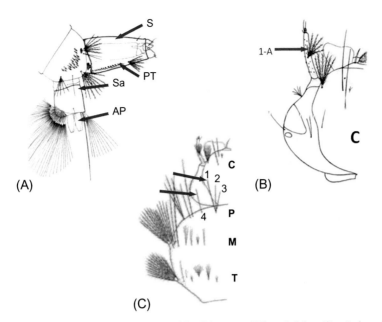

FIG. 6.4   (A) Abdominal X showing *(arrows)* anal papillae (AP), saddle (Sa), pecten (PT), and siphon (S)—*Aedes sollicitans.* (B) Head (C) showing *(arrow)* seta 1-A reaching tip of antenna—*Aedes sollicitans.* (C) Thorax showing *(arrows)* prothorax seta 1-P slightly longer than seta 4-P (*C*, head; *P*, prothorax; *M*, mesothorax; and *T*, metathorax)—*Aedes infirmatus.*

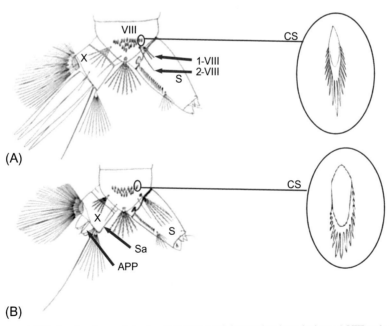

FIG. 6.5   (A) Abdominal segment VIII showing *(arrows)* single seta 2-VIII much longer than branched seta 1-VIII and a sharply pointed comb scale with median spine longer than submedian spines (CS) (*S*, siphon; *X*, abdominal segment X)—*Aedes trivittatus.* (B) Abdominal segment X showing anal papillae (APP) shorter than length of saddle (Sa); rounded comb scales (CS) with subequal median and submedian spines (*S*, siphon; *X*, abdominal segment X)—*Aedes taeniorhynchus.*

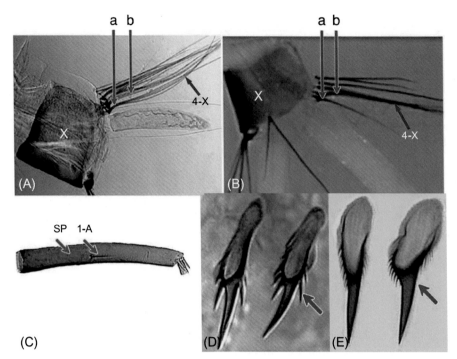

FIG. 6.6 (A) Abdomen showing ventral brush (4-X) with five pairs of setae; seta 4-a,b X branched—*Aedes aegypti*. (B) Abdomen showing ventral brush (4-X) with four pairs of setae; seta 4-a,b X single—*Aedes albopictus*. (C) Antennal seta 1-A *(arrow*; SP, spicules)—*Aedes albopictus*. (D) Comb scale (CS) with stout, submedian spines *(arrow)*—*Aedes aegypti*. (E) Comb scale (CS) without stout, submedian spines *(arrow)*—*Aedes albopictus*.

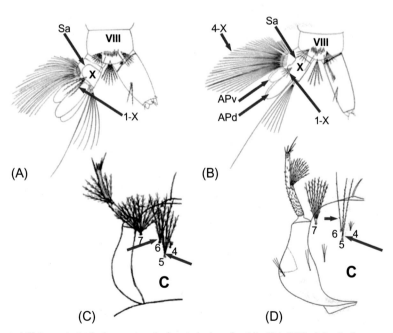

FIG. 6.7 (A) Abdominal seta 1-X *(arrow)* attached more dorsal of ventral edge of saddle (Sa) (*VIII*, abdominal segment VIII; *X*, abdominal segment X)—*Aedes zoosophus*. (B) Abdominal seta 1-X *(arrow)* attached near ventral edge of saddle (Sa); dorsal papillae (APd) and ventral papillae (APv); ventral brush (4-X) setae (*VIII*, abdominal segment VIII; *X*, abdominal segment X)—*Aedes triseriatus*. (C) Head (C) showing *(arrows)* setae 5-C and 6-C with four branches—*Aedes canadensis*. (D) Head (C) showing *(arrows)* setae 5-C and 6-C with two branches—*Aedes campestris*.

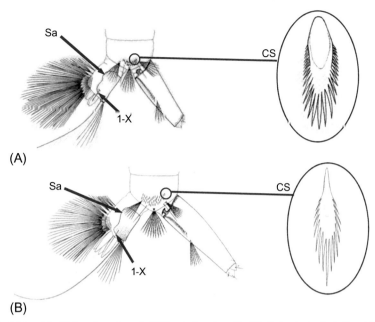

FIG. 6.8 (A) Abdominal seta 1-X, saddle (Sa), and comb scale (CS)—*Aedes dorsalis*. (B) Abdominal seta 1-X, saddle (Sa), and comb scale (CS)—*Aedes grossbecki*.

6.2   Key to species of fourth instar larvae of genus *Anopheles* (= *An.*)

1 Head. Setae 5-7-C small, single, or double (Fig. 6.9A). Abdomen. Setae 6-IV-VI plumose 2
   Head. Setae 5-7-C large, multibranched (Fig. 6.9B). Abdomen. Setae 6-IV-VI not plumose 3
2(1) Head. Setae 2-C widely separated, closer to setae 3-C than to each other (Fig. 6.9A). Abdomen. Setae 13-II-V and VII usually with three branches .... *An. barberi*
   Head. Setae 2-C close together, closer to each other than to setae 3-C. Abdomen. Setae 13-II-V and VII usually single .... *An. judithae*
3(1) Head. Seta 3-C unbranched 4
   Head. Seta 3-C with at least five branches 6
4(3) Abdomen. Setae 1-I-VII palmate, leaflets with smooth margin (Fig. 6.9C). Head. Setae 2-3-C aciculate (Fig. 6.9B) .... *An. albimanus*
   Abdomen. Setae 1-I-VII palmate, leaflets with serrate margin (Fig. 6.10A). Head. Setae 2-3-C smooth 5
5(4) Abdomen. Seta 2-IV single (Fig. 6.10A); spiracular apparatus with elongated process on caudal margin of posterolateral lobe .... *An. pseudopunctipennis*
   Abdomen. Seta 2-IV usually with two to three branches; spiracular apparatus without elongated process on caudal margin of posterolateral lobe .... *An. franciscanus*
6(3) Head. Seta 3-C with 10 or less branches (Fig. 6.11B) .... *An. atropos*
   Head. Seta 3-C with 11 or more branches (dendritic) (Fig. 6.11C) 7
7(6) Abdomen. Setae 0-IV,V with four or more branches, about same size as setae 2-IV,V (Fig. 6.10B) .... *An. crucians*
   Abdomen. Setae 0-IV,V with three or less branches, much smaller than setae 2-IV,V (Fig. 6.10C) 8
8(7) Head. Seta 2-C single, aciculate toward apex. Thorax. Seta 1-P well-developed, with three to five branches from near base (Fig. 6.11A) .... *An. walkeri*
   Head. Seta 2-C single or forked, not aciculate toward apex. Thorax. Seta 1-P weak, single or branched in distal 0.5 9

9(8) Abdomen. Setae 1-IV-VI fully palmate usually with fine marginal serrations on apical 0.5 of leaflets .... *An. bradleyi*

Abdomen. Setae 1-III-VII fully palmate with coarse marginal serrations on apical 0.5 of leaflets 10

10(9) Head. Alveoli of setae 2-C separated by more than diameter of one alveolus (Fig. 6.11C); setae 8-9-C large, usually with eight or more branches 11

Head. Alveoli of setae 2-C separated by less than diameter of one alveolus (Fig. 6.11B); setae 8-9-C small, usually with seven or less branches 12

11(10) Head. Both setae 3-C usually with sum of 63 or less branches (Fig. 6.11C). Abdomen. Segment VIII usually with anteromedian sternal plate (Fig. 6.12A) .... *An. quadrimaculatus*

Head. Both setae 3-C usually with sum of 64 or more branches. Abdomen. Segment VIII usually without antero-median sternal plate .... *An. smaragdinus*

12(10) Abdomen. Segments IV-VI with three accessory tergal plates (Fig. 6.12B). Head. Seta 1-A attached at or beyond basal 0.33 of antenna .... *An. freeborni*

Abdomen. Segments IV-VI with 1 accessory tergal plate (Fig. 6.12C). Head. Seta 1-A attached within basal 0.33 of antenna .... *An. punctipennis*

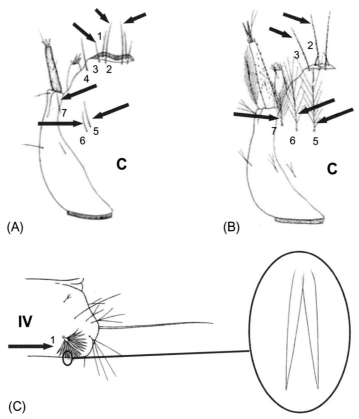

FIG. 6.9 (A) Head (C) showing *(arrows)* small, single setae 5-7-C; widely separated setae 2-C, and closer to 3-C—*Anopheles barberi*. (B) Head (C) showing *(arrows)* large, multibranched seta 5-7-C—*Anopheles albimanus*. (C) Abdominal segment IV showing *(arrow)* palmate seta 1-IV and leaflets (close-up) with smooth margins—*Anopheles albimanus*.

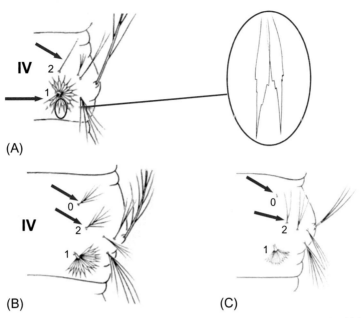

(A)

IV

(B)   (C)

FIG. 6.10 (A) Abdominal segment IV showing *(arrows)* single seta 2-IV and palmate seta 1-IV (close-up) with leaflets having serrate margins—
Anopheles pseudopunctipennis. (B) Abdomen segment IV showing *(arrows)* seta 0-IV with branches, about same size as seta 2-IV—*Anopheles crucians.*
(C) Abdomen segment IV showing *(arrows)* seta 0-IV with branches, smaller than seta 2-IV—*Anopheles punctipennis.*

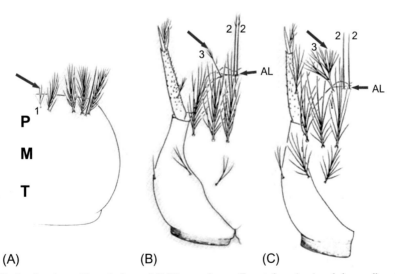

(A)   (B)   (C)

FIG. 6.11 (A) Prothorax (P) showing *(arrow)* branched seta 1-P (*M*, mesothorax; *T*, metathorax)—*Anopheles walker.* (B) Head showing *(arrow)*
branched seta 3-C—*Anopheles atropos.* (C) Head showing *(arrow)* branched or dendritic seta 3-C; alveoli (AL) of setae 2-C separated by more than
diameter of one alveolus *(arrow)*—*Anopheles quadrimaculatus.*

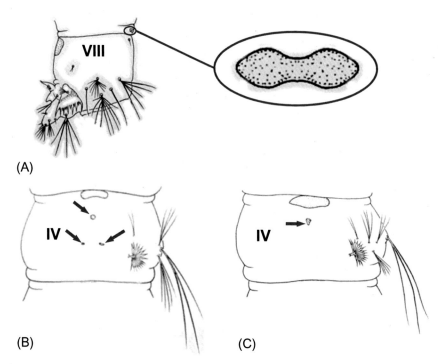

FIG. 6.12 (A) Abdominal segment VIII showing *(arrow)* anteromedian sternal plate—*Anopheles quadrimaculatus.* (B) Abdominal segment IV showing *(arrows)* three accessory tergal plates—*Anopheles freeborni.* (C) Abdominal segment IV showing *(arrow)* 1 accessory tergal plate—*Anopheles punctipennis.*

6.3 Key to species of fourth instar larvae of genus *Culex* (= *Cx.*)

   1   Head. Seta 6-C with at least three branches (Fig. 6.13A). Thorax. Prothoracic seta 3-P at least two-thirds length of seta 1-P (Fig. 6.13C) (subgenus *Culex*) 2
           Head. Seta 6-C single or double (Fig. 6.13B). Thorax. Prothoracic seta 3-P one-third or less length of seta 1-P 13
2(1)   Abdomen. Pecten spines reaching distal 0.75 of siphon, with four or five large apical spines (Fig. 6.13D) .... *Cx. interrogator*
           Abdomen. Pecten spines in basal 0.33 of siphon, without large apical spines (Fig. 6.13E) 3
3(2)   Abdomen. Siphon with mostly single, long setae, irregularly placed 4
           Abdomen. Siphon with mostly branched setae, linearly placed, with one or two pairs dorsally out of line 5
4(3)   Head. Antennal seta 1-A inserted at or before midpoint of antenna (Fig. 6.14A); distal part of antenna not slender (Fig. 6.14A) .... *Cx. restuans*
           Head. Antennal seta 1-A inserted beyond midpoint on apical third antenna (Fig. 6.14B); distal part of antenna more slender (Fig. 6.14B) .... *Cx. thriambus*
5(3)   Abdomen. Siphon with few to many short spines near apex (Fig. 6.14C) .... *Cx. coronator*
           Abdomen. Siphon without short spines near apex (Fig. 6.14D) 6
6(5)   Abdomen. Siphon with seta 1-S tufts inserted ventrally in a straight row (Fig. 6.14D) 7
           Abdomen. Siphon with seta 1-S tufts inserted ventrally and laterally not in a straight row 8
7(6)   Abdomen. Siphon with seta 1-S usually with five pairs of tufts (Fig. 6.14D); siphon length 4.5–5.5 times basal width (Fig. 6.14D) .... *Cx. tarsalis*
           Abdomen. Siphon with seta 1-S usually with six to nine pairs of tufts; siphon length at least eight times basal width .... *Cx. chidesteri*
8(6)   Abdomen. Siphon with seta 1-S with three pairs of tufts .... *Cx. declarator*
           Abdomen. Siphon with seta 1-S with four or five pairs of tufts 9
9(8)   Abdomen. Siphon length 4.0–5.0 times basal width 10
           Abdomen. Siphon length 6.0–8.0 times basal width 11
10(9)   Abdomen. Saddle with dorsoposterior aciculae larger than those at dorsomedian (Fig. 6.15A); setae 6-III, IV with three branches (Fig. 6.15B) .... *Cx. stigmatosoma*

Abdomen. Saddle with dorsoposterior aciculae not larger than those at dorsomedian; setae 6-III, IV single or double .... *Cx. quinquefasciatus*

11(9) Abdomen. Seta 1-X single (Fig. 6.15C). Thorax. Integument with fine aciculae .... *Cx. nigripalpus*

Abdomen. Seta 1-X usually double (Fig. 6.15D). Thorax. Integument smooth without distinct aciculae 12

12(11) Abdomen. Siphon with seta 1-A usually with five pairs of tufts, usually two pairs out of line dorsally ....

.... *Cx. erythrothorax*

Abdomen. Siphon with seta 1-A usually with four pairs of tufts, usually one pair out of line dorsally ....

.... *Cx. salinarius*

13(1) Abdomen. Siphon without subdorsal seta 1-S tuft, most apical 1-S tuft lateral, and with straight dorsoapical seta 2-S; pecten spines with 1–6 lateral teeth; segment X with saddle having ventral precratal tufts of seta 4-X (subgenus *Neoculex*) 14

Abdomen. Siphon with lateral or subdorsal seta 1-S tufts and with curved dorsoapical seta 2-S; pecten spines with at least 10 lateral teeth; segment X with saddle without ventral precratal tufts of seta 4-X (subgenus *Melanoconion*) 16

14(13) Head. Seta 5-C and seta 6-C about equal in length, double (Fig. 6.16A). Abdomen. Seta 1a-S narrowly separated from seta 1b-S (Fig. 6.16B) .... *Cx. arizonensis*

Head. Seta 6-C longer than seta 5-C, usually not double (Fig. 6.16C). Abdomen. Seta 1a-S widely separated from seta 1b-S (Fig. 6.16D) 15

15(14) Abdomen. Siphon six or more times longer than basal seta (seta 1a-S) .... *Cx. apicalis*

Abdomen. Siphon less than six times longer than basal seta (seta 1a-S) .... *Cx. territans*

16(13) Abdomen. Comb scale with long median spine and short lateral spines (Fig. 6.17A and B) 17

Abdomen. Comb scale rounded apically with subequal median and lateral spines (Fig. 6.17C and D) 18

17(16) Abdomen. Siphon length 4.5 or less width at base, with dorsal margin straight and ventral margin curved upward; seta 1-S with eight long ventral tufts. Head. Seta 5-C single or double .... *Cx. pilosus*

Abdomen. Siphon length at least six times width at base, with dorsal and ventral margins slightly curved upward; seta 1-S with five ventral tufts and two subdorsal tufts. Head. Seta 5-C with at least four branches .... *Cx. erraticus*

18(16) Abdomen. Siphon with small light to dark pigmented band almost at near midpoint; seta 2-S hook-like with a tooth (Fig. 6.17E); each comb scale length about 3.5 times widest width .... *Cx. peccator*

Abdomen. Siphon without dark median band; each comb scale length about 2.5 times widest width ....

.... *Cx. abominator*

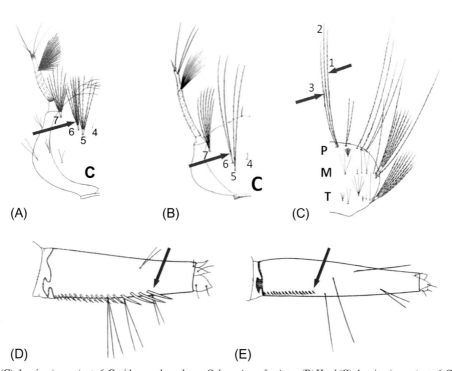

FIG. 6.13 (A) Head (C) showing *(arrow)* seta 6-C with many branches—*Culex quinquefasciatus*. (B) Head (C) showing *(arrow)* seta 6-C with two branches—*Culex arizonensis*. (C) Thorax showing *(arrows)* prothoracic setae 1-P and 2-P—*Culex erythrothorax*. (D) Siphon showing *(arrow)* pecten spines reaching distal 0.75 of siphon and four large apical spines—*Culex interrogator*. (E) Siphon showing *(arrow)* pecten spines in basal 0.33 of siphon—*Culex restuans*.

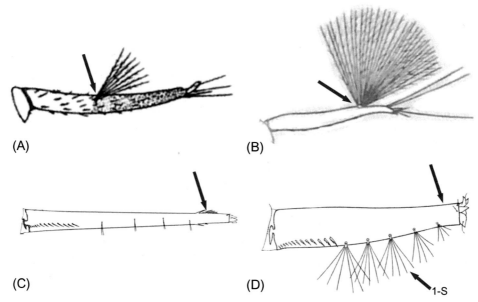

FIG. 6.14 (A) Antenna showing *(arrow)* seta 1-A—*Culex restuans.* (B) Antenna showing *(arrow)* seta 1-A—*Culex thriambus.* (C) Siphon showing *(arrow)* short spines near apex—*Culex coronator.* (D) Siphon showing *(arrows)* seta 1-S tufts without short spines near apex—*Culex tarsalis.*

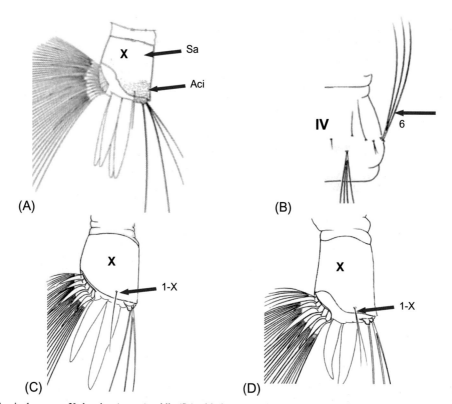

FIG. 6.15 (A) Abdominal segment X showing *(arrow)* saddle (Sa) with dorsoposterior aciculae (Aci)—*Culex stigmatosoma.* (B) Abdominal segment IV showing *(arrow)* seta 6-IV with three branches—*Culex stigmatosoma.* (C) Abdominal seta 1-X single *(arrow)*—*Culex nigripalpus.* (D) Abdominal seta 1-X double *(arrow)*—*Culex salinarius.*

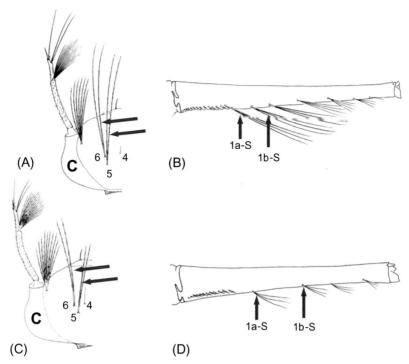

FIG. 6.16 (A) Head (C) showing *(arrows)* seta 5-C and seta 6-C—*Culex arizonensis.* (B) Siphon showing *(arrows)* seta 1a-S and seta 1b-S—*Culex arizonensis.* (C) Head (C) showing *(arrows)* seta 5-C and seta 6-C—*Culex territans.* (D) Siphon showing *(arrows)* seta 1a-S and seta 1b-S—*Culex territans.*

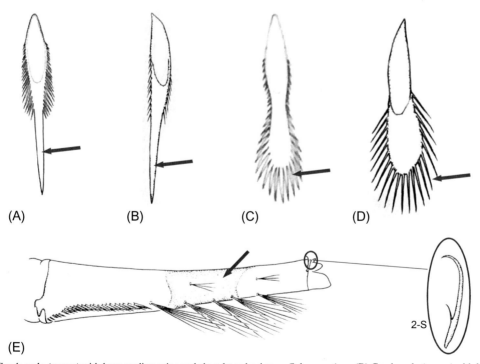

FIG. 6.17 (A) Comb scale *(arrow)* with long median spine and short lateral spines—*Culex erraticus.* (B) Comb scale *(arrow)* with long median spine and short lateral spines—*Culex pilosus.* (C) Comb scale *(arrow)* with subequal median and lateral spines—*Culex peccator.* (D) Comb scale *(arrow)* with long median and lateral spines—*Culex abominator.* (E) Siphon showing *(arrows)* dark pigmented band and close-up of 2-S—*Culex peccator.*

6.4 Key to species of fourth instar larvae of genus *Culiseta* (= Cs.)

1 Abdomen. Siphon with midventral row of about 8–14 setae (subgenus *Climacura*) (Fig. 6.18A) ....*Cs. melanura*
   Abdomen. Siphon without midventral row of setae (subgenus *Culiseta*) (Fig. 6.18B) 2
2(1) Abdomen. Seta 1-X with strong branches as long as or longer than saddle (Fig. 6.18C) .... *Cs. inornata*
   Abdomen. Seta 1-X with fine branches shorter than saddle (Fig. 6.18D) .... *Cs. incidens*

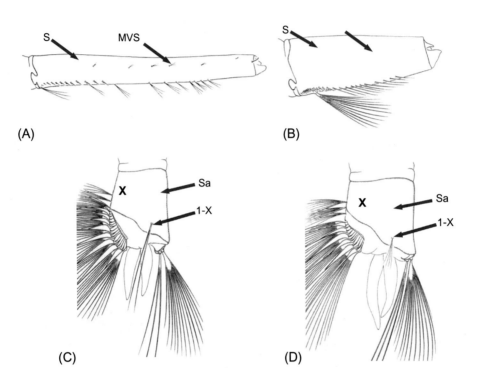

FIG. 6.18 (A) Siphon (S) with midventral row of setae (MVS, *arrow*)—*Culiseta melanura*. (B) Siphon (S) without midventral row of setae *(arrow)*—*Culiseta inornata*. (C) Abdominal segment X showing *(arrow)* seta 1-X longer than saddle (Sa)—*Culiseta inornata*. (D) Abdominal segment X showing *(arrow)* seta 1-X shorter than saddle (Sa)—*Culiseta incidens*.

6.5 Key to species of fourth instar larvae of genus *Deinocerites* (= De.)

1 Abdomen. Segment II with seta 6-II double (Fig. 6.19A); siphon with seta 1-S with two to three branches (Fig. 6.19C)
   .... *De. pseudes*
   Abdomen. Segment II with seta 6-II single (Fig. 6.19B); siphon with seta 1-S with four to six branches (Fig. 6.19D)
   .... *De. mathesoni*

6.6 Key to species of fourth instar larvae of genus *Orthopodomyia* (= Or.)

1 Abdomen. Siphon with seta 1-S usually with three or four branches (Fig. 6.20A); segment VIII without large tergal
   plate (Fig. 6.20A) .... *Or. alba*
   Abdomen. Siphon with seta 1-S usually with at least six branches (Fig. 6.20B); segment VIII with large tergal plate
   (Fig. 6.20B) 2

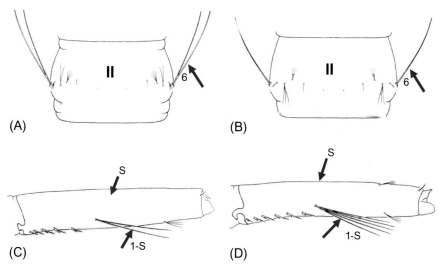

FIG. 6.19 (A) Abdominal segment II showing *(arrow)* double seta 6-II—*Deinocerites pseudes*. (B) Abdominal segment II showing *(arrow)* single seta 6-II—*Deinocerites mathesoni*. (C) Siphon (S) showing *(arrow)* with seta 1-S with two to three branches—*Deinocerites pseudes*. (D) Siphon (S) showing *(arrow)* with seta 1-S with five branches—*Deinocerites mathesoni*.

2   Abdomen. Siphon with seta 1-S with branches longer than the distance from its socket (or alveolus) to apex of siphon (Fig. 6.20C); segment X with saddle shorter than the dorsal pair of anal papillae (Fig. 6.20D) .... *Or. kummi*
Abdomen. Siphon with seta 1-S with branches shorter than the distance from its socket (or alveolus) to apex of siphon (Fig. 6.20B); segment X with saddle longer than the dorsal pair of anal papillae (Fig. 6.20E) .... *Or. signifera*

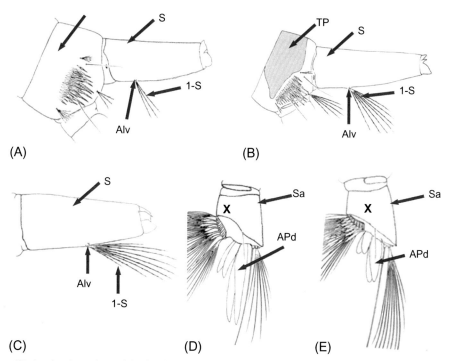

FIG. 6.20 (A) Siphon (S) showing *(arrow)* seta 1-S with three or four branches; abdominal segment VIII without large tergal plate (TP, *arrow*; Alv, alveolus)—*Orthopodomyia alba*. (B) Siphon (S) showing *(arrow)* with seta 1-S with seven branches; segment VIII with large tergal plate (TP, *arrow*; Alv, alveolus)—*Orthopodomyia signifera*. (C) Siphon (S) showing *(arrows)* seta 1-S with long branches, socket, or alveolus (Alv, *arrow*)—*Orthopodomyia kummi*. (D) Abdominal segment X showing *(arrows)* saddle (Sa) and long dorsal anal papillae (APd)—*Orthopodomyia kummi*. (E) Abdominal segment X showing *(arrows)* saddle (Sa) and short dorsal anal papillae (APd)—*Orthopodomyia signifera*.

6.7 Key to species of fourth instar larvae of genus *Psorophora* (= *Ps.*)

1   Larva very large, length more than 0.5 in. Head. Front of head truncate (Fig. 6.21A); antenna short, not reaching or barely reaching front of head (Fig. 6.21A). Abdomen. Siphon with 15 or more long pecten spines (Fig. 6.21C) (sub-genus *Psorophora*) 2

      Larva small, length 0.5 in. or less. Head. Front of head rounded (Fig. 6.21B); antenna long, reaching well beyond front of head (Fig. 6.21B). Abdomen. Siphon with eight or less small pecten spines (Fig. 6.21D) 3

2(1)   Abdomen. Seta 1-X on short saddle with three to four branches originating from base (Fig. 6.22A) ....
      .... *Ps. ciliata*

      Abdomen. Seta 1-X on saddle single or forked well beyond mid-length (Fig. 6.22B) .... *Ps. howardii*

3(1)   Head. Antenna shorter than head (Fig. 6.23C); if not, then seta 1-S with some branches at least as long as siphon. Abdomen. Siphon with seta 6-S shorter than apical diameter of siphon (subgenus *Grabhamia*) 4

      Head. Antenna equal or longer than head (Fig. 6.21B); if not, then seta 6-S subequal to apical diameter of siphon Abdomen. Siphon with seta 1-S shorter than length of siphon (subgenus *Janthinosoma*) 6

4(3)   Head. Antenna longer than head, appearing swollen or flatten. Abdomen. Siphon with seta 1-S stout, with many branches at least as long as siphon (Fig. 6.23A) .... *Ps. discolor*

      Head. Antenna shorter than head, not swollen or flatten. Abdomen. Siphon with seta 1-S small, with branches shorter than siphon (Fig. 6.23B) 5

5(4)   Head. Antenna longer than seta 5-C or seta 6-C, each seta with at least four branches (Fig. 6.23C) .... *Ps. columbiae*

      Head. Antenna shorter or as long as seta 5-C or seta 6-C, each seta single to three branches (Fig. 6.23D) ....
      .... *Ps. signipennis*

6(3)   Head. Antenna shorter than head (Fig. 6.24A). Abdomen. Siphon with seta 6-S about as long as apical diameter of siphon (Fig. 6.24B) .... *Ps. cyanescens*

      Head. Antenna as long as, or longer than head. Abdomen. Siphon with seta 6-S much shorter than apical diameter of siphon 7

7(6)   Head. Antenna shorter than or as long as median length of head 8

      Head. Antenna longer than median length of head 9

8(7)   Head. Setae 5,6-C shorter than seta 7-C (Fig. 6.24C). Abdomen. Setae 6-IV-VI with two to three branches, shorter than length of abdominal segment of origin (Fig. 6.24D); siphon with seta 1-S shorter than 2-S .... *Ps. horrida*

      Head. Setae 5,6-C longer than seta 7-C. Abdomen. Setae 6-IV-VI longer than following tergum, with variable branches (Fig. 6.24E); siphon with seta 1-S longer than 2-S .... *Ps. mathesoni*

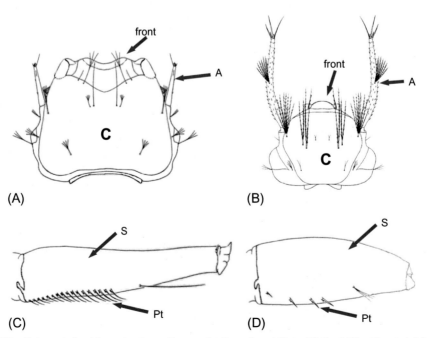

FIG. 6.21 (A) Head (C) with truncate front *(arrow)*; antenna (A, *arrow*)—*Psorophora ciliata*. (B) Head (C) with rounded front *(arrow)*; antenna (A, *arrow*)—*Psorophora ferox*. (C) Siphon (S) showing long pecten spines (Pt, *arrow*)—*Psorophora howardii*. (D) Siphon (S) showing short pecten spines (Pt, *arrow*)—*Psorophora columbiae*.

9(7) Head. Setae 5,6-C with branches nearly equal in length (Fig. 6.25A). Abdomen. Setae 6-IV-VI single or double (Fig. 6.25C) .... *Ps. ferox*

Head. Setae 5,6-C with branches not equal in length, at least one shorter or weaker (Fig. 6.25B). Abdomen. Setae 6-IV-VI with three or more branches (Fig. 6.25D) .... *Ps. longipalpus*

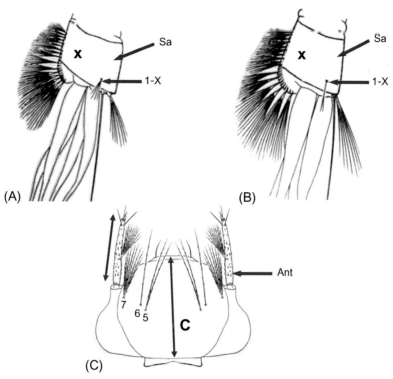

FIG. 6.22   (A) Abdominal segment X showing *(arrows)* short saddle (Sa) with branched seta 1-X—*Psorophora ciliata*. (B) Abdominal segment X showing *(arrows)* longer saddle (Sa) with single seta 1-X—*Psorophora howardii*. (C) Antenna (Ant, *arrow*), head (C)—*Psorophora signipennis*.

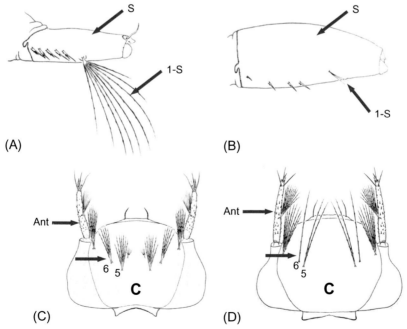

FIG. 6.23   (A) Siphon (S) showing *(arrow)* stout, long seta 1-S—*Psorophora discolor*. (B) Siphon (S) showing *(arrow)* small seta 1-S—*Psorophora columbiae*. (C) Head (C) showing antenna (Ant) longer than seta 5-C or seta 6-C—*Psorophora columbiae*. (D) Head (C) showing antenna (Ant) shorter or as long as seta 5-C or seta 6-C—*Psorophora signipennis*.

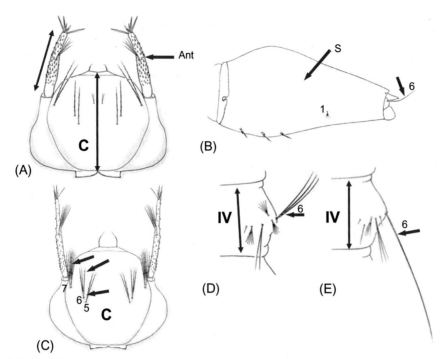

FIG. 6.24 (A) Head (C) showing *(arrow)* short antenna (Ant)—*Psorophora cyanescens*. (B) Siphon (S) showing *(arrows)* seta 1-S and seta 6-S—*Psorophora cyanescens*. (C) Head (C) showing *(arrows)* setae 5,6,7-C—*Psorophora horrida*. (D) Abdominal segment IV showing *(arrow)* seta 6-IV—*Psorophora horrida*. (E) Abdominal segment IV showing *(arrow)* seta 6-IV—*Psorophora mathesoni*.

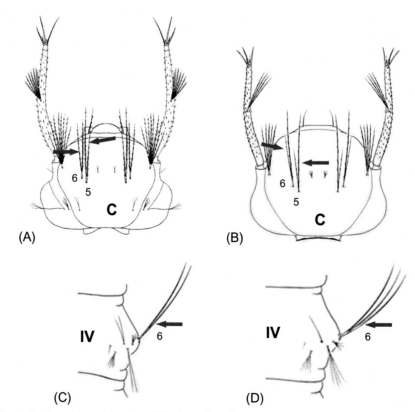

FIG. 6.25 (A) Head (C) showing *(arrows)* seta 5-C and seta 6-C—*Psorophora ferox*. (B) Head (C) showing *(arrows)* seta 5-C and seta 6-C—*Psorophora longipalpus*. (C) Abdominal setae 6-IV with two branches *(arrow)*—*Psorophora ferox*. (D) Abdomen. Setae 6-IV-VI with three branches *(arrow)*—*Psorophora longipalpus*.

6.8 Key to species of fourth instar larvae of genus *Uranotaenia* (= *Ur.*)

1 Head. Seta 5-C double or triple; seta 6-C single, thin, not spiniform (subgenus *Pseudoficalbia*) (Fig. 6.26A) ....
.... *Ur. anhydor syntheta*
Head. Setae 5,6-C single, stout, spiniform (subgenus *Uranotaenia*) (Fig. 6.26B) 2
2(1) Abdomen. Setae 6-I,II double (Fig. 6.26C). Thorax. Seta 3-P more than 0.5 length of 1-P (Fig. 6.27A) ....
.... *Ur. lowii*
Abdomen. Setae 6-I,II triple (Fig. 6.26D). Thorax. Seta 3-P much less than 0.5 length of 1-P (Fig. 6.27B) ....
.... *Ur. sapphirina*

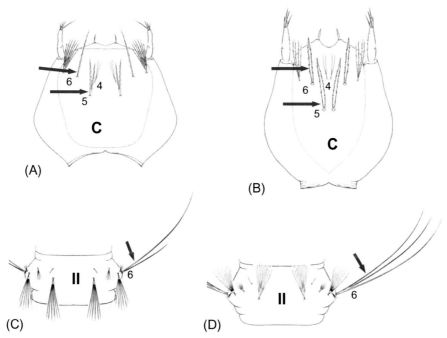

FIG. 6.26 (A) Head (C) showing *(arrows)* double seta 5-C and single seta 6-C—*Uranotaenia anhydor syntheta*. (B) Head (C) showing *(arrows)* single, spiniform seta 5-C and seta 6-C—*Uranotaenia sapphirina*. (C) Abdominal segment II showing *(arrow)* double seta 6-II—*Uranotaenia lowii*. (D) Abdominal segment II showing *(arrow)* triple seta 6-II—*Uranotaenia sapphirina*.

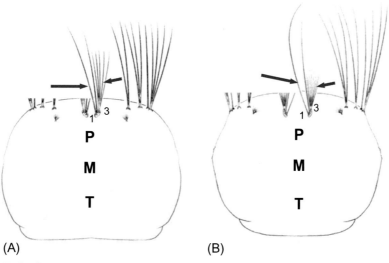

FIG. 6.27 (A) Prothorax (P) showing *(arrows)* seta 1-P and seta 3-P (*M*, mesothorax; *T*, metathorax)—*Uranotaenia lowii*. (B) Prothorax (P) showing *(arrows)* seta 1-P and seta 3-P (*M*, mesothorax; *T*, metathorax)—*Uranotaenia sapphirina*.

Acknowledgments

Great appreciations go to J. Stopper for taking photographs when she was under contract at Walter Reed Biosystematics Unit, G. White for assisting with sorting and retrieving specimens for initial photographs, and B.P. Rueda for the support and enormous help in finalizing the illustrations and photographs used in the keys. Thanks also to B.P. Rueda and C.R. Summers for their valuable comments and suggestions on this chapter.

References and further reading

Belkin, J.N., 1962. The Mosquitoes of the South Pacific (Diptera, Culicidae). vols. I & II. University of California Press, Berkely, CA. 608 pp.

Carpenter, S.J., LaCasse, W.J., 1955. Mosquitoes of North America (North of Mexico). University of California Press, Berkeley and Los Angeles, CA. 360 pp. +127 pl.

Darsie Jr., R.F., Ward, R.A., 2005. Identification and Geographical Distribution of the Mosquitoes of North America, North of Mexico. University Press of Florida, Gainesville, FL. 383 pp.

Harbach, R.E., Knight, K.L., 1980. Taxonomists' Glossary of Mosquito Anatomy. Plexus Publishing, Marlton, NJ. 415 pp.

Harbach, R.E., Knight, K.L., 1982. Corrections and additions to taxonomists' glossary of mosquito anatomy. Mosq. Syst. 13, 201–217.

Harrison, B.A., Byrd, B.A., Sither, C.B., Whitt, P.B., 2016. The Mosquitoes of the Mid-Atlantic Region: An Identification Guide. Western Carolina University, Cullowhee, NC. 201 pp.

Rueda, L.M., 2004. Pictorial keys for the identification of mosquitoes (Diptera: Culicidae) associated with dengue virus transmission. Zootaxa 589, 1–60.

Rueda, L.M., Stockwell, S.A., Pecor, J.E., Gaffigan, T., 1998. Key to the Mosquito Genera of the World (INTKEY Module). Walter Reed Biosystematics Unit, Smithsonian Institution, Washington, DC. [CD]. Also, in The Diptera Dissemination Disk – vol. 1, North American Dipterists Society, Washington, DC.

Walter Reed Biosystematics Unit (WRBU), 2018. Mosquito Identification Keys. Walter Reed Biosystematics Unit, Smithsonian Institution, Suitland, MD. Available from: http://www.wrbu.org/aors/northcom_Keys.html. (Accessed December 15, 2018).

Part II

Communities

Chapter 7

Mosquito Surveillance

Nina M. Dacko[a], Martin Reyna Nava[b], Christopher Vitek[c], Mustapha Debboun[b]

[a]*Tarrant County Public Health Environmental Health Division, Fort Worth, TX, United States,* [b]*Mosquito and Vector Control Division, Harris County Public Health, Houston, TX, United States,* [c]*The University of Texas Rio Grande Valley, Center for Vector-Borne Diseases, Edinburg, TX, United States*

Chapter Outline

7.1 Introduction: The needs and goals of mosquito surveillance

Mosquitoes transmit mosquito-borne diseases to humans and animals. They are also a nuisance, causing people discomfort and keeping them from enjoying outdoor activities. For these reasons, it is critical to understand the population dynamics of mosquitoes present in a specific geographic area, especially those of medical, veterinary, and public health importance. The ideal method by which this information could be obtained is through mosquito surveillance. Field mosquito collection should be a critical component of any strong vector control or surveillance program. Mosquito surveillance is defined as a systematic, rigorous, and continued effort to monitor mosquito populations over time to obtain information about distribution, abundance, and species composition. Surveillance data may also be used to assess the risk of mosquito-borne disease outbreaks and the need or efficacy of intervention efforts. It does not require a specified number of field sites, trapping events, or mosquitoes collected, but should be conducted on a regular schedule, at the same locations, with the same trap types. Increased data collection will result in a greater ability to estimate pest and vector species of mosquitoes before they become a problem. While the primary goal of vector control programs may be to mitigate and control both nuisance and pathogen-transmitting mosquitoes, field collections also enable analyzing the efficacy of mosquito control efforts and assessing risk of vector-borne disease transmission.

Assessing periods and locations that exhibit increased risk of disease transmission is an ancillary benefit of sustained field mosquito collections. For the purpose of field surveillance, increased risk would be associated with an observed increase in relative mosquito population size. Routine, weekly field collections over multiple years will help identify times of the year when increased mosquito activity is known. In addition, if field sites are located in multiple areas, it is possible to identify regions that show variation in mosquito activity. In addition to assessing overall risk, this information will allow predictions about the times and locations where mosquito activity may increase to help inform decisions regarding both the time and place for intervention efforts, potentially ensuring an increased efficacy of intervention efforts.

In addition to overall abundance monitoring, surveying for specific mosquito species is critical for assessing risk, as not all mosquitoes are capable of transmitting diseases. An observed increase in *Psorophora columbiae* (Dyar and Knab), for example, may result in increased biting annoyance (presumably followed by an increase in nuisance calls), but will not suggest an increased risk of West Nile virus (WNV) transmission. Field collections will enable mosquito control programs to determine which mosquitoes are most common at different times of the year and different locations.

Mosquitoes, Communities, and Public Health in Texas. https://doi.org/10.1016/B978-0-12-814545-6.00007-9

This will aid public health departments when assessing risk of disease transmission or potentially identifying human vector-borne cases. It will also enable the identification of novel or invasive species that are new in that location, which may impact the introduction of new mosquito-borne diseases or transmission of existing ones.

Lastly, working with health departments, entomologists, and epidemiologists may help identify a threshold density for certain mosquitoes as they relate to disease risk. This information may help control efforts, as interventions may be mandated once a certain mosquito species reaches a threshold abundance. The calculation of a threshold density (a certain number of mosquitoes per trap or pupae per container) would vary based on the mosquito species in question, the trap type used, and the location where trapping occurs. For example, Ricthie et al. (2004) determined that a density of less than one *Aedes aegypti* (Linnaeus) female per sticky gravid trap in Australia was considered "safe," while in Thailand, egg densities of more than two *Ae. aegypti* eggs per oviposition trap per week were correlated with dengue hemorrhagic fever (Mogi et al., 1990).

7.2 Building a surveillance program

The essential first step in developing a robust and effective surveillance program is conducting a preliminary field survey, which is the planning phase of a mosquito control program. Surveys are a component of a successful vector-monitoring program as they provide insight and allow you to plan, operate, and evaluate any mosquito surveillance and control effort. The initial survey will determine the mosquito species causing health hazard and assess the regions of jurisdiction that would be best suited for surveillance activities. In addition to surveying the population regarding mosquito activities and assessing land use or habitat characteristics that may promote mosquito breeding, field surveys may also consist of preliminary mosquito collection to identify common mosquitoes in the area. The goal of these initial surveys is to obtain preliminary data on mosquitoes, land use, breeding habitats, or other factors that may be used for identifying future surveillance sites and improving control efforts.

Helpful information collected during the planning phase of a mosquito surveillance program should include specifics on the potential area. These may include topography, land use, flood plain, and infrastructure including floodwater, storm water, septic drainage, and vegetation coverage in both rural and urban areas. In addition, knowledge of socioeconomic status, cultural customs, and ethnicity of the population may provide insight into areas of higher risk for mosquito-borne diseases. When scouting (searching for mosquito breeding habitats), one should work to identify potential factors that may influence mosquito species or abundance. While in the field, detailed notes should be taken on field sites, and potential breeding sources should be recorded. Conditions may include unkempt properties, abundant vegetation, identification of potential breeding locations, and areas prone to collecting water. It is worth noting that areas of higher socioeconomic status may also be at risk, where unmaintained yard features and malfunctioning irrigation systems are more prolific and may increase potential mosquito breeding habitats.

When conducting preliminary field surveys, it is important to understand that variations in seasonal climate produce variability in the spatial distribution and mosquito species. Data collected at one time may vary significantly from data collected at a different time of the year. Field surveys should be conducted throughout the entire year. Standard criteria and assessment methods should be followed for all potential survey sites to ensure reliability of data. If possible the same surveillance personnel should be involved to avoid collector bias. Permissions to be on private property should be obtained in advance. It is a good idea to supply a statement of intent and signed permission form that can be filed by an agency for its records. In addition to routine surveillance areas, those not under routine surveillance should be monitored periodically. Data collected may include breeding habitats, life cycles, seasonality, flight range, host preferences, resting places, public health importance, and susceptibility to insecticides. These data will enable the maintenance of an effective, sustainable control effort. Regular surveys may also identify locations that may require special attention such as areas prone to flooding.

Once locations for field surveillance are identified, it is critical to develop a long-term, continual surveillance plan. Ideally, surveillance should be sustained when mosquitoes remain active. In Texas and other southern states, surveillance should be conducted year-round. Warmer temperatures may contribute to continued activity of disease-vectoring mosquitoes. Continued surveillance from year to year is needed to monitor the variation in density and distribution of the local mosquito population. It is also important to note that surveillance results do not reflect the total mosquito population, but may provide a relative index of the population density.

Collecting sites should be mapped out using mapping software such as geographic information systems (GIS) to provide easy overall visual output. Computer programs such as Google Earth or Google Maps aid in understanding site topography. For larger mosquito control districts, the geographic area to be monitored may be divided into sections. Smaller sampling areas represent a more accurate representation of the local mosquito population. Usually, these subdivided areas are delimited by boundaries such as streets, major highways, or waterways with similar habitat features.

For ease of trap collections, trapping sites may be clustered and assigned into routes. These clusters should be set up to maximize efficiency of the surveillance program. Sites grouped in a route may comprise more than one type of trap as different traps result in different targeted collections. For example, a collector may have a light trap at one house and a gravid trap located a few houses away. Selection of trap sites should consider the target mosquito biology, both in terms of the traps utilized and the locations where the traps are placed. If traps are set on private properties, permission should be obtained in advance.

Data should be recorded on forms created for field use that are easy to understand. Collection methods and details should be determined in collaboration with any partnering agency or health department. Collections should be recorded into a database to preserve data records. Examples include Microsoft (MS) Access, MS Excel, and Structured Query Language (SQL). Web-based databases provide for more accessible record-keeping and may include additional analysis tools.

Cost is also something that must be taken into consideration when planning surveillance activities. Personnel, vehicles, traps, supplies, laboratory equipment, storage devices, and trap maintenance efforts must all be considered when estimating the budgetary needs of surveillance activities. Costs may be reduced by collaborating with external agencies. Storing and shipping samples may be taken into consideration.

A well-established systematic surveillance program comprising and following appropriate surveillance procedures will provide better decision-making for control measures either when a confirmed positive mosquito pool is detected or when the number of mosquitoes or the number of citizens' complaints increases.

7.3 Field collections: Egg and larval stages

Egg and larval collections are other useful methods for monitoring mosquito populations. If the intervention is larval control, the larval collections can be used to identify potential aquatic habitats where larval control will be most effective. Larval collection may be straightforward, depending on the types of mosquitoes being collected. However, larval collections may be difficult in certain circumstances when larval habitats are cryptic or inaccessible (Hanna et al., 1998; Montgomery and Ritchie, 2002). Larval collections for container-breeding mosquitoes such as *Ae. aegypti*, *Ae. albopictus* (Skuse), and occasionally *Culex* spp. consist of observing potential habitats in the appropriate areas, that is, around houses, cemeteries, tire shops, or other urban sites. Container monitoring can be done by pouring the water out of a container or using a turkey baster, ladle, plastic pipette, medicine dropper, or small dipper (Fig. 7.1) to extract the water. Larval counts can be conducted in the field, and larvae can be brought back to the laboratory to be identified or reared to the adult stage for complete identification. In addition, pupae can be monitored if there are known pupae thresholds for control and disease risk assessment.

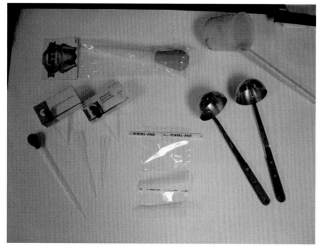

FIG. 7.1   Artificial container tools used to obtain immature mosquito stages: turkey baster, pipettes, medicine dropper, ladles, and small dipper.

In many cases, certain habitats will have containers that have a notably greater abundance of mosquito larvae than others. These containers may be referred to as essential or "key" containers (Lloyd, 2003; Ferdousi et al., 2015). Key containers may not be consistent across different habitats and ecoregions (Lloyd et al., 1992; Tun-Lin et al., 1995a,b; Morrison et al., 2004). Characteristics such as the size of container, location, method of water filling, surface area, and lid status may all influence the potential of different containers to serve as essential breeding habitats in different areas (Ferdousi et al., 2015). The only method to reliably identify such containers is through repeated systematic larval surveillance. Once these containers are identified, targeted control efforts may be utilized to reduce the production of adult mosquitoes. Removing "key" containers is not an effective control strategy as it will simply result in a new habitat or container becoming the new "key" container for larval development.

For noncontainer breeding species, larval collection primarily consists of surveying potential larval habitats through larval dipping. Larval dipping (Fig. 7.2) is a process where a cup is dipped into the water and, with a quick wrist motion, mosquito larvae (or other organisms) are scooped into the cup. The larval cup will often have a long handle (up to 2 m) for easy access to larval breeding habitats. Larvae collected can be counted and returned to the facility for identification. The selected habitat should depend on the mosquito species of interest, but any areas of standing water has the potential to support mosquito larvae, including floodplains, retention ponds, storm drains, wetlands, drainage ditches or canals, and naturally occurring ponds and streams. Once mosquito larvae are identified in a habitat, appropriate intervention efforts may be utilized.

FIG. 7.2   Regular larval dipper.

There are multiple methods or techniques to dip for larvae (Service, 1993; O'Malley, 1995). One is to quickly skim across the surface of the water with the dipping cup held at a slight angle that is particularly useful in monitoring *Anopheles* spp. The scoop method, if conducted fast enough, can collect mosquito larvae that are resting near the surface of the water. When disturbed, however, many larvae will dive below the surface; thus submerging the cup minimizes losing larvae. With this method, larvae are drawn toward the cup as it is submerged and captured as the cup is brought out of the water. It is critical to do this carefully, so larvae are not lost through spillover from excess water as you bring the cup up to the surface. Scraping may be used when there are a significant number of plants in the water, floating vegetation, or debris near the bottom of the water. Larvae may use these areas as refuges, so moving the cup through these (either along the surface or along the bottom) may collect larvae that would normally not be seen. When collecting, it is best to repeat the collection process multiple times to obtain a range of values for the number collected, then averaging them to develop an average larval index. When using multiple collections to obtain an average index, the same collection methods and number of times should be used each time to assure consistency across the estimates.

Egg collection is another method to assess mosquito presence. Gravid traps (Fig. 7.3A–E) may also allow for eggs collection. If a *Culex* mosquito approaches a gravid trap, it may land on the water's surface prior to being sucked into the collection tube. If so, it may have the opportunity to oviposit prior to being collected. In addition, these eggs may hatch prior to monitoring the trap. For this reason, water in the gravid traps should be monitored each week at a minimum, with egg rafts being removed if observed. If gravid traps are left for extended periods of time, the water should be emptied periodically to avoid rearing new mosquitoes.

FIG. 7.3   (A and B) Several variations of the CDC gravid trap, (C) Reiter-Cummings gravid trap, (D) Frommer updraft gravid trap, and (E) placement of Harris County gravid trap (HCGT).

Oviposition traps (Fig. 7.4A–C) for container-breeding mosquitoes are a particularly easy way to monitor their presence. Collection cups are primarily used for *Aedes* mosquitoes that lay their eggs on the sides of water-holding containers (Fay and Eliason, 1966; Reiter et al., 1991; Mackey et al., 2003). They are inexpensive, easy to deploy, and ideal for monitoring field mosquito populations.

Oviposition traps include an artificial substrate upon which a mosquito may lay its eggs. These traps mimic oviposition sites when placed in appropriate locations. They are filled approximately one-half to two-thirds with water. The artificial substrate can be a wide variety of substances: germination paper, cardboard, tongue depressors, or cutup wooden tomato stakes. The trap allows the mosquito to enter, lay her eggs on the substrate, and then fly away.

The container may be any water-holding vessel; often large stadium cups (473.17 mL) or similar cups are used. Containers should be of a dark color and placed in common mosquito oviposition habitats. The dark color attracts mosquitoes and potentially avoids the attention of curious citizens who may be walking by (as dark colors may be more difficult to see and may blend into the background or shadows where the cup is placed). The side of the container (approximately two-thirds from the bottom) should have a hole drilled into it, which will prevent the water rising above a certain level if it rains while the trap is in the field. If this hole is not present and the water level rises to submerge the eggs, they may hatch, impacting the estimates of the eggs collected that may be derived from field collections and potentially producing more mosquitoes that may impact local residents.

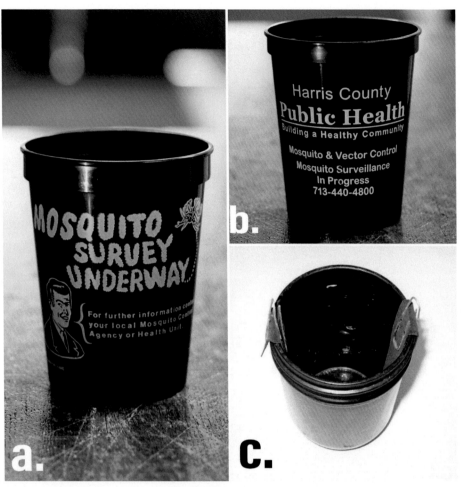

FIG. 7.4 (A and B) Oviposition cups (ovicups) or (C) little black jars (LBJ) with *red* velour strip attached in place with paper clips.

The oviposition traps can be set in the field, with the oviposition substrate removed once per week and replaced with new substrate. The water should be emptied and replaced to prevent any hatched larvae to develop, and the sides of the cup should be wiped down to make sure there aren't any eggs that may hatch when submerged. Alternatively the water in the cup can be collected to bring back with the eggs, and the larvae may be counted or identified to species.

Eggs collected from these traps are often used as a presence/absence indicator for the mosquito species. Eggs collected can be brought to the laboratory, counted, and hatched to identify the mosquito species that are ovipositing in the area. There are conflicting data regarding the use of oviposition cups to assess the relative abundance of mosquitoes. Previous data have suggested using the number of eggs per cup estimate as a threshold indicator for disease risk (Mogi et al., 1990; Wu et al., 2013), but there are limited data that suggest that egg counts correlate with actual adult populations (Mogi et al., 1990).

Oviposition data may vary significantly based on control efforts. For example, if a concerted effort is being made to empty containers and eliminate potential breeding habitats, then egg count may drastically increase as other oviposition sites become scarce (Focks, 2004). This should not be interpreted as an increased in population size, and other changing variables such as the increased or decreased presence of competing oviposition sites should be taken into consideration when analyzing the data.

7.4 Field collections: Host-seeking adults

Adult mosquito collections are often the most common field-collected ones, as they may be the easiest to implement. Different collection methods influence the species and abundance of mosquitoes collected; thus prior to identifying a potential method of collection, it is critical to determine the goals of the collections. For example, in mosquito-borne disease surveillance, often host-seeking collections are utilized to identify mosquitoes that may be actively feeding. If WNV is a primary concern, collections that will include *Culex* mosquitoes are essential. But if targeting dengue virus, *Ae. aegypti* and *Ae. albopictus* would be the primary target mosquitoes. Trap choice should be influenced by the goals of the trapping because different traps have biases in what they collect (Williams and Gingrich, 2007).

7.5 Landing rate counts

Landing rate counts have been used to measure the efficacy of a pesticide treatment for adult mosquitoes. This mostly took place before many adult mosquito surveillance traps were invented. It had been recorded as early as 1917 by Dr. T. J. Headlee in New Jersey when surveying for *Ae. sollicitans* (Walker) (Headlee, 1921; Leslie, 1985; Schmidt, 1989). Currently, it is considered a "quick and dirty" method because it's quick and easy. However, the ethics of this method are questionable when people are used as hosts, but it is still used by many mosquito control districts and agencies throughout the world. The concept is simple; two individuals enter the pretreatment area, one as a collector and the other as the attracting host (Fig. 7.5). The collector collects all the mosquitoes that land on the other individual with an aspirator (either a battery-operated flashlight or mouth aspirator (Figs. 7.6A–D) within an allotted amount of time known as the "landing rate." For example, if 28 mosquitoes landed on an individual in 1-min pretreatment and two posttreatment, then there was a 92.86% reduction in mosquito landing. This may be done with a number of sites and compared with untreated sites and pretreatment and posttreatment. It is recommended to use the same individual as the host to avoid variability in attractiveness and be used when the mosquitoes are most active such as during dusk and dawn, rather than in the middle of the day (DSHS, 2015). Frequently, these are used in post disaster events such as hurricanes or heavy rains producing flooding and when pestiferous floodwater mosquitoes reach their greatest abundance.

FIG. 7.5   Landing rate count and human collection of mosquitoes.

FIG. 7.6   (A–C) Mouth aspirators and (D) battery-operated flashlight aspirator.

7.6 Resting adult collections

Resting collections are as simple as collecting mosquitoes in places where they are resting. Mosquitoes will generally seek out wind-sheltered shaded areas during peak heat daytime hours or will seek shelter after they have taken a blood meal to produce their eggs. Fournier and Snyder (1977) noted that it is important to collect nocturnal mosquitoes at the correct time of morning when they are most likely to be inactive. These habitats may change depending on weather, species, time of year, and circadian feeding times. In less populated areas, mosquitoes may be collected in animal burrows, tree cavities, and ground vegetation (Cox, 1944; Burkett-Cadena et al., 2008); woodpiles (Nelms et al., 2013a,b); and manmade animal shelters (Gomes et al., 2013). In more populated areas, they may be collected in storm drains or catch basins (Tesh et al., 2004; Rios et al., 2006; Nelms et al., 2013a, b; Arana-Guiardia et al., 2014) corridors between or within buildings (Farajollahi et al., 2012), abandoned buildings (Andreadis et al., 2010), under bridges, or in privies (Cox, 1944; Reisen et al., 2010). In areas where there is little or no air-conditioning, mosquitoes may also be collected resting inside people's homes as was the case along the border with Mexico (Elizondo-Quiroga et al., 2006). However, it has been pointed out by Reiter et al. (2003) that the Texas lifestyle greatly reduces the probability of these mosquitoes resting in homes. It is also ideal to provide areas where it is convenient for mosquitoes to rest such as resting boxes (Fig. 7.7). They may range in size and color but are typically dark with a red color where the collector should be able to see the difference in contrast between the color and the resting mosquitoes. The advantage of utilizing resting boxes is to collect many mosquitoes that may not be attracted to certain traps and females within various stages of their gonotrophic cycle. It may also be possible to collect more males since they seek out the same resting places as do females (Unlu et al., 2014). In certain areas, one may infer what proportion of the local population is ready to oviposit or depending on the season and location and how many females are ready to enter diapause or quiescence (Nelms et al., 2013a, b). Collecting overwintering female mosquitoes may also give insight to whether a mosquito may fully enter diapause or a period of quiescence. In areas where there are many pier and beam houses or buildings resting on stilts as in Houston, TX, the need for another collection tool was warranted to collect mosquitoes that were either seeking shelter from less than favorable climatic conditions, after a blood meal, or for overwintering studies. Usually the height at which these houses are raised on their pillars is not high enough to place a regular CDC gravid trap. Thus the implementation of the under-house trap (Fig. 7.8A–C) in Harris County, TX, was incorporated. This collection device is a modification of the Harris County gravid trap (HCGV) (Fig. 7.9A–C) (Dennett et al., 2007). Such modification required laying the chimney horizontally inside the black restaurant bus box, removing the base where the chimney attaches to and using a piece of wood instead to hold the chimney leveled horizontally and keep the net horizontally straight with the rings in contact with the bottom of the bucket to go underneath the house when the trap is set. In addition, the regular liquid attractant or hay infusion cannot be used as it would cause the net to be partially submerged in the liquid, thus impacting the collection efforts. Therefore, to avoid impacting the collection efforts, another attractant, that is, carbon dioxide (CO_2) in the form of dry ice, had to be used to keep the net and collection dry. The attractant is placed on one end of the box in front of the chimney aperture, so the specimens can be sucked in by the fan into the collection bag on the other end. The bus box is pushed and retrieved from underneath houses with a hook attached to a long wooden pole that attaches to a screwed-in "U" hook on one end of the bucket.

FIG. 7.7 Resting box.

FIG. 7.8 (A) Harris County under-house trap (a modification of a CDC gravid trap) placed horizontally using dry ice (CO_2) underneath pier and beam houses (B and C).

FIG. 7.9 (A–C) Harris County gravid trap (HCGT).

7.7 Light trap collections

Collecting mosquitoes that land on humans may be inconsistent. It was suggested by Headley in 1932 that mechanical methods be introduced to reduce human-induced variability. Many crepuscular and nocturnal mosquitoes are attracted to light. The New Jersey light trap (Fig. 7.10) was one of the first light traps utilized to trap mosquitoes and was created by Mulhern in 1942 (McNelly, 1989). These traps should be placed in areas where power is readily available because they always require an outlet to be powered. Many include a photocell. In the absence of light, the trap will turn on illuminating an incandescent lightbulb and spin a fan creating negative pressure to pull insects into a collection jar. Most people will use an insecticide within the collection chamber to kill the insects that are drawn into the trap. Collections may be set daily or weekly depending on the purpose of use. Pestiferous mosquitoes, that is, *Anopheles quadrimaculatus* Say, may be collected in great numbers in these traps. They are also ideal for monitoring seasonal mosquitoes, that is, *Culiseta* in changing seasons. Because these traps need to be plugged into an outlet, their placement is limited and may inaccurately reflect the mosquito population. Moreover, they only collect mosquitoes that are attracted to light, and the specimens collected are usually dead.

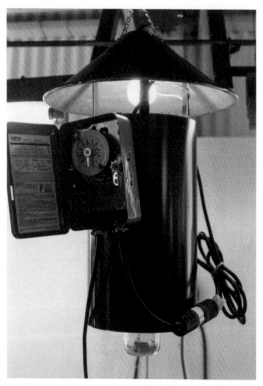

FIG. 7.10   New Jersey light trap (NJLT) used in Harris County surveillance program.

The New Jersey light trap has been used by many mosquito control districts, but because of its many limitations, other traps were designed. The Centers for Disease Control and Prevention (CDC) miniature light trap (Fig. 7.11) was introduced by Sudia and Chamberlain in 1962 (McNelly, 1989). They are compact, sturdy, and easy to set up. They utilize 6-V rechargeable or D-cell batteries, which enable the traps to be set in more versatile habitats. When the mosquito flies toward the light, negative pressure from a fan pulls it into a net connected to a tube containing the fan. When used in combination with CO_2, the CDC light trap captures a wide variety of mosquitoes (Sudia and Chamberlain, 1962; Meyer, 1991; Service, 1993; Hoekman et al., 2016). Because it attracts a wide variety of mosquitoes, it is advantageous to use it during complaint investigations when the target pest mosquitoes are not known. Light traps may be especially useful when monitoring floodplain mosquitoes after a rain or flood event such as a hurricane and for pre- and posttreatment of ground-based or aerial ultra-low volume (ULV) applications (Carney et al., 2008; Breidenbaugh et al., 2008; Chaskopoulou et al., 2011). They may also be used for surveillance of WNV in where the *Culex pipiens* complex is not the main vector (Godsey et al., 2013). The CDC gravid traps are a better option for surveillance of WNV, where *Cx. pipiens* complex mosquitoes are the main vectors for WNV (Williams and Gingrich, 2007). There are modifications of the CDC light trap for the ultraviolet (UV) light sources to be used instead of incandescent or light-emitting diode (LED) lights and modifications to size, cylinder, and net materials to hang inside storm sewers (Fig. 7.12A–F). These were created by entomologists in Harris County around the urban areas of Houston where storm drains are plentiful. Another variation of this light trap is the encephalitis virus surveillance (EVS) trap (Fig. 7.13A and B), which is always used with the addition of carbon dioxide and contains an insulated canister placed above the main unit containing the fan. Although these traps capture a large variety of mosquitoes, they specifically target host-seeking nocturnal female mosquitoes. For this reason, they may not be the best choice for monitoring diurnal mosquitoes such as some *Aedes* species or other specific mosquitoes that are not attracted to CO_2.

FIG. 7.11   CDC miniature light trap.

FIG. 7.12   (A and B) Ruggedized CDC light trap; (C) CDC light trap; (D) placement of CDC light trap in storm sewer system; and (E and F) hanging from a manhole lid.

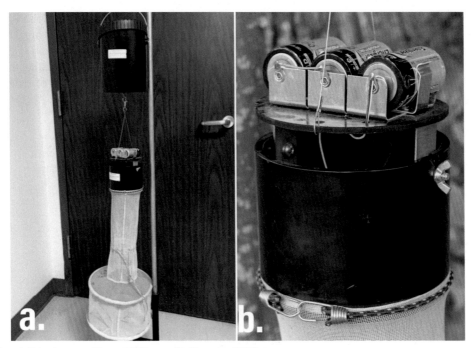

FIG. 7.13   (A) Encephalitis virus surveillance (EVS) trap. (B) Close-up of EVS trap.

7.8   Targeting *Stegomyia* mosquitoes

When Zika became a serious mosquito-borne threat in South and Central America, the popularity of traps that targeted *Aedes* subgenus *Stegomyia* (specifically *Ae. albopictus* and *Ae. aegypti*) was increased. The CDC light traps were found to be ineffective in capturing *Aedes* because of their diurnal host-seeking habits (Thurman and Thurman, 1955; Service, 1993; Heol et al., 2009; Obernauer et al., 2010). The CDC Fay-Prince traps (Fig. 7.14A and B) were not utilized often because they were cumbersome to set due to their size and bulkiness; however, some were found to contribute information to the collection of *Stegomyia* because these mosquitoes have an affinity to black and white contrast, which is one of the attractant premises of these traps. Like the CDC light traps, they are typically set out with the addition of CO_2 and use the downdraft of a fan to pull mosquitoes into the net. The Fay-Prince traps come in different forms: a bidirectional (two openings) or multidirection four openings, which lead to a small fan that will suck mosquitoes into a collection net. They have also been used to collect *Cx. quinquefasciatus* Say and *Ae. sierrensis* (Ludlow) west of the Rocky Mountains. Both sexes of these mosquitoes may be attracted to this trap. The CDC Wilton trap (Fig. 7.15A and B) was created by the CDC and the New Orleans Mosquito Control District to focus on *Ae. aegypti* and *Cx. quinquefasciatus* mosquitoes (Wilton and Kloter, 1985; Heol et al., 2009). Like the Fay-Prince trap, it utilizes color cues to attract the targeted mosquitoes and also negative pressure from a fan to pull them into a collection chamber. It is also believed to mimic tree holes or other oviposition sites or an area for mosquitoes to rest. Heol et al. (2009) reported that the mosquitoes collected in this trap were difficult to identify due to the constant high velocity of the fan.

The Biogents Sentinel (BGS) trap (Regensburg, Germany) has been the "go to" trap for the surveillance of *Stegomyia* mosquitoes (Fig. 7.16A and B). It has been shown to outperform other *Stegomyia* collection traps in some situations (Luhken et al., 2014; Springer et al., 2016). It is designed to have dark and white contrast, placed close to the ground and utilizes attractants made to mimic the scent of human sweat (Kröckel et al., 2006; Obernauer et al., 2010). There may be three lures used in this trap, which include a discontinued octanol lure, designed to mimic the smell of human feet; lactic acid, which mimics human sweat; and CO_2, which is exhaled in respiration. This combination is particularly attractive to anthropogenic mosquitoes. Both males and females of *Stegomyia* mosquitoes were captured by these traps (Unlu et al., 2014). The first iteration of this trap (BGS-I) had some design flaw making it difficult to set up and was flimsy. However, a second generation of BG Sentinel traps (BGS-II) (Fig. 7.17) fixed many problems of the first generation. They also slightly changed the design to be white in color, but Barrera et al. (2014) found that a darker color captured a greater amount of *Stegomyia* mosquitoes. The BG Sentinel traps are also used to analyze the efficiency of treatment for *Stegomyia* mosquitoes

FIG. 7.14   (A) Bidirectional and (B) multidirectional Fay-Prince traps.

FIG. 7.15   (A and B) Wilton trap.

(Farajollahi et al., 2012; Fonseca et al., 2013). Due to the occurrence of both *Ae. aegypti* and *Ae. albopictus* in most Texas counties, many mosquito control districts use BG Sentinel traps as routine surveillance for these potential disease vectors. This also enables municipalities to have an idea where their *Stegomyia* concentrations are the greatest and may give insight into which neighborhoods would be most at risk, should dengue, chikungunya, or Zika viruses become established in Texas.

FIG. 7.16 (A) Biogents Sentinel (BGS-I) trap; (B) placement of BGS-I with dry ice (CO_2).

FIG. 7.17 Biogents Sentinel (BGS-II) trap.

Another trapping device developed by Biogents AG is the BG-Mosquitaire trap (Fig. 7.18A), designed as a homeowner version of the BG Sentinel trap that is commonly used worldwide for monitoring *Ae. aegypti* and *Ae. albopictus* populations. The BG-Mosquitaire incorporates the same technology used in the BG Sentinel (in a smaller package) that attracts mosquitoes through a combination of visual ques and a lure that produces an artificial cocktail of odor components that were identified on human skin. It can be powered by 110-V house current. Carbon dioxide is optional as a supplemental attractant (Fig. 7.18B). The latest development is the BG-Pro system. This is an all-in-one modular trapping system toolbox that can be configured similarly to several different traps such as the CDC light trap (Fig. 7.19A), the EVS trap (Fig. 7.19B), the BG Sentinel trap, or the CDC gravid trap depending on the needs or objectives of the monitoring program. The BG-Pro saves cost by purchasing less equipment and allows for flexibility in collecting methods. The trap is light and can be disassembled to decrease pack size. The packing bag is insulated and can function as the dry ice container, and all parts can be packed in the water basin used for the CDC gravid trap type. Collections are less damaged due to the three-blade ventilator, which also consumes less power, increasing battery life time, and the fan blade is robust that it can be operated in saltwater. The BG-Pro trap can be operated with a 6-V or 12-V rechargeable battery. A power adapter and extension cord are available, too. In addition, it can also run on 5-V standard powerbanks (normally used to charge cell phones or tablet computers on the go), which makes it possible to be set up for days, with only one, lightweight power source.

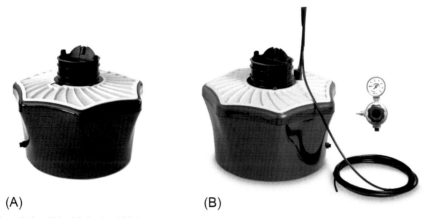

(A)   (B)

FIG. 7.18   (A) BG-Mosquitaire; (B) with dry ice (CO_2).

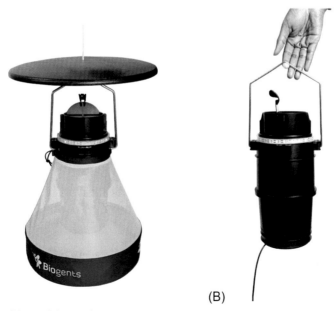

(A)   (B)

FIG. 7.19   Biogents BG-Pro system: (A) a modular trapping system toolbox that can be used as (A) CDC light trap; (B) encephalitis virus surveillance (EVS) trap; and BGS and gravid traps (latter two not pictured).

7.9 Field collections: Gravid adults

Field collections of gravid adults may be as useful as those of host-seeking adults, but there are some additional benefits related to gravid adults. The first and foremost has to do with disease surveillance activity and conducting viral testing on adult mosquitoes. The majority of nuisance and vector species of mosquitoes require a blood meal to begin the gonotrophic cycle and oogenesis. In addition, many mosquito-borne viruses have limited incidence of transovarial (or vertical) transmission (Rosen et al., 1983; Nayar et al., 1986; Nelms et al., 2013a,b; Thangamani et al., 2016; Ciota et al., 2017). As such the odds of finding an infected mosquito increase dramatically when engaging in a targeted search of female mosquitoes that have been previously blood-fed (Reiter et al., 1986; Andreadis et al., 2001; Gingrich and Casillas, 2004; Burkett, 2005; Williams and Gingrich, 2007). Blood-fed mosquitoes may be collected during aspirator collections of resting populations, but in those cases the number of blood-fed may represent a (sometimes small) portion of the entire population, and a large number of the collections may be males (Edman et al., 1968; Williams and Gingrich, 2007). The only way to ensure collecting females that have previously blood-fed is to target the collection of mosquitoes that are seeking habitats to oviposit.

Setting gravid traps (see Fig. 7.3A–D) is generally a straightforward process. Some traps may remain in the field for extended periods of times, while others must be collected daily (and often batteries must be replaced). Finding suitable locations for gravid traps depends on the mosquito species being collected. For example, *Cx. quinquefasciatus* may be collected in a wide range of habitats, including urban habitats, rural habitats, suburban habitats, and industrial habitats. *Ae. aegypti* is primarily found in or adjacent to human dwellings (including parks) or other locations with many breeding sites such as tire piles. Knowing the biology and behavior of the target mosquito species can be critical to trapping success. In addition, knowledge about the presence or absence of breeding sites can assist with trap placement or site selection. If collecting a mosquito that has limited migration distance and selecting a site with limited additional breeding habitats, there will be limited collection success. However, if a gravid trap is placed in an area with a large number of other breeding habitats, it will be "competing" with other sites for each mosquito, and the overall trap collection may be reduced. While this may not impede the goals of the trapping, it is important to recognize this effect in using gravid trapping to estimate overall abundance, as the traps may indicate fewer mosquitoes than actually present.

The most common gravid traps use some type of infusion mixture to attract mosquitoes (Reiter, 1983; Reiter et al., 1986; Millar et al., 1992; Weber and Horner, 1993; Lampman and Novak, 1996). An infusion is a simple mixture of water and some organic material that is often left to ferment over multiple days. Different organic materials may be used for this infusion, including hay, grass clippings, dried leaves, manure, and even readily available commercial products such as rabbit food pellets (Ahmadi and McClelland, 1983; Benzon and Apperson, 1988; Bentley and Day, 1989; Trexler et al., 1998; Burkett, 2005). Various studies have shown that different materials differ in the levels of efficacy for attracting different genera and species of mosquitoes (Reiter, 1983; Reisen and Meyer, 1990; Meyer, 1991; Millar et al., 1992; Walton and Workman, 1998; Reisen et al., 1999; Andreadis et al., 2001; Scott et al., 2001; Lee and Kokas, 2004; Oh and Schmaedick, 2016). Synthetic lures are also available, including synthetic hay lure from Bioquip. Likewise a synthetic mosquito oviposition pheromone that focuses primarily on attracting *Culex* mosquitoes is also available for purchase, although the pheromone is nonspecific within the genera (Iltis and Zweig, 1962; Dawson et al., 1989). *Aedes* synthetic attractant, AtrAedes, has been shown to attract gravid *Ae. aegypti* to oviposition sites and can enhance their surveillance (Eiras and Santanna, 2001; Fávaro et al., 2006; Eiras and Resende, 2009).

The most common design for gravid traps relies on a large basin in which the water/organic material mixture is placed. Above this basin a battery-operated fan creates a small suction, which creates an air flow that will pull the mosquito into a collection bin above the basin. One of the primary differences between the models is the location of the fan relative to the collection bin. In the CDC gravid trap (available from Clarke Mosquito Control, St. Charles, Illinois, and John W. Hock Company, Gainesville, FL), the fan is in front of the catch bin to force the mosquito through the fan before it is collected. This may result in a damaged, or even dead, specimen. However, in the Frommer Updraft gravid trap (Fig. 7.20) (available from John W. Hock Company) and the BioQuip gravid trap (a modification of the Reiter's gravid trap) (Fig. 7.21) (available from BioQuip, Rancho Dominguez, CA), the fan is located behind the collection bin (creating a slight vacuum in the collection bin that draws the mosquito in). This results in a greater probability that the mosquito remains alive after the collection, which may be important to test the infection status of the field-collected mosquitoes. All of these traps run on batteries, which, if they run low, reduce the speed of fan rotation and potentially allow the mosquitoes to escape (the Frommer Updraft gravid trap has a cone-shaped collection entrance designed to limit escape of mosquitoes, but without an actively blowing fan, they can still escape). As a result, traps must be monitored daily, with mosquitoes collected and batteries replaced. In addition, the basin holding the infusion water should be monitored for eggs and larvae to ensure that there is no breeding of new mosquitoes.

Placement of these traps should be in a relatively flat, somewhat protected or covered area (shade increases the probability that mosquitoes will attempt to oviposit) (Lee and Kokas, 2004; Williams and Gingrich, 2007; Oh and Schmaedick, 2016). While many of these traps were designed for *Culex* mosquitoes, they have also been shown to collect a variety of *Aedes* species as well (Lee and Kokas, 2004; Williams and Gingrich, 2007). However, if targeting container-breeding mosquitoes such as *Ae. aegypti* and *Ae. albopictus*, two additional traps are likely to produce better results.

FIG. 7.20 Frommer updraft gravid trap.

FIG. 7.21 Reiter-Cummings modified gravid trap.

The Biogents Gravid *Aedes* Trap (BG-GAT) (Fig. 7.22A and B) and the CDC autocidal gravid ovitrap (CDC-AGO) (Fig. 7.23A and B) are both specifically designed to target *Aedes* mosquitoes by replicating the preferred container oviposition habitat, although they both have been shown to successfully collect *Culex* mosquitoes as well. Both traps function on a similar design: they have a large covered bucket that holds water, with a small cylindrical tube at the top through which a mosquito must fly to get to the water. The water may be plain tap water, or it may be treated with some infusion or lure similar to the gravid traps previously described. Access to the water, however, is blocked off by some screen mesh or netting, and the sides (or in the case of the CDC-AGO trap, the entire interior cylinder) have a sticky trap-like substance that traps any landing mosquito.

One of the benefits of this trap is that it does not require batteries; thus, it can be set and monitored on a weekly (or even more sporadic) basis. However, the water may evaporate and would need to be replaced. In addition, if a mosquito doesn't land on the sticky material, it may fly out of the trap. One difficulty encountered with the sticky paper is the removal and subsequent identification of the mosquitoes collected. Sometimes, removing the mosquitoes will result in legs, wings, even the heads being torn from the mosquitoes, as well as losing a significant number of scales and setae. Some mosquito species with clear and straightforward markings (such as *Ae. albopictus*) may still be easily identified, but identification may be challenging for more cryptic or difficult-to-identify species such as many *Culex* spp. Surveillance technicians are encouraged to identify the mosquitoes while on the sticky paper rather than removing them first, although this may present challenges. In addition, the majority of specimens collected from this trap will be dead (almost certainly after removal from the sticky paper), so collecting live specimens is not likely.

FIG. 7.22   (A) BG Gravid *Aedes* Trap (GAT); (B) top view.

FIG. 7.23   (A) Autocidal Gravid Ovitrap (AGO); (B) top view.

7.10 Novel mosquito traps

Originally the creation of the In2Care mosquito trap (Fig. 7.24) was related to research to combat insecticide-resistant malaria vectors in West Africa. Knols et al. (2010) reported a major problem with insecticide resistance due to the use of insecticide-treated bed nets. Because many anopheline mosquitoes were becoming resistant to chemical control methods, novel methods to control mosquitoes with entomopathogenic fungus were developed. It was also noted that some fungal infections caused mosquitoes to be more susceptible to chemical control methods that they had become resistant to (Farenhorst et al., 2009). *Ae. aegypti* is a serious vector for various mosquito-borne diseases around the world. It is difficult to control because it uses skip-oviposition methods where a female mosquito oviposits in many different containers. Some of these containers are difficult to locate due to their cryptic nature. Researchers from the Netherlands created a trap that uses the combination of the entomopathogenic fungus, *Beauveria bassiana* (Balsamo-Crivelli) Vuillemin and a chemical additive insect growth regulator, pyriproxyfen. When the female container-breeding *Stegomyia* enters the trap to lay eggs, she lands on a strip of gauze that is impregnated with both aforementioned products, which have a slight negative charge that sticks to the slightly positively charged mosquito. The mosquito becomes infected with *B. bassiana*, which slowly kills it, and thus not only does the ovipositing mosquito dies, but also she becomes a vehicle for insecticide distribution. Thereafter, she contaminates every source she visits with pyriproxyfen until she eventually succumbs to the fungus. Some mosquito control agencies such as Manatee County, FL (Buckner et al., 2017), and Harris County Public Health Mosquito and Vector Control Division in Texas have tested them, especially where the risk for Zika and dengue transmission is high.

FIG. 7.24   In2Care trap.

The German company Biogents AG has introduced its remote mosquito-monitoring device called the "BG-Counter" (Fig. 7.25A). The device is combined with the BG Sentinel mosquito trap, which is widely used for the monitoring and surveillance of *Aedes* mosquitoes using human odor attractants and, in combination with CO_2 attractant, can also be used for monitoring other species of mosquitoes.

The BG-Counter automatically differentiates mosquitoes from larger or smaller insects and dust particles entering the trap, counts them with an accuracy of 90% (established in field tests when working with CO_2 as an attractant), and can record up to five mosquitoes per second. The patented insect sensor (Fig. 7.25B) consists of arrays of infrared LEDs and light detectors that provide reliable and sensitive detection and differentiation of mosquitoes from other objects entering the trap. The BG-Counter also samples local environmental data such as temperature, humidity, and light. It is equipped with a cellular modem communicating with a cloud-based database for storage of mosquito counts and geospatial (GPS) and environmental data. The database is automatically updated with the data constantly generated in the field. The data can be accessed, displayed, and analyzed in a cloud-based dashboard. Data can also be exported to Excel at the push of a button. The intuitive graphical user interface can be accessed from PCs and smartphones and tablets. The web application also allows to remotely switch the traps on and off in the field and set up varying time schedules to run the traps including application times of attractants. The device is designed for unattended field operation over several weeks at a time, powered by electricity from the outlet or by a battery with solar charging. It can be integrated into a trap station protected from weather (Fig. 7.25C). Maintenance consists of periodic replacement of the CO_2 tank and emptying and cleaning the trap. The BG-Counter is a mature technology, with several hundred devices deployed worldwide. It enables real-time measurements and prediction models and historical analysis of infested areas to enable surveillance programs with unprecedented data density and accuracy, overcoming labor constraints associated with manual inspection. It is a tool that can be used in place of human landing collections (HLC) or landing rates (LR) and pre- and posttreatment assessments. The effectiveness of control measures can be validated immediately.

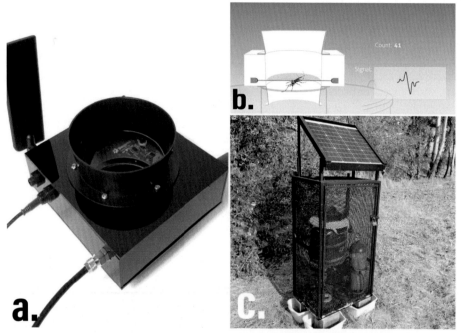

FIG. 7.25   (A) BG-Counter; (B) BG-Counter Sensor; (C) security and weather protection for solar powered BGS-II with BG-Counter and CO_2 supply.

Another type of a novel mosquito trap is the Premonition "smart" trap developed as part of Microsoft's Project Premonition in collaboration with partner academic and public health agencies (Fig. 7.26A). The goal of Project Premonition smart trap is the accurate, timely, and scalable detection and monitoring of pathogens in the environment to better predict the occurrence of arthropod-borne and other diseases. Project Premonition devices are designed to be placed in the environment where they observe flying insects and (1) identify species autonomously, (2) collect high-fidelity phenotypic data, and (3) selectively capture and store specimens while preserving whole-body morphology and nucleic acids for pathogen detection and other downstream metagenomic analyses. The Project Premonition "smart" trap efficiently attracts and collects live mosquito specimens and data about the environment from which they were collected. The initial version of the trap is composed of 64 independently actuated cells/chambers that close their doors automatically as a mosquito enters a cell/chamber (Fig. 7.26B and C). As with most traps the Premonition device uses lures such as CO_2, light, passive color, and other odorants to attract arthropods of interest. It is designed to attract mosquitoes and other insects by modulating lures in real-time as species are autonomously identified. Upon deployment the Premonition trap collects gigabytes of environmental data such as geographic location of the trap, ambient light levels at the time of mosquito capture, time of mosquito collection, temperature, humidity, wind, and barometric pressure. Onboard sensors capture phenotypic data such as wingbeat frequency to determine species identity from these data. Following collection, DNA from field-collected specimens is extracted and sequenced using next-generation genomic sequencing methods and analyzed using a state-of-the-art metagenomic analysis tool, enabling the system to confirm the identity of collected specimens to species, determine the origin of vertebrate host blood meals and detect potential pathogens. The broad-spectrum metagenomic pipeline compares the complete genetic composition of the sample against a database of over 650,000 reference genomes spanning the entire tree of life and is able to recognize vertebrates, plants, bacteria, viruses, and arthropods. Metagenomic analysis is particularly well suited for identifying novel microbial and viral threats, which would not be detected by other methods.

The Premonition "smart" trap (v.2) was deployed in Harris County/Houston, TX, in collaboration with Harris County Public Health Mosquito and Vector Control Division in the summer of 2016. In 87 experiments (trap nights), it captured over 20 GB of biotic and abiotic data, including the behavior of *Ae. aegypti* and *Ae. albopictus* and their interactions with the trap. It detected over 22,000 mosquito encounters and has been used to detect nine mosquito species, including those that vector Zika virus, dengue virus, WNV, and malaria parasites. The trap collected mosquitoes by identifying relevant public health mosquitoes with an accuracy >90%, including *Cx. quinquefasciatus*, *Ae. aegypti*, and *Ae. albopictus*. Key results and lessons learned were as follows: (1) The Premonition "smart" trap could be trained to identify mosquitoes of public health importance with high accuracy (>90%), and (2) the traps were also able to record mosquito behaviors at high resolution, allowing the construction of more accurate predictive mosquito models via statistical machine learning. These results suggested that Project Premonition could potentially improve public health interventions by providing timely predictions of disease vectors in the environment.

FIG. 7.26   (A) Microsoft's Project Premonition smart trap; (B) trap close up; (C) trap chambers/cells close up.

7.11   Using surveillance data

After data have been collected, several analyses may be conducted to reveal information about population dynamics, risk to human diseases, and efficacy of control efforts. Many GIS software programs allow statistical analyses of data and provide a visual output on a map to quickly assess the mosquito population dynamics in a determined area. Patterns and trends can be easily identified with the use of the software, and clustering or hot spots can be detected easier than by examining the data. Field surveillance is also a useful tool when determining if specific interventions or control strategies are effective and are meeting their target goals. Ideally an intervention or control effort should be undertaken with a specific goal, such as a certain percentage reduction in population or the reduction of a target mosquito species. Conducting field collections both preintervention and postintervention will determine if the goals of the intervention have been met. It is critical to conduct both pre- and postintervention trapping, as only conducting one or the other will not assess the impact (or the lack of impact) of a specific intervention effort. For example, if adulticiding is planned for a neighborhood, multiple light traps can be set to identify the pretreatment population size, and the same traps in the same locations can be set posttreatment to determine if the population size has decreased. This will determine efficacy of any specific treatment or if the intervention has achieved the objectives. Ideally, pre- and postintervention trapping should be conducted on a regular basis and also compared with a control site where the intervention has not occurred. This will determine if any observed decrease in population is due to the intervention or some other factor.

Recording data throughout time, that is, all year long and from 1 year to another, will allow for further analyses of the data. Comparison of mosquito population dynamics and any other parameter recorded such as avian data and human cases can be analyzed. Correlations can be performed and observed between these parameters and climatic data to assess the degree of health risk posed in a certain area.

7.12   Measuring mosquito abundance

Modeling population size in mosquitoes can be a difficult undertaking with field collection. Trapping data can vary widely, even with traps that are set at the same time relatively close to each other. As a result, using a small number of traps that are sporadically set is not an effective method for monitoring overall population size. While sporadic trapping may help in some areas (e.g., when identifying an invasive species), effective temporal and spatial population estimates require traps to be set in a large number of places and frequently (weekly). However, this represents a challenge to many mosquito control programs, due to resource limitations (not many traps available and not enough personnel to set the traps). It may be more valuable to identify fewer locations to set traps (possibly focusing on "hot spots" that historically have shown a high number of nuisance calls or mosquito-borne disease cases) and collect data on a weekly basis for overall population monitoring.

Assessing the absolute population size is a challenge for mosquito populations. There are three primary methods that can be used to assess absolute population size: mark and recapture, removal method, and quadrat method. Mark and recapture is conducted by using a standard method of field collection (such as a standard light trap). Specimens collected are marked with a fluorescent powder to allow their recognition when recaptured and then released back into the wild population. A second collection effort is conducted at a later date (within the lifespan of the originally collected individuals), and an estimate of the overall population size is generated using the formula:

$$\frac{M}{N} = \frac{x}{n}$$

where M is the number that was collected and marked in the first attempt, x is the number collected in the second attempt that was marked, n is the total number collected in the second attempt, and N is the estimate of the population size. Written another way the overall population size is represented as

$$N = \frac{Mn}{x}$$

The removal method also relies on multiple, sequential trapping efforts; however, samples are not released back into the wild. Instead the removal method works on the assumption that the collection size is a similar proportion of individuals from the population at each trapping interval and can calculate the population size based on the decreasing size of the collections in the field.

Both the mark and recapture method and the removal method are only effective if certain assumptions are met, that is, that the population remains static between each trapping effort (no births, no deaths, and no migration of individuals). The probability that this assumption is accurate is highly unlikely, given both the facts that mosquitoes do move around and that the population may fluctuate highly with adults dying and pupae emerging into the population. In addition and as previously mentioned, trapping results may be highly variable even within a small geographic location. Thus there are many factors that influence the collection rates of a trap, making its estimation more difficult. In many cases the mosquito population is so large that the number of mosquitoes collected may represent an extremely small percentage of the total population size, making estimates unreliable. Lastly, in the mark and recapture method, the releasing of mosquitoes back into the wild would most likely not be a popular move, regardless of the intent of such an effort.

The quadrat method is a technique used for estimating population size. However, this technique relies on the assumption that we are able to collect all mosquitoes within a given area, an assumption that is highly unlikely when it comes to field-collected mosquitoes. Therefore any estimate provided by this method is also likely to be unreliable. For any of these methods to produce a reliable estimate of actual population size, it would require a significant effort, that is, placing an extremely large number of traps in a given area to ensure that a large proportion of the total population is collected. Given the previously mentioned resource limitations, it is unlikely that this is possible.

It is significantly easier to calculate the relative population abundance than to calculate absolute population abundance. Relative population abundance is an indication that trapping efforts are assumed to consistently collect a similar proportion of the total population over time. Hence, any changes in trap collection densities would reflect changes in the actual population size, even if the population size is not known. For example, if there is a 50% increase in the number of adults collected in BG Sentinel traps, it may indicate that the actual population size parallels such increase. Estimating relative population is focused more on seeing how populations change over time rather than determining the total number of mosquito individuals in the population.

Relative population changes can be used to assess intervention techniques. If the trapping shows a 90% decrease in the number of mosquitoes collected following an intervention, it can be assumed that the intervention killed approximately 90% of the mosquitoes. In addition, if trap counts are gradually increasing over time in a certain area, it can be presumed that the mosquito populations are also increasing at the same rate in that area. Relative population estimates require multiple traps as there may be significant variability in traps even within a short distance of each other during the same trapping period. Multiple traps allow to average the collections and develop an overall estimate for the trap counts (such as eggs per cup or adults per trap). These average data are more reliable than using one or two traps and expecting them to be accurate.

While the absolute population abundance may be difficult or impossible to calculate, identifying relative changes in population abundance are ultimately just as useful for vector control and surveillance efforts. Reliably collecting abundance data over time can aid in determining the best time for and the efficacy of intervention efforts and assess the relative risk of disease transmission. Data collected based on the relative abundance of certain mosquito species can be used for identifying a threshold abundance for disease risk assessment (and any correlated increase in intervention efforts).

7.13 Calculating minimum infection rates and vector indices

Surveillance most importantly contributes information about how the local vector mosquito populations are connected to the risk of human mosquito-borne diseases. Often the minimum infection rate (MIR) is calculated when mosquito-borne diseases are detected within a population. This is simply the proportion of positive mosquito pools divided by the number of total mosquitoes tested. This rate can be given per number of mosquitoes (usually per 1000 or per 100,000). The problem with this calculation is that it makes some assumptions: (1) The detection of an infectious agent can only be assumed to be one per the number that is in the pool, and (2) mosquito pool sizes vary between 1 and x (usually 50). Gu et al. (2003) pointed out some of the issues with using the MIR as a calculation and also some issues where previous publications actually used a different calculation and suggested using the corrected calculation, the maximum likelihood estimate (MLE) (Chiang and Reeves, 1962; Gu et al., 2003):

$$MLE = \left(1 - \left(1 - Y/X\right)\right)^{1/m}$$

where Y is the number of positive pools, X is the total number of pools, and m is the pool size. The reason the MIR is problematic is because varying pool sizes must be weighted differently depending on the pool size. A positive sample from a pool of one mosquito does not have the same meaning as a positive pool sample from 50 mosquitoes. The MLE helps to weigh the importance of pool size when detecting a positive sample. Dr. Brad Biggerstaff with the CDC built an Excel Add-In to calculate the biased corrected MLE not long after this was pointed out and there have been several improved iterations since it was first released. It is also available on The R Project (n.d.), a free downloadable statistical analysis tool (https://www.r-project.org/). After calculating the MLE the abundance of mosquitoes within a population must also be taken into consideration. Surveillance for abundance must also be taken into account when assessing risk. Naturally the more vector mosquitoes that are available within a community, the larger the risk is when introducing an agent and humans. Since it is impossible to know the actual population size, one can use the average number of mosquitoes captured in one selected trap. This is simply the average number of mosquitoes per trap night. In much of Texas, it is most used with gravid traps where *Cx. quinquefasciatus* is the main vector for WNV. Since gravid traps are typically set out for only one night, one can simply calculate the average number of mosquitoes captured per trap divided by the number of total traps set. This can then be multiplied by the MLE to reveal the vector index (VI) that should be correlated with the number of human cases. This must happen over many years of collecting data and only to very local populations due to differences in climate, habitat distribution, abundance of mosquitoes, mosquito behavior, bird populations, etc. per year. Only then can these VIs be related to human disease. It must also be noted that MLEs or VIs should be calculated per species of mosquito and per one trap type. Different species of mosquitoes may have different competencies and different human biting rates, which is why they should not be combined. If two mosquito species are combined, then this should be stated before a given value so that biases may be taken into consideration.

7.14 Conclusion

Mosquito surveillance and control operations vary greatly across Texas due to varying testing capacities, funding issues, public awareness, politics, population size, and variable climatic environmental factors. Some counties may only conduct chemical control operations in reaction to reported human cases, while others have an advanced surveillance system to assess risk of vector-borne disease transmission, and others may treat in response to nuisance mosquitoes to improve the comfort of the community. At a minimum, a basic continuous collection of mosquitoes (surveillance) should be carried out to provide a sound and scientific reliable decision to apply integrated mosquito control measures effectively.

Ideally a fully developed mosquito control operation should include adult and larval surveillance; disease surveillance for potential vector species; and other integrated control methods such as larviciding, adulticiding, biological control, physical, mechanical, cultural control, education/outreach, and insecticide resistance testing. Best practices utilizing integrated pest management (IPM), integrated vector management (IVM), or integrated mosquito management (IMM) principles should be the common factor among surveillance and control programs. It is important to implement these best practices in programs across Texas for a better outcome on the battle against mosquito vectors to prevent mosquito-borne diseases.

Acknowledgments

We thank Oscar De La Garza, Jeremy Vela, Oscar Salazar, and the rest of the surveillance team at Harris County Public Health Mosquito and Vector Control Division for their assistance with the trap figures. We also thank Ethan Jackson and Mike Reddy from Microsoft Inc. for their review and photos of the premonition trap; John W. Hock Company for allowing us to use their trap photos; and Andreas Rose, Martin Geier, Scott Gordon, and Michael Weber from Biogents AG for the review and photos of the BG counter and BG-Mosquitaire and BG-Pro traps.

References and further reading

Ahmadi, A., McClelland, G.A.H., 1983. Oviposition attractants of the western treehole mosquito, *Aedes sierrensis*. Mosq. News 43 (3), 343–345.

Andreadis, T.G., Anderson, J.F., Vossbrink, C.R., 2001. Mosquito surveillance for West Nile virus in Connecticut, 2000: isolation from *Culex pipiens*, *Cx. restuans*, *Cx. salinarius*, and *Culiseta melanura*. Emerg. Infect. Dis. 7 (4), 670–674.

Andreadis, T.G., Anderson, J.F., Armstrong, P.M., Bajwa, W.I., 2010. Studies on hibernating populations of *Culex pipiens* from a West Nile virus endemic focus in New York City: parity rates and isolation of West Nile virus. J. Med. Entomol. 26 (3), 257–264.

Arana-Guiardia, R., Baak-Baak, C.M., Lorono-Pino, M.A., Machain-Willaims, C., Beaty, B.J., Eisen, L., Garcia-Rejon, J.E., 2014. Stormwater drains and catch basins as sources for production of *Aedes aegypti* and *Culex quinquefasciatus*. Acta Trop. 134, 33–42. https://doi.org/10.1016/j.actatropica.2014.01.011.

Barrera, R., Amador, M., Acevedo, V., Caban, B., Felix, G., Mackay, A.J., 2014. Use of the CDC autocidal gravid ovitrap to control and prevent outbreaks of *Aedes aegypti* (Diptera: Culicidae). J. Med. Entomol. 51 (1), 145–154.

Bentley, M.D., Day, J.F., 1989. Chemical ecology and behavioral aspects of mosquito oviposition. Annu. Rev. Entomol. 34, 401–421.

Benzon, G.L., Apperson, C.S., 1988. Reexamination of chemically mediated oviposition behavior in *Aedes aegypti* (L.) (Diptera: Culicidae). J. Med. Entomol. 25 (3), 158–164.

Bolling, B.G., Kennedy, J.H., Zimmerman, E.G., 2005. Seasonal dynamics of four potential West Nile vector species in north-central Texas. J. Vector Ecol. 30 (2), 186–194.

Breidenbaugh, M.S., Haagsma, K.A., Walker, W.W., Sanders, D.M., 2008. Post-hurricane Rita mosquito surveillance and the efficacy of air force aerial applications for mosquito control in east Texas. J. Am. Mosq. Control Assoc. 24 (2), 327–330.

Buckner, E.A., Williams, K.F., Marsicano, A.L., Latham, M.D., Lesser, C.R., 2017. Evaluating the vector control potential of the In2Care® Mosquito Trap against *Aedes aegytpi* and *Aedes albopictus* under semifield conditions in Manatee County, Florida. J. Am. Mosq. Control Assoc. 33 (3), 193–199.

Burkett, N.D., 2005. Comparative Study of Gravid-Trap Infusions for Capturing Blood-Fed Mosquitoes (Diptera: Culicidae) of the Genera *Aedes*, *Ochlerotatus*, and *Culex*. Master of Science Degree Thesis, Auburn University, Auburn, Alabama.

Burkett-Cadena, N.D., Graham, S.P., Hassan, H.K., Guyer, C., Eubanks, M.D., Katholi, C.R., Unnasch, T.R., 2008. Blood feeding patterns of potential arbovirus vectors of the genus *Culex* targeting ectothermic hosts. Am. J. Trop. Med. Hyg. 79 (5), 809–815.

Carney, R.M., Husted, S., Jean, C., Glaser, C., Kramer, V., 2008. Efficacy of aerial spraying of mosquito adulticide in reducing incidence of West Nile virus California, 2005. Emerg. Infect. Dis. 14 (5), 747–754.

Chaskopoulou, A., Latham, M.D., Pereira, R.M., Connelly, R., Bonds, J.A.S., Koehler, P.G., 2011. Efficacy of aerial ultra-low-volume applications of two novel water-based formulations of unsynergized pyrethroids against Riceland mosquitoes in Greece. J. Am. Mosq. Control Assoc. 27 (4), 414–422.

Chaves, L.F., Harrington, L.C., Keogh, C.L., Nguyen, A.N., Kitron, U.D., 2010. Blood feeding patterns of mosquitoes: random or structured? Front. Zool. 7 (3), 1–11.

Chiang, C.L., Reeves, W.C., 1962. Statistical estimation of virus infection rates in mosquito vector populations. Am. J. Hyg. 75, 377–391.

Ciota, A.T., Biolosuknia, S.M., Dylan, J.E., Kramer, L.D., 2017. Vertical transmission of Zika virus in *Aedes aegypti* and *Ae. albopictus* mosquitoes. Emerg. Infect. Dis. 23 (5), 880–882.

Cox, G.W., 1944. The Mosquitoes of Texas. Texas State Health Department. 100 pp.

Dawson, G.W., Laurence, B.R., Pickett, J.A., Pile, M.M., Wadhams, L.J., 1989. A note on the mosquito oviposition pheromone. Pest Manag. Sci. 27 (3), 277–280.

Dennett, J.A., Wuithiranyagool, T., Reyna Nava, M., Bala, A., Tesh, R.B., Parsons, I.R., Bueno Jr., R., 2007. Description and use of the Harris County gravid trap for West Nile virus surveillance 2003–06. J. Am. Mosq. Control Assoc. 23 (3), 359–362.

DSHS (Texas Department of State Health Services), 2015. DSHS Response Operating Guidelines 2015: Vector Control. https://dshs.texas.gov/commprep/response/ROG.aspx.

Edman, J.D., 1974. Host-feeding patterns of Florida Mosquitoes III *Culex (Culex)* and *Culex (Neoculex)*. J. Med. Entomol. 11 (1), 95–104.

Edman, J.D., Evans, F.D., Williams, J.A., 1968. Development of a diurnal resting box to collect *Culiseta melanura* (Coq.). Am. J. Trop. Med. Hyg. 17 (3), 451–456.

Eiras, A.E., Resende, M.C., 2009. Preliminary evaluation of the 'Dengue-MI' technology for *Aedes aegypti* monitoring and control. Cadernos de Saúde Pública 25 (Suppl. 1), S45–S58.

Eiras, A.E., Santanna, A.L., 2001. Atraentes de Oviposição de Mosquitos. Patente; Privilégio e Inovação. n. PI0106701–0. "Atraentes de Oviposição de Mosquitos". 20 de dez. de 2001.

Elizondo-Quiroga, A., Flores-Suarez, A., Elizondo-Quiroga, D., Ponce-Garcia, G., Blitvich, B.J., Contreras-Cordero, J.F., Gonzalez-Rojas, J.I., Mercado-Hernandez, R., Beaty, B.J., Fernandez-Salas, I., 2006. Host-feeding preference of *Culex quinquefasciatus* in Monterrey, Northeastern Mexico. J. Am. Mosq. Control Assoc. 22 (4), 654–661.

Farajollahi, A., Healy, S.P., Unlu, I., Gaugler, R., Fonseca, D.M., 2012. Effectiveness of Ultra Low Volume nighttime applications of an adulticide against diurnal *Aedes albopictus*, a critical vector of dengue and chikungunya viruses. PLoS One 7 (11), 1–7.

Farenhorst, M., Mouatcho, J.C., Kikankie, C.K., Brooke, B.D., Hunt, R.H., Thomas, M.B., Koekemoer, L.L., Knols, B.G., Coetzee, M., 2009. Fungal infection counters insecticide resistance in African malaria mosquitoes. PNAS 106 (41), 17443–17447.

Fávaro, E.A., Dibo, M.R., Mondini, A., Ferreria, A.C., Barbosa, A.A., Erisa, A.E., Barata, E.A., Chiaravalloti-Neto, F., 2006. Physiological state of Aedes (Stegomyia) aegypti mosquitoes captured with MosquiTRAPs in Mirasol, São Paulo, Brazil. J. Vector Ecol. 31 (2), 285–291.

Fay, R.W., Eliason, D.A., 1966. A preferred oviposition site as surveillance method for *Aedes aegypti*. Mosq. News 26 (4), 531–535.

Ferdousi, F., Yoshimatsu, S., Ma, E., Sohel, N., Wagatsuma, Y., 2015. Identification of essential containers for *Aedes* larval breeding to control dengue in Dhaka, Bangladesh. Trop. Med. Health 43 (4), 253–264.

Focks, D.A., 2004. A Review of Entomological Sampling Methods and Indicators for Dengue Vectors. Special Programme for Research and Training in Tropical Diseases, World Health Organization, Geneva, Switzerland. http://www.who.int/iris/handle/10665/68575.

Fonseca, D.M., Unlu, I., Crepeau, T., Farajollahi, A., Healy, S.P., Barlett-Healy, K., Strickman, D., Guagler, R., Hamilton, G., Kline, D., Clark, G.G., 2013. Area-wide management of *Aedes albopictus*. Part 2: gauging the efficacy of traditional integrated pest control measures against urban container mosquitoes. Pest Manag. Sci. 69, 1351–1361.

Fournier, P.V., Snyder, J.L., 1977. Introductory Manual on Arthropod-Borne Disease Surveillance Part I: Mosquito-Borne Encephalitis. Texas Department of Health Resources Bureau of Laboratories.

Gingrich, J.B., Casillas, L., 2004. Selected mosquito vectors of West Nile virus: comparison of their ecological dynamics in four woodland and marsh habitats in Delaware. J. Am. Mosq. Control Assoc. 20 (2), 138–145.

Godsey Jr., M.S., King, R.J., Burkhalter, K., Delorey, M., Colton, L., Charnetzky, D., Sutherland, G., Ezenwa, V.O., Wilson, L.A., Coffey, M., Miheim, L.E., Taylor, V.G., Palmisano, C., Wesson, D.M., Guptill, S.C., 2013. Ecology of potential West Nile virus vectors in southeastern Louisiana: enzootic transmission in the relative absence of *Culex quinquefasciatus*. Am. J. Trop. Med. Hyg. 88 (5), 986–996.

Gomes, B., Sousa, C.A., Vicente, J.L., Pinho, L., Calderon, I., Arez, E., Almeida, A.P.G., Donnelly, M.J., Pinto, J., 2013. Feeding patterns of *molestus* and *pipiens* forms of *Culex pipiens* (Diptera: Culicadae) in a region of high hybridization. Parasit. Vectors 6 (93), 1–10.

Gu, W., Lampman, R., Novak, R.J., 2003. Problems in estimating mosquito infection rates using minimum infection rate. J. Med. Entomol. 40 (5), 595–596.

Hanna, J.N., Ritchie, S.A., Merritt, A.D., van den Hurk, A.F., Phillips, D.A., Serafin, I.L., Norton, R.E., McBride, W.J., Gleeson, F.V., Poidinger, M., 1998. Two contiguous outbreaks of dengue type 2 in north Queensland. Med. J. Australia 168 (5), 221–225.

Headlee, T.J., 1921. The Mosquitoes of New Jersey and their Control. N.J. Agricultural Experiment Station Bull. 348, 229. New Brunswick, NJ.

Heol, D.F., Kline, D.L., Allan, S.A., 2009. Evaluation of six mosquito traps for collection of *Aedes albopicuts* and associated mosquitoes species in a suburban setting in north central Florida. J. Am. Mosq. Control Assoc. 25 (1), 47–57.

Hoekman, D., Springer, Y.P., Barker, C.M., Barrera, R., Blackmore, M.S., Bradshaw, W.E., Foley, D.H., Ginsberg, H.S., Hayden, M.H., Holzapeel, C.M., Juliano, S.A., Kramer, L.D., LaDeau, S.L., Livdahl, T.P., Moore, C.G., Nasci, R.S., Reisen, W.K., Savage, H.M., 2016. Designs for mosquito abundance and diversity, and phenology, sampling within the National Ecological Observatory Network. Ecosphere 7 (5), 1–13.

Iltis, W.G., Zweig, G., 1962. Surfactant in apical drop of eggs of some culicine mosquitoes. Ann. Entomol. Soc. Am. 55 (4), 409–415.

Knols, B.G.J., Burkhari, R., Farnhorse, M., 2010. Entomopathogenic fungi as the next-generation control agents against malaria mosquitoes. Future Microbiol. 5 (3), 339–341.

Kröckel, U., Rose, A., Eiras, A., Geier, M., 2006. Newtools for surveillance of adult yellow fever mosquitoes: comparison of trap catches with human landing rates in an urban environment. J. Am. Mosq. Control Assoc. 22, 229–238.

Lampman, R.L., Novak, R.J., 1996. Oviposition preferences of *Culex pipiens* and *Culex restuans* for infusion-baited traps. J. Am. Mosq. Control Assoc. 12 (1), 23–32.

Lee, J.H., Kokas, J.E., 2004. Field evaluation of CDC gravid trap attractants to primary West Nile virus vectors, *Culex* mosquitoes in New York state. J. Am. Mosq. Control Assoc. 20 (3), 248–253.

Leslie, J.B., 1985. Mosquito control: a historical view. Proc. N.J. Mosq. Control Assoc. 72, 17–21.

Lloyd, L.S., 2003. Best Practices for Dengue Prevention and Control in the Americas. Environmental Health Project. Strategic report 7.

Lloyd, L.S., Winch, P., Ortega-Canto, J., Kendall, C., 1992. Results of a community-based *Aedes aegypti* control program in Merida, Yucatan, Mexico. Am. J. Trop. Med. Hyg. 46 (6), 635–642.

Lühken, R., Pfitzner, W.P., Börstler, J., Garms, R., Huber, K., Schork, N., Steinke, S., Kiel, E., Becker, N., Tannich, E., 2014. Field evaluation of four widely used mosquito traps in Central Europe. Parasit. Vectors 7, 268–278.

Mackey, A., Amador, M., Barrera, R., 2003. An improved autocidal gravid ovitrap for the control and surveillance of *Aedes aegypti*. Parasit. Vectors 6 (1), 225.

Mari, T.R., Arunachalam, N., Rajendran, R., Satyanarayana, K., Dash, A.P., 2005. Efficacy of thermal fog application of deltacide, a synergized mixture of pyrethroids, against *Aedes aegypti*, the vector of dengue. Tropical Med. Int. Health 10 (12), 1298–1304.

McNelly, J.R., 1989. In: The CDC trap as a special monitoring tool. Proceedings of the Seventy-Sixth Annual Meeting of the New Jersey Mosquito Control Association, Inc. vol. 99. pp. 26–36.

Meyer, R.P., 1991. Urbanization and the efficiency of carbon dioxide and gravid traps for sampling *Culex quinquefasciatus*. J. Am. Mosq. Control Assoc. 7 (3), 467–470.

Millar, J.G., Chaney, J.D., Mulla, M.S., 1992. Identification of oviposition attractants for *Culex quinquefasciatus* from fermented Bermuda grass infusion. J. Am. Mosq. Control Assoc. 8 (1), 11–17.

Mogi, M., Choochote, W., Khamboonruan, C., Suwanpanit, P., 1990. Applicability of presence-absence and sequential sampling for ovitrap surveillance of *Aedes* (Diptera: Culicidae) in Chian May, Northern Thailand. J. Med. Entomol. 27 (4), 509–514.

Montgomery, B.L., Ritchie, S.A., 2002. Roof gutters: a key container for *Aedes aegypti* and *Ochlerotatus notoscriptus* (Diptera: Culicidae) in Australia. Am. J. Trop. Med. Hyg. 67 (3), 244–246.

Morrison, A.C., Gray, K., Getis, A., Astete, H., Sihuincha, M., Focks, D., Watts, D., Stancil, J.D., Olson, J.G., Blair, P., 2004. Temporal and geographic patterns of *Aedes aegypti* (Diptera: Culicidae) production in Iquitos, Peru. J. Med. Entomol. 41 (6), 1123–1142.

Nayar, J.K., Rosen, L., Knight, J.W., 1986. Experimental vertical transmission of Saint Louis encephalitis virus by Florida mosquitoes. Am. J. Trop. Med. Hyg. 35 (6), 1296–1301.

Nelms, B.M., Macedo, P.A., Kothera, L., Savage, H.M., Reisen, W.K., 2013a. Overwintering biology of *Culex* (Diptera: Culicidae) mosquitoes in the Sacramento Valley of California. J. Med. Entomol. 50 (4), 773–790.

Nelms, B.M., Fechter-Leggett, E., Carroll, B.D., Macedo, P., Kluh, S., Reisen, W.K., 2013b. Experimental and natural vertical transmission of West Nile virus by California *Culex* (Diptera: Culicidae) mosquitoes. J. Med. Entomol. 50 (2), 371–378.

Obernauer, P.J., Kaufman, P.E., Kline, D.L., Allan, S.A., 2010. Detection of the monitoring for *Aedes albopictus* (Diptera: Culicidae) in suburban and sylvatic habitats in north central Florida using four sampling techniques. Environ. Entomol. 39 (5), 1608–1616.

Oh, H.D., Schmaedick, M., 2016. Attractiveness of three gravid trap infusions for ovipositing Polynesian Tiger Mosquitoes (*Aedes polynesiensis*) in American Samoa. J. Health Dispar. Res. Pract. 9, 72–73. Sp. Ed. 1.

O'Malley, C., 1995. Seven ways to a successful dipping career. Wing Beats 6 (4), 23–24.

R Project, The R Project for Statistical Computing. https://www.r-project.org/. (Accessed December 18, 2018).

Reisen, W.K., Meyer, R.P., 1990. Attractiveness of selected oviposition substrates for gravid *Culex tarsalis* and *Culex quinquefasciatus* in California. J. Am. Mosq. Control Assoc. 6 (2), 244–250.

Reisen, W.K., Boyce, K., Cummings, R.C., Delgado, O., Gutierrez, A., Meyer, R.P., Scott, T.W., 1999. Comparative effectiveness of three adult mosquito sampling methods in habitats representative of four different biomes of California. J. Am. Mosq. Control Assoc. 15 (1), 24–31.

Reisen, W.K., Thiemann, T., Barker, C.M., Lu, H., Carroll, B., Fang, Y., Lothrop, H.D., 2010. Effects of warm winter temperature on the abundance and gonotrophic activity of *Culex* (Diptera: Culicidae) in California. J. Med. Entomol. 47 (2), 230–237.

Reiter, P., 1983. A portable, battery-powered trap for collecting gravid *Culex* mosquitoes. Mosq. News 43 (4), 496–498.

Reiter, P., Jakob, W.L.K., Francy, D.B., Mullenix, J.B., 1986. Evaluation of the CDC gravid trap for surveillance of St. Louis encephalitis vectors in Memphis, Tennessee. J. Am. Mosq. Control Assoc. 2 (2), 209–211.

Reiter, P., Amador, M.A., Colon, N., 1991. Enhancement of the CDC ovitrap with hay infusion for daily monitoring of *Aedes aegypti* populations. J. Am. Mosq. Control Assoc. 7 (1), 52–55.

Reiter, P., Lathrop, S., Bunninc, M., Biggerstaff, B., Singer, D., Tiwari, T., Barber, L., Amador, M., Thirion, J., Haynes, J., Seca, C., Mendez, J., Ramirez, B., Robinson, J., Rawlings, J., Vorndam, V., Waterman, S., Gubler, D., Clark, G., Edward, H., 2003. Texas lifestyle limits transmission of dengue virus. Emerg. Infect. Dis. 9 (1), 86–89.

Ricthie, S.A., Long, S., Smith, G., Pyke, A., Knox, T.B., 2004. Entomological investigations in a focus of dengue transmission in Cairns, Queensland, Australia by using the sticky ovitraps. J. Med. Entomol. 41 (4), 1–4.

Rios, J., Hacker, C.S., Hailey, C.A., Parsons, R.E., 2006. Demographics and spatial analysis of West Nile virus and St. Louis encephalitis in Houston, Texas. J. Am. Mosq. Control Assoc. 22 (2), 254–263.

Rosen, L., Shroyer, D.A., Tesh, R.B., Freier, J.E., Lien, J.C., 1983. Transovarial transmission of dengue viruses by mosquitoes: *Aedes albopictus* and *Aedes aegypti*. Am. J. Trop. Med. Hyg. 32 (5), 1109.

Schmidt, R.F., 1989. In: Landing rates and bite counts for nuisance evaluation. Proceedings of the Seventy-Sixth Annual Meeting of the New Jersey Mosquito Control Association, Inc. pp. 34–37.

Scott, J.J., Crans, S.C., Crans, W.J., 2001. Use of an infusion-baited gravid trap to collect adult *Ochlerotatus japonicus*. J. Am. Mosq. Control Assoc. 17 (2), 142–143.

Service, M.W., 1993. Mosquito Ecology: Field Sampling Methods, second ed. Chapman & Hall. 988 pp.

Springer, Y.P., Hoekman, D., Johnson, P.T.J., Duffy, P.A., Hufft, R.A., Barnett, D.T., Allan, B.F., Amman, B.R., Barker, C.M., Barrera, R., Beard, C.B., Beati, L., Begon, M., Blackmore, M.S., Bradshaw, W.E., Brisson, D., Calisher, C.H., Childs, J.E., Diuk-Wasser, M.A., Douglass, R.J., Eisen, R.J., Foley, D.H., Foley, J.E., Gaff, H.D., Gardner, S.L., Ginsberg, H.S., Glass, G.E., Hamer, S.A., Hayden, M.H., Hjelle, B., Holzapfel, C.M., Juliano, S.A., Kramer, L.D., Kuenzi, A.J., LaDeau, S.L., Livdahl, T.P., Mills, J.N., Moore, C.G., Morand, S., Nasci, R.S., Ogden, N.H., Ostfeld, R.S., Parmenter, R.R., Piesman, J., Reisen, W.K., Savage, H.M., Sonenshine, D.E., Swei, A., Yabsley, M.J., 2016. Tick-, mosquito-, and rodentborne parasite sampling designs for the National Ecological Observatory Network. Ecosphere 7 (5), e01271. https://doi.org/10.1002/ecs2.1271.

Sudia, W.D., Chamberlain, R.W., 1962. Battery-operated light trap, an improved model. Mosq. News 22, 126–129.

Tesh, R.B., Parsons, R., Siirin, M., Randle, Y., Sargent, C., Guzman, H., Wuldhiantagool, T., Higgs, S., Vanlandingham, D.I., Balta, A.A., Hass, K., Zerinque, B., 2004. Year-round West Nile virus activity gulf coast region, Texas and Louisiana. Emerg. Infect. Dis. 10 (9), 1649–1652.

Thangamani, S., Huang, J., Hart, C.E., Guzman, H., Tesh, R.B., 2016. Vertical transmission of Zika virus in *Aedes aegypti* mosquitoes. Am. J. Trop. Med. Hyg. 95 (5), 1169–1173.

Thurman Jr., D.C., Thurman, E.B., 1955. Report of the initial operation of a light trap in northern Thailand. Mosq. News 15, 218–224.

Trexler, J.D., Apperson, C.S., Schal, C., 1998. Laboratory and field evaluations of oviposition responses if *Aedes albopictus* and *Aedes tristeriatus* (Diptera: Culicidae) to oak leaf infusions. J. Med. Entomol. 35 (6), 967–976.

Tun-Lin, W., Kay, B.H., Barnes, A., 1995a. Understanding productivity, a key to *Aedes aegypti* surveillance. Am. J. Trop. Med. Hyg. 53 (6), 595–601.

Tun-Lin, W., Kay, B.H., Barnes, A., 1995b. The premise condition index: a tool for streamlining surveys of *Aedes aegypti*. Am. J. Trop. Med. Hyg. 53 (6), 591–594.

Unlu, I., Farajollaii, A., Rochlin, I., Crepeau, T.N., Strickman, D., Gaugler, R., 2014. Differences in male-female ratios of *Aedes albopictus* (Diptera: Culicidae) following ultra-low volume adulticide applications. Acta Trop. 137, 201–205.

Walton, W.E., Workman, P.D., 1998. Effect of marsh design of the abundance of mosquitoes in experimentally conducted wetlands in southern California. J. Am. Mosq. Control Assoc. 14 (1), 95–107.

Weber, G., Horner, T.A., 1993. The ability of *Culex pipiens* and *Culex restuans* to locate small ovisites in the field. Proc. New Jersey Mosq. Control Assoc. 1992 79, 96–103.

Wilton, D.P., Kloter, K.O., 1985. Preliminary evaluation of a black cylinder suction trap for *Aedes aegypti* and *Culex quinquefasciatus* (Diptera: Culicidae). J. Med. Entomol. 22, 113–114.

Williams, G.M., Gingrich, J.B., 2007. Comparison of light traps, gravid traps, and resting boxes for West Nile virus surveillance. J. Vector Ecol. 32 (2), 285–291.

Wu, H.H., Wang, C.Y., Teng, H.J., Lin, C., Lu, L.C., Jian, S.W., Chaing, N.T., Wen, T.H., Liu, D.P., Lin, L.J., Norris, D.E., Wu, S.H., 2013. A dengue vector surveillance by human population-stratified ovitrap survey for *Aedes* (Diptera: Culicidae) adult and egg collections in high dengue-risk areas of Taiwan. J. Med. Entomol. 50 (2), 261–269.

Chapter 8

Mosquito Control

Kyndall Dye-Braumuller, Chris Fredregill, Mustapha Debboun

Mosquito and Vector Control Division, Harris County Public Health, Houston, TX, United States

Chapter Outline

Abbreviations

AchE	acetylcholinesterase
***Ae. aegypti* (L.)**	*Aedes aegypti* (Linnaeus)
Bti	*Bacillus thuringiensis* var. *israelensis*
CDC	Centers for Disease Control and Prevention
CI	cytoplasmic incompatibility
CIM	cytoplasmic incompatibility management
DDT	dichlorodiphenyltrichloroethane
GABA	gamma aminobutyric acid
IGR	insect growth regulator
IIT	incompatible insect technique
IMM	integrated mosquito management
JH	juvenile hormone
MCWA	Office of Malaria Control in War Areas
NPDES	National Pollutant Discharge Elimination
OMWM	open marsh water management
OP	organophosphates
PAHO	Pan American Health Organization
PHS	Public Health Service
SIT	sterile insect technique
TF	thermal fog
UAS	unmanned aerial systems
ULV	ultralow volume
UNICEF	United Nations International Children's Emergency Fund
US	United States
WHO	World Health Organization
YFC	Yellow Fever Commission

Mosquitoes, Communities, and Public Health in Texas. https://doi.org/10.1016/B978-0-12-814545-6.00008-0

8.1   History of mosquito control in the United States

Globally, mosquito control has evolved over centuries, as humans have learned more about the world's deadliest animal, that is, the mosquito, and the pathogens they are capable of transmitting. Mosquitoes not only transmit pathogens but also cause severe allergic reactions, intense irritation, and even extensive blood loss to some animals such as livestock. These factors have led to the development of a multitude of techniques and approaches to control mosquitoes throughout centuries. They have been established since Sir Patrick Manson first linked mosquitoes to the transmission of human pathogens when he discovered that the house mosquito harbored and presumably—at that time—was transmitting intermediate larvae of the filarial worm, *Wuchereria bancrofti* (Manson, 1878; Chernin, 1983; Clements and Harbach, 2017). Due to the wide range of impacts mosquitoes have on humans, large quantities of resources have been dedicated to their control. For example, the United States (US) spends hundreds of millions of dollars to control mosquitoes annually (Foster and Walker, 2009). There were a few major points in US history that led to these expansive control efforts, and the first part of this chapter will describe these milestones.

8.2   Yellow fever

The well-known successes of the various yellow fever eradication campaigns in the early 1900s were a turning point for mosquito control in the Western Hemisphere (Tonn et al., 1982). Yellow fever epidemics are believed to have occurred in the Americas for centuries, with the first recorded epidemic occurring in the Yucatan Peninsula in 1648, which is believed to have been a part of a larger 1647–49 epidemic affecting multiple Caribbean islands (Soper, 1944; Monath, 1988; Gubler, 2004). For the next 150 years, yellow fever epidemics were recorded throughout South, Central, Latin, and North America, including both the Atlantic and Pacific coasts of the United States (Soper, 1958; Schleissmann, 1964; Gubler, 2004; Foster and Walker, 2009; Gould et al., 2017). Epidemics were commonly seen as far north as Boston, Massachusetts, in the United States, and these were regarded as both deadly, terrifying, and economically destructive (Schleissmann, 1964; Williams, 1964; Schleissmann and Calheiros, 1974; Patterson, 1992; Gubler, 2004). The case fatality rate for yellow fever was approximately 20%; these epidemics caused significant mortality and morbidity in the Americas (Monath, 1988; Gubler, 2004). Hundreds of workers fell ill during the construction of the Panama Canal (Fig. 8.1), and over 22,000 died before mosquito control efforts were initiated (Patterson, 1992; Güereña-Burgueño, 2002; Gould et al., 2017). Thousands of Americans died in the continental United States as well: a 1793 outbreak in Philadelphia caused 4000 deaths (Foster and Walker, 2009); New Orleans recorded more than 41,000 yellow fever deaths from 1817 to 1905 (Güereña-Burgueño, 2002).

FIG. 8.1   The Panama Canal. This image depicts the locks at the Pacific Ocean–facing portion of the Panama Canal. Thousands of laborers died due to the disease-carrying mosquitoes of the region before this engineering feat was completed. *(Photo from the CDC Public Health Image Library/Dr. Edwin P. Ewing Jr.)*

Yellow fever was a public health crisis. Requested by the Secretary of War at the time, a board of medical officers was appointed in 1900 to study yellow fever in Cuba and other Caribbean islands, which was later named the Yellow Fever Commission (YFC) (Güereña-Burgueño, 2002; Clements and Harbach, 2017). The four members comprising the commission were US Army officers Walter Reed, James Carroll, Aristides Agramonte, and Jesse Lazear (Güereña-Burgueño,

2002; Foster and Walker, 2009; Clements and Harbach, 2017). Applying the theory of Cuban physician Carlos Finlay (from 1881) that *Culex fasciatus* (now *Aedes aegypti* (L.)) carried the agent for yellow fever (Finlay, 1881; Clements and Harbach, 2017), the YFC conducted various experiments including allowing potentially infectious mosquitoes to feed on themselves and volunteers (Reed et al., 1900; Güereña-Burgueño, 2002; Clements and Harbach, 2017). Reed and the Commission proved this theory correct, and yellow fever became the first arbovirus recognized as a mosquito-borne agent (Schleissmann, 1964; Güereña-Burgueño, 2002; Gubler, 2004; Foster and Walker, 2009).

Following this transmission cycle breakthrough, the Chief Sanitary Officer in the US Army, William Gorgas, began a program to control the *Ae. aegypti* mosquito, specifically in Havana, Cuba (Schleissmann, 1964; Güereña-Burgueño, 2002; Foster and Walker, 2009). Work was done to isolate cases, eliminate mosquito breeding sites, treat inside homes, and install screens over windows and doors throughout the region (Güereña-Burgueño, 2002). Operations were targeted in urban centers, adjacent to human dwellings, and it was of key importance to have the cooperation and support of the public and local health agencies (Schleissmann, 1964). Various members of the YFC suggested that yellow fever eradication was actually feasible (Strode, 1951; Schleissmann and Calheiros, 1974). Yellow fever incidence and the number of epidemics drastically fell, and within 8 months, yellow fever was eradicated from Cuba (Soper, 1944, 1958; Schleissmann, 1964; Monath, 1988; Gubler, 2004). Gorgas then applied the same mosquito tactics in Panama in 1904, and urban yellow fever was successfully eliminated in 16 months (Schleissmann, 1964).

Due to the dramatic decrease in yellow fever incidence in these endemic centers, yellow fever eradication programs focusing on reduction of mosquito breeding habitats were adopted in additional areas, including Rio de Janeiro (Soper, 1958). Public health officials noted how there was a reduction in disease in areas where mosquito control was enacted, but also in the adjacent and surrounding cities and areas (Soper, 1958). In 1915 the Rockefeller Foundation took a heavy role in the *Ae. aegypti* control programs in the Americas based on this success (Soper, 1958; Schleissmann, 1964; Güereña-Burgueño, 2002). Following these mosquito control efforts, it appeared yellow fever would be eliminated from the Western Hemisphere, as only a small portion of Northeast Brazil was the last remaining endemic area by 1925 (Strode, 1951; Soper, 1963; Schleissmann, 1964; Schleissmann and Calheiros, 1974; Gubler, 2004). In the United States, *Ae. aegypti* mosquito control efforts successfully reduced yellow fever epidemics as well. The last recorded epidemic in the United States was in New Orleans in 1905; a significantly lower risk rate was observed for this particular outbreak, and the last recorded case of yellow fever in the United States was in 1924, which was an introduced case (Schleissmann, 1964).

Unfortunately, this victory over yellow fever did not last long. Due to a lack in the understanding of the ecology and biology of both the vector mosquito and the virus, yellow fever emerged once again in the Americas in 1928. Rio de Janiero reported yellow fever for the first time in the Americas in 20 years; soon, outbreaks were seen in isolated cities and towns in adjacent countries as well (Soper, 1958; Schleissmann, 1964). These outbreaks led to the discovery of the sylvatic, or jungle, transmission cycle of yellow fever, which is maintained by lower apes and mosquitoes other than *Ae. aegypti* (Fig. 8.2A and B) (Soper et al., 1933; Soper, 1958, 1963; Schleissmann, 1964; Schleissmann and Calheiros, 1974; Tonn et al., 1982; Gubler, 2004; Foster and Walker, 2009). The hope of yellow fever eradication through *Ae. aegypti* control faded

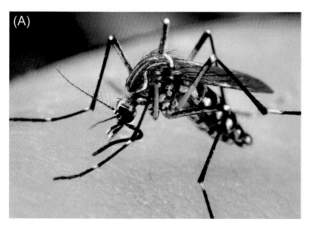

FIG. 8.2 (A) The *Aedes aegypti* (L.) mosquito. This mosquito breeds in urban environments, specifically near human dwellings. The *white* lyre shape on the dorsal portion of the thorax is a key characteristic for this species. (Photo from the CDC Public Health Image Library/Prof. Frank Hadley Collins, Dir., Cntr. for Global Health and Infectious Diseases, Univ. of Notre Dame.) (B) *Sabethes* mosquito. Compared to the *Aedes aegypti* mosquito, this *Sabethes* mosquito is much more colorful, but just as deadly. Some *Sabethes* spp. are known to vector yellow fever in the jungle, or sylvatic cycle. These mosquitoes typically undergo their larval and pupal stages in tree holes in the vast jungles of Central and South America. (Photo from the CDC Public Health Image Library/Prof. Woodbridge Foster; Prof. Franklin H. Collins.)

fast following this discovery, as jungle yellow fever outbreaks spread from Guatemala to Argentina; when it entered towns, subsequent outbreaks vectored by *Ae. aegypti* followed soon after (Soper, 1958, 1963; Tonn et al., 1982). A small break, the development of the yellow fever vaccine in the late 1930s, offered preventive relief for the virus as this was distributed widely (Schleissmann and Calheiros, 1974; Stapleton, 2004; Barrett, 2017). Urban yellow fever seemed to be controllable to a point, but jungle yellow fever control and its subsequent spillover into urban centers were seemingly uncontrollable.

Following this setback, multiple agencies; health authorities, including the Pan American Sanitary Organization (now the Pan American Health Organization); and countries met and agreed that a successful eradication program depended on preventing reinfestation from neighboring countries (Soper, 1963; Schleissmann, 1964; Tonn et al., 1982). Eradication programs were initiated around the Americas to secure a stronger hold on the viral disease. Community cleanup activities, mosquito breeding site reduction, insecticidal treatments of breeding sites and homes, and general sanitary activities were launched countrywide throughout Latin, Central, and South America in 1947 (Soper, 1963; Schleissmann, 1964; Schleissmann and Calheiros, 1974). Eradication programs in many countries were slow and uncoordinated: political instability, lack of resources, frequent infestation from surrounding areas, and overall lack of program organization contributed to limited complete *Ae. aegypti* eradication progress throughout South, Central, and Latin America (Soper, 1963; Schleissmann and Calheiros, 1974; Gubler, 2004).

The United States was the last country with *Ae. aegypti* infestations to join the eradication efforts; many Latin and Central American countries thought this apparent lack of enthusiasm was a mistake as infestation of *Ae. aegypti* mosquitoes from the United States was possible and would be detrimental to the eradication progress (Soper, 1963; Schleissmann, 1964). The US Public Health Service (PHS) and Communicable Disease Center (former Malaria Control in War Areas) formulated an eradication plan for the vector including surveillance of the *Ae. aegypti* vector population, preparedness, treatment methods, and even a pilot eradication program in 1962 (Figs. 8.3–8.5) (Hayes and Tinker, 1958; Tinker and Hayes, 1959; Tinker, 1963; Schleissmann, 1964; Morlan and Tinker, 1965). The United States successfully prevented yellow fever from causing anymore outbreaks on its soil; however, *Ae. aegypti* populations are still breeding throughout the country today.

Although urban yellow fever has been controlled to an acceptable level in Latin, South, and Central America, there is a serious risk for sylvatic yellow fever, also known as jungle yellow fever, to readily spill over into the urban cycle and cause widespread epidemics in larger, more transient populations of today (Gubler, 2004; Gould et al., 2017). Over the past couple of decades, populations of these cities have grown exponentially, mosquito control capabilities have decreased substantially, and, even though there is a vaccine available, the global supply is not nearly adequate enough for the possible size of an outbreak in these larger cities (Gubler, 2004; Roukens and Visser, 2008; Barrett, 2017; Gould et al., 2017; Paules and Fauci, 2017).

FIG. 8.3 Mosquito dipping. This photo was taken during the 1965 *Aedes aegypti* eradication program in Miami, Florida, United States. This technician is investigating this site for mosquito larvae that can breed in the multiple containers capable of holding water. *(Photo from the CDC Public Health Image Library.)*

FIG. 8.4   Mosquito spray truck. This truck is from the 1965 *Aedes aegypti* eradication program in Miami, Florida, United States. *(Photo from the CDC Public Health Image Library.)*

FIG. 8.5   Aerial application. A Ford Trimotor aircraft sprays insecticides to control mosquitoes in the Southern United States in 1930. *(Photo from the CDC Public Health Image Library.)*

8.3   Malaria

Overlapping slightly with the birth of yellow fever control programs in the early 1900s, malaria eradication in the United States was also a milestone in the history of mosquito control. Malaria is believed to have been introduced to North America with early immigrants from Europe, where it was common (Bradley, 1966; Russell, 1968). Malaria became firmly established in the United States during the slave trade, as many slaves were infected with more virulent strains of the parasite (Bradley, 1966; Zucker, 1996). It became prevalent throughout the colonies for decades, and by 1850, it had become established in every settlement except mountainous areas, deserts, and Northern New England (Faust, 1951; Bradley, 1966; Russell, 1968; Zucker, 1996; Stapleton, 2004; Foster and Walker, 2009). Malaria was a serious rural disease in the southeastern states, along with California, and was regarded as one of the most important diseases affecting public health—globally and in the United States (Schleissmann, 1964; Bradley, 1966; Zucker, 1996; Mendis et al., 2009). An estimated 600,000 cases occurred in 1914, and only about 135,000 were reported in 1934 (Bradley, 1966; PAHO, 1969; Zucker, 1996). This dramatic decrease in cases was due to a general movement of people to urban areas, malaria treatment options becoming available, increased quality in housing, drainage, etc. (Bradley, 1966; PAHO, 1969; Zucker, 1996; Stapleton, 2004; Foster and Walker, 2009). Despite the drop in cases, malaria was still a major threat to public health locally and globally. In 1935 approximately 4000 people died from malaria in the United States (Russell, 1968; Mendis et al., 2009). It was also a threat economically. Williams (1938) estimated that the morbidity, mortality, and unproductiveness caused by malaria during the 1930s were costing the US $500 million annually (Bradley, 1966).

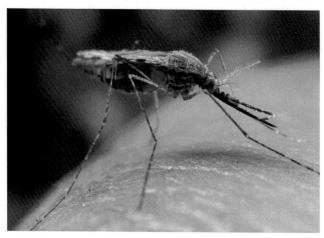

FIG. 8.6 *Anopheles quadrimaculatus* Say. This mosquito is native to the United States and was the main malaria vector when this disease was threatening the United States. *(Photo from the CDC Public Health Image Library/James Gathany.)*

Malaria control in the United States was enhanced greatly by the discovery of anopheline mosquitoes as the vectors in the early 1900s (Fig. 8.6) (Ross, 1923; Russell, 1955; Foster and Walker, 2009; Cox, 2010). Chief Sanitary Officer, William Gorgas, expanded the US Army's *Ae. aegypti* control measures in Cuba to include *Anopheles* mosquitoes in 1901. Control measures included drainage (Fig. 8.7A and B), brush and grass maintenance, soluble larvicides, quinine distribution for prophylaxis, screens, etc. (Russell, 1955; Bradley, 1966). Towns and cities throughout the United States initiated small and local efforts to control *Anopheles* mosquitoes as bed nets, screens, and draining ditches became common household items and practices (Matheson, 1941; Bradley, 1966; Russell, 1968; Stapleton, 2004).

Antimalaria programs advanced quickly during World War I (1917–18) as the US Army and PHS strived to protect civilians and soldiers. Approximately $3.25 million was spent on malaria control on military bases in the United States alone during this time (Bradley, 1966; Russell, 1968). Following World War I, various institutions became involved with malaria control and research around the country, including the Rockefeller Foundation, the Tennessee Valley Authority, and the Federal Bureau of Entomology—all working with the US PHS (Bradley, 1966). These groups encouraged local health departments to promote control activities in their jurisdictions and made strides in research—including the development of paris green as a larvicide (Barber and Hayne, 1921; Russell, 1955, 1968; Bradley, 1966). Paris green, an inorganic compound with arsenic, significantly simplified mosquito control efforts and was a great addition to the mechanical and habitat modification techniques already in place (King and Bradley, 1926; Russell, 1968; Stapleton, 2004). Malaria control programs were small and localized, but effective, throughout the United States during this time.

Eventually, a common consensus was reached throughout the nation that a carefully constructed, countrywide plan was needed for successful and permanent malaria eradication (Williams, 1937; Mountin, 1944; Bradley, 1966; Russell, 1968). As World War II began, the PHS was mainly in charge of malaria control activities again. In 1942 the PHS established the Office of Malaria Control in War Areas (MCWA) (Fig. 8.8)—the precursor to the current Centers for Disease Control and Prevention (CDC)—headquartered in Atlanta, GA (Bradley, 1966; Stapleton, 2004). This office advocated for a national coordinated mosquito control plan using larvicides (various oils, paris green, or pyrethrum emulsions) and adulticiding with dichlorodiphenyltrichloroethane (DDT) when it became available (Fig. 8.9A and B); drainage and filling were only used when necessary (Bradley, 1966; Russell, 1968; Zucker, 1996). In addition, mosquito surveillance was carefully recorded in order to measure malaria transmission risk throughout the Southern United States (Bradley, 1948, 1966). This national program not only prevented the disease from impacting the war effort but also prevented the predicted increase in malaria incidence for the 1940s (Williams, 1941; Andrews, 1951; Soper, 1963; Schleissmann, 1964; Bradley, 1966).

This national program evolved into the Extended Malaria Control Program, which aimed to eliminate malaria from the entire United States, which was approved by Congress in 1945 (Bradley, 1966; Russell, 1968). Following this, the CDC was officially founded in Atlanta, GA, charged with preventing malaria from spreading across the United States (Fig. 8.10). The mosquito control procedures from the MCWA were encouraged, and old proven methods of malaria control were not abandoned: the use of DDT for indoor applications, increased installation of screens on windows and doors, and drainage of *Anopheles* breeding habitats and larviciding until the local infection died out (Bradley, 1966). A second tier of the program consisted of epidemiologic investigations from state health departments into any and all reported or discovered malaria cases: infected patients were promptly interviewed and treated, and their homes (and neighbors'

FIG. 8.7 (A) Early malaria control. This photograph shows three malaria control health workers from the 1920s. They are about to use dynamite to remove tree stumps to create ditches to drain standing water. (Photo from the CDC Public Health Image Library.) (B) Early malaria control. This image depicts a group of workers in the 1920s digging a drainage ditch to reduce *Anopheles* mosquito breeding areas in the Southern United States. (*Photo from the CDC Public Health Image Library.*)

homes) were sprayed with DDT immediately to prevent additional autochthonous cases (Bradley, 1966; Zucker, 1996). This type of "attritional eradication" depended on the concurrent reduction of *Plasmodium* parasites in humans with mosquito control activities to a point where malaria transmission could not occur, as vector control and parasite eradication from the population were not independently feasible (Andrews and Gilbertson, 1948; Russell, 1968). The CDC shifted focus to solely malaria surveillance and prevention activities at the end of the program, and these have continued since (Bradley, 1966; Russell, 1968). This program cost approximately $27.5 million and was successful across the entire country at eradicating malaria from the United States (Andrews, 1951; Bradley, 1966; Russell, 1968). In 1952 the final year of the Extended Malaria Control Program, only 25 malaria deaths were recorded, compared to approximately 4000 deaths in 1935, and from 1961 to 1966, only four malaria deaths were recorded in the United States—all of which were imported (Bradley, 1966).

FIG. 8.8 Malaria Control in War Areas. Three public health professionals carry out tasks for the office of Malaria Control in War Areas in the 1940s. *(Photo from the CDC Public Health Image Library/K. Lord.)*

FIG. 8.9 (A) Mosquito larviciding. This man applying an insecticidal oil, which forms a layer on top of the water surface of puddles such as shown in the photo to prevent any mosquito larvae from breathing through their siphons, thus killing them. (Photo from the CDC Public Health Image Library.) (B) DDT applications. This man is spraying a chemical mixture containing DDT following a flood in Connecticut, United States. Community applications of DDT were extremely common and welcome in neighborhoods to get rid of mosquitoes and other insect pests. (Photo from the CDC Public Health Image Library.)

FIG. 8.10 The founder of CDC. Dr. Joseph W. Mountin founded the Centers for Disease Control and Prevention in 1946. He envisioned multiple "Centers of Excellence," which transformed the wartime malaria control efforts to the CDC of today. *(Photo from the CDC Public Health Image Library/ Jarrad Hogg.)*

Globally, malaria control efforts were also picking up steam around this time. In 1954 a malaria eradication program with a healthy budget was initiated from the Pan American Sanitary Conference, and in 1955 the Eighth World Health Assembly approved a resolution for 46 participating nations to work toward malaria eradication globally (Johnson, 1956; WHO, 1956; Soper, 1963; Stapleton, 2004; Mendis et al., 2009). Malaria eradication was also declared a global interest of the United Nations International Children's Emergency Fund (UNICEF) and the WHO; this was pushed so strongly in order to eradicate the parasite before the *Anopheles* vectors developed resistance to the commonly used adulticides at the time (Johnson, 1956; Soper, 1963). Because of this campaign, 37 out of 143 malaria-endemic countries (determined endemic in 1950) were free from malaria by 1978 (Wernsdorfer, 1980; Mendis et al., 2009). The global burden of malaria was drastically reduced. For example, Sri Lanka reported 18 cases in 1966, compared to 2.8 million cases in 1946 (Mendis et al., 2009).

Unfortunately, due to a combination of factors, malaria incidence and mortality began to drastically increase again around the world. The unsustainability of the global eradication campaign, parasitic resistance to chloroquine, and mosquito resistance to DDT all contributed to the decline of malaria eradication (WHO, 1969; Mendis et al., 2009). The goal of eradication was thus rejected, and malaria control became the global philosophy (Mendis et al., 2009).

8.4 Breakthroughs in mosquito control

Briefly mentioned in the earlier sections on malaria and yellow fever eradication in the United States, there were multiple breakthroughs that catapulted control efforts forward, including advancements in habitat and environmental understanding, as well as chemical control. After malaria, yellow fever, and other mosquito-borne diseases were definitively linked to their Culicine vectors, control efforts focused on mosquito habitats and behaviors (Simmons and Upholt, 1951; Russell, 1955; Soper, 1958; Bradley, 1966; Güereña-Burgueño, 2002). Mosquito netting and screening were implemented to prevent blood-feeding behavior, isolation of cases was key for preventing further local transmission, and drainage, brush maintenance, and elimination of water-holding receptacles were key tactics for habitat modification (Barber and Hayne 1921; Matheson, 1941; Simmons and Upholt, 1951; Russell, 1955, 1968; Soper, 1958, 1963; Schleissmann, 1964; Bradley, 1966; Schleissmann and Calheiros, 1974; Zucker, 1996; Güereña-Burgueño, 2002; Stapleton, 2004). Preventing contact between the vector and humans was the philosophy of the time (Andrews, 1950). Gradually, techniques targeting the mosquito vectors specifically were added after their efficacious discoveries, like paris green and various oils (Barber and Hayne, 1921; Andrews, 1950; Russell, 1955, 1968; Soper, 1963; Schleissmann, 1964; Bradley, 1966; Schleissmann and Calheiros, 1974; Zucker, 1996). Some advancements in general sanitation are also noted as having influenced the war on mosquitoes during this time. Community water sources, fire station water barrels, and water tanks for trains were all eliminated, leading to fewer *Ae. aegypti* habitats inadvertently (Schleissmann, 1964). In addition, improvements in community health and urbanization helped reduce transmission and clear *Anopheles* breeding habitats (Schleissmann, 1964).

Chemical control breakthroughs, specifically regarding adulticides, contributed to much of the mosquito control success in the United States and the rest of the world. DDT was one of the most notable adulticides that changed mosquito control forever. Although this chlorinated hydrocarbon was first synthesized in the 1800s, its insecticidal properties were not discovered until the 1930s (Berry-Cabán, 2011; Prato et al., 2013). It was found to be a potent pediculicide in the early 1940s in the United States, and tests for mosquito control were conducted shortly after (Stapleton, 2004). DDT was tested against US-native *Anopheles* mosquitoes, and its potency was extremely promising (Gahan et al., 1945; Bradley, 1966; Stapleton, 2004; Berry-Cabán, 2011). Its residual activity was of special note, as its insecticidal effects lasted for weeks or months in the environment following its application (Upholt et al., 1947; Bradley, 1966; Stapleton, 2004; Prato et al., 2013). Longer residual activity meant fewer applications, saving time and money. It also played a key role in the World War II effort by using it to protect military personnel from typhus and malaria, both arthropod-borne diseases capable of causing high morbidity and mortality in military personnel (Fig. 8.11) (Beard and ARHRC, 2005; Prato et al., 2013). In 1945 the Extended Malaria Program began widely utilizing DDT operated by the PHS, and DDT was broadly accepted and even welcomed by the general public during this campaign for mosquito control efforts indoors and outdoors for both yellow fever and malaria (Upholt et al., 1947; Paoliello, 1948; Bradley, 1966; Tonn et al., 1982; Stapleton, 2004; Eskenazi et al., 2009; Prato et al., 2013). Its use was recommended for any and all control campaigns during this time (Soper, 1963; Schleissmann, 1964; Bradley, 1966; Russell, 1968; Schleissmann and Calheiros, 1974; Zucker, 1996; Güereña-Burgueño, 2002). Following the world war, DDT remained the most used and most publicized synthetic chemical in the world (Berry-Cabán, 2011; Prato et al., 2013). It was even credited with the eradication of malaria from both the United States and Europe (Johnson, 1956; Attaran et al., 2000; Gubler, 2004; Beard and ARHRC, 2005; Dash et al., 2007). DDT's efficacy and impact on mosquito control efforts globally were so welcomed that multiple public health and mosquito control professionals declared any mosquito control program without the chemical were worthless (Simmons and Upholt, 1951; Stapleton, 2004).

Despite its extraordinary success, public health and mosquito control officials knew its potency would not last forever as its use increased annually; resistance to DDT would develop in mosquito populations at some point (Prato et al., 2013). Resistance was documented in the common house fly as early as 1946; however, this was not widely reported at the time (Metcalf, 1989; Stapleton, 2004; Prato et al., 2013). Anopheline mosquitoes also began to develop resistance to the

FIG. 8.11 DDT applications. These soldiers are demonstrating the hand sprayer application of DDT for protection against multiple pathogen-carrying arthropods. *(Photo from the CDC Public Health Image Library.)*

chlorinated hydrocarbon—many species of which were vectors of malaria (Johnson, 1956; Hemingway and Ranson, 2000; Prato et al., 2013). The eradication campaign was undoubtebly mortally wounded by this development (Johnson, 1956; Tonn et al., 1982). By 1960, over 130 species of insect pests developed some resistance to DDT, and by 1984, Metcalf (1989) demonstrated that over 200 species of insects were resistant to DDT (Stapleton, 2004).

The development of resistance to DDT was followed shortly by concern over the environmental impacts of the chemical. Its residual activity in the environment became a double-edged sword, as DDT's half-life was reported around 6 months in tropical soils (Varca and Magallona, 1994; Wandiga, 2001) and even up to 15 years in temperate soils (Ritter et al., 1995; Prato et al., 2013). Massive amounts of scientific research efforts were dedicated to investigating the deleterious effects of DDT and its metabolic products on humans, animals, and the environment, and the results did not support its continued use (Stapleton, 2004; Beard and ARHRC, 2005; Dash et al., 2007; Berry-Cabán, 2011). DDT and its metabolic products bioaccumulated readily in the environment, and animals at the top of the food chain were found to be the most impacted by it (Stapleton, 2004; Beard and ARHRC, 2005; Dash et al., 2007; Eskenazi et al., 2009; Prato et al., 2013). Eventually due to these environmental concerns, DDT was banned, or its use was heavily restricted, in most developed countries by 1970 (Casida, 1980; Beard and ARHRC, 2005; Dash et al., 2007; Eskenazi et al., 2009; Prato et al., 2013).

Coinciding with the ban and restrictions on DDT for mosquito control, another class of insecticides was introduced that advanced control efforts. In the early 1980s, synthetic pyrethroids became very important for mosquito control (Elliott and Janes, 1978; Prato et al., 2013). Based on the natural, yet unstable in sunlight and air, pyrethrins produced by flowers of the genera *Chrysanthemum* and *Tanacetum*, synthetic pyrethroids were created to mimic the insecticidal and repellent activities of these compounds over a century ago while also increasing their stability in the environment (Elliott, 1976; Elliott and Janes, 1978; Casida, 1980). In addition, these natural compounds were found to be relatively safe for humans and animals compared to the many options before, including DDT and other chlorinated hydrocarbons or organophosphorus insecticides, and were found to be metabolized easily by mammals as well (Elliott, 1976; Casida, 1980). Many of these pyrethroids demonstrated adequate efficacy against multiple mosquito species of public health importance, marking their debut as a cost-friendly and environmentally friendly alternative to the previous organochlorine and organophosphorus compounds (Darwazeh and Mulla, 1974; Mulla et al., 1975, 1978; Elliott, 1976). Four of the most important pyrethroids for public health pest control were described in the 1970s: permethrin, cypermethrin, decamethrin, and fenvalerate (Casida, 1980). Research began to continue to modify the synthetic pyrethroids to optimize low mammalian toxicity, potency, and suitable persistence to reduce toxicity to fish, honeybees, and predators (Elliott et al., 1978; Mulla et al., 1978; Casida, 1980). Although resistance to these synthetic compounds developed as well, public health professionals continued to utilize them in mosquito control with more forethought and caution than in previous years (Elliott, 1976).

There are many lessons to learn from the era of mosquito control in the United States and the world during the 1900s. Mosquito control technologies and tools have evolved significantly. As demonstrated through the many years of mosquito control discussed, chemical control is not the only tool in the public health pest control toolbox. The rest of the chapter will describe the multiple approaches and tools available for mosquito control today and how they can work together in a successful integrated mosquito management (IMM) program.

8.5 Overview of mosquito control strategies

Effective mosquito control cannot be attained using a single approach or strategy. Therefore, all responsible mosquito abatement programs are based on the principles of integrated pest management (IPM) or IMM. The IMM is not a list of specific actions to take but instead a toolbox that can be adapted and applied to the specific requirements of each situation. Integrated mosquito management "… is a comprehensive mosquito prevention and control strategy that utilizes all available mosquito control methods, either singly or in combination, to exploit the known vulnerabilities of mosquitoes to reduce their numbers while maintaining a quality environment" (AMCA, 2017). An integrated program consists of five parts: (1) background knowledge, (2) surveillance activities, (3) multitactic management, (4) continuous program evaluation, and (5) continued education. All of these steps are crucial to the success of a mosquito abatement program and the protection of humans, animals, and the environment. We will focus on the multitactic management and control of mosquitoes.

When planning control efforts in a given area, it is imperative to have the most up-to-date surveillance information and that the data meet the action threshold(s) previously set. Treating an area without surveillance data and background knowledge to back it up is irresponsible, wasteful, unsustainable, and potentially damaging to the environment. It is incumbent upon all mosquito control professionals to use the available tools judiciously and educate the public in order to minimize the impact of mosquito control efforts on the environment while preventing mosquito-borne diseases and maintaining mosquito populations at tolerable levels.

Once the decision to undertake mosquito control activities is made, one must rely on the background knowledge and surveillance to determine what life stage is to be controlled and which methods will be used to attain maximum efficacy. Common strategies are habitat modification and larval source reduction, biological control, chemical control, and public education. It is important to keep in mind that controlling the larvae of mosquitoes is always the most efficient as they are "concentrated, immobile, and accessible"; once the mosquitoes emerge and disperse, they are much more difficult to control. However, if an active mosquito-borne disease is isolated in an area, some form of adult control measures must be undertaken to kill potentially infected female mosquitoes and limit the spread of the mosquito-borne disease.

8.6 Source reduction

The most effective way to affect long-term control on mosquito populations is through the use of source reduction techniques. Since source reduction addresses the root cause of a mosquito issue (primarily larval habitats), if thoroughly conducted and maintained, it can reduce or eliminate the need for other control strategies. As the name implies, source reduction is physically reducing the availability of sources of mosquito production, which can be as simple as homeowners eliminating artificial containers on their property all the way to wetland modification and management practiced by some agencies along the coasts. The type of source reduction depends on the species of mosquito requiring control. For example, reducing the sources of *Aedes albopictus* (Skuse) in which a homeowner can conduct much of the source reduction, will look quite different from efforts for *Aedes sollicitans* (Walker), which will require water level and vegetation management and possibly ditching of salt marshes.

The actions that residents can take are the primary focus of the education and outreach messaging performed by public health authorities from the Centers for Disease Control and Prevention to the local mosquito control organization and health departments.

The most common actions residents or property owners can take to reduce sources of mosquitoes are as follows:

(1) Keep property free of unwanted artificial containers, tires, trash, and debris
(2) Drain buckets or containers (drill holes or overturn to prevent future water holding)
(3) Cover/screen rain collection devices
(4) Change water in pet bowls and bird baths twice per week (every 3 days)
(5) Keep roof rain gutters clean and flowing
(6) Fill in low spots in lawn or tree holes with sand
(7) Report or fix water leaks immediately
(8) Maintain swimming pools and/or septic systems if applicable
(9) Do not sweep debris/lawn waste into storm drain systems
(10) Eliminate thick/overgrown vegetation (adult harborage)

The actions typically undertaken by organized mosquito control to reduce the sources of mosquitoes include the following:

(1) Conducting property inspections and showing residents "cryptic" or hidden breeding sites
(2) Collaborating with other agencies to dispose/recycle discarded tires
(3) Working with infrastructure departments to ensure proposed water management projects will not produce mosquitoes (either drain completely or harbor predators) including vegetation management
(4) Removing trash/debris from ditches to allow draining
(5) Working with authorities for water management techniques to prevent mosquito production on dredge spoil sites and marshes

Larger area-wide actions are often undertaken in areas producing mosquito populations too large to be handled by individuals and may often require the joint efforts of several mosquito organizations. Many times, these efforts can be as simple as properly grading sites ensuring proper drainage of areas holding water unnaturally or controlling vegetation that is impeding the natural flow of water.

One technique used in the northeast and certain portions of the Gulf Coast is open marsh water management (OMWM) tactics. The OMWM involves creating radial ditches from more permanent pools in salt marshes (Wolfe, 1996). The idea behind these tactics is to concentrate mosquito larvae and predators into refuge areas to be preyed upon during periods of low water and allow predators to expand out of the refuge and into the flooded areas much faster when waters are high. The OMWM and all habitat modifications in salt marsh and spoil sites should be exercised with great caution as salt marshes and wetlands are habitats and home to many species, which must be factored in before any steps are taken.

Source reduction is extremely effective and efficient when executed fully, but unfortunately, it is very time-consuming for homeowners and labor intensive for mosquito control organizations making it very difficult to maintain long term, particularly in large urban areas of varying socioeconomic statuses. Therefore, source reduction remains a critical first step in any IMM program but is far from a silver bullet and must be incorporated with other available control tools and techniques.

8.7 Biological control

Biological control, or using a mosquito's natural enemies against it, includes a variety of organisms, which fall broadly into the following categories: vertebrates (fish), invertebrates (predacious mosquitoes, copepods), fungi, nematodes, bacteria, and viruses. The exploration for new biocontrol agents is ongoing, but for the purposes of this chapter, we will discuss the biological agents commonly used in Texas. When using aquatic biological control organisms in an area, it is best to use nonprotected species, native to the area, and only release them at sites that are contained (abandoned swimming pools, rain barrels, containers, etc.) and do not connect to natural bodies of water. The reason for the strict release standards is that naturally occurring bodies of water likely already have existing populations of predators; it is not advisable to waste resources by applying additional predators. Careful application can control mosquito populations while limiting environmental disruption and potential legal liability. Always consider the environmental implications before releasing a biocontrol agent into the environment.

Historically, one of the most common "traditional" biological control agents used is the mosquitofish: *Gambusia affinis* (Western mosquitofish) and *Gambusia holbrooki* (Eastern mosquitofish). *Gambusia* now exist on every continent except Antarctica after being introduced for mosquito control purposes (Pyke, 2008). They are highly adaptable, fast-reproducing, voracious predators, which make them a successful biocontrol candidate. Unfortunately, *Gambusia* are not mosquito specific, but are aggressive and can become an environmental detriment in naïve habitats, reducing or eliminating existing mosquito predators and negatively affecting endangered species (Cheng et al., 2018). More recent data have also shown that in different locales and temperatures, native freshwater fish may outperform *Gambusia* in controlling mosquito populations (Lawrence and Hamilton, 2016). Thus, always consider the effects that larvivorous fish may have on an environment and consult with local wildlife authorities before releasing them in a particular habitat or jurisdiction.

The mosquitoes in the genus *Toxorhynchites* (Theobald) are those that are predacious as larvae and do not bite/take a blood meal as adults (Carpenter and Lacasse, 1955). The species most commonly encountered locally in Texas is *Toxorhynchites rutilus septentrionalis* (Dyar and Knab). *Toxorhynchites* will lay eggs in a multitude of natural and artificial containers, often the same containers inhabited by *Ae. aegypti* and *Ae. albopictus* (Focks, 2007). Adult *Toxorhynchites* females do not feed on blood but instead on nectar from flowers, making them appealing candidates for biological control. Adult female *Toxorhynchites* releases were tested several times in Louisiana in the 1980s and showed some promise when combined with ultralow volume adulticiding activities (Focks et al., 1985, 1986), although cannibalism has been reported as a drawback, possibly limiting their efficacy (Focks et al., 1982; Schreiber, 2007). Schreiber and Jones (1994) were unable to obtain satisfactory colonization in natural containers in Florida. Many of the "Tox" rearing and release strategies have been discontinued due to the difficulty in rearing and limited success affecting control in an area.

Bacillus thuringiensis var. *israelensis* (*Bti*) is a bacteria that produces a protoxin that is highly toxic when ingested by mosquito larvae. The *Bti* bacteria has the advantage of being highly specific to nematocerous flies especially mosquitoes (Floore, 2006). It has been used as a larvicide for decades due to its relatively low cost, variety of formulations, fast-acting nature, and lack of notable resistance. The *Bti* can be combined with other slower-acting residual larvicides to achieve a quick kill. It has also been combined with products to mitigate resistance to other actives such as *B. sphaericus* (Ahmed et al., 2018; De Mendonça Santos et al., 2019). The drawback of *Bti* is its short-term residual in field environments (Maracombe et al., 2018), but some new formulation technologies are able to extend its release and mitigate the "flash release" of many prior products.

Lysinibacillus sphaericus formerly known as *B. sphaericus* is another commonly applied biological control method targeting primarily *Culex, Anopheles*, and *Aedes* species (Berry, 2012). The advantage of *L. sphaericus* is that it has a recycling effect allowing a prolonged control period in the environment while demonstrating no significant effects on cohabiting nontarget organisms (Floore, 2006; Derua et al., 2018). There are many formulations of this product including "toss its," which can be applied by hand or water-dispersible granules, which can be mixed with water and applied over wide areas with hand tanks, backpack sprayers, or vehicle-mounted mist blowers. Some disadvantages of *L. sphaericus* are as follows: It costs more than *Bti*, and instances of resistance have been noted in populations of *Culex* mosquitoes (Su et al., 2018). However, if insecticide resistance levels of local mosquito populations are monitored, the product can still be an effective tool to control larval populations.

Spinosad was first labeled for mosquito control in 2007. It is an insecticide derived from a fungal soil, actinomycete, through fermentation (Mertz and Yao, 1990) and affects the nicotinic acetylcholine and gamma aminobutyric acid (GABA) receptors. It has been formulated into granules, tablets, and emulsifiable concentrates. Mosquitoes can be exposed through either contact or ingestion, and although mortality may take up to 72 h, there are limited nontarget effects (Hertlein et al., 2010).

Another group of organisms used in biological mosquito control are the bacteria in the genus *Wolbachia*. This group of bacteria is maternally inherited and obligately intracellular within their host organisms, which include a large majority of arthropods and some nematodes (O'Neill et al., 1997; Werren, 1997; Bourtzis and O'Neill, 1998; Bandi et al., 1998; Dobson et al., 1999, Popovici et al., 2010). These bacteria were first described in *Culex pipiens* Linnaeus mosquitoes (Hertig, 1936). *Wolbachia* bacteria are estimated to inhabit up to 60% of all terrestrial insects, an estimated 52% of aquatic insects, and between 40% and 66% of all arthropods (Hilgenboecker et al., 2008; Zug and Hammerstein, 2012; Weinert et al., 2015; Sazama et al., 2017; Truitt et al., 2018). *Wolbachia* is considered the most abundant intracellular bacteria, and its biological success can be attributed to the fact that they can impact host reproduction in various ways in order to increase their own vertical transmission (eggs from an infected female) (Werren, 1997; Dobson et al., 1999, 2004; Werren et al., 2008; Gavotte et al., 2009; De Barro et al., 2011). Due to this phenomenon, sometimes, *Wolbachia* are referred to as reproductive parasites (Werren et al., 2008). *Wolbachia* provide infected female insects with a reproductive advantage over uninfected females (Dobson et al., 2004; De Barro et al., 2011). Examples of these mechanisms include cytoplasmic incompatibility (CI), parthenogenesis, male killing, and feminization of males (O'Neill et al., 1997; Werren, 1997; Bourtzis and O'Neill, 1998; Dobson et al., 1999; Hurst et al., 1999; Sinkins, 2004; Popovici et al., 2010). *Wolbachia* has been found in additional tissues other than those involved with reproduction in both *Drosophila* flies and mosquitoes (Min and Benzer, 1997; Dobson et al., 1999).

Cytoplasmic incompatibility is the reduction in brood hatching when an infected male and uninfected female mate or the male and female mating pair have different *Wolbachia* strains (Dobson et al., 2002; Gavotte et al., 2009; O'Connor et al., 2012; Lees et al., 2015; Mains et al., 2016). The infected male sperm is modified so that it is no longer able to fertilize uninfected eggs or those with a different strain of bacteria (Sinkins, 2004; Zabalou et al., 2004; Moretti and Calvitti, 2013; Hoffmann et al., 2014). Thus, uninfected female mosquitoes are at a reproductive advantage compared to uninfected females due to their reproductive success and fitness advantage (Dobson et al., 2004). The CI can be unidirectional, where infected males mate with uninfected females, or bidirectional, where both males and females are infected with different *Wolbachia* strains (Sinkins, 2004; Dobson et al., 2004; Zabalou et al., 2004). *Wolbachia* bacteria are known to only cause the CI phenotype in mosquitoes, and for this reason, they have been used considerably for various control techniques (Sinkins, 2004). As CI occurs throughout a population of mosquitoes, the amount of eggs that hatch will be reduced, therefore suppressing the vector population (Sinkins, 2004). An additional phenotype has also been described in *Ae. aegypti* where *Wolbachia* infection in females causes a "bendy proboscis" where they are incapable of penetrating human skin and taking a blood meal (Turley et al., 2009).

One way to harness the impacts of CI in vector mosquito populations is to release CI-capable males, that is, males that are infected with a strain of *Wolbachia* that the wild-type females are not infected with (Dobson et al., 2002; Moretti and Calvitti, 2013). Approximately 50 years ago, the first example of *Wolbachia*-induced CI was successfully utilized for control of the *Cx. pipiens* mosquito (Laven, 1967). The suppressed population result will depend on the threshold release rate of males, which is determined by the reproductive rate and type of competition in the wild population (Dobson et al., 2002). This exploitation of CI for mosquito control is called the incompatible insect technique (IIT) (Boller et al., 1976; Chambers et al., 2011; Moretti and Calvitti, 2013). This technique does not pose a health threat as male mosquitoes do not bite and do not vector any pathogens and the *Wolbachia* will not become established in the wild mosquito population since it is maternally inherited (O'Connor et al., 2012).

If *Wolbachia*-infected females are released into the wild population, the maternally inherited bacteria could potentially invade and become established in the environment (Xi et al., 2005; Mains et al., 2016). This has been suggested as another possible mosquito control strategy apart from IIT—called cytoplasmic incompatibility management (CIM); modeling has demonstrated that this could significantly reduce the vector population and keep it at very low levels (Dobson et al., 2002; Mains et al., 2016). The CIM strategy has not been explored as extensively as IIT and poses additional safety concerns, which will be addressed below.

Research has been ongoing for decades to determine the efficacy of the IIT for mosquito control. One of the first concerns addressed was whether or not *Wolbachia*-infected male mosquitoes could successfully compete in wild target populations for mates, that is, male fitness (Brelsford and Dobson, 2011; Moretti and Calvitti, 2013). Multiple studies have concluded that artificially *Wolbachia*-infected male mosquitoes are not significantly different from the wild-type males in regard to longevity (Dobson et al., 2004; Moretti and Calvitti, 2013), sexual compatibility (Chambers et al., 2011), overall fitness (Dobson et al., 2004), insemination capacity (Moretti and Calvitti, 2013), and competitiveness for mating (Chambers et al., 2011; Moretti and Calvitti, 2013). Age was also determined to not impact the success of CI, that is, male *Wolbachia*-infected *Ae. albopictus* mosquitoes as old as 60 days produced complete sterility when crossed with uninfected females (Kittayapong et al., 2002).

Multiple groups have also demonstrated that *Wolbachia* bacteria will successfully invade laboratory mosquito populations for future release and control opportunities. Complete or nearly complete sterility has been observed between *Ae. albopictus* mosquitoes in strictly laboratory reared colonies (Kambhampati et al., 1993; Sinkins et al., 1995; Dobson et al., 2001; Calvitti et al., 2012). In addition, Kittayapong et al. (2002) observed almost complete sterility after crossing laboratory-reared with field-collected *Ae. albopictus* mosquitoes. Xi et al. (2005) microinjected *Ae. aegypti* with wAlbB *Wolbachia* infection from *Ae. albopictus* and reported complete CI success (0% egg hatch) and high maternal transmission rates with 0% maternal transmission failure. *Wolbachia* strains transferred from nonmosquitoes have also shown complete CI in their new mosquito hosts in the laboratory (Zabalou et al., 2004).

In addition to the first CI field trial from the 1960s (Laven, 1967), additional field studies have successfully proven the efficacy of CI for mosquito control. Significant reductions in target mosquito populations have been demonstrated from various male release field trials, along with no observed *Wolbachia* strain establishment into the environment (Atyame et al., 2011, 2015; Hoffmann et al., 2011, 2014; O'Connor et al., 2012; Mains et al., 2016).

Another benefit to *Wolbachia* infection in vector mosquito populations is its ability to interfere with pathogen transmission (McGraw and O'Neill, 2013; Sinkins, 2013; Hoffmann et al., 2014). This discovery opened a new avenue of research and control to use the bacterium for strict viral or pathogen interference to reduce or eliminate mosquito-borne disease (Popovici et al., 2010). Protection against establishment, replication, and transmission of *Plasmodium*, Chikungunya virus, and dengue virus has been documented in *Aedes* mosquitoes (Moreira et al., 2009a; Iturbe-Ormaetxe et al., 2011; Sinkins, 2013; Bull and Turelli, 2013; Frentiu et al., 2014; Lees et al., 2015); additional pathogen interference phenotypes include *Plasmodium* development inhibition in *Anopheles* (Kambris et al., 2010), reduced filarial competence in *Anopheles* (Kambris et al., 2009), reduced yellow fever virus replication (van den Hurk et al., 2012), West Nile virus host resistance in *Culex* (Glaser and Meola, 2010), and blockage of circulating Zika virus in *Ae. aegypti* (Dutra et al., 2016). The Eliminate Dengue Program is utilizing this characteristic of the bacterium to help combat dengue virus transmission in several countries including Australia and Indonesia (Moreira et al., 2009a; Caragata et al., 2016). For further reading on pathogen interference and *Wolbachia*, refer to Iturbe-Ormaetxe et al. (2011).

As mosquito research and control methods utilizing *Wolbachia* became more established, scientists and government agencies worked to also address any safety and environmental concerns that emerged regarding this strategy. Existing policy and regulation did not exist for *Wolbachia*-based mosquito control strategies before these were initiated (De Barro et al., 2011). Questions included whether or not *Wolbachia* could be transferred to humans through the bite of an infected female mosquito, how would artificial *Wolbachia* strains impact the environment, and how likely was it for *Wolbachia* to establish itself in the environment (Popovici et al., 2010). *Wolbachia* has not been detected in mosquito saliva (Moreira et al., 2009b); in other animals such as humans, mammals, birds, reptiles, and fish; or in food products for consumption (Popovici et al., 2010). In addition, when *Wolbachia*-infected mosquitoes bite humans, no antigens are injected into the human host during the blood-feeding process (Popovici et al., 2010). Lastly, precautions and additional research are still underway to prevent the accidental release of any female mosquitoes into the environment. This is due to the possibility of unwanted population replacement and establishment of *Wolbachia* bacteria in the environment and the ability of female mosquitoes to bite and spread pathogens (Dobson et al., 2002; Chambers et al., 2011; Lees et al., 2015; Mains et al., 2016).

8.8 Chemical control

Chemical control of mosquitoes becomes necessary only when other naturalistic methods have been tried or are not feasible, that is, when a confirmed mosquito-borne disease is circulating in an area or extremely high adult mosquito counts are affecting human quality of life (landing rate counts >50 per min). The primary responsibility of a pesticide applicator is to know when and where to apply insecticides and apply them in a manner that effectively controls target mosquito populations while limiting the hazard for the applicator, the community, and the environment. Overreliance on chemical control strategies will eventually lead to insecticide resistance in mosquito populations if not used in conjunction with other tactics. For this reason, chemical control should be used to manage mosquito populations when only necessary or required. Insecticide resistance monitoring and management should also be used with any type of chemical control measures to ensure that the products used are currently and will continue to be effective on target mosquito populations.

Before applying any pesticide, read and become familiar with the label. The golden rule in pesticide applications is, "the label is the law" (this includes the earlier stated biorational pesticides). The label will state, in plain language, everything an applicator will need to know regarding that product including if the product is restricted use or state-limited use insecticide, active ingredients, application rates, application instructions, personal protective equipment required, precautionary statements, and storage and disposal information. A restricted use or state-limited use pesticide will require a licensed applicator and an application record that is maintained for two or more years after the application. Products applied directly to water sources or "point source" applications to water sources may also be regulated by the National pollutant Discharge Elimination System (NPDES) under the Clean Water Act. The applicator is responsible for maintaining and calibrating application equipment and keeping current with the laws, regulations, and any changes made by the regulatory agencies responsible for pesticide applications.

The chemical control of mosquitoes can be divided into two categories: larviciding and adulticiding. Larvicides are applied over or directly to water containing larvae through a variety of means or may be applied via ultralow volume (ULV) or mist blowing technologies allowing the material to drift and deposit in the water. Adulticides are primarily applied as space sprays in the form of ULV cold fogs (aerosols) or thermal fogs (smokes) and, in certain specific situations, residual barrier sprays. The applications can also be made by either air- or ground-based equipment either piloted or, in some cases, remotely operated unmanned aerial systems (UAS). Recently, developments have also begun using autodissemination techniques, which use the adult mosquito to transfer the products to each other and larval habitats.

Chemical control of larvae in Texas is primarily limited to the biorational pesticides already mentioned earlier, insect growth regulators (IGRs), and larvicidal oils or surface films. These materials may be formulated into dry formulations, tablets, briquettes, pellets, granules, or liquids. The briquettes, tablets, etc. are often referred to as "toss its" as they are premeasured doses of larvicides that can be tossed or applied to a specific quantity of water by hand. Granular applications usually require the use of a spreader or seeder in order to disperse the granules evenly. The liquid formulations may be applied neat or mixed with water and applied using handheld pump up sprayers or vehicle-mounted sprayers, which on occasion may include mist blowers or ULV application.

IGRs are insecticides that mimic hormones in insects and interfere with their normal growth and development. There are three IGRs used in mosquito control, that is, methoprene; pyriproxyfen, which are juvenile hormone (JH) mimics; and diflubenzuron, which is a chitin synthesis inhibitor. In Texas the JH mimics are used for mosquito control, and diflubenzuron is a restricted use product used in agricultural applications due to its effects on other nontarget invertebrates in aquatic environments.

Methoprene has been used since the mid-1970s and has been used widely due to its low toxicity to mammals and short environmental persistence, and the development of resistance is rare under normal field conditions (Henrick, 2007). It is only effective against mosquito larvae at specific points in the development cycle and, therefore, is ineffective against mosquitoes in the pupal stage (Staal, 1975). Methoprene is also slower acting than other materials based on its mode of action and the primary measure of efficacy is not larval mortality but a lack of adult emergence, which can be hard to measure in a situation where a continuous breeder, that is, *Cx. quinquefasciatus* (Say) is found.

Pyriproxyfen is the other mosquito IGR available in Texas for mosquito control. The primary formulation for this product in Texas is a granule that has been shown to have a 3-week residual in catch basins in California (Mian et al., 2017). It is effective against a wide range of mosquitoes at very low concentrations (Unlu et al., 2017), making it an ideal candidate for autodissemination strategies. Autodissemination strategies consist of contaminating a mosquito with pyriproxyfen and allowing it to contaminate other females and/or ovipositional sites to obtain control in cryptic habitats. Devine et al. (2009) used a dust formulation for autodissemination, similar to the newer autodissemination traps that use a combination of pyriproxyfen, *Beauveria bassiana*, and an ovipositional attractant for control of adult and larval mosquitoes.

Larvicidal oils have been used for many decades, beginning with the use of kerosene or other low-quality fuel oils. The mechanisms are still unclear whether they suffocate, reduce surface tension, or are toxic from inhalation (Lloyd et al., 2018). The new oils spread quickly, rarely last more than 72h, and offer quick control of larvae and pupae in areas with minimal vegetation. They may also serve as an ovipositional deterrent for *Ae. aegypti* (Hall et al., 2017). Despite these new refinements, oils are not mosquito specific and may kill other nontarget small arthropods and invertebrates as well, especially if the label is not followed.

Adult mosquito control is less efficient than controlling the larvae because the adults disperse out into the environment, whereas larvae are relegated to their aquatic habitats making them "concentrated, immobile, and accessible." Another limitation to adulticiding is that the products used are contact insecticides and need to impinge on mosquitoes out flying in the environment to control them. However, adulticiding is necessary to quickly break the transmission cycle of a mosquito-borne disease in a given area. The remaining classes of mosquito adulticides are the organophosphates, which are acetylcholinesterase (AchE) inhibitors and the pyrethroids, which interact with the voltage-gated sodium channels.

Traditionally, adulticides were applied using a thermal fog (TF) technique in which a low concentration of insecticide was diluted with a large amount of diluent (oil), which was passed through a nozzle and injected into combusted fuel at temperatures between 316°C and 649°C. The mixture is then vaporized, and the resultant mixture is pushed out of the machine where the cooler external temperature causes it to condense into a "fog" (Pratt and Littig, 1974). The droplet size of a TF may be from 0.1 to 50 μm (Brown, 1968) and may be applied by ground using either a handheld or vehicle-mounted machine but not aerially as the droplets are too small and light to settle to the ground appropriately. The TF applies large volumes of low concentrated insecticides (30–120 gal per hour), which can penetrate vegetation and obstacles (Rose, 2001) but can obscure vision and at times pose a traffic hazard.

Ultralow volume is the process of generating an aerosol by forcing material through a small orifice and blowing high-pressure air to cleave and distribute the drops instead of heating the drops by combusting kerosene or diesel. The material can be applied via air- or ground-based equipment. The material applied is a small volume of highly concentrated insecticide, which generally produces a drop of 5–25 μm for ground-based mosquito control (Mount et al., 1996) typically with a flow rate of 0.5–6 oz./minute depending on speed and application rates. The most effective droplet size for controlling mosquitoes is between 8 and 15 μm (Mount, 1998). The drops applied by aerial ULV tend to be larger (>45 μm) to allow them to settle more accurately but not too large as to damage automotive paint. The ULV droplet size is controlled by the pressure and flow and is larger than that of TF equipment which is determined by flow and temperature. The larger drops do not penetrate vegetation as well or stay aloft as long as a thermal fog but also do not lead to the obscured visibility and reduce the environmental impact and cost by applying smaller amounts of material in an area. For further review on ULV and TF, the US Armed Forces Pest Management Board (2011) has a full technical description of equipment.

The organophosphates (OP) used in America currently are malathion and naled. Malathion has been used since the 1950s in agriculture and mosquito control, although the application rates for agriculture are much higher. Mosquito control applies malathion primarily by ground but a small percentage of aerial malathion treatments remain (EPA, 2016). The primary complaint about malathion is its odor, which causes concerns among some residents. Pietrantonio et al. (2000) reported high levels of resistance to malathion in Harris County, Texas, but the susceptibility was restored in local populations by resting (refraining to use) the material for several years (Dennett and Debboun, 2017). Naled historically has been applied via ground but, more recently, is primarily applied via aerial to cover large areas in time of disease risk. Aerial spraying has become controversial in recent years, and naled has not been excluded from that controversy even though it has been thoroughly studied and vetted by the EPA and countless other researchers (Duprey et al., 2008). Another drawback of naled is its corrosiveness to application equipment that should be considered before applying. Despite the controversy, the risk of aerial application of naled is minimal if the material is applied in accordance to the label.

The pyrethroids are insecticides derived from and or synthesized from the products derived from a species of chrysanthemum. They have filled the void left by several OPs that have been removed from the market (Palmquist et al., 2012). Due to their low mammalian toxicity and fast-acting "knockdown," pyrethroids can be found almost everywhere from homeowners' pest control products to agriculture to professional mosquito control products. The ones commonly used in mosquito control are pyrethrins, permethrin, Sumithrin, prallethrin, etofenprox, and, most recently, deltamethrin. Unfortunately, the registration was not renewed for resmethrin; therefore it is no longer for sale for mosquito control. These products may or may not be synergized with piperonyl butoxide to increase efficacy. Pyrethrins and pyrethrum are derived from the flower itself and therefore can be expensive and hard to obtain depending on the harvest. The pyrethroids have a similar chemical structure to pyrethrin but have been synthesized and modified. All of these products are effective on insects at low doses and can also be very toxic to fish and aquatic invertebrates (Antwi and Reddy, 2015). They are generally applied by ground but still are applied via air in some locales.

The fact that there are only two classes of adulticides remaining stresses the importance for judicious usage and insecticide resistance monitoring and the need for new chemistries. There are a number of ways to test local populations of mosquitoes for insecticide resistance levels including CDC bottle bioassays, field cage tests, and PCR to name a few.

8.9 Genetic control

As the world develops, it is obvious that one method of control will not stay effective forever. Insecticide and drug resistance have both been an issue with every tool added to the mosquito control toolbox; their relatively fast emergence is of special concern (Coleman and Alphey, 2004; Wilke et al., 2009; Gabrieli et al., 2014). Unfortunate and unforeseen environmental impacts from the use of certain chemicals have also been under scrutiny following the advent of DDT and others (Alphey et al., 2007; Wilke et al., 2009). The overall failure to achieve global malaria and yellow fever eradication and the grueling fight with dengue, inadequate technology, and unsustainable chemical resistance in mosquitoes have all stirred interest in developing additional techniques that would potentially outlast these and future failures dependent on chemicals and other previously mentioned control tactics (Coleman and Alphey, 2004; Wilke et al., 2009; Alphey et al., 2010; Macias et al., 2017). Specifically, advancements in genetic mosquito control have been sought after to solve this control crisis

(Alphey et al., 2007, 2010, 2013). The WHO broadly defines genetic control as "the use of any condition or treatment that can reduce the reproductive potential of noxious forms by altering or replacing the hereditary material," with "noxious forms" referring to the modified vectors (WHO, 1964).

Current genetic mosquito control approaches fall into one of two strategies: (1) population suppression, which is the reduction or elimination of a vector population, and (2) population replacement, meaning the existing vector population is replaced by strains or species of the vector that are refractory to the pathogen of concern or unable to transmit this pathogen (WHO, 1964; Coleman and Alphey, 2004; Wilke et al., 2009; Alphey et al., 2013). These strategies have become well developed over the past few decades since their discovery. In the 1940s, x-ray–induced chromosomal translocations were found to cause sterility in offspring in certain pest insects (Muller, 1927; Serebrovsky, 1969; Klassen and Curtis, 2006), and in the 1950s, these ideas were successfully applied to the screwworm, *Cochliomyia hominivorax* (Coquerel), a monogamous and destructive agricultural pest (Knipling, 1955). Coincidentally, work was also being conducted on tsetse flies during the 1940s, which demonstrated that when two species mated from different regions, some of the offspring were sterile (Vanderplank, 1944, 1947, 1948).

The first genetic manipulations of mosquitoes were demonstrated in the late 1990s through the integration of transgenes using transposons, which are genes that have the ability to mobilize within an organism's genome (Coates et al., 1998; Jasinskiene et al., 1998; Macias et al., 2017). These first transgenic mosquitoes introduced the age of modern mosquito genome engineering (Gabrieli et al., 2014).

The sterile insect technique (SIT) is one of the most successful and well-known genetic techniques used for control of pest insects, established through the previously mentioned experiments using sterile insects from Muller (1927), Serebrovsky (1969), Knipling (1955), and Vanderplank (1944). A classic example of the population suppression strategy, SIT relies on the release of large numbers of sterile male insects that will compete with wild males to mate, and the subsequent offspring will be sterile (i.e., nonviable), thereby reducing female reproductive capacity and, ultimately, reducing the target population (Knipling, 1955, 1979, 1998; Krafsur, 1998; Klassen, 2006; Alphey et al., 2010). If enough sterile males are released over a sufficient period of time, the target insect population will collapse—failing to self-replicate—yielding possible elimination (Coleman and Alphey, 2004; Alphey et al., 2007, 2010; Wilke et al., 2009; Gabrieli et al., 2014). Sterilization can be accomplished through either irradiation (ionizing radiation) or chemical sterilants (Alphey et al., 2010; Gabrieli et al., 2014; de Araújo et al., 2018). The SIT strategies have been used successfully in multiple countries for various types of agricultural pests including fruit fly species (Enkerlin, 2006), moth species (Bloem et al., 2006), and additional screwworm fly outbreaks (Lindquist et al., 1992; Vargas-Terán et al., 1994). These programs have validated this technique in large-scale, long-term, and sustainable control programs (Alphey et al., 2007, 2010).

The SIT has also been utilized to combat mosquitoes since the 1960s, with successes demonstrated from the 1970s to 1980s (Lofgren et al., 1974; Benedict and Robinson, 2003). Documented mosquito species for these past SIT field trials included many vector species including *Ae. aegypti*, *Ae. albopictus* (Skuse), *Culex pipiens* Linnaeus, *Cx. quinquefasciatus* Say, *Anopheles gambiae* Giles, and *An. albimanus* Wiedemann (Benedict and Robinson, 2003). Although most of these trials did not result in long-term control, researchers in favor of SIT believe that the recent advancements in technology have improved this technique for mosquitoes (Benedict and Robinson, 2003; Wilke et al., 2009). In addition to the release itself, entomological surveillance of the vector population through standardized trapping is also conducted throughout mosquito SIT programs in order to monitor the target vector population and the released males' survival and distribution (Alphey et al., 2010). For mosquito releases, males are desired for two reasons: (1) male mosquitoes do not bite, and sterile female mosquitoes have the ability to still transmit pathogens, and (2) released females might distract the sterile males from seeking wild females for mating (Coleman and Alphey, 2004; Alphey et al., 2007, 2013; Wilke et al., 2009).

There are several advantages for utilizing the SIT for mosquito control. First, SIT is species specific, as released males will only seek out females of the same species to mate; this means that this technique is very environmentally friendly as minimal nontarget effects are seen (Coleman and Alphey, 2004; Alphey et al., 2007, 2010, 2013). Since no toxic chemicals are used, the only impacts expected are to be indirect: as the population of a vector (or pest) species is reduced or locally eliminated (Alphey et al., 2010). In addition, the released males will disperse and find wild females without human intervention, and this is particularly useful when control is needed in areas that are difficult to access like private property, dangerous areas, etc. (Coleman and Alphey, 2004; Alphey et al., 2007, 2010, 2013). A third advantage to using SIT is its cost-effectiveness profile. As more sterile males are released, the program becomes more efficient as the population of wild males wanes and the population of released sterile males increase (Knipling, 1955; Coleman and Alphey, 2004; Klassen and Curtis, 2006; Alphey et al., 2010). Lastly, because the sterilized males are not delivering any genes or pathogen to the females, there is no opportunity for a foreign entity to be spread into the environment (Wilke et al., 2009).

Despite its strong success and many advantages, there are several disadvantages for using the SIT as well. Sterilants, whether irradiation or chemical, can have adverse impacts on the fitness of the male mosquitoes, and this can cause poor mating performance, thereby decreasing the efficacy of the program (Coleman and Alphey, 2004; Alphey et al., 2007, 2013; Wilke et al., 2009; Gabrieli et al., 2014). Andreasen and Curtis (2005) demonstrated that irradiating mosquitoes as pupae caused severe abnormalities in the adults, making them nonviable for SIT. In addition, if the pathogen of concern for the control program is being transmitted by multiple vector species, the species specificity aspect of SIT might not be adequate to impact the disease spread, and a more broad-spectrum approach may be needed (Alphey et al., 2010, 2013). Another disadvantage depends on the customer: if immediate population reduction is desired, the customer will be disappointed as SIT impacts the following generation after a release (the offspring of the released males), which could take weeks or months (Alphey et al., 2010). Lastly, there is no efficient method of separating males and females in the rearing facility—making this aspect of the program costly and time inefficient (Coleman and Alphey, 2004; Alphey et al., 2007, 2010; Wilke et al., 2009). More specifically, to date, an *Aedes* mass-rearing facility capable of providing male mosquitoes without the risk of accidental female contamination does not exist (de Araújo et al., 2018).

All genetic mosquito control methods have stirred up controversy with the public, but SIT has generally not been met with an immense amount of push-back, unless wrong information is spread to the public to cause mass hysteria. In India in 1975, an SIT mosquito control field trial supported by the WHO had to be completely canceled before it began due to negative publicity and a failure on the program officials to inform the public of the facts in a timely manner (Dyck et al., 2006b; Alphey et al., 2010). The program staff for this field trial was falsely accused of working on biological warfare weapons, and a proper clarification of the science and program was not made, thus causing the entire program to be derailed (Nature, 1975; WHO, 1976; Dyck et al., 2006b).

For further reading on SIT and mosquitoes, Alphey et al. (2010), Dyck et al. (2006a), and Benedict and Robinson (2003) are recommended.

Another popular, albeit more controversial, type of mosquito control population suppression strategy is Release of Insects carrying a Dominant Lethal (RIDL). This strategy is similar to SIT; however instead of irradiation, this technique utilizes a strain of insects homozygous for a dominant lethal genetic mechanism (Alphey et al., 2002; Wilke et al., 2009). The lethal gene can cause either all mosquito or all female mosquito progeny to die, depending on the strain; this is also called a female-specific lethal (Alphey and Andreasen, 2002; Alphey et al., 2002; Coleman and Alphey, 2004; Wilke et al., 2009; Oxitec, 2018a).

In order to create this system, two colonies are reared: one colony for production and one for release (Wilke et al., 2009; de Araújo et al., 2018). The dominant lethal transgene is inserted into mosquito larvae, along with a fluorescent marker, and these mosquitoes will be mass reared into the homozygous RIDL strain (Wilke et al., 2009). A suppressor is supplied to the larval mosquitoes, which suppresses the RIDL system, allowing the females to survive (Wilke et al., 2009; Alphey et al., 2013; de Araújo et al., 2018). Once the colony for production is created, the release colony will be created. For release, the females are removed by either genetic sex separation or mechanical separation methods (Wilke et al., 2009; Alphey et al., 2007, 2013; de Araújo et al., 2018). The all-male homozygous release colony will be released for control measures. The offspring of the RIDL males crossed with wild females are heterozygous for the dominant lethal gene—which leads to the eventual death of all F1 females (Alphey and Andreasen, 2002; Coleman and Alphey, 2004; Alphey et al., 2007; Wilke et al., 2009). As wild female offspring die due to the lethal mechanism—because they lack the suppressor in the field—the mosquito population will eventually be reduced due to the decrease in reproductive capacity (Heinrich and Scott, 2000; Thomas et al., 2000; Wilke et al., 2009). The F1 males will be heterozygous for the female-specific lethal, and thus half of their eventual female offspring (F2) will also die (Alphey et al., 2007). Some experts even categorize RIDL as a subtype of the traditional SIT strategy (Alphey et al., 2007).

The use of transgenic insects, specifically repressible RIDL systems, has been around and in practice since the early 2000s. The first models to show success were *Drosophila* and the Mediterranean fruit fly (Thomas et al., 2000; Gong et al., 2005). In a true agricultural model, transgenic strains of the pink bollworm were first described in 2000 (Peloquin et al., 2000), and after years of laboratory and field experiments, releases were first conducted in 2006 (Simmons et al., 2007). This transgenic insect release was shown to be equal or superior to that of the traditional SIT strain (Simmons et al., 2007; Wilke et al., 2009).

The advantages of utilizing the RIDL technique are numerous. First, utilizing the specialized female-specific lethal genetic mechanism means that this system can double as both a control mechanism in the field and a sex separation tool, which is lacking in other techniques such as SIT (Thomas et al., 2000; Alphey, 2002; Alphey and Andreasen, 2002; Alphey et al., 2013). A lethal phenotype allows a large number of females to be removed simultaneously, with a severe reduction in errors (Alphey et al., 2010). Second, RIDL allows the user to control the time of death of the offspring, compared to using irradiation, which kills the affected mosquitoes as embryos (Alphey et al., 2013). Allowing larval competition in the field is predicted to substantially increase the effectiveness of the control program (Wilke et al., 2009; Alphey et al., 2013). Irradiation in general causes significant damage to the mosquitoes, whereas the RIDL mechanism avoids the need for this costly step (Coleman and Alphey, 2004). Thus, many experts believe that RIDL is a more effective system for population control of mosquito vectors, and this has been predicted through mathematical modeling as well (Thomas et al., 2000; Gould and Schliekelman, 2004; Alphey et al., 2007; Atkinson et al., 2007). Lastly, the use of a repressor serves as a fail-safe for any escapees, meaning any mosquitoes that are accidentally released are sterile due to this system (Benedict and Robinson, 2003; Wilke et al., 2009).

One of the most successful examples of RIDL is the OX513A *Ae. aegypti* strain, first described by Phuc et al. (2007) as LA513A, a product of Oxitec, Ltd. (Oxford, UK), also known as the Friendly mosquito. This strain contains a dominant, inheritable gene, which results in death before the female mosquito larva reaches adulthood through a positive feedback mechanism (Gabrieli et al., 2014; Patil et al., 2018). When the larvae are supplemented with tetracycline, the repressor, the larvae are not killed and allowed to reach adulthood for rearing purposes (Gabrieli et al., 2014; Patil et al., 2018). The males that are reared and eventually released will mate with wild *Ae. aegypti* females, producing unviable offspring (Gabrieli et al., 2014). This strain demonstrated a 95%–97% penetrance of lethality at the larval/pupal barrier when initially tested (Phuc et al., 2007; Wilke et al., 2009). From 2009 to 2010, the first field releases of OX513A strain *Ae. aegypti* were conducted in the Cayman Islands. These demonstrated that the RIDL males were competitive with wild males, and over 80% suppression was achieved in the target population following sustained releases (Harris et al., 2011, 2012; Alphey et al., 2013; Gabrieli et al., 2014). In 2010, OX513A strain *Ae. aegypti* was released in a forested area of Malaysia, which showed that these males had comparable longevity and dispersal to unmodified males (Lacroix et al., 2012; Alphey et al., 2013; Gabrieli et al., 2014). Recently, the OX513A strain was evaluated in the field in Brazil through multiple sustained releases (Alphey et al., 2013; Gabrieli et al., 2014). These releases demonstrated a reduction in the local *Ae. aegypti* population by 95% based on adult trap data; the authors concluded that this level of suppression would likely be sufficient to prevent a dengue epidemic in areas with similar prevalence (Carvalho et al., 2014, 2015).

Another female-specific approach to RIDL, called fsRIDL, involves the disruption of the female mosquito's ability to fly through a severe impairment on the flight muscles (Fu et al., 2010; Labbé et al., 2012). Without the repressor for this specific system, the female mosquito flight muscle cells die, resulting in their inability to fly; males are unaffected by the transgene—thus the cohort for releases will be all male as well (Fu et al., 2010). This approach will also provide population suppression with a built-in late-acting, sex-specific lethal, which allows rearing and sexing prior to the release of the mosquitoes (Thomas et al., 2000; Atkinson et al., 2007; Phuc et al., 2007; Fu et al., 2010; Facchinelli et al., 2013; Gabrieli et al., 2014; Macias et al., 2017). Field cage tests did not show complete suppression of the target population with this strain, and it is suggested that fsRIDL is not suitable for large-scale releases, but more research is needed (Facchinelli et al., 2013; Gabrieli et al., 2014).

Despite its many advantages and success stories, RIDL is still faced with one of the most costly (by time and resources) steps: sex separation in the laboratory. Even though genetic sex separation and mechanical separation are available, the production of two types of colonies (for production and for release) can slow down the process as females still need to be eliminated (de Araújo et al., 2018). Carvalho et al. (2015) spent 5 days a week separating males from females in the laboratory for colony maintenance and release to achieve the needed mass numbers of transgenic males; even with this dedication, females are still able to slip through the separation process (de Araújo et al., 2018). Research is currently being conducted to navigate around mosquito sex sorting before male release; one major development is the discovery of the gene that initiates male mosquito development, also known as Nix (Hall et al., 2015; de Araújo et al., 2018). Small RNAi-mediated techniques including silencing and knockdown of specific genes have been researched, tentatively demonstrating reductions in female mosquito fecundity and lifespan, providing highly male-biased populations; however, no complete sex conversion has been achieved (Mysore et al., 2015; Whyard et al., 2015). Complete sex conversion was accomplished in a species of medfly, leading to the sex reversal of all females to fully fertile and viable males—indicating that a transgenic strain is now possible to produce male-only progeny—a promising step for mosquito control as well (Pane et al., 2005; Saccone et al., 2007; Salvemini et al., 2007).

In order to address the genetic sex separation issue, Oxitec, Ltd. (Oxford, UK), announced in the summer of 2018 that it is transitioning from the first-generation self-limiting mosquito to a second-generation technology Friendly *Aedes* mosquito, called the OX5034 strain (Oxitec, 2018b). Oxitec reports that the new strain is both male selecting and self-limiting, unlike any other mosquito control technology available (Oxitec, 2018a,b). As a completely male-selecting strain, only male mosquitoes survive; thus this eliminates the risk of releasing female mosquitoes and the need for sex sorting during the rearing process (Oxitec, 2018b). In addition, another type of gene has been introduced to the strain for added benefits: insecticide susceptibility genes (Oxitec, 2018b). A released male and wild female mating will produce all males, half carrying the female-lethal gene and the other half carrying the insecticide susceptibility gene—thus suppressing the mosquito population while also decreasing resistance to insecticides in the wild population (Oxitec, 2018b). With the possibility of an all-male production, egg-based releases are potentially in the future for this innovative technology (Oxitec, 2018b). They have received approval in Brazil for field releases of their second-generation *Ae. aegypti* mosquitoes, and more research will be done (Oxitec, 2018a).

The population replacement strategy for mosquito control does not require large sustained releases of transgenic insects; instead, this strategy focuses on creating mosquitoes that are refractory to pathogen transmission and mate with wild mosquitoes (Alphey et al., 2002, 2013). This phenotype spreads throughout the wild population without additional aid, reducing the number of mosquitoes capable of transmitting the pathogen of concern (Coleman and Alphey, 2004; Alphey et al., 2002, 2013). This has been described in *Aedes* spp. to suppress dengue virus replication. The RNAi has been used for this specific purpose to inhibit the virus in both the midgut and the salivary glands of the mosquito, which has demonstrated a strong block in virus transmission (Franz et al., 2006; Wilke et al., 2009; Mathur et al., 2010; Alphey et al., 2013; Gabrieli et al., 2014; Macias et al., 2017). The salivary gland modification reduced viral titers more than five times the control mosquitoes (Mathur et al., 2010). Unfortunately, the midgut-targeted refractory line was lost after 13 generations, suggesting this line suffered significant fitness costs (Franz et al., 2009).

Anopheles mosquitoes have also been made refractory to carrying the malaria parasite by manipulating specific antibodies, peptides, or cell signaling inside the mosquitoes using synthetic antimalarial factors (Ito et al., 2002; Arrighi et al., 2002; Kim et al., 2004; Wilke et al., 2009; Corby-Harris et al., 2010; Meredith et al., 2011; Isaacs et al., 2012; Sumitani et al., 2013; Gabrieli et al., 2014; Macias et al., 2017). The cell signaling disruption was especially successful in reducing not only the number of malaria parasites per infected mosquito but also the duration of mosquito infectivity (Corby-Harris et al., 2010). A bee venom phospholipase was even successfully used to inhibit invasion by the *Plasmodium* parasites in the mosquito midgut by Moreira et al. (2002). These strains of refractory mosquitoes showed little to no fitness costs in the *Anopheles*, suggesting strong implications for future malaria control efforts.

Because the population replacement strategy does not require sustained releases of large numbers of insects, the genetic manipulations created in the refractory insects need to be carried throughout the natural population (Gabrieli et al., 2014). This concept of spreading the refractory traits is called gene drive. Artificial gene drive systems were first developed in systems other than disease vectors, for example, in *Drosophila melanogaster* (Chen et al., 2007; Akbari et al., 2013; Gabrieli et al., 2014). Homing endonuclease genes (HEGs), double-stranded DNases (DNA-cutting molecules) that target large asymmetric recognition sites in target genomes and cut them causing shredding of the target DNA, are used for this purpose (Gabrieli et al., 2014). By inducing specific HEGs and target mosquito regulatory regions into the mosquito population, it will spread throughout the desired population (Windbichler et al., 2011; Gabrieli et al., 2014). This was successfully done in *Anopheles* mosquitoes in a laboratory colony, demonstrating the proof of concept in a disease vector system (Windbichler et al., 2011). In addition, new technologies have been explored utilizing CRISPR/Cas9, which has the ability to cut multiple genomic sequences at a time and copying the transgene into all cleaved loci, enabling the transgenic strains to be spread even faster and throughout more members of the target population (Gabrieli et al., 2014; Gantz et al., 2015; Hammond et al., 2016; Macias et al., 2017; Mitchell and Catteruccia, 2017; Marshall and Akbari, 2018). Unfortunately, the reduction in vectorial capacity from the refractory lines of genetically altered mosquitoes most likely causes a reduction in fitness cost (Coleman and Alphey, 2004; Wilke et al., 2009; Mitchell and Catteruccia, 2017; Marshall and Akbari, 2018).

More research needs to be done in order to make gene drivers as efficient and ecologically sound as possible for spread throughout the wild-type population. Little is known about how long foreign DNA can survive in a wild-type population and successfully spread as well. Hanemaaijer et al. (2018) examined two *Anopheles* sister species' genomic relationships in Mali over 25 years and found introgression within a very small portion of genes (0.26%). The frequency of genetic introgression was inconsistent throughout observed generations and suggested that this was limited to specific genes under positive selection, meaning that a gene drive system would need to be very specific and monitored closely (Hanemaaijer et al., 2018). Coleman and Alphey (2004) described the perfect gene drive system for population replacement as having the following traits: (1) perfect linkage between the drive mechanism and refractory gene; (2) ability to achieve complete population replacement; (3) controllable spread; (4) recallable, that is, able to be eliminated from the population; (5) replaceable and reusable for future traits; and (6) nontransferable to nontarget species.

The release of genetically modified mosquitoes into the environment—whether through population suppression or replacement—has caused a significant amount of controversy regarding environmental and safety concerns. Although many of the safety and environmental concerns are unlikely, researchers must address every scenario. Some of the scenarios of concern are horizontal transfer of the genetically modified transgenes to wild mosquitoes, spread of the transgenes to nontarget insects, the inevitable breach of national borders by released insects, and unintended ecological impacts (Alphey et al., 2007; Wilke et al., 2009; David et al., 2013; Gabrieli et al., 2014; Macias et al., 2017). The first two concerns are of special importance regarding population replacement and gene drive systems (Oye et al., 2014). Because mosquitoes do not respect national borders, it is unknown what consequences would occur if and when released genetically engineered mosquitoes cross into another country where that technology was not approved (Macias et al., 2017). Even further, if the modified genome (driven or not through the wild population) causes adverse consequences, the question must be asked, "Can the technology be reversed or removed once it is released?" (Marshall and Akbari, 2018). Another concern is the potential of removing mosquitoes from the ecosystem and what this would do to other species' equilibrium (Coleman and Alphey, 2004; Alphey et al., 2010). Special confined and high containment laboratories and testing facilities are required for genetically engineered insect research to avoid accidental releases (Gabrieli et al., 2014). In addition to these specialized research facilities, open and early conversations are needed between the regulatory agencies, the public, and the scientific community in addition to full disclosure for nonprofits in targeted areas, the media, politicians, and health professionals (Coleman and Alphey, 2004; Alphey et al., 2010; Oye et al., 2014).

It is also worth mentioning that genetic mosquito control techniques are not the silver bullet to mosquito control. Many of the techniques described work exceptionally well with additional integrated mosquito control tools such as larviciding, adulticiding, habitat modification, etc. (Alphey et al., 2007, 2013; Gabrieli et al., 2014; Oxitec, 2018a). For example, larviciding—which focuses on the immature stages of mosquitoes—would go well with most of the genetic mosquito control techniques, which focus on the adult stages (Alphey et al., 2010). In fact, the combination of techniques in an integrated vector management program has been proven to be more effective than one technique by itself, and they work additively with each other (Wilke et al., 2009; Alphey et al., 2010; Macias et al., 2017; de Araújo et al., 2018). As technology continues to advance, additional tools will be integrated, and the toolbox will be expanded to encompass integrated vector-borne disease management as well (Alphey et al., 2010, 2013). With the combination of various mosquito control techniques available, a truly integrated system can have great success.

8.10 Public engagement

Public education and outreach efforts are one of the most important facets of any meaningful mosquito control program and should be ongoing throughout the year and not just a stop gap measure after mosquito populations or mosquito-borne diseases become an issue. The "public" consists of all of the stakeholders in an area including, but not limited to, residents, visitors, elected officials, media, and special interest groups (ASTHO, 2005). The biggest benefit of public inclusion is fostering buy-in and support for the actions taken by the mosquito control organization as well as their assistance in mosquito control efforts. This is accomplished through maintaining an open, transparent control program in which actions are visible, the mosquito control plan is open for review, questions and complaints are taken seriously and handled quickly, and a presence is maintained in the community. Mosquito control's presence in a community is maintained through public appearances (media, meetings, town halls, etc.), as well as sending inspectors to residences and working with partners to extend the reach through schools and medical networks.

Public assistance usually comes in the form of allowing traps to be placed on and conducting source reduction on their properties, using personal protective measures (repellents, wearing long sleeves, and avoiding dawn and dusk), and convincing neighbors to do the same. These actions may seem small, but they are a tremendous benefit in keeping mosquito populations low, especially for *Ae. aegypti* and *Ae. albopictus* populations, which occur close to homes and in backyards.

Historically, education and outreach consisted of distributing educational materials to residents at either neighborhood meetings or other outreach events. Although these actions are important in getting the message out; there have been numerous advances thanks primarily to technology and social media. Currently, there are mosquito control organizations that can share mosquito testing results as soon as they are confirmed and show the public planned treatment areas, and the public is now allowed to report mosquito breeding habitats from applications on a mobile phone. This allows the public much more access to mosquito control, which is rewarded with continued support of the program.

8.11 Cost of integrated mosquito management

The cost of conducting an effective mosquito control program is one of the most complicated issues. A true IMM program is made up of many different components and each of those components will require capital. Since IMM is not a rigid set of rules but a set of guidelines to adapt to each situation, mosquito control professionals have the flexibility to consider the costs of various resources in their locale and adjust accordingly. The largest cost in many mosquito control agencies is labor (salary and overtime) and insecticides. Across the country, there are mosquito control organizations that have budgets in excess of $20,000,000 USD and some with a budget of a few thousand dollars. The disparity can be even greater internationally for some low-resourced and rural countries and territories. Thus, it is up to the mosquito control professional to plan and maximize the reach and effect of those dollars in order to protect the public health. In Harris County, the Mosquito and Vector Control Division (HCPH MVCD) costs approximately $1.67 per capita. Out of the $1.67, HCPHMVCD is able to provide comprehensive mosquito surveillance, virological testing, inspection services, education and outreach, insecticide resistance management, and adulticiding operations. The breadth of this program is afforded partially due to the population density of Harris County. Areas with higher population densities share costs and therefore bring down the per capita cost of protecting individuals when compared to low population density areas (Worral and Fillinger, 2011). The phrase, "An ounce of prevention is worth a pound of cure," is especially true in public health and mosquito control. A study of the costs of a West Nile virus outbreak of 163 human cases in Sacramento, California, found that the cost of treating patients exceeded the costs of emergency mosquito control 3:1 and that if emergency spraying prevented only 15 cases, the costs would have been offset (Barber et al., 2010). As discussed earlier, chemical adulticiding is the least efficient method of controlling mosquitoes and often one of the most expensive, which is why it is generally implemented as the last measure after action thresholds are met. If adulticiding is required, targeted or selective treatments as opposed to area-wide treatments will also save money (Konradsen et al., 1999). Conducting mosquito control under fluctuating budgets stresses the importance of continuously evaluating current strategies to ensure that funds are maximized. Things to be considered are the cost and availability of products and methods used, availability of labor and equipment, environmental conditions, population density, stakeholder engagement, and assistance from NGO or other outside groups. With effective planning and cost calculations, an effective mosquito control can be established or maintained through mosquito-borne disease outbreaks or fluctuating budgets.

8.12 Conclusion

As multiple political and environmental factors continue to influence mosquito control globally, the United States and Texas will continue to utilize mosquito control strategies that make sense and are safe for people and the environment. Policy and regulations will need to be continuously updated and practical, as the newest and innovative technologies emerge for control. It is also important to remember that the many milestones and strides in integrated mosquito control have not been accomplished through one method alone. There are multiple strategies and tools that must be used in conjunction with one another to combat mosquito-borne diseases, because there is no one silver bullet to defeat them all.

Acknowledgments

The authors thank their families (Anthwan Omar Braumuller, Kim and Lee Vincent) for their patience and support when writing portions of the chapter at home, on the weekends, and during paternity leave. We also thank our coworkers at Harris County Public Health Mosquito and Vector Control Division for their encouragement and Jimmy Mains from MosquitoMate, Inc. for his assistance with locating articles for the Biological and Genetic Control sections.

References and further reading

Ahmed, K.M., Abdulah, S.H., Mohammed, N., 2018. Evaluation of some insecticides against *Culex pipiens*, the dominant mosquito species in Abha city. Int. J. Rural. Dev. Environ. Health Res. 2 (2), 4–17.

Akbari, O.S., Matzen, K.D., Marshall, J.M., Huang, H., Ward, C.M., Hay, B.A., 2013. A synthetic gene drive system for local, reversible modification and suppression of insect populations. Curr. Biol. 23 (8), 671–677.

Alphey, L., 2002. Re-engineering the sterile insect technique. Insect Biochem. Mol. Biol. 32 (10), 1243–1247.

Alphey, L., Andreasen, M., 2002. Dominant lethality and insect population control. Mol. Biochem. Parasitol. 121 (2), 173–178.

Alphey, L., Beard, C.B., Billingsley, P., Coetzee, M., Crisanti, A., Curtis, C., Eggleston, P., Godfray, C., Hemingway, J., Jacobs-Lorena, M., James, A.A., Kafatos, F.C., Mukwaya, L.G., Paton, M., Powell, J.R., Schneider, W., Scott, T.W., Sina, B., Sinden, R., Sinkins, S., Spielman, A., Toure, Y., Collins, F.H., 2002. Malaria control with genetically manipulated insect vectors. Science 298 (5591), 119–121.

Alphey, L., Nimmo, D., O'Connell, S., Alphey, N., 2007. Insect population suppression using engineered insects. In: Aksoy, S. (Ed.), Transgenesis and the Management of Vector-Borne Disease. Springer, New York, pp. 93–103.

Alphey, L., Benedict, M., Bellini, R., Clark, G.G., Dame, D.A., Service, M.W., Dobson, S.L., 2010. Sterile-insect methods for control of mosquito-borne diseases: an analysis. Vector Borne Zoonotic Dis. 10 (3), 295–311.

Alphey, L., McKemey, A., Nimmo, D., Neira Oviedo, M., Lacroix, R., Matzen, K., Beech, C., 2013. Genetic control of *Aedes* mosquitoes. Pathog. Glob. Health 107 (4), 170–179.

American Mosquito Control Association (AMCA), 2017. Best Practices for Integrated Mosquito Management: A Focused Update. https://cdn.ymaws.com/www.mosquito.org/resource/resmgr/docs/Resource_Center/Training_Certification/12.21_amca_guidelines_final_.pdf.

Andreasen, M.H., Curtis, C.F., 2005. Optimal life stage for radiation sterilization of *Anopheles* males and their fitness for release. Med. Vet. Entomol. 19 (3), 238–244.

Andrews, J.M., 1950. Advancing frontiers in insect vector control. Am. J. Public Health 40 (4), 409–416.

Andrews, J.M., 1951. The eradication program in the U.S.A. J. Nat. Malaria Soc. 10 (2), 99–123.

Andrews, J.M., Gilbertson, W.E., 1948. Blueprint for malaria eradication in the United States. J. Nat. Malaria Soc. 7 (3), 167–170.

Antwi, F., Reddy, V.P., 2015. Toxicological effects of pyrethroids on non-target aquatic insects. Environ. Toxicol. Pharmacol. 40 (3), 915–923.

Armed Forces Pest Management Board (AFPMB), 2011. Dispersal of Ultra Low Volume (ULV) Insecticides by Cold Aerosol and Thermal Fog Ground Application Equipment. Technical Guide No. 13, Armed Forces Pest Management Board, Silver Spring, MD.

Arrighi, R.B., Nakamura, C., Miyake, J., Hurd, H., Burgess, J.G., 2002. Design and activity of antimicrobial peptides against sporogonic-stage parasites causing murine malarias. Antimicrob. Agents Chemother. 46 (7), 2104–2110.

Association of State and Territorial Health Officials (ASTHO), 2005. Public Health Confronts the Mosquito: Developing Sustainable State and Local Mosquito Control Programs. http://www.astho.org/Programs/Environmental-Health/Natural-Environment/confrontsmosquito/. Washington, DC.

Atkinson, M.P., Su, Z., Alphey, N., Alphey, L.S., Coleman, P.G., Wein, L.M., 2007. Analyzing the control of mosquito-borne diseases by a dominant lethal genetic system. Proc. Natl. Acad. Sci. U. S. A. 104 (22), 9540–9545.

Attaran, A., Liroff, R., Maharaj, R., 2000. Doctoring malaria, badly: the global campaign to ban DDT. BMJ Br. Med. J. 321 (7273), 1403–1405.

Atyame, C.M., Pasteur, N., Dumas, E., Tortosa, P., Tantely, M.L., Pocquet, N., Licciardi, S., Bheecarry, A., Zumbo, B., Weill, M., Duron, O., 2011. Cytoplasmic incompatibility as a means of controlling *Culex pipiens quinquefasciatus* mosquito in the islands of the south-western Indian Ocean. PLoS Negl. Trop. Dis. 5 (12), e1440.

Atyame, C.M., Cattel, J., Lebon, C., Flores, O., Dehecq, J.S., Weill, M., Gouagna, L.C., Tortosa, P., 2015. *Wolbachia*-based population control strategy targeting *Culex quinquefasciatus* mosquitoes proves efficient under semi-field conditions. PLoS One 10 (3), e0119288.

Bandi, C., Anderson, T.J., Genchi, C., Blaxter, M.L., 1998. Phylogeny of *Wolbachia* in filarial nematodes. Proc. R. Soc. B Biol. Sci. 265 (1413), 2407–2413.

Barber, M.A., Hayne, T.B., 1921. Arsenic as a larvicide for anopheline larvae. Public Health Rep. 36, 3027–3034.

Barber, L.M., Scheiller III, J.J., Peterson, K.D., 2010. Economic cost analysis of West Nile virus outbreak, Sacramento County, California, USA, 2005. Emerg. Infect. Dis. 16 (3), 480–486.

Barrett, A.D., 2017. Yellow fever live attenuated vaccine: a very successful live attenuated vaccine but still we have problems controlling the disease. Vaccine 35 (44), 5951–5955.

Beard, J.Australian Rural Health Research Collaboration (ARHRC), 2005. DDT and human health. Sci. Total Environ. 355 (2006), 78–89.

Benedict, M.Q., Robinson, A.S., 2003. The first releases of transgenic mosquitoes: an argument for the sterile insect technique. Trends Parasitol. 19 (8), 349–355.

Berry, J., 2012. The bacterium, *Lysinibacillus sphaericus*, as an insect pathogen. J. Invertebr. Pathol. 109 (1), 1–10.

Berry-Cabán, C.S., 2011. DDT and silent spring: fifty years after. J. Mil. Vet. Health 19 (4), 19–24.

Bloem, K.A., Bloem, S., Carpenter, J.E., 2006. Impact of moth suppression/eradication programmes using the sterile insect technique or inherited sterility. In: Dyck, V.A., Hendrichs, J., Robinson, A.S. (Eds.), Sterile Insect Technique: Principles and Practice in Area-Wide Integrated Pest Management. Springer, Dordrecht, pp. 677–700.

Boller, E.F., Russ, K., Vallo, V., Bush, G.L., 1976. Incompatible races of European cherry fruit fly, *Rhagoletis cerasi* (Diptera: Tephritidae), their origin and potential use in biological control. Entomol. Exp. Appl. 20 (3), 237–247.

Bourtzis, K., O'Neill, S., 1998. "*Wolbachia*" infections and arthropod reproduction. Bioscience 48 (4), 287–293.

Bradley, G.H., 1948. Mosquito control in the U.S. In: Abstracts. International Congress on Tropical Medicine and Malaria (4th: Washington, DC). pp. 131. vol. 56, No. 4th Congr.

Bradley, G.H., 1966. A review of malaria control and eradication in the U.S. Mosq. News 26 (4), 462–470.

Brelsfoard, C.L., Dobson, S.L., 2011. *Wolbachia* effects on host fitness and the influence of male aging on cytoplasmic incompatibility in *Aedes polynesiensis* (Diptera: Culicidae). J. Med. Entomol. 48 (5), 1008–1015.

Brown, A.W.A., 1968. Principles of dispersal in ground equipment and inspections for mosquito control. J. Am. Mosq. Control Assoc. Bull 2, 11–19.

Bull, J.J., Turelli, M., 2013. *Wolbachia* versus dengue: evolutionary forecasts. Evol. Med. Public Health 1, 197–207.

Calvitti, M., Moretti, R., Skidmore, A.R., Dobson, S.L., 2012. *Wolbachia* strain wPip yields a pattern of cytoplasmic incompatibility enhancing a *Wolbachia*-based suppression strategy against the disease vector *Aedes albopictus*. Parasit. Vectors 5 (1), 254.

Caragata, E.P., Dutra, H.L., Moreira, L.A., 2016. Exploiting intimate relationships: controlling mosquito-transmitted disease with *Wolbachia*. Trends Parasitol. 32 (3), 207–218.

Carpenter, S.J., Lacasse, W.J., 1955. Mosquitoes of North America (North of Mexico). University of California Press, London, England.

Carvalho, D.O., Nimmo, D., Naish, N., McKemey, A.R., Gray, P., Wilke, A.B., Marrelli, M.T., Virginio, J.F., Alphey, L., Capurro, M.L., 2014. Mass production of genetically modified *Aedes aegypti* for field releases in Brazil. JOVE-J. Vis. Exp. 83, e3579.

Carvalho, D.O., McKemey, A.R., Garziera, L., Lacroix, R., Donnelly, C.A., Alphey, L., Malavasi, A., Capurro, M.L., 2015. Suppression of a field population of *Aedes aegypti* in Brazil by sustained release of transgenic male mosquitoes. PLoS Negl. Trop. Dis. 9 (7), e0003864.

Casida, J.E., 1980. Pyrethrum flowers and pyrethroid insecticides. Environ. Health Perspect. 34, 189–202.

Chambers, E.W., Hapairai, L., Peel, B.A., Bossin, H., Dobson, S.L., 2011. Male mating competitiveness of a *Wolbachia*-introgressed *Aedes polynesiensis* strain under semi-field conditions. PLoS Negl. Trop. Dis. 5 (8), e1271.

Chen, C.H., Huang, H., Ward, C.M., Su, J.T., Schaeffer, L.V., Guo, M., Hay, B.A., 2007. A synthetic maternal-effect selfish genetic element drives population replacement in *Drosophila*. Science 316 (5824), 597–600.

Cheng, Y., Xiong, W., Tao, J., He, D., Chen, K., Cheng, Y., 2018. Life-history traits of the invasive mosquitofish (*Gambusia affinis* Baird and Girard, 1853) in the central Yangtze River, China. BioInvasions Rec. 7 (3), 309–318.

Chernin, E., 1983. Sir Patrick Manson's studies on the transmission and biology of filariasis. Rev. Infect. Dis. 5 (1), 148–166.

Clements, A.N., Harbach, R.E., 2017. History of the discovery of the mode of transmission of yellow fever virus. J. Vector Ecol. 42 (2), 208–222.

Coates, C.J., Jasinskiene, N., Miyashiro, L., James, A.A., 1998. Mariner transposition and transformation of the yellow fever mosquito, *Aedes aegypti*. Proc. Natl. Acad. Sci. 95 (7), 3748–3751.

Coleman, P.G., Alphey, L., 2004. Genetic control of vector populations: an imminent prospect. Trop. Med. Int. Health 9 (4), 433–437.

Corby-Harris, V., Drexler, A., De Jong, L.W., Antonova, Y., Pakpour, N., Ziegler, R., Ramberg, F., Lewis, E.E., Brown, J.M., Luckhart, S., Riehle, M.A., 2010. Activation of Akt signaling reduces the prevalence and intensity of malaria parasite infection and lifespan in *Anopheles stephensi* mosquitoes. PLoS Pathog. 6 (7), e1001003.

Cox, F.E.G., 2010. History of the discovery of the malaria parasites and their vectors. Parasit. Vectors 3 (1), 5.

Da Silva Carvalho, K., Crespo, M.M., Araújo, A.P., da Silva, R.S., de Melo-Santos, M.A., de Oliveira, C.M., Silva-Filha, M.H., 2018. Long-term exposure of *Aedes aegypti* to *Bacillus thuringiensis svar. israelensis* did not involve altered susceptibility to this microbial larvicide or to other control agents. Parasit. Vectors 11, 673.

Darwazeh, H.A., Mulla, M.S., 1974. Biological activity of organophosphorus compounds and synthetic pyrethroids against immature mosquitoes. Mosq. News 34, 151–154.

Dash, A.P., Raghavendra, K., Pillai, M.K.K., 2007. Resurrection of DDT: a critical appraisal. Indian J. Med. Res. 126 (1), 1.

David, A.S., Kaser, J.M., Morey, A.C., Roth, A.M., Andow, D.A., 2013. Release of genetically engineered insects: a framework to identify potential ecological effects. Ecol. Evol. 3 (11), 4000–4015.

de Araújo, H.R.C., Kojin, B.B., Capurro, M.L., 2018. Sex determination and *Aedes* population control. Parasit. Vectors 11 (2), 644.

De Barro, P.J., Murphy, B., Jansen, C.C., Murray, J., 2011. The proposed release of the yellow fever mosquito, *Aedes aegypti* containing a naturally occurring strain of *Wolbachia pipientis*, a question of regulatory responsibility. J. Verbrauch. Lebensm. 6, 33–40. Suppl. 1.

De Mendonça Santos, E.M., de Melo Chalegre, K.D., de Albuquerque, A.L., Regis, L., Fontes de Oliveira, C.M., Silva-Filha, M.H., 2019. Frequency of resistance alleles to *Lysinibacillus sphaericus* in a *Culex quinquefasciatus* population treated with a *L. sphaericus/Bti* biolarvicide. Biol. Control. 132, 95–101.

Dennett, J.A., Debboun, M., 2017. Case study: surveillance and control operations in Harris County, Texas. Pest Control Technology 86–92.

Derua, Y.A., Kahindi, S.C., Mosha, F.W., Kweka, E.J., Atieli, H.E., Wang, X., Zhou, G., Lee, M.C., Githeko, A.K., Yan, G., 2018. Microbial larvicides for mosquito control: impact of long lasting formulations of *Bacillus thuringiensis var. israelensis* and *Bacillus sphaericus* on non-target organisms in western Kenya highlands. Ecol. Evol. 8 (15), 7563–7573.

Devine, G.J., Perea, E.Z., Killeen, G.F., Stancil, J.D., Clark, S.J., Morrison, A.C., 2009. Using adult mosquitoes to transfer insecticides to *Aedes aegypti* larval habitats. Proc. Natl. Acad. Sci. U. S. A. 106 (28), 11530–11534.

Dobson, S.L., Bourtzis, K., Braig, H.R., Jones, B.F., Zhou, W., Rousset, F., O'Neill, S.L., 1999. *Wolbachia* infections are distributed throughout insect somatic and germ line tissues. Insect Biochem. Mol. Biol. 29 (2), 153–160.

Dobson, S.L., Marsland, E.J., Rattanadechakul, W., 2001. *Wolbachia*-induced cytoplasmic incompatibility in single-and superinfected *Aedes albopictus* (Diptera: Culicidae). J. Med. Entomol. 38 (3), 382–387.

Dobson, S.L., Fox, C.W., Jiggins, F.M., 2002. The effect of *Wolbachia*-induced cytoplasmic incompatibility on host population size in natural and manipulated systems. Proc. R. Soc. B Biol. Sci. 269 (1490), 437–445.

Dobson, S.L., Rattanadechakul, W., Marsland, E.J., 2004. Fitness advantage and cytoplasmic incompatibility in *Wolbachia* single-and superinfected *Aedes albopictus*. Heredity 93 (2), 135–142.

Duprey, Z., Rivers, S., Luber, G., Becker, A., Blackmore, C., Barr, D., Weerasekera, G., Kieszak, S., Flanders, W.D., Rubin, C., 2008. Community aerial mosquito control and naled exposure. J. Am. Mosq. Control Assoc. 24 (1), 42–46.

Dutra, H.L.C., Rocha, M.N., Dias, F.B.S., Mansur, S.B., Caragata, E.P., Moreira, L.A., 2016. *Wolbachia* blocks currently circulating Zika virus isolates in Brazilian *Aedes aegypti* mosquitoes. Cell Host Microbe 19, 771–774.

Dyck, V.A., Hendrichs, J., Robinson, A.S. (Eds.), 2006a. Sterile Insect Technique: Principles and Practice in Area-Wide Integrated Pest Management. Springer, Dordrecht.

Dyck, V.A., Fernandez, E.E.R., Flores, J.R., Teruya, T., Barnes, B., Riera, P.G., Lindquist, D., Reuben, R., 2006b. Public relations and political support in area-wide integrated pest management programmes that integrate the sterile insect technique. In: Dyck, V.A., Hendrichs, J., Robinson, A.S. (Eds.), Sterile Insect Technique: Principles and Practice in Area-Wide Integrated Pest Management. Springer, Dordrecht, pp. 547–559.

Elliott, M., 1976. Properties and applications of pyrethroids. Environ. Health Perspect. 14, 3–13.

Elliott, M., Janes, N.F., 1978. Synthetic pyrethroids—a new class of insecticide. Chem. Soc. Rev. 7 (4), 473–505.

Elliott, M., Janes, N.F., Potter, C., 1978. The future of pyrethroids in insect control. Annu. Rev. Entomol. 23 (1), 443–469.

Enkerlin, W.R., 2006. Impact of fruit fly control programmes using the sterile insect technique. In: Dyck, V.A., Hendrichs, J., Robinson, A.S. (Eds.), Sterile Insect Technique: Principles and Practice in Area-Wide Integrated Pest Management. Springer, Dordrecht, pp. 651–676.

Environmental Protection Agency (EPA), 2016. Malathion. EPA. 18, Nov, https://www.epa.gov/mosquitocontrol/malathion#q1.

Eskenazi, B., Chevrier, J., Rosas, L.G., Anderson, H.A., Bornman, M.S., Bouwman, H., Chen, A., Cohn, B.A., de Jager, C., Henshel, D.S., Leipzig, F., Leipzig, J.S., Lorenz, E.C., Snedeker, S.M., Stapleton, D., 2009. The Pine River statement: human health consequences of DDT use. Environ. Health Perspect. 117 (9), 1359–1367.

Facchinelli, L., Valerio, L., Ramsey, J.M., Gould, F., Walsh, R.K., Bond, G., Robert, M.A., Lloyd, A.L., James, A.A., Alphey, L., Scott, T.W., 2013. Field cage studies and progressive evaluation of genetically-engineered mosquitoes. PLoS Negl. Trop. Dis. 7 (1), e2001.

Faust, E.C., 1951. The history of malaria in the U.S. Am. Sci. 39, 121–130. 1.

Finlay, C., 1881. El mosquito hipoteticamente considerado como agente de trasmision de la fiebre amarilla. An. Real Acad. Cienc. Méd. Fisicas Nat. Habana 18, 147–169.

Floore, T.G., 2006. Mosquito larval control practices: past and present. J. Am. Mosq. Control Assoc. 22 (3), 527–533.

Focks, D.A., Sackett, S.R., Bailey, D.L., 1982. Field experiments on the control of *Aedes aegypti* and *Culex quinquefasciatus* by *Toxorhynchites rutilus rutilus* (Diptera: Culicidae). J. Med. Entomol. 19 (3), 336–339.

Focks, D.A., Sackett, S.R., Bailey, D.L., Dame, D.A., 1985. Effect of weekly releases of *Toxorhynchites amboinensis* (Doleschall) on *Aedes aegypti* (L.) (Diptera: Culicidae) in New Orleans, Louisiana. J. Econ. Entomol. 78 (3), 622–626.

Focks, D.A., Sackett, S.R., Kloter, K.O., Dame, D.A., Carmichael, G.T., 1986. The integrated use of *Toxorhynchites amboinensis* and ground-level ULV insecticide application to suppress *Aedes aegypti* (Diptera: Culicidae). J. Med. Entomol. 23 (5), 513–519.

Focks, D.A., 2007. *Toxorhynchites* as biocontrol agents. J. Am. Mosq. Control Assoc. 23 (sp2), 118–127.

Foster, W.A., Walker, E.D., 2009. Mosquitoes (Culicidae). In: Mullen, G.R., Durden, L.A. (Eds.), Medical and Veterinary Entomology. second ed. Elsevier, Oxford, pp. 207–259.

Franz, A.W., Sanchez-Vargas, I., Adelman, Z.N., Blair, C.D., Beaty, B.J., James, A.A., Olson, K.E., 2006. Engineering RNA interference-based resistance to dengue virus type 2 in genetically modified *Aedes aegypti*. Proc. Natl. Acad. Sci. U. S. A. 103 (11), 4198–4203.

Franz, A.W., Sanchez-Vargas, I., Piper, J., Smith, M.R., Khoo, C.C.H., James, A.A., Olson, K.E., 2009. Stability and loss of a virus resistance phenotype over time in transgenic mosquitoes harbouring an antiviral effector gene. Insect Mol. Biol. 18 (5), 661–672.

Frentiu, F.D., Zakir, T., Walker, T., Popovici, J., Pyke, A.T., van den Hurk, A., McGraw, E.A., O'Neill, S.L., 2014. Limited dengue virus replication in field-collected *Aedes aegypti* mosquitoes infected with *Wolbachia*. PLoS Negl. Trop. Dis. 8 (2), e2688.

Fu, G., Lees, R.S., Nimmo, D., Aw, D., Jin, L., Gray, P., Berendonk, T.U., White-Cooper, H., Scaife, S., Phuc, H.K., Marinotti, O., 2010. Female-specific flightless phenotype for mosquito control. Proc. Natl. Acad. Sci. 107 (10), 4550–4554.

Gabrieli, P., Smidler, A., Catteruccia, F., 2014. Engineering the control of mosquito-borne infectious diseases. Genome Biol. 15 (11), 535.

Gahan, J.B., Travis, B.V., Morton, F.A., Lindquist, A.W., 1945. DDT as a residual-type treatment to control *Anopheles quadrimaculatus*: practical tests. J. Econ. Entomol. 38 (2), 231–235.

Gantz, V.M., Jasinskiene, N., Tatarenkova, O., Fazekas, A., Macias, V.M., Bier, E., James, A.A., 2015. Highly efficient Cas9-mediated gene drive for population modification of the malaria vector mosquito Anopheles stephensi. Proc. Natl. Acad. Sci. 112 (49), E6736–E6743.

Gavotte, L., Mercer, D.R., Vandyke, R., Mains, J.W., Dobson, S.L., 2009. *Wolbachia* infection and resource competition effects on immature *Aedes albopictus* (Diptera: Culicidae). J. Med. Entomol. 46 (3), 451–459.

Glaser, R.L., Meola, M.A., 2010. The native *Wolbachia* endosymbionts of *Drosophila melanogaster* and *Culex quinquefasciatus* increase host resistance to West Nile virus infection. PLoS One 5, e119877.

Gong, P., Epton, M.J., Fu, G., Scaife, S., Hiscox, A., Condon, K.C., Condon, G.C., Morrison, N.I., Kelly, D.W., Dafa'alla, T., Coleman, P.G., 2005. A dominant lethal genetic system for autocidal control of the Mediterranean fruitfly. Nat. Biotechnol. 23 (4), 453.

Gould, F., Schliekelman, P., 2004. Population genetics of autocidal control and strain replacement. Annu. Rev. Entomol. 49 (1), 193–217.

Gould, E., Pettersson, J., Higgs, S., Charrel, R., de Lamballerie, X., 2017. Emerging arboviruses: why today? One Health 4, 1–13.

Gubler, D.J., 2004. The changing epidemiology of yellow fever and dengue, 1900 to 2003: full circle? Comp. Immunol. Microbiol. Infect. Dis. 27 (5), 319–330.

Güereña-Burgueño, F., 2002. The centennial of the Yellow Fever Commission and the use of informed consent in medical research. Salud Publica Mex. 44 (2), 140–144.

Hall, A.B., Basu, S., Jiang, X., Qi, Y., Timoshevskiy, V.A., Biedler, J.K., Sharakhova, M.V., Elahi, R., Anderson, M.A., Chen, X.G., Sharakhov, I.V., 2015. A male-determining factor in the mosquito *Aedes aegypti*. Science 348 (6240), 1268–1270.

Hall, M.T., Briley, A.C., Lindroth, E.J., Fajardo, J.D., Cilek, J.E., Richardson, A.G., 2017. A small-scale investigation into the effects of a larvicidal oil on oviposition site preference by *Aedes aegypti*. J. Am. Mosq. Control Assoc. 33 (4), 355–357.

Hammond, A., Galizi, R., Kyrou, K., Simoni, A., Siniscalchi, C., Katsanos, D., Gribble, M., Baker, D., Marois, E., Russell, S., Burt, A., 2016. A CRISPR-Cas9 gene drive system targeting female reproduction in the malaria mosquito vector *Anopheles gambiae*. Nat. Biotechnol. 34 (1), 78–83.

Hanemaaijer, M.J., Collier, T.C., Chang, A., Shott, C.C., Houston, P.D., Schmidt, H., Main, B.J., Cornel, A.J., Lee, Y., Lanzaro, G.C., 2018. The fate of genes that cross species boundaries after a major hybridization event in a natural mosquito population. Mol. Ecol. 27 (24), 4978–4990.

Harris, A.F., Nimmo, D., McKemey, A.R., Kelly, N., Scaife, S., Donnelly, C.A., Beech, C., Petrie, W.D., Alphey, L., 2011. Field performance of engineered male mosquitoes. Nat. Biotechnol. 29 (11), 1034–1037.

Harris, A.F., McKemey, A.R., Nimmo, D., Curtis, Z., Black, I., Morgan, S.A., Oviedo, M.N., Lacroix, R., Naish, N., Morrison, N.I., Collado, A., 2012. Successful suppression of a field mosquito population by sustained release of engineered male mosquitoes. Nat. Biotechnol. 30 (9), 828–830.

Hayes Jr., G.R., Tinker, M.E., 1958. The 1956-1957 status of *Aedes aegypti* in the U.S. Mosq. News 18 (3), 253–257.

Heinrich, J.C., Scott, M.J., 2000. A repressible female-specific lethal genetic system for making transgenic insect strains suitable for a sterile-release program. Proc. Natl. Acad. Sci. U. S. A. 97 (15), 8229–8232.

Hemingway, J., Ranson, H., 2000. Insecticide resistance in insect vectors of human disease. Annu. Rev. Entomol. 45 (1), 371–391.

Henrick, C.A., 2007. Methoprene. J. Am. Mosq. Control Assoc. 23 (Suppl. 2), 225–239.

Hertig, M., 1936. The Rickettsia, *Wolbachia pipientis* (gen. et sp. n.) and associated inclusions of the mosquito, *Culex pipiens*. Parasitology 28 (4), 453–486.

Hertlein, M.B., Mavrotas, C., Jousseaume, C., Lysandro, M., Thompson, G.D., Jany, W., Ritchie, S.A., 2010. A review of spinosad as a natural product for larval mosquito control. J. Am. Mosq. Control Assoc. 26 (1), 67–87.

Hilgenboecker, K., Hammerstein, P., Schlattmann, P., Telschow, A., Werren, J.H., 2008. How many species are infected with *Wolbachia*? A statistical analysis of current data. FEMS Microbiol. Lett. 281 (2), 215–220.

Hoffmann, A., Montgomery, B.L., Popovici, J., Iturbe-Ormaetxe, I., Johnson, P.H., Muzzi, F., Greenfield, M., Durkan, M., Leong, Y.S., Dong, Y., Cook, H., 2011. Successful establishment of *Wolbachia* in *Aedes* populations to suppress dengue transmission. Nature 476 (7361), 454–U107.

Hoffmann, A.A., Iturbe-Ormaetxe, I., Callahan, A.G., Phillips, B.L., Billington, K., Axford, J.K., Montgomery, B., Turley, A.P., O'Neill, S.L., 2014. Stability of the wMel *Wolbachia* infection following invasion into *Aedes aegypti* populations. PLoS Negl. Trop. Dis. 8 (9), e3115.

Hurst, G.D., Jiggins, F.M., von der Schulenburg, J.H.G., Bertrand, D., West, S.A., Goriacheva, I.I., Zakharov, I.A., Werren, J.H., Stouthamer, R., Majerus, M.E., 1999. Male–killing *Wolbachia* in two species of insect. Proc. R. Soc. B Biol. Sci. 266 (1420), 735–740.

Isaacs, A.T., Jasinskiene, N., Tretiakov, M., Thiery, I., Zettor, A., Bourgouin, C., James, A.A., 2012. Transgenic *Anopheles stephensi* coexpressing single-chain antibodies resist Plasmodium falciparum development. Proc. Natl. Acad. Sci. 109 (28), E1922–E1930.

Ito, J., Ghosh, A., Moreira, L.A., Wimmer, E.A., Jacobs-Lorena, M., 2002. Transgenic anopheline mosquitoes impaired in transmission of a malaria parasite. Nature 417 (6887), 452–455.

Iturbe-Ormaetxe, I., Walker, T., O'Neill, S.L., 2011. *Wolbachia* and the biologican control of mosquito-borne disease. EMBO Rep. 12 (6), 508–518.

Jasinskiene, N., Coates, C.J., Benedict, M.Q., Cornel, A.J., Rafferty, C.S., James, A.A., Collins, F.H., 1998. Stable transformation of the yellow fever mosquito, *Aedes aegypti*, with the Hermes element from the housefly. Proc. Natl. Acad. Sci. 95 (7), 3743–3747.

Johnson, D.R., 1956. The challenge of global malaria eradication. Mosq. News 16 (3), 208–211.

Kambhampati, S., Rai, K.S., Burgun, S.J., 1993. Unidirectional cytoplasmic incompatibility in the mosquito, *Aedes albopictus*. Evolution 42 (2), 673–677.

Kambris, Z., Cook, P.E., Phuc, H.K., Sinkins, S.P., 2009. Immune activiation by life-shortening *Wolbachia* and reduced filarial competence in mosquitoes. Science 326, 134–136.

Kambris, Z., Blagborough, A., Pinto, S., Blagrove, M., Godfray, H., Sinden, R., Sinkins, S., Vernick, K., 2010. *Wolbachia* stimulates immune gene expression and inhibits *Plasmodium* development in *Anopheles gambiae*. PLoS Pathog. 6, 215–220.

Kim, W., Koo, H., Richman, A.M., Seeley, D., Vizioli, J., Klocko, A.D., O'brochta, D.A., 2004. Ectopic expression of a cecropin transgene in the human malaria vector mosquito *Anopheles gambiae* (Diptera: Culicidae): effects on susceptibility to *Plasmodium*. J. Med. Entomol. 41 (3), 447–455.

King, W.V., Bradley, G.H., 1926. Airplane Dusting in the Control of Malaria Mosquitoes. U.S. Department of Agriculture, pp. 1–16. Dept. Circular 367.

Kittayapong, P., MongKalangoon, P., Baimai, V., O'Neill, S.L., 2002. Host age effect and expression of cytoplasmic incompatibility in field populations of *Wolbachia*-superinfected *Aedes albopictus*. Heredity 88 (4), 270–274.

Klassen, W., 2006. Area-wide integrated pest management and the sterile insect technique. In: Dyck, V.A., Hendrichs, J., Robinson, A.S. (Eds.), Sterile Insect Technique: Principles and Practice in Area-Wide Integrated Pest Management. Springer, Dordrecht, pp. 39–68.

Klassen, W., Curtis, C.F., 2006. History of the sterile insect technique. In: Dyck, V.A., Hendrichs, J., Robinson, A.S. (Eds.), Sterile Insect Technique: Principles and Practice in Area-Wide Integrated Pest Management. Springer, Dordrecht, pp. 3–38.

Knipling, E.F., 1955. Possibilities of insect control or eradication through the use of sexually sterile males. J. Econ. Entomol. 48 (4), 459–462.

Knipling, E.F., 1979. The Basic Principles of Insect Population Suppression and Management. Agriculture Handbook, No. 512, USDA, Washington, DC.

Knipling, E.F., 1998. Role of parasitoid augmentation and sterile insect techniques in areawide management of agricultural insect pests. J. Agr. Entomol. 15, 273–301.

Konradsen, F., Steele, P., Perera, D., Van der Hoek, W., Amerasinghe, F.P., 1999. Cost of malaria control in Sri Lanka. Bull. World Health Organ. 77 (4), 301–309.

Krafsur, E.S., 1998. Sterile insect technique for suppressing and eradicating insect population: 55 years and counting. J. Agr. Entomol. 15 (4), 303–317.

Labbé, G.M., Scaife, S., Morgan, S.A., Curtis, Z.H., Alphey, L., 2012. Female-specific flightless (fsRIDL) phenotype for control of *Aedes albopictus*. PLoS Negl. Trop. Dis. 6 (7), e1724.

Lacroix, R., McKemey, A.R., Raduan, N., Wee, L.K., Ming, W.H., Ney, T.G., AA, S.R., Salman, S., Subramaniam, S., Nordin, O., Angamuthu, C., 2012. Open field release of genetically engineered sterile male *Aedes aegypti* in Malaysia. PLoS One 7 (8), e42771.

Laven, H., 1967. A possible model for speciation by cytoplasmic isolation in the *Culex pipiens* complex. Bull. World Health Organ. 37 (2), 263–266.

Lawrence, C., Hamilton, R., 2016. Experimental evidence indicates that native freshwater fish outperform introduced *Gambusia* in mosquito suppression when water temperature is below 25°C. Hydrobiologia 766, 357–364.

Lees, R.S., Gilles, J.R., Hendrichs, J., Vreysen, M.J., Bourtzis, K., 2015. Back to the future: the sterile insect technique against mosquito disease vectors. Curr. Opin. Insect Sci. 10, 156–162.

Lindquist, D.A., Abusowa, M., Hall, M.J.R., 1992. The New World screwworm fly in Libya: a review of its introduction and eradication. Med. Vet. Entomol. 6 (1), 2–8.

Lloyd, A.M., Connelly, C.R., Carlson, D.B. (Eds.), 2018. Florida Coordinating Council on Mosquito Control. Florida Mosquito Control: The State of the Mission as Defined by Mosquito Controllers, Regulators, and Environmental Managers. University of Florida, Institute of Food and Agricultural Sciences, Florida Medical Entomology Laboratory, Vero Beach, FL, pp. 70–71.

Lofgren, C.S., Dame, D.A., Breeland, S.G., Weidhaas, D.E., Jeffery, G., Kaiser, R., Ford, H.R., Boston, M.D., Baldwin, K.F., 1974. Release of chemosterilized males for the control of *Anopheles albimanus* in El Salvador. Am. J. Trop. Med. Hyg. 23 (2), 288–297.

Macias, V., Ohm, J., Rasgon, J., 2017. Gene drive for mosquito control: where did it come from and where are we headed? Int. J. Environ. Res. Public Health 14 (9), 1006.

Mains, J.W., Brelsfoard, C.L., Rose, R.I., Dobson, S.L., 2016. Female adult *Aedes albopictus* suppression by *Wolbachia*-infected male mosquitoes. Sci. Rep. 6, 33846.

Manson, P.M., 1878. On the development of *Filaria sanguinis hominis* and on the mosquito considered as a nurse. Zool. J. Linn. Soc.-Lond. 14 (75), 304–311.

Maracombe, S., Chonephetsarath, S., Thammavong, P., Brey, P.T., 2018. Alternative insecticides for larval control of the dengue vector *Aedes aegypti* in Lao PDR: insecticide resistance and semi-field trial study. Parasit. Vectors 11, 616.

Marshall, J.M., Akbari, O.S., 2018. Can CRISPR-based gene drive be confined in the wild? A question for molecular and population biology. ACS Chem. Biol. 13 (2), 424–430.

Matheson, R., 1941. The role of anophelines in the epidemiology of malaria. In: A Symposium on Human Malaria. American Association for the Advancement of Science, Washington, DC, pp. 157–162.

Mathur, G., Sanchez-Vargas, I., Alvarez, D., Olson, K.E., Marinotti, O., James, A.A., 2010. Transgene-mediated suppression of dengue viruses in the salivary glands of the yellow fever mosquito, *Aedes aegypti*. Insect Mol. Biol. 19 (6), 753–763.

McGraw, E.A., O'Neill, S.L., 2013. Beyond insecticides: new thinking on an ancient problem. Nat. Rev. Microbiol. 11 (3), 181–193.

Mendis, K., Rietveld, A., Warsame, M., Bosman, A., Greenwood, B., Wernsdorfer, W.H., 2009. From malaria control to eradication: the WHO perspective. Trop. Med. Int. Health 14 (7), 802–809.

Meredith, J.M., Basu, S., Nimmo, D.D., Larget-Thiery, I., Warr, E.L., Underhill, A., McArthur, C.C., Carter, V., Hurd, H., Bourgouin, C., Eggleston, P., 2011. Site-specific integration and expression of an anti-malarial gene in transgenic *Anopheles gambiae* significantly reduces *Plasmodium* infections. PLoS One 6 (1), e14587.

Mertz, F.P., Yao, R.C., 1990. *Saccharopolyspora spinosa* sp. nov. isolated from soil collected in a sugar mill rum still. Int. J. Syst. Bacteriol. 40 (1), 34–39.

Metcalf, R.L., 1989. Insect resistance to insecticides. Pestic. Sci. 26 (4), 333–358.

Mian, L.S., Dhillon, M.S., Dodson, L., 2017. Field evaluation of pyriproxyfen against mosquitoes in catch basins in southern California. J. Am. Mosq. Control Assoc. 33 (2), 145–147.

Min, K.T., Benzer, S., 1997. *Wolbachia*, normally a symbiont of *Drosophila*, can be virulent, causing degeneration and early death. Proc. Natl. Acad. Sci. U. S. A 94 (20), 10792–10796.

Mitchell, S.N., Catteruccia, F., 2017. Anopheline reproductive biology: impacts on vectorial capacity and potential avenues for malaria control. Cold Spring Harb. Perspect. Med. https://doi.org/10.1101/cshperspect.a025593.

Monath, T.P., 1988. Yellow fever. In: Monath, T.P. (Ed.), The Arboviruses: Ecology and Epidemiology. vol. V. CRC Press, Boca Raton, pp. 139–231.

Moreira, L.A., Ito, J., Ghosh, A., Devenport, M., Zieler, H., Abraham, E.G., Crisanti, A., Nolan, T., Catteruccia, F., Jacobs-Lorena, M., 2002. Bee venom phospholipase inhibits malaria parasite development in transgenic mosquitoes. J. Biol. Chem. 277 (43), 40839–40843.

Moreira, L.A., Iturbe-Ormaetxe, I., Jeffery, J.A., Lu, G., Pyke, A.T., Hedges, L.M., Rocha, B.C., Hall-Mendelin, S., Day, A., Riegler, M., Hugo, L.E., 2009a. A *Wolbachia* symbiont in *Aedes aegypti* limits infection with dengue, Chikungunya, and *Plasmodium*. Cell 139 (7), 1268–1278.

Moreira, L.A., Saig, E., Turley, A.P., Ribeiro, J.M., O'Neill, S.L., McGraw, E.A., 2009b. Human probing behavior of *Aedes aegypti* when infected with a life-shortening strain of *Wolbachia*. PLoS Negl. Trop. Dis. 3, e568.

Moretti, R., Calvitti, M., 2013. Male mating performance and cytoplasmic incompatibility in a *wPip Wolbachia* trans-infected line of *Aedes albopictus* (*Stegomyia albopicta*). Med. Vet. Entomol. 27 (4), 377–386.

Morlan, H.B., Tinker, M.E., 1965. Distribution of *Aedes aegypti* infestations in the United States. Am. J. Hyg. 14 (6), 892–899.

Mount, G.A., 1998. A critical review of ultralow-volume aerosols of insecticide applied with vehicle-mounted generators for adult mosquito control. J. Am. Mosq. Control Assoc. 14, 305–334.

Mount, G.A., Biery, T.L., Haile, D.G., 1996. A review of ultra-low volume aerial sprays of insecticide for mosquito control. J. Am. Mosq. Control Assoc. 12, 601–618.

Mountin, J.W., 1944. A program for the eradication of malaria from continental United States. J. Nat. Malaria Soc. 3, 69–73.

Mulla, M.S., Darwazeh, H.A., Majori, G., 1975. Field efficacy of some promising mosquito larvicides and their effects on nontarget organisms. Mosq. News 36 (3), 251–256.

Mulla, M.S., Navvab-Gorjrati, H.A., Darwazeh, H.A., 1978. Biological activity and longevity of new synthetic pyrethroids against mosquitoes and some nontarget insects. Mosq. News 38 (1), 90–96.

Muller, H.J., 1927. Artificial transmutation of the gene. Science 66 (1699), 84–87.

Mysore, K., Sun, L., Tomchaney, M., Sullivan, G., Adams, H., Piscoya, A.S., Severson, D.W., Syed, Z., Duman-Scheel, M., 2015. siRNA-mediated silencing of doublesex during female development of the dengue vector mosquito *Aedes aegypti*. PLoS Negl. Trop. Dis. 9 (11), e0004213.

Nature, 1975. Opinion—Oh, New Dehli; Oh, Geneva. Nature 256 (5516), 355–357.

O'Connor, L., Plichart, C., Sang, A.C., Brelsfoard, C.L., Bossin, H.C., Dobson, S.L., 2012. Open release of male mosquitoes infected with a *Wolbachia* biopesticide: field performance and infection containment. PLoS Negl. Trop. Dis. 6 (11), e1797.

O'Neill, S.L., Hoffman, A.A., Werren, J.H., 1997. Influential Passengers: Inherited Microorganisms and Arthropod Reproduction. Oxford University Press, Oxford.

Oxitec, 2018a. Oxitec to Apply New Generation of Self-Limiting Mosquito Technology to Malaria-Spreading Mosquitoes. Oxford, UK, https://www.oxitec.com/oxitec-to-apply-new-generation-of-self-limiting-mosquito-technology-to-malaria-spreading-mosquitoes/.

Oxitec, 2018b. Transitioning Friendly™ Self-limiting Mosquitoes to 2nd Generation Technology Platform, Paving Way to New Scalability, Performance and Cost Breakthroughs. Oxford, UK, https://www.oxitec.com/2nd-generation-platform/.

Oye, K.A., Esvelt, K., Appleton, E., Catteruccia, F., Church, G., Kuiken, T., Lightfoot, S.B.Y., McNamara, J., Smidler, A., Collins, J.P., 2014. Regulating gene drives. Science 345 (6197), 626–628.

Palmquist, K., Salatas, J., Fairbrother, A., 2012. Pyrethroid insecticides: use, environmental fate, and ecotoxicology. In: Perveen, F. (Ed.), Insecticides—Advances in Integrated Pest Management. IntechOpen, https://doi.org/10.5772/29495. Available from: https://www.intechopen.com/books/insecticides-advances-in-integrated-pest-management/pyrethroid-insecticides-use-environmental-fate-and-ecotoxicology.

Pan American Health Organization (PAHO), 1969. Report for Registration of Malaria Eradication From U.S. of America. Pan American Health Organization, Pan American Sanitary Bureau, Regional Office of the World Health Organization.

Pane, A., De Simone, A., Saccone, G., Polito, C., 2005. Evolutionary conservation of *Ceratitis capitata* transformer gene function. Genetics 171 (2), 615–624.

Paoliello, A., 1948. Control of yellow fever and other mosquito-borne diseases. Bol. Oficina Sanit. Panam. 27 (11), 1005–1044.

Patil, P.B., Gorman, K.J., Dasgupta, S.K., Reddy, K.V., Barwale, S.R., Zehr, U.B., 2018. Self-limiting OX513A *Aedes aegypti* demonstrate full susceptibility to currently used insecticidal chemistries as compared to Indian wild-type *Aedes aegypti*. Psyche: J. Entomol. 2018, 1.

Patterson, K.D., 1992. Yellow fever epidemics and mortality in the U.S., 1693–1905. Soc. Sci. Med. 34 (8), 855–865.

Paules, C.I., Fauci, A.S., 2017. Yellow fever—once again on the radar screen in the Americas. N. Engl. J. Med. 376 (15), 1397–1399.

Peloquin, J.J., Thibault, S.T., Staten, R., Miller, T.A., 2000. Germ-line transformation of pink bollworm (Lepidoptera: Gelechiidae) mediated by the *piggyBac* transposable element. Insect Mol. Biol. 9 (3), 323–333.

Phuc, H.K., Andreasen, M.H., Burton, R.S., Vass, C., Epton, M.J., Pape, G., Fu, G., Condon, K.C., Scaife, S., Donnelly, C.A., Coleman, P.G., 2007. Late-acting dominant lethal genetic systems and mosquito control. BMC Biol. 5 (1), 11.

Pietrantonio, P.V., Gibson, G., Nawrocki, S., Carrier, F., Knight Jr., W.P., 2000. Insecticide resistance status, esterase activity, and electromorphs from mosquito populations of *Culex quinquefasciatus* Say (Diptera: Culicidae), in Houston (Harris County), Texas. J. Vector Ecol. 25 (1), 74–89.

Popovici, J., Moreira, L.A., Poinsignon, A., Iturbe-Ormaetxe, I., McNaughton, D., O'Neill, S.L., 2010. Assessing key safety concerns of a *Wolbachia*-based strategy to control dengue transmission by *Aedes* mosquitoes. Mem. Inst. Oswaldo Cruz 105 (8), 957–964.

Prato, M., Polimeni, M., Giribaldi, G., 2013. DDT as an Anti-Malaria Tool: The Bull in the China Shop of the Elephant in the Room? Intech Open Access Publisher, pp. 333–364.

Pratt, H.D., Littig, K.S., 1974. Insecticide Application Equipment for the Control of Insects of Public Health Importance. U.S. Department of Health Education and Welfare Public Health Service, Center for Disease Control and Prevention, Atlanta, Georgia.

Pyke, G.H., 2008. Plague minnow or mosquito fish? A review of the biology and impacts of introduced *Gambusia* species. Annu. Rev. Ecol. Evol. Syst. 39 (1), 171–791.

Reed, W., Carroll, J., Agramonte, A., Lazear, J.W., 1900. The etiology of yellow fever—a preliminary note. Public Health Pap. Rep. 26, 37–53.

Ritter, L., Solomon, K.R., Forget, J., Stemeroff, M., O'Leary, C., 1995. A Review of Selected Persistent Organic Pollutants. International Programme on Chemical Safety (IPCS). PCS/95.3, 9 65. World Health Organization, Geneva, p. 66.

Rose, R.I., 2001. Pesticides and public health: integrated methods of mosquito management. Emerg. Infect. Dis. 7 (1), 17.

Ross, R., 1923. Memoirs: With a Full Account of the Great Malaria Problem and its Solution. John Murray, London.

Roukens, A.H., Visser, L.G., 2008. Yellow fever vaccine: past, present and future. Expert. Opin. Biol. Ther. 8 (11), 1787–1795.

Russell, P.F., 1955. Man's mastery of malaria. In: Man's Mastery of Malaria. Oxford University Press, London.

Russell, P.F., 1968. The U.S. and malaria: debits and credits. Bull. N. Y. Acad. Med. 44 (6), 623–653.

Saccone, G., Pane, A., De Simone, A., Salvemini, M., Milano, A., Annunziata, L., Mauro, U., Polito, L.C., 2007. New sexing strains for Mediterranean fruit fly *Ceratitis capitata*: transforming females into males. In: Vreysen, M.J.B., Robinson, A.S., Hendrichs, J. (Eds.), Area-Wide Control of Insect Pests. Springer, Dordrecht, pp. 95–102.

Salvemini, M., Robertson, M., Aronson, B., Atkinson, P., Polito, L.C., Saccone, G., 2007. *Ceratitis capitata* transformer-2 gene is required to establish and maintain the autoregulation of Cctra, the master gene for female sex determination. Int. J. Dev. Biol. 53 (1), 109–120.

Sazama, E.J., Bosch, M.J., Shouldis, C.S., Ouellette, S.P., Wesner, J.S., 2017. Incidence of *Wolbachia* in aquatic insects. Ecol. Evol. 7 (4), 1165–1169.

Schleissmann, D.J., 1964. The *Aedes aegypti* eradication program of the U.S. Mosq. News 24 (2), 124–132.

Schleissmann, D.J., Calheiros, L.B., 1974. A review of the status of yellow fever and *Aedes aegypti* eradication programs in the Americas. Mosq. News 34, 1–9.

Schreiber, E.T., 2007. Toxorhynchites. J. Am. Mosq. Control Assoc. 23 (Suppl. 2), 129–132.

Schreiber, E.T., Jones, C.J., 1994. Evaluation of inoculative releases of *Toxorhynchites splendens* (Diptera: Culicidae) in urban environments in Florida. Environ. Entomol. 23 (3), 770–777.

Serebrovsky, A.S., 1969. On the possibility of a new method for the control of insect pests. Sterile-male technique for eradication or control of harmful insects. In: Proceedings of a Panel on Application of the Sterile-Male Technique for the Eradication or Control of Harmful Species of Insects, Organised by the Joint FAO/IAEA Division of Atomic Energy in Food and Agriculture and Held in Vienna, Austria, pp. 123–237.

Simmons, S.W., Upholt, W.M., 1951. Disease control with insecticides: a review of the literature. Bull. World Health Organ. 3 (4), 535–556.

Simmons, G.S., Alphey, L.S., Vasquez, T., Morrison, N.I., Epton, M.J., Miller, E., Miller, T.A., Staten, R.T., 2007. Potential use of a conditional lethal transgenic pink bollworm *Pectinophora gossypiella* in area-wide eradication or suppression programmes. In: Vreysen, M.J.B., Robinson, A.S., Hendrichs, J. (Eds.), Area-Wide Control of Insect Pests. Springer, Dordrecht, pp. 119–123.

Sinkins, S.P., 2004. *Wolbachia* and cytoplasmic incompatibility in mosquitoes. Insect Biochem. Mol. Biol. 34 (7), 723–729.

Sinkins, S.P., 2013. *Wolbachia* and arbovirus inhibition in mosquitoes. Future Microbiol. 8, 1249–1256.

Sinkins, S.P., Braig, H.R., O'Neill, S.L., 1995. *Wolbachia pipientis*: bacterial density and unidirectional cytoplasmic incompatibility between infected populations of *Aedes albopictus*. Exp. Parasitol. 81 (3), 284–291.

Soper, F.L., 1944. Yellow fever. In: Bercovitz, Z.T. (Ed.), Clinical and Tropical Medicine. Paul B. Hoeber, Inc., New York, pp. 391–420.

Soper, F.L., 1958. The 1957 status of yellow fever in the Americas. Mosq. News 18 (3), 203–216.

Soper, F.L., 1963. The elimination of urban yellow fever in the Americas through the eradication of *Aedes aegypti*. Am. J. Public Health 53 (1), 7–16.

Soper, F.L., Penna, H., Cardoso, E., Serafim Jr., J., Frobisher Jr., M., Pinheiro, J., 1933. Yellow fever without *Aedes aegypti*. Study of a rural epidemic in the Valle do Chanaan, Espirito Santo, Brazil, 1932. Am. J. Hyg. 18 (3), 555–587.

Staal, G.B., 1975. Insect growth regulators with juvenile hormone activity. Annu. Rev. Entomol. 20, 417–460.

Stapleton, D.H., 2004. Lessons of history? Anti-malaria strategies of the International Health Board and the Rockefeller Foundation from the 1920s to the era of DDT. Public Health Rep. 119 (2), 206–215.

Strode, G.K. (Ed.), 1951. Yellow Fever. McGraw-Hill, New York, pp. 1–710.

Su, T., Thieme, J., White, G.S., Lura, T., Mayerle, N., Faraji, A., Cheng, M.L., Brown, M.Q., 2018. High resistance to *Bacillus sphaericus* and susceptibility to other common pesticides in *Culex pipiens* (Diptera: Culicidae) from Salt Lake City, UT. J. Med. Entomol. 56 (2), 506–513.

Sumitani, M., Kasashima, K., Yamamoto, D.S., Yagi, K., Yuda, M., Matsuoka, H., Yoshida, S., 2013. Reduction of malaria transmission by transgenic mosquitoes expressing an antisporozoite antibody in their salivary glands. Insect Mol. Biol. 22 (1), 41–51.

Thomas, D.D., Donnelly, C.A., Wood, R.J., Alphey, L.S., 2000. Insect population control using a dominant, repressible, lethal genetic system. Science 287 (5462), 2474–2476.

Tinker, M.E., 1963. The urban nature of *Aedes aegypti*. Proc. N.J. Mosq. Ext. Assoc. 50, 363–371.

Tinker, M.E., Hayes Jr., G.R., 1959. The 1958 *Aedes aegypti* distribution in the United States. Mosq. News 19 (2), 73–78.

Tonn, R.J., Figueredo, R., Uribe, L.J., 1982. *Aedes aegypti*, yellow fever and dengue in the Americas. Mosq. News 42 (4), 497–501.

Truitt, A.M., Kapun, M., Kaur, R., Miller, W.J., 2018. *Wolbachia* modifies thermal preference in *Drosophila melanogaster*. Environ. Microbiol. https://doi.org/10.1111/1462-2920.14347.

Turley, A.P., Moreira, L.A., O'Neill, S.L., McGraw, E.A., 2009. *Wolbachia* infection reduces blood-feeding success in the dengue fever mosquito, *Aedes aegypti*. PLoS Negl. Trop. Dis. 3 (9), e516.

Unlu, I., Suman, D.S., Wang, Y., Klingler, K., Faraji, A., Gaugler, R., 2017. Effectiveness of autodissemination stations containing pyriproxyfen in reducing immature *Aedes albopictus* populations. Parasit. Vectors 10 (1), 139.

Upholt, W.M., Gaines, T.B., Simmons, S.W., Arnold, E.H., 1947. The experimental use of DDT in the control of yellow fever mosquito *Aedes aegypti* (L). Public Health Rep. 186, 90–96.

van den Hurk, A.F., Hall-Mendelin, S., Pyke, A.T., Frentiu, F.D., McElroy, K., Day, A., Higgs, S., O'Neill, S.L., 2012. Impact of *Wolbachia* on infection with Chikungunya and yellow fever viruses in the mosquito vector *Aedes aegypti*. PLoS Negl. Trop. Dis. 6 (11), e1892.

Vanderplank, F.L., 1944. Hybridization between *Glossina* species and suggested new method for control of certain species of tsetse. Nature 154 (3915), 607–608.

Vanderplank, F.L., 1947. Experiments in the hybridisation of tsetse-flies (*Glossina*, Diptera) and the possibility of a new method of control. Trans. Roy. Ent. Soc. London 98 (1), 1–18.

Vanderplank, F.L., 1948. Experiments in cross-breeding tsetse-flies (*Glossina* species). Ann. Trop. Med. Parasitol. 42 (2), 131–152.

Varca, L.M., Magallona, E.D., 1994. Dissipation and degradation of DDT and DDE in Phillippine soil under field conditions. J. Environ. Sci. Health B 29 (1), 25–35.

Vargas-Terán, M., Hursey, B.S., Cunningham, E.P., 1994. Eradication of the screwworm from Libya using the sterile insect technique. Parasitol. Today 10 (3), 119–122.

Wandiga, S.O., 2001. Use and distribution of organochlorine pesticides. The future in Africa. Pure Appl. Chem. 73 (7), 1147–1155.

Weinert, L.A., Araujo-Jnr, E.V., Ahmed, M.Z., Welch, J.J., 2015. The incidence of bacterial endosymbionts in terrestrial arthropods. Proc. R. Soc. B Biol. Sci. 282 (1807), 20150249.

Wernsdorfer, W.H., 1980. The importance of malaria in the world. In: Kreier, J.P. (Ed.), Malaria. vol. 1. Academic Press, New York, pp. 1–93.

Werren, J.H., 1997. Biology of *Wolbachia*. Annu. Rev. Entomol. 42 (1), 587–609.

Werren, J.H., Baldo, L., Clark, M.E., 2008. *Wolbachia*: master manipulators of invertebrate biology. Nat. Rev. Microbiol. 6 (10), 741.

Whyard, S., Erdelyan, C.N., Partridge, A.L., Singh, A.D., Beebe, N.W., Capina, R., 2015. Silencing the buzz: a new approach to population suppression of mosquitoes by feeding larvae double-stranded RNAs. Parasit. Vectors 8 (1), 96.

Wilke, A.B.B., Nimmo, D.D., St John, O., Kojin, B.B., Capurro, M.L., Marrelli, M.T., 2009. Mini-review: genetic enhancements to the sterile insect technique to control mosquito populations. AsPac. J. Mol. Biol. Biotechnol. 17 (3), 65–74.

Williams, L.L., 1941. The anti-malaria program in North America. In: A Symposium on Human Malaria, With Special Reference to North America and the Caribbean Region. vol. 15, pp. 365–370.

Williams, L.L., 1964. Pesticides: a contribution to public health. Am. J. Public Health 54 (Suppl. 1), 32–36.

Williams Jr., L.L., 1937. A plan for statewide control of malaria. Public Health Service Document B-2534: February.

Williams Jr., L.L., 1938. Economic importance of malaria control. New Jersey Mosq. 25, 148–151.

Windbichler, N., Menichelli, M., Papathanos, P.A., Thyme, S.B., Li, H., Ulge, U.Y., Hovde, B.T., Baker, D., Monnat, R.J., Burt, A., Crisanti, A., 2011. A synthetic homing endonuclease-based gene drive system in the human malaria mosquito. Nature 473 (7346), 212–215.

Wolfe, R.J., 1996. Effects of open marsh water management on selected tidal marsh resources: a review. J. Am. Mosq. Control Assoc. 12 (4), 701–712.

World Health Organization (WHO), 1964. Genetics of Vectors and Insecticide Resistance: Report of a WHO Scientific Group. World Health Organization Technical Report Series No. 268. Geneva, Switzerland.

World Health Organization (WHO), 1969. Re-examination of the global strategy of malaria eradication. In: A Report by the Director-General to the 22nd World Health Assembly, 30 May 1969.

World Health Organization (WHO), 1976. WHO-supported collaborative research projects in India: the facts. WHO Chron. 30, 131–139.

World Health Organization (WHO), 1956. Expert committee on malaria. WHO Expert Committee on Malaria, Athens, 20–28 June 1956: Sixth Report.

World Health Organization (WHO), 2012. Handbook for Integrated Vector Management.

Worral, E., Fillinger, U., 2011. Large-scale use of mosquito larval source management for malaria control in Africa: a cost analysis. Malar. J. 10 (1), 338.

Xi, Z., Khoo, C.C., Dobson, S.L., 2005. *Wolbachia* establishment and invasion in an *Aedes aegypti* laboratory population. Science 310 (5746), 326–328.

Zabalou, S., Riegler, M., Theodorakopoulou, M., Stauffer, C., Savakis, C., Bourtzis, K., 2004. *Wolbachia*-induced cytoplasmic incompatibility as a means for insect pest population control. Proc. Natl. Acad. Sci. 101 (42), 15042–15045.

Zucker, J.R., 1996. Changing patterns of autochthonous malaria transmission in the U.S.: a review of recent outbreaks. Emerg. Infect. Dis. 2 (1), 37–43.

Zug, R., Hammerstein, P., 2012. Still a host of hosts for *Wolbachia*: analysis of recent data suggests that 40% of terrestrial arthropod species are infected. PLoS One 7 (6), e38544.

Chapter 9

Mosquito Species of Neighboring States of Mexico

Aldo I. Ortega-Morales[a], Martin Reyna Nava[b]

[a]Parasitology Department, Agricultural Autonomous University Antonio Narro, Laguna Unit, Torreón, Coahuila, México, [b]Mosquito and Vector Control Division, Harris County Public Health, Houston, TX, United States

Chapter outline

Abbreviations

Ad.	*Aedeomyia*
Ae.	*Aedes*
An.	*Anopheles*
Cq.	*Coquillettidia*
Cs.	*Culiseta*
Cx.	*Culex*
De.	*Deinocerites*
Hg.	*Haemagogus*
Li.	*Limatus*
Lt.	*Lutzia*
Ma.	*Mansonia*
Or.	*Orthopodomyia*
Ps.	*Psorophora*
s.l.	*sensu lato*
s.s.	*sensu stricto*
Sa.	*Sabethes*
SLEV	Saint Louis encephalitis virus
Tx.	*Toxorhynchites*
Ur.	*Uranotaenia*
US	United States
VEEV	Venezuelan equine encephalitis virus
WEEV	Western equine encephalitis virus
WNV	West Nile virus
WRBU	Walter Reed Biosystematics Unit
Wy.	*Wyeomyia*

Mosquitoes, Communities, and Public Health in Texas. https://doi.org/10.1016/B978-0-12-814545-6.00009-2

9.1 Introduction

The physiography of the United States is complex. Historically, several attempts with the intention of delimiting each and every one of the physiographic regions that comprise the country have been made. Currently, eight major physiographic regions are recognized: the Appalachian Highlands, the Atlantic Plain, the Interior Highlands, the Interior Plains, the Intermontane Plateaus, the Laurentian Uplands, the Pacific Mountain, and the Rocky Mountain System. Each physiographic region has a distinct type of landscape, vegetation type, orography, and evolutionary history, and most biodiversity distribution is restricted to one or some types of physiographic region. Texas is divided into three regions: the Atlantic Plain (Coastal Plain), the Interior Plains (Central Lowland and Great Plains), and Intermontane Plateaus (Basin and Range). Some spread to the south across the international border of Mexico, where they receive different names (Atlantic Plain, Coastal Plain of North Gulf; Interior Plains, Grand North American Plains; Intermontane Plateaus, Sierra Madre Oriental and North Sierras and Plains). Although the physiographic regions have different names in the United States and Mexico, they share the same environmental conditions that result in some mosquito species distributed in both countries or restricted to a region. Texas borders northeastern Mexico, which includes the states of Coahuila (151,595 km^2); Nuevo Leon (64,156 km^2); Tamaulipas (80,249 km^2); and Chihuahua, which is located in the northwest of the country and the largest state of Mexico (247,460 km^2) (Fig. 9.1).

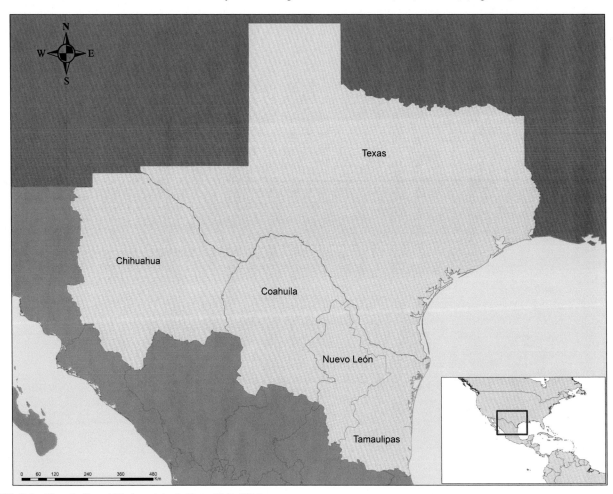

FIG. 9.1 Map of adjacent Mexican states to Texas, United States.

Knowing the physiography of a particular region is important to understand the distribution patterns of mosquitoes. In the United States, approximately 193 species of mosquitoes have been recorded and 87 in Texas. In Mexico, 235 species of mosquitoes have been recorded, although this number needs to be updated. In Texas and adjacent Mexican states, a total of 119 species have been recorded with some species present in both geopolitical regions. Of those, 100 are present in the adjacent Mexican states where some species are present in more than one Mexican state (22 species in Chihuahua, 49 in Coahuila, 64 in Nuevo Leon, and 81 in Tamaulipas) and 87 in Texas. Only 68 species occur in both regions, that is, the adjacent Mexican states to Texas and in Texas. Thus, of the total 119 species, 19 are only found in Texas and 32 only in the Mexican states adjacent to Texas (Table 9.1).

TABLE 9.1 List of 100 mosquito species present in adjacent Mexican states to Texas and 87 in Texas.

	Species	Present in any of four adjacent Mexican states	Chihuahua	Coahuila	Nuevo Leon	Tamaulipas	Texas
1	Anopheles atropos	●	●	●	●	●	✓
2	An. barberi	●	●	●	●	●	✓
3	An. bradleyi	✓	●	●	●	✓	✓
4	An. crucians	✓	●	✓	✓	✓	✓
5	An. eiseni	✓	●	●	✓	✓	●
6	An. franciscanus	✓	✓	✓	●	●	✓
7	An. freeborni	●	●	●	●	●	✓
8	An. judithae	●	✓	✓	✓	●	✓
9	An. pseudopunctipennis	✓	✓	✓	✓	✓	✓
10	An. punctipennis	✓	✓	●	✓	✓	✓
11	An. quadrimaculatus	✓	●	●	●	●	✓
12	An. smaragdinus	●	●	●	●	●	✓
13	An. walkeri	●	●	●	✓	✓	✓
14	An. vestitipennis	✓	●	✓	●	✓	●
15	An. albimanus	✓	●	●	✓	✓	✓
16	An. argyritarsis	✓	●	●	●	✓	●
17	Aedeomyia squamipennis	✓	●	●	✓	✓	●
18	Aedes vexans	✓	✓	✓	✓	✓	✓
19	Ae. epactius	✓	●	✓	✓	✓	✓
20	Ae. quadrivittatus	✓	●	✓	✓	✓	●
21	Ae. muelleri	✓	●	●	✓	●	✓
22	Ae. amateuri	●	●	●	●	✓	●
23	Ae. atlanticus	●	●	●	✓	●	✓
24	Ae. bimaculatus	✓	●	●	✓	✓	✓
25	Ae. campestris	●	●	●	●	●	✓
26	Ae. canadensis	✓	✓	●	●	●	✓

Continued

TABLE 9.1 List of 100 mosquito species present in adjacent Mexican states to Texas and 87 in Texas.

	Species	Present in any of four adjacent Mexican states	Chihuahua	Coahuila	Nuevo Leon	Tamaulipas	Texas
27	Ae. dorsalis	✓	✓	●	✓	●	✓
28	Ae. dupreei	✓	●	●	✓	●	✓
29	Ae. fulvus	●	●	●	●	●	✓
30	Ae. grossbecki	●	●	●	●	●	✓
31	Ae. infirmatus	●	●	●	●	●	✓
32	Ae. mitchellae	✓	✓	●	●	✓	✓
33	Ae. nigromaculis	✓	●	✓	✓	●	✓
34	Ae. scapularis	✓	●	✓	✓	✓	✓
35	Ae. sollicitans	✓	✓	✓	✓	✓	✓
36	Ae. sticticus	✓	●	✓	●	●	✓
37	Ae. taeniorhynchus	✓	●	✓	✓	✓	✓
38	Ae. thelcter	✓	✓	✓	●	✓	✓
39	Ae. thibaulti	●	●	●	●	●	✓
40	Ae. tormentor	●	●	●	●	●	✓
41	Ae. trivittatus	✓	✓	✓	✓	✓	✓
42	Ae. amabilis	✓	●	●	●	●	●
43	Ae. brelandi	✓	●	✓	●	✓	✓
44	Ae. hendersoni	●	●	✓	✓	●	✓
45	Ae. lewnielseni	✓	●	✓	●	●	●
46	Ae. podographicus	✓	●	✓	●	✓	●
47	Ae. triseriatus	✓	●	✓	✓	✓	✓
48	Ae. zoosophus	✓	●	✓	✓	✓	✓
49	Ae. aegypti	✓	✓	✓	✓	✓	✓
50	Ae. albopictus	✓	●	✓	✓	✓	✓
51	Haemagogus equinus	✓	●	●	✓	✓	✓
52	Psorophora columbiae	✓	●	✓	✓	✓	✓

#	Species	1	2	3	4	5	6
53	*Ps. discolor*	●	✓	✓	✓	✓	✓
54	*Ps. signipennis*	✓	✓	✓	✓	✓	✓
55	*Ps. albipes*	●	●	●	●	✓	✓
56	*Ps. cyanescens*	●	✓	✓	✓	✓	✓
57	*Ps. ferox*	●	●	✓	✓	✓	✓
58	*Ps. horrida*	●	●	●	●	✓	●
59	*Ps. longipalpus*	●	✓	●	●	✓	●
60	*Ps. mathesoni*	●	●	●	●	✓	●
61	*Ps. mexicana*	●	●	●	✓	✓	✓
62	*Ps. ciliata*	●	✓	✓	✓	✓	✓
63	*Ps. cilipes*	●	●	✓	✓	●	●
64	*Ps. howardii*	●	●	✓	✓	✓	✓
65	*Ps. stonei*	●	✓	●	✓	✓	●
66	*Culex restrictor*	●	●	✓	✓	✓	●
67	*Cx. bidens*	●	✓	✓	✓	✓	●
68	*Cx. coronator*	✓	✓	✓	✓	✓	✓
69	*Cx. declarator*	●	✓	✓	✓	✓	✓
70	*Cx. chidesteri*	●	✓	✓	●	✓	✓
71	*Cx. erythrothorax*	✓	✓	✓	●	✓	✓
72	*Cx. interrogator*	●	✓	✓	✓	✓	✓
73	*Cx. nigripalpus*	●	✓	✓	✓	✓	✓
74	*Cx. quinquefasciatus*	✓	✓	✓	✓	✓	✓
75	*Cx. restuans*	✓	✓	✓	✓	✓	✓
76	*Cx. salinarius*	●	✓	✓	✓	✓	✓
77	*Cx. stigmatosoma*	✓	✓	✓	✓	✓	✓
78	*Cx. tarsalis*	●	✓	✓	✓	✓	✓
79	*Cx. thriambus*	✓	✓	✓	✓	✓	✓
80	*Cx. abominator*	●	●	●	●	●	●

Continued

TABLE 9.1 List of 100 mosquito species present in adjacent Mexican states to Texas and 87 in Texas.

	Species	Present in any of four adjacent Mexican states	Chihuahua	Coahuila	Nuevo Leon	Tamaulipas	Texas
81	Cx. educator	✓	●	●	●	✓	●
82	Cx. erraticus	✓	●	●	✓	✓	✓
83	Cx. inhibitator	✓	●	●	●	✓	●
84	Cx. iolambdis	✓	●	●	●	✓	●
85	Cx. peccator	✓	●	●	●	✓	✓
86	Cx. pilosus	✓	●	●	●	✓	✓
87	Cx. taeniopus	✓	●	●	●	✓	●
88	Cx. rejector	✓	●	✓	●	✓	●
89	Cx. apicalis	✓	●	✓	●	●	✓
90	Cx. arizonensis	✓	●	✓	●	●	✓
91	Cx. territans	✓	✓	●	●	✓	●
92	Cx. corniger	✓	●	✓	●	✓	●
93	Cx. lactator	✓	●	●	●	✓	✓
94	Deinocerites mathesoni	✓	●	●	●	✓	✓
95	De. pseudes	✓	●	●	●	✓	●
96	Lutzia bigoti	✓	●	●	●	✓	✓
97	Culiseta melanura	✓	●	●	●	●	✓
98	Cs. incidens	✓	●	●	●	●	✓
99	Cs. inornata	✓	●	●	●	●	●
100	Cs. particeps	✓	●	✓	●	●	●
101	Coquillettidia nigricans	✓	●	✓	●	●	✓
102	Cq. perturbans	✓	●	●	✓	●	●
103	Mansonia indubitans	✓	●	●	●	●	●
104	Ma. dyari	✓	●	●	✓	●	●
105	Ma. titillans	✓	●	●	✓	✓	✓
106	Orthopodomyia alba	✓	✓	✓	✓	●	✓
107	Or. kummi	✓	●	●	✓	●	✓

No.	Species						
108	Or. signifera	✓	●	✓	●	✓	✓
109	Limatus durhamii	✓	●	●	●	✓	●
110	Sabethes chloropterus	✓	●	●	●	✓	●
111	Wyeomyia mitchellii	✓	●	●	●	✓	●
112	Toxorhynchites grandiosus	✓	●	●	●	✓	●
113	Tx. moctezuma	✓	●	●	✓	✓	✓
114	Tx. rutilus	●	●	●	●	●	✓
115	Uranotaenia syntheta	✓	●	●	✓	✓	✓
116	Ur. coatzacoalcos	✓	●	●	✓	●	●
117	Ur. geometrica	✓	●	●	●	✓	●
118	Ur. lowii	✓	●	✓	✓	✓	✓
119	Ur. sapphirina	✓	●	✓	●	✓	✓
Totals	119	100	22	49	64	81	87

● Not present; ✓ Present.

Many species of mosquitoes found in the United States occur in northeastern Mexico and reach their southernmost distributional limits in the states adjacent to Texas. Nearctic mosquitoes that occur in Texas and the adjacent Mexican states, but do not extend much farther into Mexico, are *Aedes (Ochlerotatus) canadensis* (Theobald), *Ae. (Och.) dorsalis* (Meigen), *Ae. (Och.) mitchellae* (Dyar), *Ae. (Och.) sticticus* (Meigen), *Ae. (Och.) thelcter* Dyar, *Culex (Neoculex) territans* Walker, *Culiseta (Climacura) melanura* (Coquillett), and *Cs. (Culiseta) incidens* (Thomson). A second group of mosquitoes is the Neotropical group of species that extend from south/middle America to the northeastern Mexican states adjacent to Texas, but do not reach the United States. Mosquitoes in this group are *Anopheles (Anopheles) eiseni* Coquillett, *An. (Ano.) vestitipennis* Dyar and Knab, *An. (Nyssorhynchus) argyritarsis* Robineau-Desvoidy, *Aedeomyia (Aedeomyia) squamipennis* (Lynch Arribálzaga), *Ae. (Howardina) quadrivittatus* (Coquillett), *Ae. (Protomacleaya) amabilis* Schick, *Ae. (Pro.) podographicus* Dyar and Knab, *Psorophora (Janthinosoma) albipes* (Theobald), *Ps. (Psorophora) cilipes* (Fabricius), *Ps. (Pso.) stonei* Vargas, *Cx. (Anoedioporpa) restrictor* Dyar and Knab, *Cx. (Culex) bidens* Dyar and Knab, *Cx. (Melanoconion) educator* Dyar and Knab, *Cx. (Mel.) inhibitator* Dyar and Knab, *Cx. (Microculex) rejector* Dyar and Knab, *Cx. (Phenacomyia) corniger* Theobald, *Cx. (Phc.) lactator* Dyar and Knab, *Lutzia (Lutzia) bigoti* (Bellardi), *Coquillettidia (Rhynchotaenia) nigricans* (Coquillett), *Mansonia (Mansonia) indubitans* Dyar and Shannon, *Limatus durhamii* Theobald, *Sabethes (Sabethoides) chloropterus* (von Humboldt), *Toxorhynchites (Lynchiella) grandiosus* (Williston), *Uranotaenia (Uranotaenia) coatzacoalcos* Dyar and Knab, and *Ur. (Ura.) geometrica* Theobald. Finally a third group of mosquitoes is the Neotropical group that reach their northernmost distributional limit in Texas are *Ae. (Och.) bimaculatus* (Coquillett), *Haemagogus (Haemagogus) equinus* Theobald, and *Ps. (Jan.) mexicana* (Bellardi). *Aedes (Och.) amateuri* Ortega and Zavortink and *Ae. (Pro.) lewnielseni* Ortega and Zavortink were discovered and described recently (Ortega-Morales et al., 2019), and they are endemic for the northeastern Mexico.

The 19 mosquito species found in Texas but not in adjacent Mexican states (*although some species have been reported in other states of Mexico) are *An. (Ano.) atropos* Dyar and Knab*, *An. (Ano.) barberi*, *An. (Ano.) freeborni* Aitken*, *An. (Ano.) judithae* Zavortink*, *An. (Ano.) smaragdinus* Reinert, *An. (Ano.) walkeri* Theobald*, *Ae. (Och.) atlanticus* Dyar and Knab, *Ae. (Och.) campestris* Dyar and Knab*, *Ae. (Och.) fulvus* (Wiedemann)*, *Ae. (Och.) grossbecki* Dyar and Knab, *Ae. (Och.) infirmatus* Dyar and Knab*, *Ae. (Och.) thibaulti* Dyar and Knab, *Ae. (Och.) tormentor* Dyar and Knab*, *Ae. (Pro.) hendersoni* Cockerell, *Ps. (Jan.) horrida* (Dyar and Knab), *Ps. (Jan.) longipalpus* Randolph and O'Neill*, *Ps. (Jan.) mathesoni* Belkin and Heinemann, *Cx. (Mel.) abominator* Dyar and Knab, and *Tx. (Lyn.) rutilus* (Coquillett). These mosquitoes are discussed in more detail in Chapter 2 of this book.

The Culicidae mosquitoes have been subject to nomenclatural changes in recent years, especially members of the tribe Aedini. Based on phylogenetic studies, many of the subgenera of *Aedes* were elevated to generic level, resulting in many changes of specific epithets. As a result, many taxonomists, entomologists, researchers, and public health personnel around the world preferred not to accept the changes proposed by this classification system, since it resulted in confusion for some major medically important species such as *Ae. aegypti = Stegomyia aegypti* or *Ae. albopictus = Stegomyia albopicta*. In this chapter the classification proposed by Wilkerson et al. (2015) available in the Walter Reed Biosystematics Unit (WRBU, 2018) is used.

9.2 How to read this chapter

The mosquito species are presented taxonomically as follows: Culicidae subfamilies (Anophelinae and Culicinae), tribes of Culicinae (Aedeomyiini, Aedini, Culicini, Culisetini, Orthopodomyiini, Mansoniini, Sabethini, Toxorhynchitini, and Uranotaeniini), genera, subgenera, and species. In each tribal section, general information about distribution, bionomics, and taxonomy is provided. The species list includes all of the 100 species recorded in the adjacent Mexican states including 68 mosquito species that have been previously recorded in Texas and the adjacent Mexican states with their general information and their distribution in the Mexican states adjacent to Texas and 32 mosquito species that occur only in the Mexican states adjacent to Texas. For each species the description, bionomics, distribution, and medical importance are provided. The descriptions provide general diagnostic characters and are not intended to be used as an identification key. The reader must have previously identified the genus using appropriate identification keys, including the larval chaetotaxy that is commonly used in this section. The references used for the description for each species in this chapter are (Dyar, 1928) *Cx. bidens* (adult female), *Cx. inhibitator* (larvae only), and *Cx. rejector*; (Lane, 1953) *Ma. indubitans*, and *Tx. grandiosus*; (Vargas, 1956) *Ps. stonei*; (Bram, 1967) *Cx. bidens* (larvae only); (Schick, 1970) *Ae. amabilis*; (Sirivanakarn, 1982) *Cx. inhibitator* (adult female only, placed in the inhibitator group); (Clark-Gil and Darsie, 1983) *An. eiseni, An. vestitipennis, An. argyritarsis, Ad. squamipennis, Ae. quadrivittatus, Ae. podographicus, Ps. albipes* (adult female only), *Cx. restrictor, Cx. educator, Cx. iolambdis, Cx. taeniopus, Lt. bigoti, Cq. nigricans, Ma. dyari, Li. durhamii, Sa. chloropterus, Ur. coatzacoalcos,* and *Ur. geometrica*; (Strickman and Pratt, 1989) *Cx. corniger* and *Cx. lactator*; (Darsie and Ward, 2005)

Cs. particeps, Wy. mitchellii; and (Ortega-Morales et al., 2019) *Ae. amateuri* and *Ae. lewnielseni*. The bionomics includes general information about the type of aquatic habitat where the species are commonly found and the known feeding preferences of adult females. The distribution is where each mosquito species has previously been recorded in Mexico and it also specifies in which Mexican state adjacent to Texas the species occurs. The known medical importance of each species is mentioned in the last section, although it is very possible that many species can participate actively in cycles of transmission of diseases to humans and/or animals that are presently unknown.

9.3 Family Culicidae Meigen

Subfamily Anophelinae Grassi

Genus Anopheles *Meigen*

Mosquitoes of the subfamily Anophelinae, also known as anophelines, are divided into three genera: *Anopheles*, *Bironella* Theobald, and *Chagasia* Cruz. Only the genera *Anopheles* and *Chagasia* are present in America. In the United States, only the genus *Anopheles* with 22 species placed into two subgenera *Anopheles* Meigen and *Nyssorhynchus* Blanchard has been reported. In Texas, 13 anopheline species have been reported, while in the Mexican states adjacent to Texas, 10 have been reported. The larvae of the anophelines are characterized by the absence of a respiratory siphon, which is common in the rest of the mosquito groups, and by the resting position that is parallel to the surface of the water. Adult females can be distinguished by having the palps as long as the proboscis. Adults of both sexes are also characterized by having a rounded scutellum and active during the night and dusk, and many bite humans. Historically, *Anopheles* has received much attention by entomologists because many species are vectors of *Plasmodia* that cause malaria. In Texas and adjacent Mexican states, seven *Anopheles* species have been recorded, including two complexes: *An. crucians* and *An. quadrimaculatus, sensu lato* (*s.l.*), a taxonomic category that means "in a broad sense," while *sensu stricto* (*s.s.*) "in a narrow sense."

(1) *An. (Ano.) bradleyi* King. This species belongs to the *An. crucians* complex. Its distribution is restricted to a coastal strip along the Atlantic Ocean and the Gulf of Mexico from Middle America to New York. Distribution in Mexico: Tamaulipas.

(2) *An. (Ano.) crucians* (*s.l.*). The group comprising the *An. crucians* complex includes two described species: *An. bradleyi* King and *An. georgianus* King (there are at least five additional undescribed species). Distribution: Coahuila, Nuevo Leon, and Tamaulipas.

(3) *An. (Ano.) franciscanus* McCracken. Distribution: Chihuahua, Coahuila, Nuevo Leon, and Tamaulipas.

(4) *An. (Ano.) pseudopunctipennis* Theobald. In Mexico, it is the main vector of malaria in regions 600 m above sea level and is most common during the dry season. Distribution: Chihuahua, Coahuila, Nuevo Leon, and Tamaulipas.

(5) *An. (Ano.) punctipennis* (Say). Distribution: Chihuahua, Coahuila, Nuevo Leon, and Tamaulipas.

(6) *An. (Ano.) quadrimaculatus* Say. Five species comprise the *An. quadrimaculatus* complex: *An. diluvialis* Reinert, *An. inundatus* Reinert, *An. maverlius* Reinert, *An. quadrimaculatus* (*s.s.*) Say, and *An. smaragdinus* Reinert. All are found in the United States except *An. quadrimaculatus* (*s.s.*) that occurs from Canada to Middle America and Puerto Rico. Members of this complex were historically involved in the transmission of malaria to humans in the United States. Distribution: Nuevo Leon and Tamaulipas.

(7) *An. (Nys.) albimanus* Wiedemann. Distribution: Coahuila, Nuevo Leon, and Tamaulipas.

The species of *Anopheles* found in the Mexican state of Nuevo Leon and Tamaulipas but absent in Texas are *An. eiseni* Coquillett, *An. vestitipennis* Dyar and Knab, and *An. argyritarsis* Robineau-Desvoidy.

(8) *An. (Ano.) eiseni* Coquillett (Fig. 9.2).

Description: Larvae: setae 3-C single or double, setae 5–7-C plumose, setae 1-I, II well developed, setae 6-III plumose, setae 6-VI very short, spiracular lobe without long process. Adults: vein C dark scaled, hind leg entirely dark scaled, and hind tibia with broad apical band of white scales.

Bionomics: Its larvae are commonly found in tree holes, coconut shells, bromeliad axils, bamboo internodes, ponds with aquatic vegetation, ditches, marshes, and artificial containers such as discarded tires, tin cans, and buckets, always with complete shade and leaves at bottom. Adults feed outdoors on sylvan animals and rarely enter houses and bite humans.

Distribution: It is a Neotropical forest species that occurs in the tropical and subtropical regions of the southeastern states of Mexico and reaches its northernmost distributional limit in the states of Nuevo Leon and Tamaulipas.

Medical importance: The *Plasmodium* that causes malaria has been found in wild mosquitoes of *An. eiseni*; however, it is not considered an important vector of malaria.

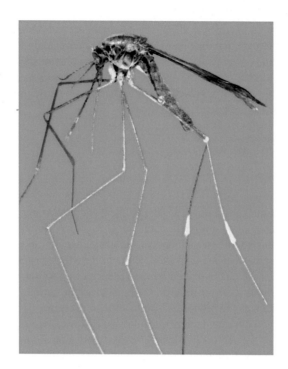

FIG. 9.2   *Anopheles eiseni.*

(9) *An. (Ano.) vestitipennis* Dyar and Knab (Fig. 9.3).

FIG. 9.3   *Anopheles vestitipennis.*

Description: Larvae: setae 3-C with fewer than 15 branches, some branches subdivided, setae 5–7-C plumose, setae 9–12-P branched, setae 6-IV, V with 2 or more branches, and spiracular lobe without long process. Adults: hind tibia with spots of white scales, hind tarsomeres with rings of white scales and abdominal terga II–VII without prominent lateral tuffs.
Bionomics: Its larvae develop in freshwater bodies such as ponds, swamps, and marshes, always with aquatic vegetation and full or partial shade. Adult females are persistent human biters on humans at dusk and night. It is very common during the rainy season, and the populations are very abundant in the swampy regions of the tropical states of southeastern Mexico.
Distribution: It has been recorded mostly in southeastern states of Mexico, and its northernmost distributional limit is the swampy regions of southern Tamaulipas.
Medical importance: It is one of the most important malaria vectors in Middle America. In Mexico, it is probably a vector in sylvan regions of Chiapas.

(10) *An. (Nys.) argyritarsis* Robineau-Desvoidy (Fig. 9.4).

FIG. 9.4   *Anopheles argyritarsis.*

Description: Larvae: same chaetotaxy as *An. eiseni* but with the seta 6-VI long. Adults: Segments 3–5 of the fore tarsomeres covered with white scales and mesanepimeron without scales.

Bionomics: Its larvae develop in margins of small streams in sun or partial shade and in rain pools, always with abundant emergent vegetation such as grass. Females feed on mammals, but are not attracted to humans, and also do not enter human dwellings.

Distribution: It occurs in several countries of South and Middle America and reaches its northernmost distributional limit in the swampy regions of southern Tamaulipas.

Medical importance: The *Plasmodium* causing malaria has been detected in wild-caught females of *An. argyritarsis*; however, since it rarely feeds on humans, it is not considered an important malaria vector.

9.4   Subfamily Culicinae Meigen

Tribe Aedeomyiini Theobald

Genus Aedeomyia *Theobald*

The tribe Aedeomyiini is monobasic, including only one genus: *Aedeomyia*, which includes seven species. This tribe occurs in the tropical regions of Africa, Australia, South Asia, and South and Middle America. Very little is known about the biology of this group of mosquitoes. Larvae usually spend a lot of time submerged in their aquatic habitats, such as marshes, ponds, and other stagnant water, always with abundant floating aquatic vegetation. In Mexico, only *Ad. squamipennis* (Lynch Arribálzaga) has been reported.

(11) *Ad. squamipennis* (Lynch Arribálzaga)

Description: Larvae: antennae very large, strongly curved, longer than head; siphon pilose. Adults: antennal segments very short, presence of broad pale scales on pedicel, antenna, and clypeus; wing covered with pale and brown broad scales, hind femora with apical scale tufts.

Bionomics: Larvae of *Ad. squamipennis* commonly occur in swamps with abundant aquatic vegetation with partial shade and have also been found in stream margins, where they are commonly found in association with other groups of mosquitoes, that is, members of the tribe Mansoniini. Females feed predominantly on birds.

Distribution: It occurs in sylvan regions of many countries of South and Middle America and reaches its northernmost distributional limit in Mexico in the swampy regions of southern Tamaulipas and in Florida, US.

Medical importance: It is not considered to be a vector of human diseases.

9.5 Tribe Aedini Neveu-Lemaire

The tribe Aedini includes the genera *Aedes* Meigen, *Armigeres* Theobald, *Eretmapodites* Theobald, *Haemagogus* Williston, *Heizmannia* Ludlow, *Opifex* Hutton, *Psorophora* Robineau-Desvoidy, *Udaya* Thurman, *Verrallina* Theobald, and *Zeugnomyia* Leicester. Three genera inhabit America: *Aedes*, *Haemagogus*, and *Psorophora*. This group of mosquitoes has great importance worldwide because it includes many species of vector pathogens that cause mosquito-borne diseases in domestic animals and humans. Many are transmitted by species of the genus *Aedes*, such as *Ae. aegypti* (Linnaeus) and *Ae. albopictus* (Skuse), which are the main vectors of yellow fever, dengue, Zika, and chikungunya, that is, diseases that have caused epidemic outbreaks in many countries of South and Central America. In Mexico, both *Ae. aegypti* and *Ae. albopictus* are the most important mosquitoes of public health and receive the most interest by medical entomologists and public health officials. In addition, they are vectors of urban yellow fever. The genus *Haemagogus* includes species that are vectors of sylvatic yellow fever, and some species of *Psorophora* have been associated with the transmission of Venezuelan Equine encephalitis virus (VEEV). In the Mexican states adjacent to Texas, 37 Aedini species are found and grouped in 3 genera and 11 subgenera: *Aedes* (*Aedimorphus*, *Georgecraigius*, *Howardina*, *Lewnielsenius*, *Ochlerotatus*, *Protomacleaya*, and *Stegomyia*); *Haemagogus* (*Haemagogus*); and *Psorophora* (*Grabhamia*, *Janthinosoma*, and *Psorophora*).

(12) *Ae. (Aedimorphus) vexans* (Meigen). Distribution: Chihuahua, Coahuila, Nuevo Leon, and Tamaulipas.

(13) *Ae. (Georgecraigius) epactius* Dyar and Knab. This is a very common species in most Mexican territory during the dry season. Distribution: Coahuila, Nuevo Leon, and Tamaulipas.

(14) *Ae. (Lewnielsenius) muelleri* Dyar. Distribution: Coahuila and Nuevo Leon. This is the first record of this species in Coahuila.

(15) *Ae. (Och.) bimaculatus* (Coquillett). This is a very uncommon species in northeastern Mexico, and very few occurrence records have been published. Distribution: Nuevo Leon and Tamaulipas.

(16) *Ae. (Och.) canadensis* (Theobald). Distribution: Chihuahua.

(17) *Ae. (Och.) dorsalis* (Meigen). Distribution: Chihuahua and Nuevo Leon.

(18) *Ae. (Och.) dupreei* (Coquillett). Distribution: Nuevo Leon.

(19) *Ae. (Ochlerotatus) mitchellae* (Dyar). Distribution: Chihuahua and Tamaulipas.

(20) *Ae. (Och.) nigromaculis* (Ludlow). This is an uncommon species in northeastern Mexico and has been recorded only few times. Distribution: Coahuila and Nuevo Leon.

(21) *Ae. (Och.) scapularis* (Rondani). This species is very similar to *Ae. infirmatus* Dyar and Knab, another species that occurs in the Texas region. Distribution: Coahuila, Nuevo Leon, and Tamaulipas.

(22) *Ae. (Och.) sollicitans* (Walker). Distribution: Chihuahua, Coahuila, Nuevo Leon, and Tamaulipas.

(23) *Ae. (Och.) sticticus* (Meigen). This is an uncommon mosquito in Mexico, with only two records known from the extreme north near the border with the United States. Distribution: Chihuahua and Coahuila.

(24) *Ae. (Och.) taeniorhynchus* (Wiedemann). This is a very common species in all coastal regions of Mexico. Distribution: Coahuila, Nuevo Leon, and Tamaulipas.

(25) *Ae. (Och.) thelcter* Dyar. Its distribution is restricted to northern Mexico and the Rio Grande Valley in Texas. Distribution: Chihuahua and Tamaulipas.

(26) *Ae. (Och.) trivittatus* (Coquillett). Distribution: Chihuahua, Coahuila, Nuevo Leon, and Tamaulipas.

(27) *Ae. (Pro.) brelandi* Zavortink. This species inhabits oak woodlands from Texas to Querétaro state in Mexico. Distribution: Coahuila, Nuevo Leon, and Tamaulipas. This is the first record of this species in Coahuila.

(28) *Ae. (Pro.) triseriatus* (Say). Distribution: Chihuahua, Coahuila, Nuevo Leon, and Tamaulipas.

(29) *Ae. (Pro.) zoosophus* Dyar and Knab. It is distributed from the state of Kansas to Guerrero state in Mexico. Distribution: Coahuila, Nuevo Leon, and Tamaulipas.

(30) *Ae. (Stegomyia) aegypti* (L.). This is the best-known mosquito in Mexico because it is the main vector of mosquito-borne diseases such as dengue, Zika, chikungunya, and yellow fever. Distribution: Chihuahua, Coahuila, Nuevo Leon, and Tamaulipas.

(31) *Ae. (Stg.) albopictus* (Skuse). It is a vector of Zika, dengue, chikungunya, and yellow fever. Distribution: Coahuila, Nuevo Leon, and Tamaulipas.

(32) *Haemagogus (Hag.) equinus* Theobald. This is the only species of *Haemagogus* that occurs in northeastern Mexico and the United States, where it has only been recorded in Cameron County, Texas. Distribution: Nuevo Leon and Tamaulipas.

(33) *Ps. (Grabhamia) columbiae* (Dyar and Knab). Distribution: Coahuila, Nuevo Leon, and Tamaulipas.

(34) *Ps. (Gra.) discolor* (Coquillett). This is an uncommon species in northeastern Mexico. Since it is very similar to *Ps. columbiae*, it is possible that both species have been confused historically in northeastern Mexico. Distribution: Coahuila, Nuevo Leon, and Tamaulipas.

(35) *Ps. (Gra.) signipennis* (Coquillett). The arid and semiarid region of northeastern Mexico is where it occurs most frequently. Distribution: Chihuahua, Coahuila, Nuevo Leon, and Tamaulipas.

(36) *Ps. (Jan.) cyanescens* (Coquillett). Distribution: Coahuila, Nuevo Leon, and Tamaulipas.

(37) *Ps. (Jan.) ferox* (von Humboldt). This is one of the most common species of *Psorophora* in northeastern Mexico during the rainy season. Distribution: Nuevo Leon, and Tamaulipas.

(38) *Ps. (Jan.) mexicana* (Bellardi). Distribution: Tamaulipas.

(39) *Ps. (Pso.) ciliata* (Fabricius). This species and *Ps, howardii* are the largest hematophagous mosquitoes in Texas. Distribution: Coahuila, Nuevo Leon, and Tamaulipas.

(40) *Ps. (Pso.) howardii* Coquillett. Its biology is very similar to *Ps. ciliata*. Distribution: Nuevo Leon and Tamaulipas.

The species of Aedini found in the Mexican states adjacent to Texas but absent in Texas are *Ae. quadrivittatus* (Coquillett), *Ae. amateuri* Ortega and Zavortink, *Ae. amabilis* Schick, *Ae. lewnielseni* Ortega and Zavortink, *Ae. podographicus* Dyar and Knab, *Ps. albipes* (Theobald), *Ps. cilipes* (Coquillett), and *Ps. stonei* Vargas.

(41) *Ae. (How.) quadrivittatus* (Coquillett) (Fig. 9.5).

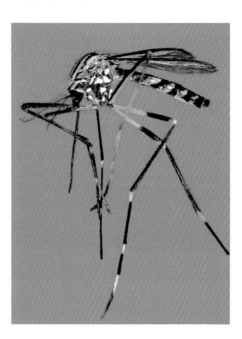

FIG. 9.5   *Aedes quadrivittatus.*

Description: Larvae: anal segment not completely encircled by saddle, segment VIII with small subapical spinules on comb scale, long spines in posterior margin of saddle, and setae 4-X with six pairs of setae. Adults: scutum with longitudinal golden lines, mid femur without spots of silvery scales on the anterior surface.

Bionomics: It develops in the leaf axils of epiphytic bromeliads, and larvae move from one axil to another by crawling and have also been collected from tree holes. Adults are diurnal, and females are readily attracted to humans.

Distribution: Like most members of the subgenus *Howardina*, it is distributed only where epiphytic bromeliads grow, such as cloud forests and oak forests, as well as tropical forests. It has been reported from Panama to the southern part of Coahuila and Nuevo Leon and the middle part of Tamaulipas in the Biosphere Reserve "El Cielo," where it reaches its northernmost distributional limit. This is the first record of this species in Coahuila.

Medical importance: Although some viruses have been detected in females of *Ae. quadrivittatus*, it is not considered a vector of diseases to humans.

(42) *Ae. (Och.) amateuri* Ortega and Zavortink (Fig. 9.6).

Description: Larvae: anal segment completely encircling by saddle; siphon moderately pigmented, spiculose thoracic, and abdominal integument; setae 13-VI short, multiple, and similar to 13-II; comb scales more than 15, in small patch of 2 or 3 irregular rows; comb scales evenly fringed; the median spinule not enlarged; and pecten teeth evenly spaced in straight row to middle of siphon. Adults: presence of white tarsal bands; the hind tarsus having white scales at both the base and apex of the tarsomeres and tarsomere 5 entirely white scaled; shape of the subspiracular scale patch, which has a prominent dorsal extension toward the postpronotum.

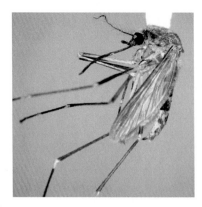

FIG. 9.6 *Aedes amateuri.*

Bionomics: The immature stages have been collected from ponds at ground level without aquatic vegetation. Females have been collected biting humans at dusk and daytime.

Distribution: *Aedes amateuri* is endemic to Mexico and has only been collected from forested regions of the states of Nuevo Leon and Tamaulipas.

Medical importance: The medical importance of *Ae. amateuri* is unknown.

(43) *Ae. (Pro.) amabilis* Schick.

Description: Larvae: unknown. Adults: The female has an unusual pattern of scales on scutum with longitudinal dorsocentral and acrostichal lines of white scales.

Bionomics: As with most members of the subgenus *Protomacleaya*, its larvae occur mostly in the holes of living trees. Adult females are active during the day and can bite humans and are persistent by keeping biting even when the host is moving.

Distribution: It is endemic to Mexico and has only been collected in the states of Veracruz, Nuevo Leon, and Tamaulipas. The tropical forest of the Biosphere Reserve "El Cielo" and southern of Nuevo Leon is the northernmost distributional limit of this species.

Medical importance: Nothing is known about the medical importance of *Ae. amabilis.*

(44) *Ae. (Pro.) lewnielseni* (Fig. 9.7) Ortega and Zavortink.

FIG. 9.7 *Aedes lewnielseni.*

Description: Larva: double or triple head hair 6-C; setae 1 and 13-I–VI with attenuate branches of setae; strong, single seta 2-II–VI; long seta 4a-X of the ventral brush. Adults: scutum entirely white scaled, palpus slightly shorter than proboscis, base of hind tarsomere 1 entirely dark scaled, remigium of wing entirely dark scaled.

Bionomics: Its larvae have been collected from a large tree hole in a living oak tree (*Quercus* sp.). Adult females were collected biting humans during daytime.

Distribution: *Ae. lewnielseni* has been collected only in one site of the Grand Folded Sierra of the Sierra Madre Oriental of Nuevo Leon.

Medical importance: Its medical importance is unknown.

(45) *Ae. (Pro.) podographicus* Dyar and Knab (Fig. 9.8).

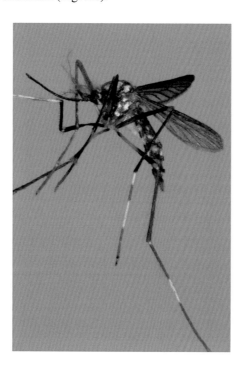

FIG. 9.8   *Aedes podographicus.*

Description: Because of its wide distribution from South America to Mexico and the variable pattern and coloring of scales of adults, it could represent a complex of species. Larvae: anal segment not completely encircled by saddle, setae 1 of siphon attached distal to the apex of pecten, spinules of comb scales small, saddle with some spicules on the margin. Adults: without acrostichal setae, hind tarsomeres with apical and basal pale-scaled bands on segments 1–2, and hind femur with dark-scaled basal band incomplete.

Bionomics: Its larvae have been found in a variety of natural and artificial containers such as tree holes, bamboo internodes, discarded tires, buckets, cattle drinking troughs, and water tanks. Adults are diurnal, and females can bite humans and monkeys.

Distribution: It is possibly the most common species of the subgenus *Protomacleaya* in Middle America and coastal regions of Mexico. It has been reported from Ecuador and Venezuela to Mexico. The tropical region of the Biosphere Reserve "El Cielo" represents its northernmost distributional limit.

Medical importance: It has not been incriminated in any disease transmission cycle to humans. However, since it is very abundant in tropical regions, highly anthropophagus, and also bites monkeys, it could be involved in the transmission of yellow fever in the jungle regions.

(46) *Ps. (Jan.) albipes* (Theobald) (Fig. 9.9).

Description: Larvae unknown. Adults with wings covered with all dark scales, hind tarsomeres 4 and 5 white scaled, scutum with median longitudinal dark stripe with patches of golden scales at sides, mid lobe of scutellum with dark scales, and pedicel of antenna with pale scales.

Bionomics: Adult females are very aggressive mosquitoes that bite humans. They prefer to bite in shaded places and can bite even when the host is moving. In southeastern Mexico, it is a very troublesome pest because of the high populations after summer rains.

FIG. 9.9 *Psorophora albipes.*

Distribution: It is very similar to other species of *Psorophora* present in Texas such as *Ps. horrida*, *Ps. mathesoni*, and *Ps. longipalpus*. Although the record of *Ps. albipes* in Tamaulipas has not been corroborated, its proximity of southern Tamaulipas with regions of the state of Veracruz where *Ps. albipes* occurs, it is considered a valid record. In Mexico the swampy region of southern Tamaulipas is its northernmost distributional limit.

Medical importance: It is not considered a vector of mosquito-borne diseases.

(47) *Ps. (Pso.) cilipes* (Fabricius)

Description: Larvae: head quadrate, siphon index 2.0 or less, seta 1-S and 1-X single, Pecten extending to apex of siphon. Adults: large species, scutum with unscaled areas, pleuron largely covered with pale scales, tarsi entirely dark scaled.

Bionomics: Immatures of *Ps. cilipes* have been collected in shallow temporary ground pools, cattle trails with rainwater, and small temporary ground pools among tree roots. Adults have been collected while attempting to bite humans.

Distribution: *Ps. cilipes* has been collected in several countries of South and Middle America. In Mexico, it has been found in Mexico state, Quintana Roo, and Nuevo Leon as its northernmost distributional limit.

Medical importance: Its medical importance is unknown.

(48) *Ps. (Pso.) stonei* Vargas

Description: Very large species. The adults have green reflections on abdominal tergum. Additional diagnostic characters are found in male genitalia.

Bionomics: Not much is known about the biology of *Ps. stonei*; however, its larvae develop in ground pools.

Distribution: It was collected from ground pools in Sinaloa state and was recorded in Tamaulipas (Díaz-Nájera and Vargas, 1973), and this record was repeated (Ortega-Morales et al., 2015), but it is very probable that this record corresponds to misidentified *Ps. howardii*. Since there is no definitive evidence of the presence or absence of *Ps. stonei* in Tamaulipas in Mexico mosquito collections, it remains a valid record for this state.

Medical importance: Nothing is known about its medical importance.

9.6 Tribe Culicini Meigen

The tribe Culicini is divided into four genera: *Culex* L., *Deinocerites* Theobald, *Galindomyia* Stone and Barreto, and *Lutzia* Theobald. Except for *Galindomyia*, all genera occur in North America. *Lutzia* reaches its northernmost distributional limit in northeastern Mexico and does not spread in the United States. The most representative genus of the tribe Culicini is *Culex*, which has a cosmopolitan distribution. *Culex* is a very important genus because many species have been incriminated as vectors of a number of arboviruses such as Western Equine encephalitis virus (WEEV), VEEV, Saint Louis encephalitis virus (SLEV), West Nile virus (WNV), and lymphatic filariasis. The taxonomy of the subgenus *Culex* is complex because there are some species complexes such as *Cx. pipiens*, *Cx. coronator*, *Cx. restuans*, and *Cx. salinarius*. The subgenus *Melanoconion* is another group with complex taxonomy since many of them have incomplete descriptions and some are based on misidentified specimens. The genus *Deinocerites* is commonly known to have the "crab-hole mosquitoes" because their larvae develop in water in holes of some crab species and thus is commonly found near coastal regions. The genus *Lutzia* is a tropical group of mosquitoes, and their larvae are predators of other mosquito larvae and other small aquatic insects. There are 30 Culicini species recorded in the adjacent Mexican states to Texas and are divided into three genera and seven subgenera: *Culex* (*Anoedioporpa*, *Culex*, *Melanoconion*, *Microculex*, *Neoculex*, and *Phenacomyia*), *Deinocerites*, and *Lutzia* (*Lutzia*). Twenty of those are recorded in Texas:

(49) *Cx. (Cux.) chidesteri* Dyar. Distribution: Nuevo Leon.

(50) *Cx. (Cux.) coronator (s.s.)* Dyar and Knab. Distribution: Chihuahua, Coahuila, Nuevo Leon, and Tamaulipas.

(51) *Cx. (Cux.) declarator* Dyar and Knab. It is an uncommon species in northeastern Mexico. Distribution: Coahuila, Nuevo Leon, and Tamaulipas.

(52) *Cx. (Cux.) erythrothorax* Dyar. It is very common in the Mexico Valley but rare in the northeast of the country. Distribution: Chihuahua, Coahuila, and Nuevo Leon.

(53) *Cx. (Cux.) interrogator* Dyar and Knab. Distribution: Coahuila, Nuevo Leon, and Tamaulipas.

(54) *Cx. (Cux.) nigripalpus* Theobald. Distribution: Coahuila, Nuevo Leon, and Tamaulipas.

(55) *Cx. (Cux.) quinquefasciatus* Say. This is one of the most common species in Mexico, including the northeast. Distribution: Chihuahua, Coahuila, Nuevo Leon, and Tamaulipas.

(56) *Cx. (Cux.) restuans* Theobald. Although *Cx. restuans* has been reported to bite humans, in Mexico, it has never been associated with humans. Distribution: Chihuahua, Coahuila, Nuevo Leon, and Tamaulipas.

(57) *Cx. (Cux.) salinarius* Coquillett. This is an uncommon mosquito in northeastern Mexico. Distribution: Coahuila, Nuevo Leon, and Tamaulipas.

(58) *Cx. (Cux.) stigmatosoma* Dyar. This is a very common species of *Culex* in northeastern Mexico. Distribution: Chihuahua, Coahuila, Nuevo Leon, and Tamaulipas.

(59) *Cx. (Cux.) tarsalis* Coquillett. During the wintertime, it is one of the most common mosquitoes in northeastern Mexico. Distribution: Chihuahua, Coahuila, Nuevo Leon, and Tamaulipas.

(60) *Cx. (Cux.) thriambus* Dyar. This is a common species in the mountain regions, where the climate is colder than in coastal or tropical regions. Distribution: Chihuahua, Coahuila, Nuevo Leon, and Tamaulipas.

(61) *Cx. (Mel.) erraticus* (Dyar and Knab). Distribution: Nuevo Leon, and Tamaulipas.

(62) *Cx. (Mel.) peccator* Dyar and Knab. Distribution: Tamaulipas.

(63) *Cx. (Mel.) pilosus* (Dyar and Knab). Distribution: Tamaulipas.

(64) *Cx. (Ncx.) apicalis* Adams. Distribution: Coahuila.

(65) *Cx. (Ncx.) arizonensis* Bohart. Distribution: Coahuila and Nuevo Leon.

(66) *Cx. (Ncx.) territans* Walker. Distribution: Chihuahua and Tamaulipas.

(67) *Deinocerites mathesoni* Belkin and Hogue. The genus *Deinocerites* is characterized by the length of the antennae of the adult females that is longer than the proboscis. Distribution: Tamaulipas.

(68) *De. pseudes* Dyar and Knab. Distribution: Tamaulipas.

The species of the tribe Culicini found in the Mexican states adjacent to Texas but absent from that state are *Cx. restrictor* Dyar and Knab, *Cx. bidens* Dyar and Knab, *Cx. educator* Dyar and Knab, *Cx. inhibitator* Dyar and Knab, *Cx. iolambdis* Dyar, *Cx. taeniopus* Dyar and Knab, *Cx. rejector* Dyar and Knab, *Cx. corniger* Theobald, *Cx. lactator* Dyar and Knab, and *Lt. bigoti* (Bellardi).

(69) *Cx. (Anoedioporpa) restrictor* Dyar and Knab (Fig. 9.10).

Description: Larvae: setae 4-X with six pairs, setae 2-VIII attached within oval plate. Adult: female with acrostichal setae extending from anterior scutum to prescutellar area.

Bionomics: It is a common tree-hole dweller in the tropical regions of southeastern Mexico. Its larvae have also been found in artificial containers such as discarded tires, always with abundant submerged leaves and colored water. Not much is known about the biology of adults; however, females have been collected at night using CDC light traps baited with octenol in southeastern Mexico.

Distribution: Nuevo Leon and Tamaulipas states represent its northernmost distributional limit.

Medical importance: Its medical importance is unknown.

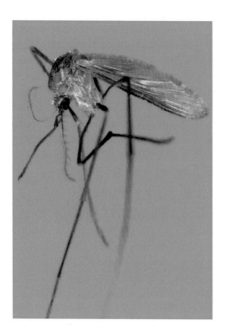

FIG. 9.10 *Culex restrictor.*

(70) *Cx. (Cux.) bidens* Dyar and Knab

Description: Larvae: pecten spines restricted to the basal third of siphon, siphon without apical spines, thorax minutely aciculate, three siphonal tufts beyond the pecten, siphon index 6.0, anal papillae longer than the length of the anal segment, basal tuft on siphon attached well beyond pecten, seta 6-C triple. Adults: vertex of the head with narrow curved scales, abdomen light brown and without metallic reflections, hind tarsomeres with pale bands and fourth segment dark scaled in the middle, proboscis without white ring.

Bionomics: Its larvae have been found in a variety of aquatic habitats such as ponds, swamps, and marshes, always with abundant emergent vegetation and partial or total shade.

Distribution: In northern Mexico, it has been found in the states of Coahuila, Nuevo Leon, and Tamaulipas.

Medical importance: The feeding preferences and medical importance of *Cx. bidens* are unknown, although it is suspected to be a vector of VEEV.

(71) *Cx. (Mel.) educator* Dyar and Knab

Description: Larvae: Abdominal setae 7-I single, siphon index less than 8.0, setae 4-P double, comb scales with prominent apical spine and small lateral spinules. Adults with hind tarsomeres entirely dark scaled, scutum with dark-brown scales, central area of occiput with wide ovate scales, mesanepimeron dark brown and without scales, upper mesokatepisternum with three or fewer broad scales, abdominal sterna entirely pale scaled, second seta of mid coxa larger than either first or third setae.

Bionomics: Its larvae have been collected from swamps with abundant aquatic vegetation. Adult females have been collected at night using CDC light traps. Its feeding preferences are unknown.

Distribution: In Mexico, it has been collected in the swampy region of southern Tamaulipas state, which is its northernmost distributional limit.

Medical importance: Its medical importance is unknown.

(72) *Cx. (Mel.) inhibitator* Dyar and Knab

Description: Larvae: comb scales in a triangular patch of three rows, siphon index 3.0, with tufts in straight line, basal tuft long and the rest progressive shorter, thorax pilose, setae 4 and 5-C branched. Adults: females are large or medium mosquitoes, with decumbent scales on vertex entirely broad and usually pale along eye margin, dark toward center, scales of the veins of the wing usually broad ovate or squamous, upper corner of mesokatepisternum without scales or microsetae, integument of scutum light brown, mesopostnotum and lower part of mesokatepisternum without microsetae.

Bionomics: Its larvae have been collected in southeastern Mexico in shaded swamps with abundant aquatic vegetation. Adult females have been collected at night using CDC light traps near the aquatic habitat of the larvae.

Distribution: *Cx. inhibitator* occurs in several countries and islands of the Neotropical region. In Mexico the swampy region of southern Tamaulipas state is its northernmost distributional limit.

Medical importance: Its medical importance is unknown

(73) *Cx. (Mel.) iolambdis* Dyar

Description: Larvae: setae 7-I single, siphon index less than 8.0, setae 4-P single, thorax and abdomen moderately aculeate, setae 7-P triple, setae 8-P double, comb scales long, abdomen smooth on segments I–IV. Adults: hind tarsomeres all dark scaled, scutum covered with dark-brown scales, mesanepimeron without patch of scales but with 4–5 hairlike to ligulate scales.

Bionomics: Its larvae have been collected from crab holes, swamps in mangrove areas, stream sides with total shade, and ground pools with abundant aquatic vegetation. Adults have been collected using aspirators after they were disturbed in their resting places on the vegetation. Adult females feed on wild birds, frogs, turtles, and other reptiles.

Distribution: This is a Neotropical species that has been reported in several countries of Middle and South America. In the northeastern Mexico, it has been collected in Tamaulipas.

Medical importance: Although VEEV has been isolated from *Cx. iolambdis*, it is not considered its vector.

(74) *Cx. (Mel.) taeniopus* Dyar and Knab

Description: Larvae: setae 7-I double, setae 1-M branched, setae 6-I, II double, siphon index 5.0. Adults: hind tarsomeres pale banded, dorsal corner of mesokatepisternum with patch of scales, femora all dark scaled.

Bionomics: Its larvae have been collected from crab holes, swamps, mangrove ponds, sluggish streams, and pools besides the streams. Adult females have been collected at night using CDC light traps baited with octenol and using Shannon traps baited with horses. This is one of the few members of subgenus *Melanoconion* that is attracted to humans with the intention of biting.

Distribution: In Mexico, it has been recorded in some states, including Tamaulipas, where it reaches its northernmost distributional limit.

Medical importance: *Cx. taeniopus* has been incriminated as a vector of VEEV.

(75) *Cx. (Mcx.) rejector* Dyar and Knab (Fig. 9.11).

FIG. 9.11 *Culex rejector.*

Description: Larvae: comb scales in a triangular patch and with very long spines, setae 6-III, IV double. Adults: hind tarsomeres with narrow white rings, thoracic markings scarcely silvery, palpi of male dark.

Bionomics: Its larvae are commonly found in the axils of epiphytic bromeliads that are filled with rainwater. It is also commonly found in association with other phytotelmata mosquitoes such as *Aedes (Howardina)* spp. and *Wyeomyia* spp. Nothing is known about the feeding preferences of the female.

Distribution: In Mexico the Biosphere Reserve "El Cielo" in Tamaulipas is its northernmost distributional limit. It is also reported in Coahuila state, but this record is presumably erroneous.

Medical importance: Since the feedings patterns of *Cx. rejector* are currently unknown, not much is known about its medical importance.

(76) *Cx. (Phc.) corniger* Theobald (Fig. 9.12).

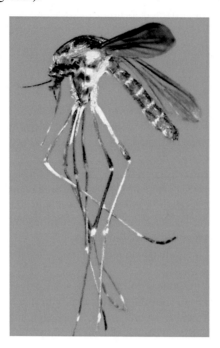

FIG. 9.12 *Culex corniger.*

Description: Adults: dark integument with greater contrast between light and dark areas. Both *Cx. corniger* and *Cx. lactator* are extremely similar in larval and adult life stages. For additional characteristics to separate this species, see Strickman and Pratt (1989).

Bionomics: Its larvae develop in a number of aquatic habitats such as tree holes, bamboo internodes, ground pools, stream margins, flower bracts, artificial containers, and rock holes. The adults have been collected resting in vegetation and biting humans, but the feeding preference for humans is not well documented.

Distribution: It has been recorded in several countries of the Neotropical region. In the Mexican states adjacent to Texas, *Cx. corniger* has been reported in Coahuila and Tamaulipas, where it reaches its northernmost distributional limit.

Medical importance: Its medical importance is unknown.

(77) *Cx. (Phc.) lactator* Dyar and Knab (Fig. 9.13).

FIG. 9.13 *Culex lactator.*

Description: Adults: distinct light bands on the proximal hind tarsomeres and less distinct bands on tarsomeres 4 and 5 than *Cx. corniger*.

Bionomics: Its larvae have been collected from sunlit waters of ground pools, stream margins, lake margins, and artificial containers. Adults have been collected from light traps and resting in vegetation. The feeding preferences of *Cx. lactator* are unknown.

Distribution: In Mexico, it has been recorded in Tamaulipas state, which is its northernmost distributional limit.

Medical Importance. Its medical importance is unknown

(78) *Lt. (Lut.) bigoti* (Bellardi) (Fig. 9.14).

FIG. 9.14 *Lutzia bigoti.*

Description: Larvae: pecten of siphon extending distally from the middle of siphon. Adults with vein Cu with dark scales extending to fork of Cu_1 and Cu_2, subcostal pale spot on wing extending to subcostal vein.

Bionomics: In Mexico, its larvae have been commonly collected from artificial containers such as discarded tires, flower vases, and tree holes and are predators of other mosquito larvae. Adults have been collected from vegetation and shaded resting places, and females feed on small domestic animals.

Distribution: In the Mexican states adjacent to Texas, *Lt. bigoti* has been collected in Nuevo Leon and Tamaulipas as its northernmost distributional limit.

Medical importance: Its medical importance is unknown.

9.7 Tribe Culisetini Belkin

The tribe Culisetini is a monobasic group of mosquitoes that includes only the genus *Culiseta* with 37 species that are distributed in all zoogeographic regions except in South America. Most species of the tribe Culisetini occur in the Palearctic, Nearctic, and Australasian regions, and some also occur in the Afrotropical and Oriental regions (Harbach, 2018a). They are very similar to other mosquitoes in the tribes Aedini, Culicini, and Orthopodomyiini. Most are large, strong, light brown– or dark brown–colored mosquitoes. Many are ornithophagous, and some bite mammals. In North America, eight species occur while in Texas three: *Cs. melanura* (Coquillett), *Cs. incidens* (Thomson), and *Cs. inornata* (Williston).

(79) *Cs. (Cli.) melanura* (Coquillett): Distribution: Nuevo Leon.

(80) *Cs. (Cus.) incidens* (Thomson): Distribution: Coahuila.

(81) *Cs. (Cus.) inornata* (Williston): Distribution: Coahuila, Nuevo Leon, and Tamaulipas.

The only species of the tribe Culisetini found in the Mexican states adjacent to Texas but absent in Texas is *Cs. particeps* (Adams).

(82) *Cs. (Cus.) particeps* (Adams)

Description: Larvae: setae 1-S longer than rest of setae of siphon, setae 6-C with fewer branches than 5-C, n and saddle with aciculae dorsoposteriorly. Adults: abdominal terga with distinct basal bands, hind tarsomeres with broad pale bands, and femora with narrow subapical band of pale scales.

Bionomics: Its larvae develop in cold and shaded pools with green filamentous algae and dead tree leaves at bottom of the pool and in artificial containers such as discarded tires with cold water. Adult females are attracted to humans at night and dusk but prefer to feed on wild birds.

Distribution: Coahuila, Nuevo Leon, and Tamaulipas.

Medical importance: *Cs. particeps* is not known to be involved in the transmission of any human pathogens.

9.8 Tribe Mansoniini Belkin

The tribe Mansoniini includes the genera *Coquillettidia* Dyar and *Mansonia* Blanchard. Most species of the tribe Mansoniini occur in the Afrotropical and Nearctic regions. The main characteristic of the tribe Mansoniini is found in the larvae, which have an attenuated siphon modified for piercing the tissue of aquatic plants and attach themselves to submerged aquatic plants to obtain oxygen. Adult females of some species are troublesome in outdoor areas, since they are very persistent biters. In Texas and north of Mexico, five species of the tribe Mansoniini are found: *Cq. nigricans* (Coquillett); *Cq. perturbans* (Walker); *Ma. dyari* Belkin, Heinemann, and Page; *Ma. indubitans* Dyar and Shannon; and *Ma. titillans* (Walker). With the exception of *Ma. indubitans*, all species of the genus *Mansonia* mentioned here are recorded in the United States.

(83) *Cq. (Rhy.) perturbans* (Walker). Distribution: Coahuila.

(84) *Ma. (Man.) titillans* (Walker). Distribution: Nuevo Leon and Tamaulipas.

The species of the tribe Mansoniini found in the Mexican states adjacent to Texas but absent from Texas are *Cq. nigricans*, *Ma. dyari*, and *Ma. indubitans*.

(85) *Cq. (Coq.) nigricans* (Coquillett) (Fig. 9.15).

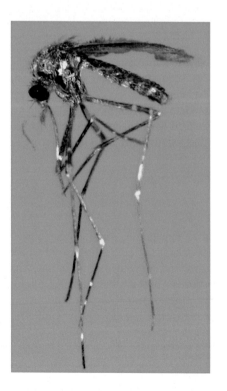

FIG. 9.15 *Coquillettidia nigricans.*

Description: Larvae: setae 2–3-A short, part of antennae distal to their point of attachment longer than part basal to it. Adults: Wings covered with all dark scales, tibiae dark on anterior surface except for the preapical pale-scaled band.

Distribution: In Mexico the swampy region of southern Tamaulipas state is its northernmost distributional limit.

Medical importance: Its medical importance is unknown.

(86) *Ma. (Man.) dyari* Belkin, Heinemann, and Page (Fig. 9.16).

FIG. 9.16 *Mansonia dyari.*

Description: Larvae: setae 4-X with three pairs of fanlike attached to grid, comb scales boarder, with several subequal spinules. Adults: apex of abdominal tergum VII without row of short dark spiniforms, ventral surface of proboscis with broad pale-scaled stripe in middle.

Bionomics: Its larvae develop in swamps and ponds with abundant aquatic vegetation. Adult females are mostly active during nighttime and at dusk and can be collected at night using CDC light traps.

Distribution: In the United States, it has been recorded in Florida and Georgia, while in the Mexican states adjacent to Texas, *Ma. dyari* has been recorded in Nuevo Leon state.

Medical importance: The medical importance of *Ma. dyari* is unknown, but it could be involved in zoonotic cycles of WNV.

(87) *Ma. (Man.) indubitans* Dyar and Shannon

Description: Larvae: setae 4-X with four pairs of fanlike setae attached to grid, comb scales slender, with single spine apically. Adults: apex of abdominal tergum VII with row of short dark spiniforms, ventral surface of proboscis predominantly dark scaled, lightly to moderately speckled with pale scales.

Bionomics: The aquatic habitats and bionomics of *Ma. indubitans* are the same as *Ma. dyari*.

Distribution: In Mexico the swampy region of southern Tamaulipas state is its northernmost distributional limit.

Medical importance: Its medical importance is unknown.

9.9 Tribe Orthopodomyiini Belkin, Heinemann, and Page

The tribe Orthopodomyiini is a monobasic group of mosquitoes that includes only the genus *Orthopodomyia* containing 36 species distributed in the Neotropical, Palearctic, Oriental, Afrotropical, Nearctic, and Australian regions. In North America, there are three species: *Or. alba* Baker, *Or. kummi* Edwards, and *Or. signifera* (Coquillett) all found in Texas and described in Chapter 2.

(88) *Or. alba* Baker. Distribution: Coahuila and Nuevo Leon.

(89) *Or. kummi* Edwards. Distribution: Nuevo Leon.

(90) *Or. signifera* (Coquillett). The habitats of its larvae are similar to those of *Or. alba*, where both species can be found in the same tree hole where populations overlap. Distribution: Coahuila and Tamaulipas.

9.10 Tribe Sabethini Blanchard

The tribe Sabethini is a group of tropical mosquitoes that occur in the Neotropical, Oriental, Afrotropical, and Australasian regions. The only genus of the tribe Sabethini found in the United States is *Wyeomyia*, and it includes *Wyeomyia mitchellii* (Theobald) distributed in Florida and Georgia, *Wy. smithii* (Coquillett) distributed from southern Canada to the southeastern United States, and *Wy. vanduzeei* Dyar and Knab distributed in southern Florida. None of these species have been recorded in Texas. In North America, Sabethini includes the genera *Johnbelkinia* Zavortink, *Limatus* Theobald, *Sabethes* Robineau-Desvoidy, *Shannoniana* Lane and Cerqueira, *Trichoprosopon* Theobald, and *Wyeomyia* Theobald, which are all present in the Mexican states adjacent to Texas except *Johnbelkinia*, *Shannoniana*, and *Trichoprosopon*. The species of Sabethini found in the Mexican state of Tamaulipas but absent in Texas are *Li. durhamii* Theobald, *Sa. chloropterus* (von Humboldt), and *Wy. mitchellii* (Theobald)

(91) *Li. durhamii* Theobald (Fig. 9.17).

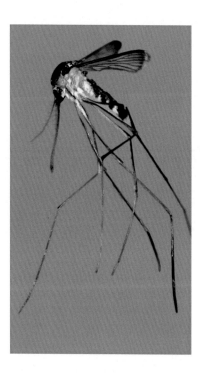

FIG. 9.17 *Limatus durhamii.*

> **Description:** Larvae: setae 4-X are with 3–4 branches, and siphon index is 2.5–3.0. Adults: purple scales on the dorsal surface of scutum and abdominal terga have dark dorsal scales in incised pattern.
> **Bionomics:** Its larvae can be found in a variety of aquatic habitats such as tree holes, bamboo internodes, axils of terrestrial plants, and artificial containers such as discarded tires and flower vases. In Mexico, *Li. durhamii* is the most common Sabethini species in artificial containers. Adult females are forest mosquitoes that rest in shaded vegetation and are attracted to humans.
> **Distribution:** It has a widespread distribution in the Neotropical region, from Argentina to northeastern Mexico. In the Biosphere Reserve "El Cielo" in Tamaulipas, *Li. durhamii* reaches its northernmost distributional limit.
> **Medical importance:** Even though the yellow fever virus has been isolated from wild *Li. durhamii* mosquitoes, its importance as a vector is unknown.

(92) *Sa. (Sbo.) chloropterus* (von Humboldt) (Fig. 9.18).

> **Description:** Larvae: setae 1-X double, siphon index 3.0. Adults: proepisternal setae present and mid legs without paddles formed by long scales.

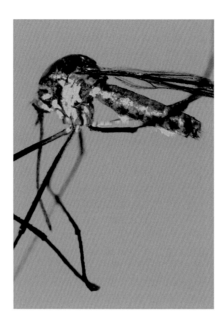

FIG. 9.18 *Sabethes chloropterus.*

Bionomics: Its larvae have been found in tree holes and discarded tires with colored water. Adult females have been collected during the day using CDC light trap and while approaching humans.

Distribution: In Mexico, it has been reported in the Biosphere Reserve "El Cielo" in Tamaulipas, where it reaches its northernmost distributional limit.

Medical importance: It has been involved as a potential vector of sylvatic yellow fever.

(93) *Wy. (Wyo.) mitchellii* (Theobald) (Fig. 9.19).

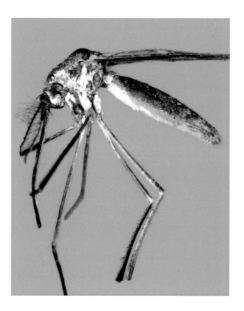

FIG. 9.19 *Wyeomyia mitchellii.*

Description: Larvae: setae 1–3-X single, and setae 4-X with 7 or more branches. Adults: antepronotal lobes with bluish to purple scales, occiput with ocular line with pale scales, and postpronotum with board pale scales.

Bionomics: Its larvae are commonly found in axils of epiphyte bromeliads at both ground and canopy levels and less frequently in other types of phytotelmata and artificial containers. Adults are fragile mosquitoes that rest in shaded vegetation, active during daytime, and attracted to humans.

Distribution: In Mexico, it reaches its northernmost distributional limit in the Biosphere Reserve "El Cielo" in Tamaulipas state.

Medical importance: Its medical importance is unknown.

9.11 Tribe Toxorhynchitini Lahille

The tribe Toxorhynchitini is a monobasic group of mosquitoes, including only the genus *Toxorhynchites*, which has a pantropical distribution. Adult females do not feed on blood; both sexes are completely phytophagous, feeding only on plant juices and other plant exudates. The larvae are carnivorous and extremely aggressive predators that feed on other mosquito larvae and other aquatic insects and can even feed on smaller larvae of the same species. In America, two subgenera of *Toxorhynchites* are found: *Ankylorhynchus* and *Lynchiella* of which only the latter is found in North America. In Mexico, three species have been found. In the adjacent Mexican states to Texas, two species are found: *Tx. grandiosus* (Williston) and *Tx. moctezuma* (Dyar and Knab). The latter species is also found in Texas along with *Tx. rutilus* (Coquillett), which has not been recorded in Mexico, but it could be present in the northeastern Mexico. Since *Toxorhynchites* does not feed on blood, it is of no medical importance. The species of the tribe Toxorhynchitini found in the Mexican state of Tamaulipas but absent in Texas is *Tx. grandiosus* (Williston).

(94) *Tx. (Lyn.) grandiosus* (Williston)
   Description: It's a large mosquito. Larvae have anal papillae very narrow and longer than segment X. Adults: abdominal segments without tufts, and tibiae are mostly golden.
   Bionomics: In Tamaulipas, it was collected as larvae from a discarded tire with colored water and abundant submerged leaves feeding on larvae of *Ae. podographicus* and *Cx. thriambus*. The larvae were reared to obtain the adult stages but died in the fourth instar stage. Its adult biology is unknown.
   Distribution: In Mexico, it has been recorded in the states of Guerrero and Tamaulipas. In the Biosphere Reserve "El Cielo" in Tamaulipas, it reaches its northernmost distributional limit.

(95) *Tx. (Lyn.) moctezuma* (Dyar and Knab)
   Distribution: In the Mexican states adjacent to Texas, it has been recorded in Tamaulipas, but it may reach more northern regions into the United States than Cameron County, Texas, and even be sympatric with *Tx. rutilus*.

9.12 Tribe Uranotaeniini Lahille

The tribe Uranotaeniini is another monobasic group of mosquitoes, including only the genus *Uranotaenia*. This genus has a worldwide distribution, with species occurring in tropical regions such as the Neotropical, Afrotropical, and Oriental regions and warm regions of Europe, southern United States, Australia, and Middle East. It is absent on several oceanic islands. It includes 270 species grouped in two subgenera: *Pseudoficalbia* and *Uranotaenia* (Harbach, 2018b). In Mexico, 10 species of *Uranotaenia* have been recorded and 5 in the adjacent Mexican states to Texas while 3 in Texas.

(96) *Ur. (Pseudoficalbia) anhydor* Dyar. It is the only species of the subgenus *Pseudoficalbia* that occurs in Mexico and the United States. Its larvae develop in sunlight ponds and swamps with or without aquatic vegetation. Its adult female feeding preferences and the medical importance are unknown. Distribution: Nuevo Leon and Tamaulipas.

(97) *Ur. (Ura.) lowii* Theobald. Distribution: Coahuila, Nuevo Leon, and Tamaulipas.

(98) *Ur. (Ura.) sapphirina* (Osten Sacken). *Distribution*: Coahuila and Tamaulipas.

   The two species of the tribe Uranotaeniini found in the Mexican state of Tamaulipas but absent in Texas are *Ur. coatzacoalcos* Dyar and Knab and *Ur. geometrica* Theobald.

(99) *Ur. (Ura.) coatzacoalcos* Dyar and Knab (Fig. 9.20).
   Description: Larvae: setae 6-I, II double, setae 14-P multibranched. Adults: medium-sized species; hind tarsomeres with segments 1, 2 and part of 3 dark scaled; remainder of tarsi entirely pale scaled; line of blue scales above root of wing extend to anterior margin of paratergite; scutal integument dark brown.

FIG. 9.20   *Uranotaenia coatzacoalcos.*

Bionomics: Its larvae develop in swamps and pools at ground level with abundant emerging vegetation. In Mexico, it has been found in artificial containers with colored water and abundant leaves at bottom. Adults have been collected resting in vegetation.

Distribution: This is a Neotropical species. In Mexico, it has been found in some southeastern states, and Nuevo Leon is its northernmost distributional limit.

Medical importance: Its medical importance is unknown.

(100) *Ur. (Ura.) geometrica* Theobald (Fig. 9.21).

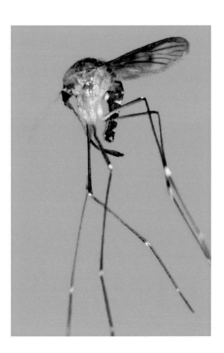

FIG. 9.21   *Uranotaenia geometrica.*

Description: Larvae: setae 6-I, II triple, and 14 P single. Adults: Hind tarsomeres with bands of pale scales, tarsomeres 4 and 5 entirely white scaled, and scutum with median spot of bluish scales just before prescutellar area.

Bionomics: Its larvae develop in sunlight aquatic habitats with abundant aquatic vegetation and green algae, such as swamps, ponds, and less frequently artificial containers such as flower vases and discarded tires with colored water and abundant leaves at bottom. Adult females have been collected using CDC light traps and approaching humans, but are not known to bite humans.

Distribution: It has been recorded in several countries of South and Middle America, The Biosphere Reserve "El Cielo" in Tamaulipas is its northernmost distributional limit.

Medical importance: Its medical importance is unknown.

Acknowledgments

We thank Quetzaly K. Siller-Rodriguez for her assistance in preparing Fig. 9.1, Rafael Vazquez Marroquin for providing adult mosquito pictures, and Thomas J. Zavortink and Mustapha Debboun for reviewing this chapter.

References and further reading

Bram, R.A., 1967. Classification of *Culex* Subgenus *Culex* in the New World (Diptera: Culicidae). Smithsonian Institution, Washington.

Clark-Gil, S., Darsie, R.F., 1983. The mosquitoes of Guatemala, their identification, distribution and bionomics. Mosq. Syst. 15, 151–284.

Darsie, R.F., Ward, R.A., 2005. Identification and Geographical Distribution of the Mosquitoes of North America, North of Mexico. University Press of Florida.

Díaz-Nájera, A., Vargas, L., 1973. Mosquitos mexicanos, distribución geográfica actualizada. Rev. Invest. Salud Publica 33, 111–125.

Dyar, H., 1928. The Mosquitoes of the Americas. Carnegie Institution of Washington.

Harbach, R., 2018a. Culisetini Tribe. Mosquito Taxonomic Inventory. http://mosquito-taxonomic-inventory.info/simpletaxonomy/term/6196. (Accessed March 12, 2018).

Harbach, R., 2018b. Mansoniini Tribe. Mosquito Taxonomic Inventory. http://mosquito-taxonomic-inventory.info/simpletaxonomy/term/6213. (Accessed March 12, 2018).

Lane, J., 1953. Neotropical Culicidae. University of Sao Paulo, Sao Paulo.

Ortega-Morales, A.I., Zavortink, T.J., Huerta-Jiménez, H., Sánchez-Ramos, F.J., Valdés-Perezgasga, M.T., Reyes-Villanueva, F., Siller-Rodríguez, Q.K., Fernández-Salas, I., 2015. Mosquito records from Mexico: the mosquitoes (Diptera: Culicidae) of Tamaulipas state. J. Med. Entomol. 52, 171–184.

Ortega-Morales, A.I., Zavortink, T.J., Garza-Hernández, J.A., Siller-Rodríguez, K.Q., Fernández-Salas, I., 2019. The mosquitoes (Diptera: Culicidae) of Nuevo Leon, with descriptions of two new species. PLoS ONE . (In press).

Schick, R.X., 1970. Mosquito studies (Diptera, Culicidae), the Terrens group of *Aedes* (*Finlaya*). Contrib. Am. Entomol. Inst. 5 (3), 1–158.

Sirivanakarn, S., 1982. A review of the systematics and proposed scheme of internal classification of the New World subgenus *Melanoconion* of *Culex* (Diptera, Culicidae). Mosq. Syst. 14, 265–332.

Strickman, D., Pratt, J., 1989. Redescription of *Cx. corniger* Theobald and elevation of *Culex* (*Culex*) *lactator* Dyar and Knab from synonymy based on specimens from Central America (Diptera: Culicidae). Proc. Entomol. Soc. Wash. 91, 551–574.

Vargas, L., 1956. Psorophora (Psorophora) stonei n. sp. Rev. Inst. Salubr. Enferm. Trop. 2, 15–16.

Wilkerson, R.C., Linton, R.M., Fonseca, D.M., Schultz, T.R., Price, D.C., Strickman, D.A., 2015. Making mosquito taxonomy useful: a stable classification of tribe Aedini that balances utility with current knowledge of evolutionary relationships. PLoS ONE 10, 1–26.

WRBU (Walter Reed Biosystematic Unit), 2018. Systematic Catalog of Culicidae. http://www.mosquitocatalog.org/. (Accessed 13 March 2018).

Chapter 10

Invasive Mosquito Species and Potential Introductions

Daniel Strickman

Global Health Program, Bill & Melinda Gates Foundation, Seattle, WA, United States

Chapter Outline

Abbreviations

Ae.	*Aedes*
An.	*Anopheles*
Cx.	*Culex*
EICA	evolution of increased competitive ability
US	United States

10.1 Biogeography of species invasion with respect to mosquitoes

Given a long enough time frame, the distribution of species is never stable. Changing conditions and changing genetic makeups combine to restrict or expand the geographical extent of a population. Such changes can occur over a small range and short time, such as typical movement of *Psorophora columbiae* (Dyar and Knab) or *Aedes taeniorhynchus* (Wiedemann) from their larval habitats to sources of blood meals and back to oviposition sites by gravid females (Service, 1997). In the course of such movement, populations may expand to areas that are temporarily suitable, expanding the range of that species until conditions change again to unfavorable (e.g., *Culex quinquefasciatus* Say, Keyghobadi et al., 2006). The subtropical climate of southern Texas is particularly favorable to these expansions and contractions, as series of mild winters are punctuated occasionally by northern weather that moves in unrestricted by geographical barriers to the north. Such conditions created bizarre circumstances for mosquitoes in 1984 (Strickman, 1988). Against this general backdrop is an accelerating redistribution of species globally, brought about by human transport (Mooney and Cleland, 2001). That transport creates repeated opportunities for species to become established in new places, affecting genetics of both the invading species and the native species with which they interact.

The genetics of invasion can be problematic for a species, as informed by detailed studies of invasive plants. The new range of a species presents its own challenges, and studies of gene flow have shown that the rate for neutral markers is greater than that for expressed, phenotypic markers (Xu et al., 2010). That difference represents selection for those phenotypes that are adaptive for the new environment. Establishment of new, adaptive genotypes can occur rapidly when unusual alleles do not encounter the variety of competitive types present in the home range (Dlugosch and Parker, 2008). Development of new, unique genetic combinations for a species in its invaded range can also be enhanced by isolation of smaller populations in a mosaic of suitable habitats (Holderegger and Wagner, 2008). Hybridization of populations can also produce new combinations of traits, as populations merge across a mosaic of habitats or as additional individuals are introduced from the home range to the invaded territory (Schierenbeck and Ellstrand, 2009). Selection can tend toward "evolution of increased competitive ability" or EICA (Franks et al., 2011). In plants the trend toward EICA results in invasive species that are more capable of displacing native species. Displacement of indigenous flora by invasive species has been blamed for serious damage to biodiversity (Lowe et al., 2004). Invasive plants can also contribute to a "pathogenic landscape" that favors harmful mosquitoes (Mack and Smith, 2011; Stone et al., 2018).

Whether due to accidental introduction or changes in conditions that make an area suitable, there have been more than a few examples of introduction of mosquitoes to North America (Table 10.1). In general the phases of a successful species invasion involve transport to a location in the new range, establishment (i.e., colonization) of a reproductively viable population at that site, spread from that site to others, and finally the negative impact on humans or the environment that separates

TABLE 10.1 Introduced mosquitoes in the US since 1500.

Species	From	To	First introduction	Conveyance	Comment	References
Aedes aegypti	Mexico	Arizona	1994	Ground transport?	Previously present 1931–46 and 1951, but established in Tucson since 1994	Engelthaler et al. (1997)
Aedes aegypti		California	2013		In two counties with current efforts to eradicate	Mulligan (2013)
Aedes aegypti	Europe or Africa	Western Hemisphere	1490s	Ships	Suspected in Haiti 1495 documented in Yucatan 1648	Lounibos (2002)
Aedes albopictus	Asia	California	2001	Lucky bamboo	First introductions seemed to disappear but reemerged in 2011 with same genetic signature in Los Angeles County	Zhong et al. (2013)
Aedes albopictus	Japan	Texas, United States	1985	Tires	Previously introduced to United States, but not established	Lounibos (2002)
Aedes albopictus	Japan	Texas, United States	1985			Mouchet et al. (1995)
Aedes atropalpus	Eastern United States	Midwestern United States	After 1972	Tires	Species expanded suitable larval sites from rock pools to include tires	Lounibos (2002)
Aedes bahamensis	Bahamas	Southern Florida	1986	Tires	Present in only two counties	Lounibos (2002)
Aedes japonicus	United States	Ontario, Canada	2001	Ground transport	Established	Thielman and Hunter (2006)
Aedes japonicus	Japan	United States	1998	Tires	Japan probably source of original two introductions	Lounibos (2002), Fonseca et al. (2001)
Aedes notoscriptus	Australia	California	2015	Unknown	In three counties as of 2018	Peterson and Campbell (2015); website San Mateo County Mosquito and Vector Control District 2018

TABLE 10.1 Introduced mosquitoes in the US since 1500.—cont'd

Species	From	To	First introduction	Conveyance	Comment	References
Aedes togoi	Asia	British Columbia and Washington	1960s	Ships and tires		Lounibos (2002)
Culex biscayensis	Caribbean	Florida		Bromeliads	Possibly an indigenous species, present in only one location	Lounibos (2002)
Culex coronator	Southwest United States	Georgia, United States	2006		Prior to 2003, only in south central, southwestern, and north central Texas, southern New Mexico, and southeastern Arizona. 2003 found in Oklahoma, Louisiana, Mississippi; 2005 in Florida	Kelly et al. (2008)
Culex pipiens	Europe	North America	1500s	Ships	Species considered originally to be from Africa	Farajollahi et al. (2011)
Culex quinquefasciatus	Asia	Western Hemisphere	Before 1800	Ships	Uncertainty about origin and timing, traditionally thought to be African but now thought that African populations from America. Possibly from Southeast Asia	Farajollahi et al. (2011)

invasive species from those only expanding their ranges (Lowe et al., 2004). The process of genetic selection and adaptation in the new range may be slower than that of plants, requiring hundreds or thousands of generations for a trait to become fixed in the population (Kiszewski and Spielman, 1998). The need for a significant number of introduced individuals to provide sufficient genetic variation for establishment of a viable population may account for the relatively small number of mosquitoes that have become established. Undoubtedly, many species have been introduced at one time or another but have not become established (Lounibos, 2002). The rapid expansion of *Anopheles arabiensis* Patton in Brazil may have been an exception. Introduced on rapid warships traveling from Dakar, Senegal, to Natal, Brazil, in 1930 (Soper, 1966), this species quickly established itself in this arid part of the country. It spread rapidly along river systems, especially taking advantage of agricultural practices that created small pools suitable for its larvae (Causey et al., 1943; Soper and Wilson, 1943). A deadly epidemic of malaria ensued, and *An. arabiensis* was not completely eliminated from northeastern Brazil until 1939.

Movement of mosquito species into new ranges can also occur between regions of the same land mass. A great example is *Ae. atropalpus* (Coquillett). Its larvae are commonly found in rock pools near rivers in the Middle Atlantic States, the type locality being just outside Washington, DC. In the early 1970s, oviposition expanded from rock pools to artificial containers (Lounibos, 2002). As a consequence, *Ae. atropalpus* expanded its range to the Midwest. Eventually, it even invaded Europe (Scholte et al., 2009; Medlock et al., 2012). Another example is *Cx. coronator* Dyar and Knab, which expanded its range in the opposite direction from Arizona, New Mexico, and Texas to Oklahoma, Louisiana, and Mississippi by 2003; Florida by 2005; and Georgia by 2006 (Kelly et al., 2008).

Some species appear to take advantage of chances to expand their ranges by adjusting elevation in accordance with latitude. For example, *Ae. impiger* (Walker) is an Arctic species that lives much further south at high elevations in Colorado (Baker, 1961). *Cx. restuans* Theobald lives at high elevations in Central America (Strickman and Darsie, 1988), as does

Ae. epactius Dyar and Knab in Mexico (Lozano-Fuentes et al., 2012). Just as important, many high-elevation species in Mexico and Central America do not occur lower in the United States, despite elevations where the climate might be suitable (*Ae. quadrivittatus* (Coquillett) (Berlin, 1969), *Cx. pseudostigmatosoma* Strickman (Strickman, 1989), and *An. eiseni* Coquillett (Wilkerson and Strickman, 1990)).

Climate change (Portier et al., 2010) is likely to affect mosquitoes in many different ways (Table 10.2). This will affect the distribution, ecology, behavior, and medical importance of mosquitoes already in Texas. For example, species restricted to the southern part of the state might move further north. Some mosquitoes restricted by seasonality of temperature, in particular, may be able to have more generations per year. Species that are marginal vectors may become better able to transmit certain pathogens. Climate change will also lead to an increase in human influence (Ross, 1964) on mosquito distribution through influence on the availability of many kinds of larval sites (Service, 1993). We could expect an increase in the following factors that influence the ranges of mosquitoes: forest fragmentation and new areas of agricultural activity (Reiter and LaPointe, 2007), irrigation and the presence of people in an area where they had been sparse (Wimberly et al., 2008), and an increase in water-holding containers (Walker et al., 2011). As the population distributions of mosquitoes shift, a given area may become the home to an assemblage of species that is new, with only limited coevolutionary adaptation between them. Those biological relationships of predation and niche specialization in the aquatic habitat might be more of

TABLE 10.2 Potential influences of climate change on mosquitoes (Epstein et al., 1998; Dukes and Mooney, 1999; Gubler et al., 2001; Parmesan, 2006; De la Rocque et al., 2008; Gage et al., 2008; Holderegger and Wagner, 2008; Lafferty, 2009; Hoffmann and Sgrò, 2011; Peterson and Lieberman, 2012; Chen et al., 2013).

Effects directly on the mosquito
  Developmental
    Direct physiological effects
    Developmental period
    Longevity and survival
  Symbiotic
    Susceptibility to infection
    External incubation period
  Behavioral
    Host selection
    Blood feeding rate
Effects on groups of individuals and populations
  Genetic
    Population genetic structure
    Genetic adaptation
    Hybridization
  Phenological
    Seasonality
    Generations per year
    Timing of life cycle events
    Influence of weather events
  Populational
    Population growth
    Mass outbreaks
Indirect effects on mosquitoes
  Geographic
    Barriers formed by areas of unsuitable habitat
    Population relationships to El Niño-Southern Oscillation cycle
  Habitat structure
    Landscape
    Destruction of high-altitude habitats
    Habitat fragmentation
    Ecosystem shifts caused by changes in flora
  Ecological
    Distribution of blood meal hosts
    Biodiversity
    Conditions favoring colonizing species (*r*-strategists)
    Conditions that favor most adaptable pathogens and symbionts

a challenge to those species than if conditions had remained the same. In one manner of thinking, shifts in the distributions of indigenous insects in response to environmental changes will create large areas of what are essentially disturbed habitats that are often the most susceptible to invasive species because competition from native species is relatively less than in a habitat with strong ecological associations between fauna and flora.

10.2 Invasive mosquito species in Texas

Texas is home to three of the four most important invasive mosquito species, namely, *Ae. aegypti* (L.), *Ae. albopictus* (Skuse), and *Cx. quinquefasciatus*. The species not in Texas is *Cx. pipiens* Linnaeus, though northern populations of *Cx. quinquefasciatus* are undoubtedly heavily hybridized with *Cx. pipiens* (Barr, 1957; Farajollahi et al., 2011), and there is some evidence that these species started as one population in North America (Huang et al., 2011). Both *Cx. quinquefasciatus* and *Ae. aegypti* have been in North America for a long time, presumably brought from the Old World early during the colonization period (Ross, 1964).

Ae. albopictus was probably introduced multiple times but did not become established until the 1980s, first discovered in abundance in Houston (Sprenger and Whuithiranyagool, 1986; Moore and Mitchell, 1997) in 1985. It was almost certainly introduced in tires as part of the bulk trade in used tires. The robust trade in tires introduced multiple genetic types from Asia, as did the importation of an ornamental plant normally shipped in water. Its spread across the eastern half of the United States was rapid, and some of those populations made their way to Italy (Urbanelli et al., 2000), to Mexico probably directly from Texas in the 1990s (Bonizzoni et al., 2013), and elsewhere. Remarkably, completely new genetic types of the species emerged just 15 years in the United States in response to local selective forces (Medley, 2010). In the laboratory, mixed larval populations of *Ae. albopictus* and *Ae. triseriatus* (Say), a Texas native, increase the vector competence of the latter species for La Crosse virus by at least 50% (Bevins, 2008). The mechanism for this increase may be production of larger individuals as survival decreases and remaining larvae have access to more nutrition.

Ae. aegypti was abundant in southern Texas prior to the introduction of *Ae. albopictus,* and its range extended throughout the state, except perhaps for the extreme northwestern corner of the Panhandle (https://www.cdc.gov/zika/vector/range.html, accessed 19 January 2019). Competition, reproductive interference, and/or introduction of mosquito pathogens from *Ae. albopictus* reduced the abundance of *Ae. aegypti* in most locations (Tripet et al., 2011), but recently there are indications that the latter species is becoming more numerous (M. Debboun, personal communication). Currently, the range of *Ae. aegypti* is expanding within the United States (Pless et al., 2017), possibly including from Texas populations. There was a time when international public health entities attempted to eliminate this important species from the Western Hemisphere, successfully eliminating it from 15 Latin American countries (Soper, 1963) and Brazil (Soper et al., 1943).

Texas may be susceptible to many other mosquito introductions. Mosquitoes are most diverse in the tropics (Foley et al., 2007), and many tropical genera are poorly represented among invasive species. This may suggest that there is a bank of species poised to increase their distributions through either random opportunity or environmental change. The distribution of mosquito groups in Texas is more similar to the tropics than to temperate areas (global summary from Rueda (2008)). The ratio of the number of species in the tribe Aedini to Culicini is much greater in the Nearctic (2.74) than in the Neotropics (0.72). The ratio in Texas (0.82) is more similar to the tropical portion of the Western Hemisphere. On the other hand the ratio of species in subfamily Culicinae to Anophelinae in Texas (4.87 if the likely distribution of *An. maverlius* Reinert and *An. diluvialis* Reinert (Levine et al., 2004) and *An. crucians* Wiedemann is included) is different from other regions (world = 6.14, Nearctic = 7.48, and Neotropics = 7.55).

Worldwide, there are a number of species that appear to be expanding their distributions currently. Of these, *Ae. japonicus* (Theobald), *Ae. koreicus* (Edwards), and *Ae. atropalpus* seem more adapted to cooler climates. *Cx. coronator* is expanding its range, but it is indigenous to Texas (Connelly et al., 2016). Possibly more likely are *An. stephensi* Liston, a container-developing malaria vector from South Asia (Faulde et al., 2014), and *Ae. notoscriptus* (Skuse). The latter species expanded its range of habitats in Australia from only rock pools to include artificial containers (Russell, 1986), which may have facilitated its introduction and establishment in California (Peterson and Campbell, 2015). More broadly, many mosquito species live in nonhumid tropical or subtropical climates outside of the Western Hemisphere; however, few have shown a tendency to expand their distributions. In addition, many of these would not be a notable medical threat even if they did become established in Texas. The question is whether any of these species might change habitat preference to something commonly transported (like *Ae. notoscriptus*) or simply have the "good" fortune to be transported in sufficient genetic quantity (like *An. arabiensis* to Brazil).

Looking closer to home, there are some species in the border states of Mexico (Martinez Palacios, 1987) that do not occur in Texas. *An. albimanus* Wiedemann and *Haemagogus equinus* Theobald might qualify in this category, though they have been found occasionally north of the border. The other species are listed in Table 10.3. For potential medical

TABLE 10.3 Species of mosquitoes occurring in Mexican states bordering Texas, but not yet in Texas.

Species	Mexican states		
	Coahuila	Nuevo Léon	Tamaulipas
Aedes serratus	X		
Anopheles vestitipennis			X
Anopheles argyritarsis			X
Culex restrictor			X
Culex corniger			X
Culex educator			X
Culex inhibitator			X
Culex iolamdis			X
Culex reevesi	X	X	X
Culiseta impatiens	X		
Culiseta particeps	X	X	X
Psorophora pruinosa	X		
Psorophora stonei			X

Mexican species were extracted by Martinez Palacios (1987) and Ortega-Morales et al. (2015). Note that all mosquito species in Chihuahua also occur in Texas.

importance, *An. albimanus* and *An. vestitipennis* Dyar and Knab could be important as vectors of malaria and *Ae. serratus* as a significant floodwater pest. *Haemagogus equinus* is a significant vector of sylvatic yellow fever; however, the absence of monkeys in Texas would seem to make this an unlikely problem.

10.3 Some implications for action

A discussion of the actions that should be taken against invasive mosquito species would result in opinions varying between no action and a comprehensive national defensive force. Of course, we are much closer to the former than the latter currently, but some efforts are made. The publication of this book may be the occasion for such a discussion with respect to Texas, which would seem to be more susceptible than other states to new introductions and establishment.

The principles of insect pest management apply equally to management of invasive species. Intelligent pest control can be divided into four elements: risk assessment, surveillance, control, and sustainment. Risk assessment is the process of scholarship, experience, and thought that leads to a logical evaluation of the likelihood of events, leading to prioritization of effort and selection of what those efforts should be. It is the essential justification for basic research leading to understanding of species and ecology. Understanding the factors that will affect the ability of a given species to successfully establish populations and to predict the likely damage from that species is clearly important (Juliano and Philip Lounibos, 2005). Surveillance is detection and measurement of populations with the purposes of targeting control efforts, monitoring the effectiveness of those efforts, and archiving accurate information for historical comparisons. Control is, of course, about limiting the damage from populations of mosquitoes. Hard experience during the use of chemical insecticides for the last 80 years has shown that control activities must be carefully conducted for both personal and environmental safety, which usually coincide with the most economical use of those tools. Sustainment is the part of a program that maintains the gains against a pest by watchful monitoring and preparation for targeted response.

In some ways, Texas is better prepared than many states to detect and respond to new invasive mosquito species. It is home to knowledgeable experts on mosquitoes at both academic and governmental institutions who contribute to risk assessment. One of five federally funded institutions guided by the US Centers for Disease Control and Prevention for training and coordination of integrated vector management is in Texas, the Western Gulf Center of Excellence for Vector-Borne Diseases (https://www.utmb.edu/wgcvbd). Its network of cooperators will contribute greatly to the state's ability to detect and identify new infestations early. Major mosquito abatement efforts, like the one housed in Harris County Public Health, Houston, Texas, are a connection to professional expertise on large-scale mosquito control and provide local, intensive surveillance (e.g., Nava and Debboun, 2016). The US Navy, Army, and Air Force all have dedicated entomological efforts in Texas that connect the state to mobile, skilled mosquito control.

Possibly the weakest link in Texas' defense against new invasive mosquito species is in the area of surveillance. The experience with *Ae. albopictus* was an important example of how late detection of an infestation resulted in the near impossibility of preventing its spread. There is no systematic sampling program across the state, though some locations like Corpus Christi and Houston have programs. Central funding of such an effort seems unlikely, but there have been suggestions of how a combination of institutions and citizen involvement might create one (Crowl et al., 2008). Europe has initiated a system that seems to work well and that has summarized the relative importance of all the current invasive species (Medlock et al., 2015). Accurate and timely surveillance combined with good biological knowledge could target existing mosquito control assets to the correct areas at the correct times, assuming that resources and legal requirements were prepared in advance. To a large extent, getting such a system in place only means a small amount of reorganization and cooperation among existing resources.

10.4 Conclusion

Invasive species biology is a well-developed field, especially informed by experience with invasive species of plants. Many of the same genetic principles probably apply to mosquitoes, implying that an introduced species must have a minimum genetic pool for successful establishment. Following establishment, rapid evolutionary change is a possibility as a species adapts to its new environment in the absence of some of the natural competition and predation it experienced in its home range. The result can be a damaging species that threatens health and biodiversity.

Texas is probably particularly susceptible to invasion by mosquitoes. A number of species are present to the south that might easily cross the border if conditions allow. The Texas environment is also very variable both seasonally and spatially, creating a mosaic of habitats, any one of which might be suitable for an introduced species. Climate change and the resulting changes in human ecology will challenge native faunal and floral assemblages, making them more susceptible to invasion by nonnative species.

Texas has a good start on preparation for invasive mosquito species. The state has been able to take advantage of new federal emphasis on this problem, and there are centers of expertise. The main challenge is early detection through systematic surveillance systems, which are currently scattered. Reorganization and coordination of existing surveillance resources, possibly with the addition of public participation, could go a long way toward creating a network of detection that would detect invasive species before they have a chance to adapt to their new range.

References and further reading

Baker, M., 1961. The altitudinal distribution of mosquito larvae in the Colorado Front Range. Trans. Am. Entomol. Soc. 87 (4), 231–246.

Barr, A.B., 1957. The distribution of *Culex pipiens* and *C.p. quinquefasciatus* in North America. Am. J. Trop. Med. Hyg. 6 (1), 153–165.

Berlin, O.G.W., 1969. Mosquito studies (Diptera: Culicidae) XII. A revision of the neotropical subgenus *Howardina* of *Aedes*. Contrib. Am. Entomol. Inst. 4 (2), 1–190.

Bevins, S.N., 2008. Invasive mosquitoes, larval competition, and indirect effects on the vector competence of native mosquito species (Diptera: Culicidae). Biol. Invasions 10 (7), 1109–1117.

Bonizzoni, M., Gasperi, G., Chen, X., James, A.A., 2013. The invasive mosquito species *Aedes albopictus*: current knowledge and future perspectives. Trends Parasitol. 29 (9), 460–468.

Causey, O.R., Deane, L.M., Deane, M.P., 1943. Ecology of *Anopheles gambiae* in Brazil. Am. J. Trop. Med. 23 (1), 73–94.

Chen, C.C., Jenkins, E., Epp, T., Waldner, C., Curry, P.S., Soos, C., 2013. Climate change and West Nile virus in a highly endemic region of North America. Int. J. Environ. Res. Public Health 10 (7), 3052–3071.

Connelly, C.R., Alto, B.W., O'Meara, G.F., 2016. The spread of *Culex coronator* (Diptera: Culicidae) throughout Florida. J. Vector Ecol. 41 (1), 195–199.

Crowl, T.A., Crist, T.O., Parmenter, R.R., Belovsky, G., Lugo, A.E., 2008. The spread of invasive species and infectious disease as drivers of ecosystem change. Front. Ecol. Environ. 6 (5), 238–246.

De la Rocque, S., Rioux, J.A., Slingenbergh, J., 2008. Climate change: effects on animal disease systems, and implications for surveillance and control. Rev. Sci. Tech. Off. Int. Des Epiz. 27 (2), 340–354.

Dlugosch, K.M., Parker, I.M., 2008. Invading populations of an ornamental shrub show rapid life history evolution despite genetic bottlenecks. Ecol. Lett. 11 (7), 701–709.

Dukes, J.S., Mooney, H.A., 1999. Does global change increase the success of biological invaders? Trends Ecol. Evol. 14 (4), 135–139.

Engelthaler, D.M., Fink, T.M., Levy, C.E., Leslie, M.J., 1997. The reemergence of *Aedes aegypti* in Arizona. Emerg. Infect. Dis. 3 (2), 241–242.

Epstein, P.R., Diaz, H.F., Elias, S., Grabherr, G., Graham, N.E., Martens, W.J.M., Susskind, J., 1998. Biological and physical signs of climate change: focus on mosquito-borne diseases. Bull. Am. Meteorol. Soc. 79 (3), 409–417.

Farajollahi, A., Fonseca, D.M., Kramer, L.D., Kilpatrick, A.M., 2011. "Bird biting" mosquitoes and human disease: a review of the role of Culex pipiens complex mosquitoes in epidemiology. Infect. Genet. Evol. 11 (7), 1577–1585.

Faulde, M.K., Rueda, L.M., Khaireh, B.A., 2014. First record of the Asian malaria vector *Anopheles stephensi* and its possible role in the resurgence of malaria in Djibouti, Horn of Africa. Acta Trop. 139, 39–43.

Foley, D.H., Rueda, L.M., Wilkerson, R.C., 2007. Insight into global mosquito biogeography from country species records. J. Med. Entomol. 44 (4), 554–567.

Fonseca, D.M., Campbell, S., Crans, W.J., Mogi, M., Miyagi, I., Toma, T., … Wilkerson, R.C., 2001. *Aedes (Finlaya) japonicus* (Diptera: Culicidae), a newly recognized mosquito in the United States: analysis of genetic variation in the United States and putative source populations. J. Med. Entomol. 38 (2), 135–146.

Franks, S.J., Wheeler, G.S., Goodnight, C., 2011. Genetic variation and evolution of secondary compounds in native and introduced populations of the invasive plant *Melaleuca quinquenervia*. Evolution 66 (5), 1398–1412.

Gage, K.L., Burkot, T.R., Eisen, R.J., Hayes, E.B., 2008. Climate and vectorborne diseases. Am. J. Prev. Med. 35 (5), 436–450.

Gubler, D.J., Reiter, P., Ebi, K.L., Yap, W., Nasci, R., Patz, J.A., 2001. Climate variability and change in the United States: potential impacts on vector- and rodent-borne diseases. Environ. Health Perspect. 109 (Suppl. 2), 223–233.

Hoffmann, A.A., Sgrò, C.M., 2011. Climate change and evolutionary adaptation. Nature 470 (7335), 479–485.

Holderegger, R., Wagner, H.H., 2008. Landscape genetics. Bioscience 58 (3), 199–207.

Huang, S., Molaei, G., Andreadis, T.G., 2011. Reexamination of *Culex pipiens* hybridization zone in the eastern United States by ribosomal DNA-based single nucleotide polymorphism markers. Am. J. Trop. Med. Hyg. 85 (3), 434–441.

Juliano, S.A., Philip Lounibos, L., 2005. Ecology of invasive mosquitoes: effects on resident species and on human health. Ecol. Lett. 8 (5), 558–574.

Kelly, R., Mead, D., Harrison, B.A., 2008. Discovery of *Culex coronator* Dyar and Knab (Diptera: Culicidae) in Georgia. Proc. Entomol. Soc. Wash. 110 (1), 258–260.

Keyghobadi, N., La Pointe, D., Fleischer, R.C., Fonseca, D.M., 2006. Fine-scale population genetic structure of a wildlife disease vector: the southern house mosquito on the island of Hawaii. Mol. Ecol. 15 (13), 3919–3930.

Kiszewski, A.E., Spielman, A., 1998. Spatially explicit model of transposon-based genetic drive mechanisms for displacing fluctuating populations of anopheline vector mosquitoes. J. Med. Entomol. 35 (4), 584–590.

Lafferty, K.D., 2009. The ecology of climate change and infectious diseases. Ecology 90 (4), 888–890.

Levine, R.S., Peterson, A.T., Benedict, M.Q., 2004. Geographic and ecologic distributions of the *Anopheles gambiae* complex predicted using a genetic algorithm. Am. J. Trop. Med. Hyg. 70 (2), 105–109.

Lounibos, L.P., 2002. Invasions by insect vectors of human disease. Annu. Rev. Entomol. 47 (1), 233–266.

Lowe, S., Browne, M., Boudjelas, S., DePoorter, M., 2004. 100 of the World's Worst Invasive Alien Species: A Selection From the Global Invasive Species Database. The Invasive Species Specialist Group, Species Survival Commission, World Conservation Union. Retrieved from: www.k-state.edu/withlab/consbiol/IUCN_invaders.pdf.

Lozano-Fuentes, S., Welsh-Rodriguez, C., Hayden, M.H., Tapia-Santos, B., Ochoa-Martinez, C., Kobylinski, K.C., … Eisen, L., 2012. *Aedes (Ochlerotatus) epactius* along an elevation and climate gradient in Veracruz and Puebla States, México. J. Med. Entomol. 49 (6), 1244–1253.

Mack, R., Smith, M., 2011. Invasive plants as catalysts for the spread of human parasites. NeoBiota 9, 13–29.

Martinez Palacios, F.C., 1987. Los Mosquitoes de Mexico (Diptera: Culicidae) Taxonomia, Distribucion Geographica y Su Importancia en Salud Publica. Thesis, Universidad Nacional Autonoma de Mexico, Facultad de Ciencias, Biologia.

Medley, K.A., 2010. Niche shifts during the global invasion of the Asian tiger mosquito, *Aedes albopictus* Skuse (Culicidae), revealed by reciprocal distribution models. Glob. Ecol. Biogeogr. 19 (1), 122–133.

Medlock, J.M., Hansford, K.M., Schaffner, F., Versteirt, V., Hendrickx, G., Zeller, H., Van Bortel, W., 2012. A review of the invasive mosquitoes in Europe: ecology, public health risks, and control options. Vector Borne Zoonotic Dis. 12 (6), 1–13.

Medlock, J.M., Hansford, K.M., Versteirt, V., Cull, B., Kampen, H., Fontenille, D., … Schaffner, F., 2015. An entomological review of invasive mosquitoes in Europe. Bull. Entomol. Res. 105 (6), 637–663.

Mooney, H.A., Cleland, E.E., 2001. The evolutionary impact of invasive species. Proc. Natl. Acad. Sci. 98 (10), 5446–5451.

Moore, C.G., Mitchell, C.J., 1997. *Aedes albopictus* in the United States. Ten-year presence and public health implications. Emerg. Infect. Dis. 3 (3), 329–334.

Mouchet, J., Giacomini, T., Julvez, J., 1995. La diffusion anthropique des arthropods vecteurs de maladie dans le monde. Cahiers Santé 5 (5), 293–298.

Mulligan, S., 2013. *Aedes aegypti* (Yellow Fever) Mosquito Found in Fresno County. County of Fresno, Department of Public Health.

Nava, M.R., Debboun, M., 2016. A taxonomic checklist of the mosquitoes of Harris County, Texas. J. Vector Ecol. 41 (1), 190–194.

Ortega-Morales, A.I., Zavortink, T.J., Huerta-Jiménez, H., Sánchez-Rámos, F.J., Valdés-Perezgasga, M.T., Reyes-Villanueva, F., … Fernandez-Salas, I., 2015. Mosquito records from Mexico: the mosquitoes (Diptera: Culicidae) of Tamaulipas State. J. Med. Entomol. 52 (2), 171–184.

Parmesan, C., 2006. Ecological and evolutionary responses to recent climate change. Annu. Rev. Ecol. Evol. Syst. 37, 637–669.

Peterson, A.T., Campbell, L.P., 2015. Global potential distribution of the mosquito *Aedes notoscriptus*, a new alien species in the United States. J. Vector Ecol. 40 (1), 191–194.

Peterson, A.T., Lieberman, B.S., 2012. Species' geographic distributions through time: playing catch-up with changing climates. Evol. Educ. Outreach 5 (4), 569–581.

Pless, E., Gloria-Soria, A., Evans, B.R., Kramer, V., Bolling, B.G., Tabachnick, W.J., Powell, J.R., 2017. Multiple introductions of the dengue vector, *Aedes aegypti* into California. PLoS Negl. Trop. Dis. 11 (8), e0005718.

Portier, C.J., Thigpen, T.K., Carter, S.R., Dilworth, C.H., Grambsch, A.E., Gohlke, J., … Whung, P.Y., 2010. A Human Health Perspective on Climate Change: A Report Outlining the Research Needs on the Human Health Effects of Climate Change. Environmental Health Perspectives. Retrieved from: http://www.niehs.nih.gov/health/materials/a_human_health_perspective_on_climate_change_full_report_508.pdf.

Reiter, M.E., LaPointe, D.A., 2007. Landscape factors influencing the spatial distribution and abundance of mosquito vector *Culex quinquefasciatus* (Diptera: Culicidae) in a mixed residential-agricultural community in Hawaii. J. Med. Entomol. 44 (5), 861–868.

Ross, H.H., 1964. The colonization of temperate North America by mosquitoes and man. Mosquito News 24 (2), 103–118.

Rueda, L.M., 2008. Global diversity of mosquitoes (Insecta: Diptera: Culicidae) in freshwater. Hydrobiologia 595 (1), 477–487.

Russell, R.C., 1986. Larval competition between the introduced vector of dengue fever in Australia, *Aedes aegypti* (L.), and a native container-breeding mosquito, *Aedes notoscriptus* (Skuse) (Diptera: Culicidae). Aust. J. Zool. 34 (4), 527–534.

Schierenbeck, K.A., Ellstrand, N.C., 2009. Hybridization and the evolution of invasiveness in plants and other organisms. Biol. Invasions 11 (5), 1093–1105.

Scholte, E.J., Den Hartog, W., Braks, M., Reusken, C., Dik, M., Hessels, A., 2009. First report of a North American invasive mosquito species *Ochlerotatus atropalpus* (Coquillett) in the Netherlands, 2009. Euro Surveill. 14 (45), 429–433.

Service, M.W., 1993. Mosquitoes (Culicidae). In: Lave, R.P., Crosskey, R.W. (Eds.), Medical Insects and Arachnids. Chapman and Hall, New York, pp. 120–240.

Service, M.W., 1997. Mosquito (Diptera: Culicidae) dispersal—the long and short of it. J. Med. Entomol. 34 (6), 579–588.

Soper, F.L., 1963. The elimination of urban yellow fever in the Americas through the eradication of *Aedes aegypti*. Am. J. Public Health Nations Health 53 (1), 7–16.

Soper, F.L., 1966. Paris green in the eradication of *Anopheles gambiae*: Brazil, 1940; Egypt, 1945. Mosquito News 26 (4), 470–476.

Soper, F.L., Wilson, D.B., 1943. *Anopheles gambiae* in Brazil, 1930–1940. The Rockefeller Foundation, New York.

Soper, F.L., Wilson, D.B., Lima, S., Antunes, W.S., 1943. The Organization of Permanent Nation-Wide Anti-*Aedes aegypti* Measures in Brazil. The Rockefeller Foundation, New York.

Sprenger, D., Whuithiranyagool, T., 1986. The discovery and distribution of *Aedes albopictus* in Harris County, Texas. J. Am. Mosq. Control Assoc. 2 (2), 217–219.

Stone, C.M., Witt, A.B.R., Walsh, G.C., Foster, W.A., Murphy, S.T., 2018. Would the control of invasive alien plants reduce malaria transmission? A review. Parasit. Vectors 11 (1), 76. https://doi.org/10.1186/s13071-018-2644-8.

Strickman, D., 1988. Rate of oviposition by *Culex quinquefasciatus* in San Antonio, Texas, during three years. J. Am. Mosquito Control Assoc. 4, 339–344.

Strickman, D., 1989. *Culex pseudostigmatosoma, Cx. yojoae*, and *Cx. aquarius*: New Central American species in the subgenus *Culex* (Diptera: Culicidae). Mosq. Syst. 21 (3), 143–177.

Strickman, D., Darsie Jr., R.F., 1988. The previously undetected presence of *Culex restuans* (Diptera: Culicidae) in Central America, with notes on identification. Mosq. Syst. 20 (1), 21–27.

Thielman, A., Hunter, F.F., 2006. Establishment of *Ochlerotatus japonicus* (Diptera: Culicidae) in Ontario, Canada. J. Med. Entomol. 43 (2), 138–142.

Tripet, F., Lounibos, L.P., Robbins, D., Moran, J., Nishimura, N., Blosser, E.M., 2011. Competitive reduction by satyrization? Evidence for interspecific mating in nature and asymmetric reproductive competition between invasive mosquito vectors. Am. J. Trop. Med. Hyg. 85 (2), 265–270.

Urbanelli, S., Bellini, R., Carrieri, M., Sallicandro, P., Celli, G., 2000. Population structure of *Aedes albopictus* (Skuse): the mosquito which is colonizing Mediterranean countries. Heredity 84 (3), 331–337.

Walker, K.R., Joy, T.K., Ellers-Kirk, C., Ramberg, F.B., 2011. Human and environmental factors affecting *Aedes aegypti* distribution in an arid urban environment. J. Am. Mosq. Control Assoc. 27 (2), 135–141.

Wilkerson, R.C., Strickman, D., 1990. Illustrated key to the female anopheline mosquitoes of Central America and Mexico. J. Am. Mosq. Control Assoc. 6 (1), 7–34.

Wimberly, M.C., Hildreth, M.B., Boyte, S.P., Lindquist, E., Kightlinger, L., 2008. Ecological niche of the 2003 West Nile virus epidemic in the northern Great Plains of the United States. PLoS One 3 (12), e3744. https://doi.org/10.1371/journal.pone.0003744.

Xu, C.Y., Julien, M.H., Faterni, M., Girod, C., Van Klinken, R.D., Gross, C.L., Novak, S.J., 2010. Phenotypic divergence during the invasion of *Phyla canescens* in Australia and France: evidence for selection-driven evolution. Ecol. Lett. 13 (1), 32–44.

Zhong, D., Lo, E., Hu, R., Metzger, M.E., Cummings, R., Bonizzoni, M., Yan, G., 2013. Genetic analysis of invasive *Aedes albopictus* populations in Los Angeles County, California and its potential health impact. PLoS One 8 (7), e68586. https://doi.org/10.1371/journal.pone.0068586.

Part III

Public Health

Chapter 11

Mosquito-Borne Diseases

Dagne Duguma[a], Leopoldo M. Rueda[b], Mustapha Debboun[a]

[a]*Mosquito and Vector Control Division, Harris County Public Health, Houston, TX, United States,* [b]*Walter Reed Biosystematics Unit, Department of Entomology, Smithsonian Institution, Suitland, MD, United States*

Chapter Outline

11.1 Introduction

Mosquito-borne diseases debilitate humans and other animals worldwide with most of the casualties occurring in tropical and subtropical regions. For example, malaria alone infected 219 million humans and killed 435,000 people in 2017 with the vast majority (>90%) of malaria-associated human deaths occurring in sub-Saharan African countries causing significant economic and social burden (WHO, 2017).

The mosquito was one of the first known arthropods to transmit arthropod-borne pathogens causing arthropod-borne human diseases. Human lymphatic filariasis, caused by a filarial nematode, *Wuchereria bancrofti* (Cobbold), is an important arthropod-borne disease that was first identified to be transmitted by the female mosquito, *Culex quinquefasciatus* Say by Patrick Manson in 1877 (Gubler, 2009). Manson documented the development of *W. bancrofti* in mosquitoes, which gave rise to many discoveries incriminating mosquitoes and other arthropods as vectors of human and animal diseases during the subsequent decades. While historically some of the mosquito-borne diseases such as yellow fever (YF) and malaria significantly affected the United States (US), mainly the southern states, including Texas, these diseases have been largely managed over the past century with a combination of approaches, including improved socioeconomics, public health, and vector control services (Moreno-Madriñán and Turell, 2018). However, there continues to be a serious concern for emergence of new pathogens (Zika, Mayaro fever (MF), chikungunya (CHIK), etc.) and resurgence of the already established mosquito-borne diseases due to changes in several factors such as environmental, climatic, and demographic changes (Gubler, 1998; Morse, 2001; Hotez, 2018). For example, *Aedes albopictus* (Skuse), which is indigenous to Asia and is one of the vectors of chikungunya, dengue, and Zika viruses, was discovered breeding in discarded used tires in Houston, Texas, in 1985 (Sprenger and Wuithiranyagool, 1986).

The concern of acquiring new pathogens from various parts of the world is greater today than 60 years ago. For example, Texas has the top three megacities in the United States that have strong global links to other metropolitan cities worldwide, including the disease-endemic countries. This global connection enhances an increased risk of acquiring new emerging mosquito-borne diseases. For example, in the past 5 years, the two more recent emerging mosquito-borne diseases, chikungunya and Zika, which caused epidemics in South America, have been introduced to Texas (Hotez, 2016).

Mosquitoes, Communities, and Public Health in Texas. https://doi.org/10.1016/B978-0-12-814545-6.00011-0

Mosquitoes transmit many viral, protozoan, and filarial nematode pathogens causing human and animal diseases. The importance and status of mosquito-borne diseases present or relevant to the state of Texas are briefly discussed in this chapter.

11.2 Mosquito-borne viral diseases

Mosquito-borne viral diseases of serious concern in Texas include West Nile fever (WNF), Saint Louis encephalitis (SLE), California encephalitis (CE), Eastern equine encephalitis (EEE), Western equine encephalitis (WEE), dengue (DEN), chikungunya (CHIK), and Zika (ZIK) (http://www.dshs.texas.gov). In addition to YF, which historically impacted residents of Texas, there are many other mosquito-borne viruses including Barmah Forest (BAM), Batai (BAT), Babanki (BBK), Bouboui (BOU), Bunyamwera (BUN), chikungunya (CHIK), Cache Valley (CV), dengue fever (DENF), Edge Hill (EH), Everglades (EYE), Getah (GETV), Gan Gan (GG), Highland J (HJ), Ilheus (ILH), Jamestown Canyon (JC), Japanese encephalitis (JE), Kedougou (KED), La Crosse (LAC), Lebombo (LEB), Murray Valley encephalitis (MVE), Nyando (NDO), Ngari (NRI), Oropouche (ORO), Oriboca (ORI), Orungo (ORU), Pongola (PGA), O'nyong-nyong (ONN), Ross River (RR), Rocio (ROC), Rift Valley fever (RVF), Semliki Forest (SF), Sindbis (SIN), Spondweni (SPO), Tahyna (TAH), Tensaw (TEN), Trivittatus (TVT), Uganda S (UGS), Venezuelan equine encephalitis (VEE), Wesselsbron (WSL), and Wyeomyia (WYO) that can be potentially transmitted by mosquitoes in Texas. Table 11.1 shows the list of mosquito species from Texas, with associated human pathogens or mosquito-borne diseases as reported elsewhere.

TABLE 11.1 List of mosquito species from Texas, with associated human pathogens or diseases as reported elsewhere.

Species	Pathogen/disease[a] (references)
Aedes (Aedimorphus) vexans (Meigen)	WNV (Fontenille et al., 1998); Potosi (Wozniak et al., 2001); SLE, WEE (Turell et al., 2005a,b); Chaoyang (Wang et al., 2009); Banna (Liu et al., 2010); EEE, TAHV, TVTV (CDC, 2015)
Aedes (Georgecraigius) epactius Dyar and Knab	WNV (CDC, 2012)
Aedes (Ochlerotatus) canadensis (Theobald)	WEE (Huang, 1972); RVFV (Gargan et al., 1988); EEEV (Fontenille et al., 1998); WNV (Foster and Walker, 2009); HJV (Armstrong and Andreadis, 2010); JCV (CDC, 2012)
Aedes (Ochlerotatus) dorsalis (Meigen)	WEE (Carpenter and LaCasse, 1955); Liaoning (Tao et al., 2003); Banna (Liu et al., 2010); WNV (CDC, 2012)
Aedes (Ochlerotatus) fulvus (Wiedemann)	EEEV, WNV, WYOV (CDC, 2015)
Aedes (Ochlerotatus) grossbecki Dyar and Knab	WNV (CDC, 2012)
Aedes (Ochlerotatus) infirmatus Dyar and Knab	EEEV, Keystone, Tensaw (Wellings et al., 1972); TVTV, WNV (CDC, 2012, 2015)
Aedes (Ochlerotatus) mitchellae (Dyar)	EEEV, TENV (CDC, 2015)
Aedes (Ochlerotatus) nigromaculis (Ludlow)	WNV (CDC, 2012)
Aedes (Ochlerotatus) scapularis (Rondani)	WB, YFV (Arnell 1976); ILHV (Pauvolid-Corrêa et al., 2013); SLEV, VEEV, WYOV (CDC, 2015)
Aedes (Ochlerotatus) sollicitans (Walker)	RVFV (Gargan et al., 1988); CVV (Andreadis et al., 2014); EEEV, VEEV, WNV (CDC, 2012, 2015)
Aedes (Ochlerotatus) sticticus (Meigen)	EEEV, JCV, TAHV, WNV (CDC, 2012, 2015)
Aedes (Ochlerotatus) taeniorhynchus (Wiedemann)	RVFV (Armstrong and Andreadis, 2010); CVV, EEEV, EYEV, ORIV, TENV, VEEV, WYOV, WNV (CDC, 2012, 2015)
Aedes (Ochlerotatus) trivittatus (Coquillett)	EEEV (Armstrong and Andreadis, 2010); LACV, TVTV, WNV (CDC, 2012)
Aedes (Ochlerotatus) vigilax (Skuse)	CVV, POTV (Mitchell et al., 1998); JCV, LACV, RVFV (Gargan et al., 1988); EEEV (Armstrong and Andreadis, 2010); WNV, BAMV, EHV, GGV, KOKV, RRV, SINV (CDC, 2012, 2015)

TABLE 11.1 List of mosquito species from Texas, with associated human pathogens or diseases as reported elsewhere.—cont'd

Species	Pathogen/disease[a] (references)
Aedes (Protomacleaya) triseriatus (Say)	
Aedes (Stegomyia) aegypti (Linnaeus)	CHIK, DENV, ORIIV, VEEV, WNV, YFV (Christophers, 1960); ZIKV (Ali et al., 2017)
Aedes (Stegomyia) albopictus (Skuse)	JE, WEEV (Huang, 1972); EEEV (Niebylski et al., 1992); LACV; CVV (Mitchell et al., 1998); CHIK (Pagès et al., 2009); DENV, WNV, YFV (CDC, 2015)
Anopheles (Anopheles) freeborni Aitken	MAL (Carpenter and LaCasse, 1955)
Anopheles (Anopheles) pseudopunctipennis Theobald	MAL (Horsfall, 1955)
Anopheles (Anopheles) punctipennis (Say)	MAL (Foster and Walker, 2009)
Anopheles (Anopheles) quadrimaculatus Say	MAL (Foster and Walker, 2009)
Anopheles (Nyssorhynchus) albimanus (Wiedemann)	MAL (Foster and Walker, 2009)
Coquillettidia (Coquillettidia) perturbans (Walker)	EEEV (Turell et al., 2005a)
Culex (Culex) nigripalpus (Theobald)	EEEV, SLE, WNV (Turell et al., 2005a)
Culex (Culex) quinquefasciatus (Say)	WB, WEEV, SLEV (Carpenter and LaCasse, 1955; Turell et al., 2005a)
Culex (Culex) restuans Theobald	SLEV, WNV (Turell et al., 2005a)
Culex (Culex) tarsalis Coquillett	SLEV, WEEV (Turell et al., 2005a); WNV (Hayes et al., 2005)
Mansonia (Mansonia) titillans (Walker)	VEEV (Gilyard, 1944)
Psorophora (Grabhamia) columbiae (Dyar and Knab)	Potosi (Wozniak et al., 2001); WNV (CDC, 2012)
Psorophora (Grabhamia) discolor (Coquillett)	VEEV (CDC, 2015)
Psorophora (Janthinsoma) cyanescens (Coquillett)	VEEV (CDC, 2015)
Psorophora (Janthinsoma) ferox (von Humboldt)	EEEV (Armstrong and Andreadis, 2010); WNV (CDC, 2012); CVV (Andreadis et al., 2014); ORIV, SLEV, WYOV (CDC, 2015)
Psorophora (Psorophora) ciliata (Fabricius)	TENV (Wozniak et al., 2001); WNV (CDC, 2012);VEEV (CDC, 2015)
Psorophora (Psorophora) howardii Coquillett	WNV (CDC, 2012)

[a]Pathogen/disease names and abbreviations: BAMV, Barmah Forest virus; BATV, Batai virus; BBKV, Babanki virus; BOUV, Bouboui virus; BUNV, Bunyamwera virus; CHIKV, chikungunya virus; CVV, Cache Valley virus; DENV, dengue virus; EEEV, Eastern equine encephalitis virus; EHV, Edge hill virus; EYEV, Everglades virus; GETV, Getah virus; GGV, Gan Gan virus; HJV, Highland J virus; ILHV, Ilheus virus; JCV, Jamestown Canyon virus; JE, Japanese encephalitis; KEDV, Kedougou virus; LACV, La Crosse virus; LEBV, Lebombo virus; MAL, Malaria; MVEV, Murray Valley encephalitis virus; NDOV, Nyando virus; NRIV, Ngari virus; ORIV, Oriboca virus; ORUV, Orungo virus; PGAV, Pongola virus; POTV, Potosi virus; RRV, Ross River virus; RVFV, Rift Valley fever virus; SFV, Semliki Forest virus; SINV, Sindbis virus; SLEV, St. Louis encephalitis virus; SPOV, Spondweni virus; TAHV, Tahyna virus; TENV, Tensaw virus; TVTV, Trivittatus virus; UGSV, Uganda S virus; VEEV, Venezuelan virus Equine Encephalitis; WEEV, Western equine encephalitis virus; WNV, West Nile virus; WB, Wuchereria bancrofti; WSLV, Wesselsbron virus; WYOV, Wyeomyia virus; YFV, yellow fever virus; ZIKV, Zika virus.

11.2.1 Yellow fever

Cause, transmission, morbidity, and geographic distribution

YF is a viral hemorrhagic fever caused by the *Alphavirus* of family Flaviviridae and is transmitted by *Ae. aegypti* (Linnaeus). YF exhibits three transmission cycles: sylvatic (forest type), intermediate, and urban (Barrett and Higgs, 2007). Primates are the main definitive hosts in the sylvatic cycle, whereas humans are the main definitive hosts in the urban cycle. The disease is maintained by both primates and humans in the urban and sylvatic interface (intermediate cycle). About 200,000 YF cases occur annually, mainly in Africa and South America, and about 30,000 human deaths per year worldwide with over 90% of deaths occurring in Africa (Barrett and Higgs, 2007). It is endemic to tropical regions of Africa and was introduced to the Americas in the 1600s (Tabachnick and Powell, 1979). YF devastated the southern United States throughout the 18th and 19th centuries (Hotez, 2016;Moreno-Madriñán and Turell, 2018). During the Spanish American War in 1889,

YF inflicted more casualties to the US soldiers than the war itself. Carlos Finlay in 1881 speculated that mosquitoes were culpable for transmitting YF. However, it was in 1900 that the US Army Yellow Fever Commission was led by Major Walter Reed and that the transmission of YF by the mosquito, *Ae. aegypti*, was experimentally demonstrated in Cuba. Thousands of Texans suffered and died of YF in Galveston and San Antonio, Texas, during the YF epidemics between 1839 and 1867 (https://tshaonline.org/handbook/online/articles/sme01, Moreno-Madriñán and Turell, 2018). While a local epidemic of such magnitude has not occurred in Texas in the last century, its risk of introduction from travel cases have and still exist due to increased globalization and trade with YF endemic countries. No YF travel cases have been reported in Texas from 2004 to 2017 (TDSHS, 2018). Risk factors, including widespread presence of the principal YF vector, *Ae. aegypti* in the majority of Texas counties (see Chapter 2 for its distribution), and increased travel from and to endemic regions are still a major concern about the reintroduction of YF in Texas.

Symptoms and diagnosis

YF shows acute symptoms including fever, chills, vomiting, nausea, muscle pain, and headache. Diagnosis during early stages of infection is difficult, but polymerase chain reaction (PCR) can be used to detect the virus in blood and urine samples (WHO, 2009). Testing of serum for virus-specific level IgM-ELISA and IgG-ELISA and neutralizing antibodies are used to confirm its infection at later stages (WHO, 2009).

Treatment and prevention

There is no antiviral drug therapy or treatment available for YF to date. However, YF vaccines, including YF-Vax (Sanofi Pasteur vaccine) in the United States, 17DD in Brazil, and 17DD-204 in many other parts of the world, have been effectively used to immunize people against YF (Barrett and Higgs, 2007).

Prevention that includes the use of insect repellents and proper clothing to prevent mosquitoes from biting, breeding habitat source reduction, and aerial spraying for adults has been shown to manage its epidemics, especially in the Americas. Emerging novel technologies such as sterile insect techniques using genetically modified mosquitoes and *Wolbachia*-based sterilization techniques are undergoing field trials that are hoped to be widely implemented in years to come to reduce the mosquito-borne disease burden worldwide.

11.2.2 Dengue fever (DF) and dengue hemorrhagic fever (DHF)

Dengue ("O'nyong-nyong fever," "dengue-like disease," and "breakbone fever") is caused by RNA virus belonging to the genus *Flavivirus* in the family Flaviviridae. There are four serotypes of dengue virus (DENV1, DENV2, DENV3, and DENV4) with each causing dengue infection. It is common in Asia, Eastern and Western Africa, Polynesia, Micronesia, Caribbean region, and Central and Northern South America. They cause DF and the more severe DHF. Dengue is one of the few mosquito-borne diseases known to be transmitted only to humans by *Aedes* mosquitoes mainly, *Ae. aegypti* and *Ae. albopictus*. About 3.9 billion people in more than 125 countries are at risk of dengue infection, about 500,000 requiring hospitalization due to dengue, and about 20,000 deaths occur due to severe dengue annually (WHO, 2018).

Dengue was once a major mosquito-borne disease in Texas. For example, dengue outbreak occurred in 1922 with more than 500,000 cases, and it remained endemic to some parts of Texas with occasional outbreaks (Rawlings et al., 1998). The autochthonous (local) dengue transmission was reported in Houston, Texas, between 2003 and 2005 (Murray et al., 2013) and in Brownsville, Texas, in 2005 (Ramos et al., 2008). According to the Texas Department of State Health Services (TDSHS, 2018), 232 dengue cases were reported in Texas in 2017 of which 66% were associated with travel, 23% locally acquired, and 1% from unknown sources (Fig. 11.1). The majority of reported dengue cases in Texas were from travel to and from dengue disease–endemic countries that included Mexico (43%) and India (35%), and no locally transmitted dengue cases were reported in 2017 (TDSHS, 2018). Between 1980 and 2017 a total of 634 dengue cases were reported in Texas with majority of the cases associated with travel and with at least one case every year since 1991 (Fig. 11.1.). Texas is one of the three contiguous states with Mexico where dengue outbreak is common, and the widespread presence of primary vectors, *Ae. aegypti* and *Ae. albopictus*, makes it vulnerable for local dengue transmission (Reiter et al., 2003).

Symptoms and diagnosis

Dengue causes fever chills, cephalalgia, retro-ocular pain, photophobia, muscle and joint pains, nausea, vomiting, and a sore throat. It also causes acute and benign febrile diseases. Diagnosis of dengue involves the detection of the virus, DNA virus, antigens, antibodies, or a combination of these techniques (WHO, 2009). Similar to YF diagnosis, dengue nucleic acid or antigen detection can be used during its early stage, whereas serology is considered the best diagnostic method

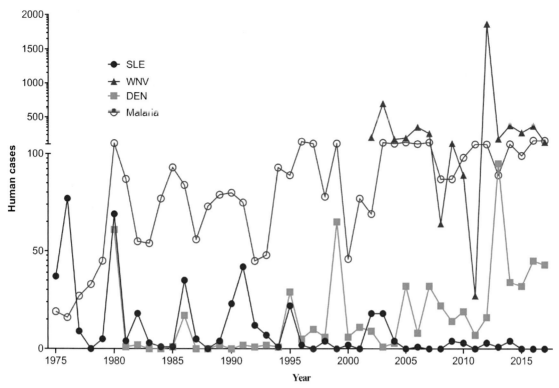

FIG. 11.1 Human cases of Saint Louis encephalitis, West Nile virus, dengue, and malaria in Texas. Only available statewide data obtained from Texas Department of State Health Services is included. The majority of dengue and malaria cases are travel associated. West Nile virus was first reported in Texas in 2002. *(Source: Zoonosis Control Branch, Texas Department of State Health Services.)*

at a later stage (WHO, 2009). The serum samples can also be tested using antidengue immunoglobulin M (IgM) by IgM antibody-capture enzyme-linked immunosorbent assay (MAC-ELISA) and antidengue IgG by IgG-ELISA and mixed antigens (Chungue et al., 1989; Innis et al., 1989).

Treatment and prevention

Currently, there is no effective antiviral drug therapy or approved vaccine available for dengue. However, there have been several dengue vaccines undergoing trials in South America and Asia (Whitehead, 2016). The live attenuated recombinant tetravalent dengue vaccine CYD-TDV (Dengvaxia) developed and licensed in 20 countries by Sanofi Pasteur uses the attenuated YF virus 17D vaccine strain as the replication backbone (Wilder-Smith et al., 2018). Because the virus does not provide full protection, its use is currently limited to those already infected by dengue. Another promising tetravalent dengue vaccine (TetraVax-DV) TV005, developed by National Institute of Allergy and Infectious Diseases, targets both serotypes of DENV, and Phase I trial has been completed (https://clinicaltrials.gov/). Vector control, which includes the use of personal protective measures (insect repellents and proper clothing), breeding habitat source reduction, education, and aerial spraying for adults, has been the primary methods of dengue prevention. The emerging vector control technologies such as sterile insect techniques using genetically modified mosquitoes and *Wolbachia*-based biological control are promising strategies that are undergoing field trials to minimize the dengue disease burden.

11.2.3 Chikungunya fever

Chikungunya ("to walk bent over") is a tropical disease caused by the RNA virus (CHIKV) of the genus *Alphavirus* in the family Togaviridae. It widely occurs in sub-Saharan Africa, Southeast Asia, India, and the Philippines. It was first discovered in 1952 in Tanzania and transmitted by arboreal *Aedes* mosquitoes (*Ae. africanus* Theobald and *Ae. furcifer* group species) in rural areas. *Ae. aegypti* and *Ae. albopictus* are associated with outbreak of CHIKV in urban areas. It is a zoonotic disease occurring commonly in primates such as green monkeys, baboons, and chimpanzees in Africa (Acha and Szyfres, 1987; Weaver, 2014). While it is usually an endemic rural disease, CHIK has occasional epidemic cycles in urban centers with more susceptible human populations. The disease first arrived in the Americas in 2013 to the Caribbean

TABLE 11.2 Chikungunya and Zika cases in Texas from 2004 to 2018.

Disease	2014	2015	2016	2017	2018[a]
Chikungunya	114	55	20	15	7
Zika	NR	8	315	55	3

NR, not reportable.
[a]Cases reported to the TDSHS but have not yet been verified.
Source: TexasZika.org.

islands. In Texas, 211 human cases between 2014 and 2018 were reported with all, but one was acquired from traveling abroad (Table 11.2). There was one case suspected of autochthonous transmission with a person having no history of travel to chikungunya-endemic countries, and risk factors conducive for an outbreak of CHIK are present in Texas, especially in the southern border where *Aedes* vectors and poor sanitation conditions are prevalent (Hotez, 2016).

Symptoms and diagnosis

Chikungunya fever (CHIKF) causes similar symptoms as dengue, including fever, rash, headache, muscle pain, joint pain or joint swelling, and other diseases such as Ross River, Mayaro, Sindbis, and O'nyong-nyong fevers (Tesh, 1982). Diagnosis of CHIKF involves detection of the virus, virus nucleic acid, antigens, antibodies, or a combination of these techniques (WHO, 2009). Similar to YF virus diagnosis, nucleic acid or antigen detection of CHIKV can be used for its early stage, whereas serology is the best method to diagnose it at a later stage (WHO, 2009).

Treatment and prevention

There is no vaccine or treatment against chikungunya infection. Prevention methods are similar to other *Aedes*-borne viruses, as indicated in Sections 11.2.1 and 11.2.2.

11.2.4 Zika (ZIK)

The Zika virus (ZIKV) of the *Flavivirus* genus (family Flaviviridae) was first discovered in Uganda in 1947 from monkeys and was later identified from humans in Uganda and Tanzania in 1952. The ZIKV was also isolated from *Ae. africanus* collected from Zika forest in Uganda (https://www.who.int/emergencies/zika-virus/timeline/en/). It causes a neurological condition called microcephaly, which is associated with smaller heads in infants than the average head size of normal children of the same age and sex. The ZIKV is primarily transmitted by *Ae. aegypti* mosquito but has been reported to be transmitted through unprotected sex (without proper use of condoms) and other means including from mother to babies and blood transfusion (Musso et al., 2014; Ali et al., 2017). Zika outbreak in the United States was primarily restricted to two states, that is, Florida and Texas. In 2015, TDSHS reported eight Zika cases (seven from Harris and one from Fort Bend counties, Texas) for the first time. The first local mosquito-borne transmission occurred in 2016 in Brownsville, TX, located along the US-Mexico border (Hall et al., 2017). A total of 315 Zika cases were reported in 2016, 55 in 2017, and 3 in 2018 (Table 11.2). A total of 381 Zika cases were reported in Texas between 2015 and 2018 of which 11 cases were locally transmitted by *Ae. aegypti* mosquitoes in Cameron and Hidalgo counties in Texas. Four Zika cases were transmitted from mother to child before birth and two via sexual transmission (TexasZika.org). Zika remains to be a serious disease of public health importance because of the presence of conducive environments for both its confirmed and suspected mosquito vectors in Texas. A study showed that a 1% Zika attack rate could cost the public more than 1 billion dollars in associated economic cost (direct medical cost and productivity losses) in Texas and five other southern states including, Florida, Louisiana, Georgia, Mississippi, and Alabama (Lee et al., 2017).

Symptoms and diagnosis

The majority of people (about 80%) who have ZIKV do not develop symptoms, but some develop symptoms of fever, rash, conjunctivitis, muscle and joint pain, malaise, and headache that last 2–7 days. Zika has also been associated with Guillain-Barré syndrome, a neurological disorder in which the part of peripheral nervous system is mistakenly attacked by its body's immune system (NIH, 2018). The methods used to test ZIKV are the RNA nucleic acid test (NAT) on paired serum and urine sample, Triplex real-time RT-PCR assay, serological test using Zika virus–specific IgM and neutralizing antibodies, and Zika MAC-ELISA techniques.

Treatment and prevention

There is no vaccine or treatment against Zika infection. Prevention methods are similar to other *Aedes*-borne viruses, as indicated in Sections 11.2.1 and 11.2.2. Since Zika has been known to be sexually transmitted, it is highly recommended to properly use condoms (Polen et al., 2018).

11.2.5 West Nile fever

The WNF is caused by West Nile virus (WNV) in the genus *Flavivirus* of the family Flaviviridae. The virus was first iso lated from humans in 1937 in Uganda (Hayes and Gubler, 2006). It causes occasional outbreaks in Africa, Asia, Australia, and some parts of Europe. It was first reported in the United States in 1999 from New York and in 2002 from Texas. It has an enzootic transmission cycle with wild birds known to be its major reservoirs and transmitted by *Culex* mosquitoes, mainly *Cx. quinquefasciatus* in Texas. In a study conducted in Harris County, Texas, *Ae. albopictus* (Skuse) was found to be the second most important mosquito vector for WNV infection (Dennett et al., 2007). American robins are mainly involved in the transmission cycle and maintenance of WNV in Texas and the United States, although house sparrows are considered more susceptible to WNV (Colpitts et al., 2012). The WNV infection of blue jays throughout the year was found to be correlated with infections in humans in Harris County and is considered indicators of WNV infection (Dennett et al., 2007). Humans, horses, sheep, and cattle are accidental and dead-end hosts for WNV (Acha and Szyfres, 1987) (Fig. 11.2).

West Nile virus transmission cycle

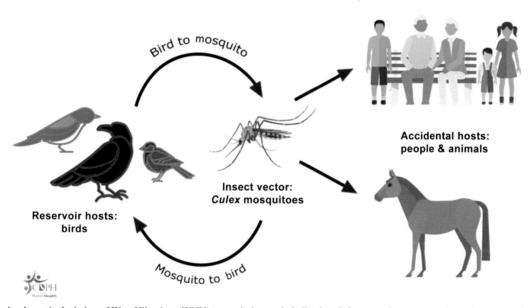

FIG. 11.2 A schematic depiction of West Nile virus (WNV) transmission cycle indicating *Culex* mosquito vectors and vertebrate hosts (birds, human, and horses). Although several mosquito species are known to transmit WNV, *Culex quinquefasciatus* is the primary vector of WNV in Texas. Reservoir and virus amplification bird hosts of WNV in Texas include blue jays, house sparrows, and American robins. Humans and horses are dead-end hosts. *(Source: http://www.westnile.ca.gov/.)*

WNF is the most common mosquito-borne disease infecting humans, birds, and other mammals in Texas and causes significant economic and social burden (Nolan et al., 2013). For example, an estimated productivity loss and medical cost of $47.6 million (range $14.5–$140.7 million) was incurred during the 2012 outbreak of WNV in Texas (Table 11.3). This estimated account does not include the millions of dollars spent to control mosquito vectors. While the disease remained to be endemic over the last 17 years, the major outbreak with 1864 human cases including 844 West Nile neuroinvasive disease (WNND) and 1024 WNF occurred in 2012 in Texas (Fig. 11.1). The majority of cases during the 2012 outbreak were from northern Texas counties (Murray et al., 2013). A total of 5412 cases were reported between 2002 and 2017 in Texas, and it is the third state following California and Colorado in the number of cases (CDC, 2018a).

Local variations in the WNV outbreak have been reported in Texas, and factors influencing these variations are less well understood. For example, a local outbreak of WNV (about 139 human cases and two deaths) occurred in Harris County, Texas, in 2014, unlike in other counties where the number of cases was almost similar or comparable with the previous

TABLE. 11.3 Estimated medical cost and productivity loss of West Nile virus diseases (neuroinvasive (WNND) form and West Nile fever (WNF)) in Texas.

Year	Total[a]	WNND[a]	Cost of WNND per patient[b]			Cost of all WNND patients[b]			WNF
			Average	Range		Average	Range		
2002	202	202	$43,213	$12,260	$130,254	$8,728,943	$2,476,429	$26,311,389	0
2003	702	431	$44,335	$12,578	$133,638	$8,955,668	$5,421,109	$57,597,819	271
2004	176	119	$45,189	$12,820	$136,212	$19,476,455	$1,525,612	$16,209,208	57
2005	195	128	$46,531	$13,201	$140,257	$5,537,189	$1,689,728	$17,952,896	67
2006	354	233	$48,385	$13,727	$145,847	$6,193,332	$3,198,414	$33,982,279	121
2007	260	170	$49,390	$14,012	$148,874	$11,507,805	$2,382,045	$25,308,570	90
2008	64	40	$51,504	$14,612	$155,246	$8,755,636	$584,471	$6,209,848	24
2009	115	93	$51,519	$14,616	$155,293	$2,060,764	$1,359,301	$14,442,204	22
2010	89	77	$52,872	$15,000	$159,370	$4,917,083	$1,154,994	$12,271,494	12
2011	27	20	$53,735	$15,245	$161,971	$4,137,567	$304,894	$3,239,415	7
2012	1868	844	$55,307	$15,691	$166,709	$1,106,130	$13,242,900	$140,702,160	1024
2013	183	113	$56,189	$15,941	$169,368	$47,423,145	$1,801,319	$19,138,528	70
2014	379	253	$57,076	$16,193	$172,042	$6,449,560	$4,096,723	$43,526,555	126
2015	275	196	$57,025	$16,178	$171,888	$14,427,262	$3,170,910	$33,690,050	79
2016	370	252	$57,808	$16,400	$174,248	$11,330,319	$4,132,863	$43,910,541	118
2017	135	87	$59,253	$16,810	$178,604	$14,931,748	$1,462,493	$15,538,588	48
2018	131	96	$60,480	$17,158	$182,302	$5,806,062	$1,647,199	$17,501,037	35

[a] West Nile virus.

[b] The table was adapted from previous studies (Murray et al. 2013; Barber et al., 2010). The EEC associated with West Nile virus was calculated by the methodology established by Barber et al. (2010), and extrapolated using CPI inflation calculator (https://www.bls.gov/data/inflation_calculator.htm) for the different years.

Cost of WNF per patient[b]			Cost of all WNF patients[b]			Cost of both WNF and WNND patients[b]		
Average	Range		Average	Range		Average	Range	
$1087	$1048	$1147	$0	$0	$0	$8,728,943	$2,476,429	$26,311,389
$1115	$1048	$1177	$302,105	$283,889	$318,888	$9,257,773	$5,704,998	$57,916,707
$1136	$1095	$1199	$64,767	$62,442	$68,365	$19,541,222	$1,588,054	$16,277,572
$1170	$1128	$1235	$78,390	$75,576	$82,745	$5,615,579	$1,765,304	$18,035,641
$1217	$1173	$1284	$147,212	$141,927	$155,391	$6,340,545	$3,340,341	$34,137,669
$1242	$1197	$1311	$111,769	$107,757	$117,978	$11,619,574	$2,489,802	$25,426,548
$1295	$1249	$1367	$31,081	$29,965	$32,808	$8,786,717	$614,436	$6,242,655
$1295	$1249	$1367	$28,499	$27,476	$30,074	$2,089,264	$1,386,777	$14,472,278
$1329	$1282	$1367	$15,953	$15,381	$16,409	$4,933,036	$1170,374	$12,287,903
$1351	$1303	$1403	$9458	$9118	$9823	$4,147,025	$314,012	$3,249,238
$1391	$1341	$1426	$1,424,036	$1,372,908	$1,460,429	$2,530,166	$14,615,808	$142,162,588
$1413	$1362	$1491	$98,898	$95,348	$104,393	$47,522,043	$1,896,668	$19,242,921
$1435	$1384	$1515	$180,828	$174,336	$190,874	$6,630,387	$4,271,059	$43,717,429
$1434	$1382	$1514	$113,275	$109,209	$119,568	$14,540,537	$3,280,118	$33,809,618
$1454	$1401	$1534	$171,519	$165,362	$181,047	$11,501,838	$4,298,225	$44,091,589
$1490	$1436	$1573	$71,515	$68,947	$75,488	$15,003,263	$1,531,441	$15,614,076
$1521	$1466	$1605	$53,226	$51,315	$56,183	$5,859,288	$1,698,514	$17,557,220

years (Martinez et al., 2017). The changes in WNV strain or the introduction of novel WNV strain from different geographic areas may have caused the differences in outbreaks among counties in Texas (Mann et al., 2013). Studies showed disparities in fatal and nonfatal WNV infection cases by gender and race from 2002 to 2012 (Murray et al., 2013; Philpott et al., 2019).

Symptoms and diagnosis

About 80% of WNV-infected humans remain asymptomatic; however, some may develop debilitating illness or even death (Hayes and Gubler, 2006). The WNF causes febrile illness in about 20% of WNV-infected patients. In <1% of WNV-infected patients, a severe WNND typified with encephalitis, meningitis, and flaccid paralysis can occur and can be severe in older adults that may result in death (Hayes and Gubler, 2006; CDC, 2018a). Other symptoms include acute onset of fever; headache; malaise; fatigue; and ocular, muscular, and articular pain as well as gastrointestinal upset (Hayes and Gubler, 2006). Diagnosis of WNV involves serum or cerebrospinal fluid to detect WNV-specific IGM antibodies and detect viral RNA (CDC, 2018a).

Treatment and prevention

There is no treatment or vaccine available against WNV infection in humans. The primary method of WNV prevention is through effective control of mosquito vectors (see Control Chapter 8 of this book). In addition, the use of mosquito repellents on human skin and proper wear of clothing will reduce the risk of acquiring WNV.

11.2.6 St. Louis encephalitis

SLE virus belongs to the genus *Flavivirus* (family Flaviviridae). It was first isolated in St. Louis, Missouri, the United States in 1933. The SLE is found in both North America and South America and known to be transmitted by the *Culex* mosquitoes at various geographic areas. It is considered the disease of equines with transmission cycle between mosquitoes and vertebrates, primarily birds and horses (Fig. 11.3). The primary vector of SLE in Texas and other southern US areas is *Cx. quinquefasciatus*. Epidemics of SLE occurred numerous times in Texas, including in Houston and Corpus Christi in 1964 (Luby et al., 1967) and Dallas, Texas, in 1966 (Hopkins et al., 1975). During the epidemic years the SLE virus was isolated from various birds including blue jays, mocking birds, pigeons, house sparrows, and domestic geese (Luby et al., 1967). Other SLE outbreaks that occurred in 1976 and 1980 in Harris County/Houston, Texas, had the incidence per 100,000 humans significantly lower than the 1964 outbreak (Bell et al., 1981).

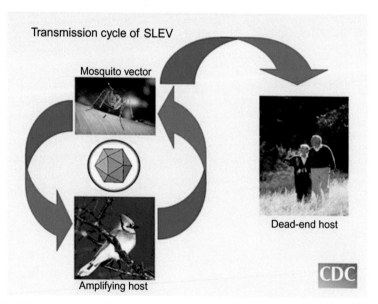

FIG. 11.3 Transmission cycle of Saint Louis encephalitis virus (SLEV) in mosquito and vertebrate hosts. *(Source: Centers for Disease Control and Prevention (CDC).)*

According to the TDSHS a total of 439 SLE cases in Texas were reported between 1975 and 2017 (TDSHS, 2018). Although large outbreaks have not occurred in recent years, SLE remains endemic in Texas (Fig. 11.1). Over the past two decades, about 1.3 cases per year have been reported compared with about 14.3 cases per year for the previous three decades (Fig. 11.1). The decline in SLE over the last two decades coincided with the introduction and persistence of WNV in Texas.

It is unknown whether the viral pathogens of the two diseases might have interfered with one another in the vector populations and impacted the transmission of SLE or other factors were probably involved including its evolution (Beard et al., 2016).

The Harris County Public Health Mosquito and Vector Control Division in Houston, Texas, were founded as result of the 1964 SLE outbreak. It has since become the primary organization to survey and prevent SLE, WNV, dengue, and other emerging and endemic mosquito-borne diseases in Harris County/Houston.

Symptoms and diagnosis

During SLE epidemics, patients may experience sustained high fever, disorientation, confusion, delirium, lethargy, stupor and or comma, and neurologic disorders (Day, 2001). Diagnosis of SLE can be done by testing the serum, nucleic acid amplification, histopathology with immunohistochemistry, and/or virus culture.

Treatment and prevention

There is no treatment or vaccine available against SLE infection in humans. The primary method of SLE prevention is through effective integrated mosquito management (IMM) of mosquito vectors. In addition, the use of personal protective measures such as the use of mosquito repellents on skin and proper wear of clothing will reduce the risk of acquiring SLE.

11.2.7 Mayaro fever

Mayaro fever (MF) is caused by a single-stranded positive RNA virus in the genus *Alphavirus* of family Togaviridae, and it is mainly found in South and Central America (Acha and Szyfres, 1987). It is also considered an emerging mosquito-borne disease. The MF virus was first isolated from five febrile rural workers in Trinidad and Tobago in 1954 and also from a bird in Louisiana, the United States (Acha and Szyfres, 1987; Mota et al., 2015). MF is transmitted primarily by *Haemagogus* mosquitoes but *Aedes* mosquitoes have also been found to be its competent vectors (Tesh et al., 1999). The hosts include primates, birds, and other vertebrates including reptiles (Mota et al., 2015; Lednicky et al., 2016). While MF is not currently present in Texas, the presence of *Aedes* vectors that are often associated with urban dwellings increase the risk of its introduction.

Symptoms and diagnosis

MF causes several symptoms similar to other arboviral diseases, including pyrexia, frontal cephalalgia, conjunctival congestion, photophobia, myalgia, and occasionally arthralgia and can be mistaken for infection by DENV or other arboviruses (Acha and Szyfres, 1987; Mota et al., 2015). During an outbreak, MF symptoms resemble that of CHIKV infection. Although the short viremia period of the virus can be problematic, the best method to diagnose MF infection is by isolating the virus from blood and conducting nucleic acid test to detect it (Mota et al., 2015).

Prevention and treatment

There is no treatment or vaccine available against MF infection in humans. However, there are some vaccine candidates in the developmental stages (Mota et al., 2015). The primary method of MF prevention is through effective IMM of the mosquito vectors. In addition, the use of mosquito repellents on the skin and proper wear of clothing will reduce the risk of acquiring MF.

11.2.8 Eastern equine encephalitis

Eastern equine encephalitis (EEE) is caused by a single-stranded, positive sense RNA virus in the genus *Alphavirus* (family Togaviridae) and maintained principally by *Culiseta melanura* (Coquillett) and some avian hosts in aquatic habitats. It is considered a rare illness with most of the cases reported in Florida, Georgia, Massachusetts, and New Jersey (CDC, 2019a). Several mosquito species in the genera *Aedes*, *Coquillettidia*, and *Culex* have been incriminated as vectors of EEEV. Its fatality rate is high and about 90% higher than WEE or VEE in certain areas. The horses and humans are considered dead-end hosts.

Symptoms and diagnosis

EEE causes systemic or encephalic symptoms (CDC, 2019a). The systemic infection is characterized by chills, fever, malaise, arthralgia, and myalgia. Among some people the symptoms can be asymptomatic. The mortality rate of EEE is about

30% and death often occurs within 2–10 days after the onset of symptoms. The diagnosis of EEE involves isolating the virus from the brain or cerebrospinal tissue and other serological tests.

Prevention and treatment

There is no human vaccine available for EEE infection or antiviral therapy. Similar to many other encephalitis viruses, the best mechanism to reduce the risk of EEE human infection is through proper use of personal protective measures including application of repellents, wearing long sleeve shirts and pants, and covering the full body with clothing during dusk and dawn when the mosquito activity is higher. Integrated mosquito management that reduces mosquito vector populations in areas suspected of EEE activity will also reduce the risk of human infection.

11.2.9 Western equine encephalitis

WEE belongs to the genus *Alphavirus* (family Togaviridae). It was identified in 1930 from a horse and is mainly considered a disease of equines. It is an important disease of humans and horses along with EEE and VEE diseases (Sherman and Weaver, 2010). Its geographic distribution is limited to the Americas ranging from Canada to Uruguay (Acha and Szyfres, 1987). The primary vector of WEE is *Culex tarsalis* Coquillett, and wild birds and humans are incidental hosts. Both humans and horses are dead-end hosts. Other vectors involved in the transmission of WEE include species of *Culiseta* and *Aedes* mosquitoes.

Symptoms and diagnosis

The symptoms of WEE infection range from mild flu-like illness to encephalitis, coma, or death, and it can cause a severe and permanent neurological damage for patients who survive acute infection of WEEV (Sherman and Weaver, 2010). Successful diagnosis includes its isolation or serological testing from brain tissue of dead humans or horses (Acha and Szyfres, 1987).

Prevention and treatment

No vaccine is available to prevent WEE in humans. Integrated mosquito management approach involving habitat modification and use of biological and chemical pesticides is considered the primary preventive mechanism to prevent the transmission of the WEE from infected *Cx. tarsalis*. In addition, proper use of clothing, mosquito repellents, and screens is also recommended to minimize the risk of WEE infection. A formalin-inactivated chick embryo vaccine has been available for horses for many years.

11.2.10 Venezuelan equine encephalitis

VEE virus is in the genus *Alphavirus* (family Togaviridae). The VEE is a zoonotic disease first discovered in horses in 1930s in South America and is considered to be native to the Americas, including North and South Americas (Acha and Szyfres, 1987; Weaver et al., 2004; Weaver and Barrett, 2004). The VEEV was first isolated in 1938 from the dead equine brain in Venezuela (Weaver et al., 2004). While the primary mode of transmission for VEE involves mosquitoes to humans and equines, it is also known to be transmitted by accidental inhalation of the virus in the laboratory or by other biting flies (Acha and Szyfres, 1987). It is transmitted primarily by the mosquito species in the genus *Culex* (*Melanconion*), but during outbreaks, floodwater, *Aedes* and *Psorophora* mosquitoes have also been incriminated (Acha and Szyfres, 1987; Weaver et al., 2004; Weaver and Barrett, 2004). Weaver et al. (2004) provided the complete list of mosquito species, ticks, and other flies associated with the transmission of VEE. Acha and Szyfres (1987) and Weaver et al. (2004) provided the relevant account of VEE. Rodent species in the genera *Sigmodon*, *Oryzomys*, *Zygodontomys*, *Heteromys*, *Peromyscus*, and *Peomyscus* are considered reservoirs during enzootic transmission cycle of VEEV (Weaver et al., 2004).

In 1971, there was a VEE outbreak including 88 reported human cases in southern Texas counties including, Cameron, Hidalgo, and Corpus Christi area (Bowen et al., 1976). The epidemics originated in Central America in 1969 and progressed north to the United States through Texas over the next 2 years (Zehmer et al., 1974). It remains to be a concern especially in the US-Mexico border that includes vast areas in Texas where movement of people from the VEE-endemic areas is common (Adams et al., 2012). In addition, the ubiquitous presence of the two mosquito vector species, *Ae. sollicitans* (Walker) and *Ae. taeniorhynchus* (Widemann) in Texas, may influence the outbreak of VEE.

Symptoms and diagnosis

VEE causes fever similar to influenza, chills, myalgia, cephalalgia, and frequently nausea, vomiting, and diarrhea. Its diagnosis involves isolating the virus and conducting serological tests (Acha and Szyfres, 1987).

Prevention and treatment

There is no licensed human vaccine or effective antiviral therapy or treatment that is currently available against VEE infection (Weaver et al., 2004). However, vaccinating equines during an outbreak is considered effective to reduce VEE epidemic (Weaver et al., 2004). Area-wide mosquito pesticide applications can reduce the VEEV infection during an outbreak. In addition, limiting movement of equines during VEE outbreak might reduce its introduction into new areas (Weaver et al., 2004). As with many other mosquito-borne diseases, personal protective measures, including proper wear of clothing, mosquito repellents containing DEET concentration $\leq 35\%$ for adults and $\leq 10\%$ for children, and avoiding mosquito bites will reduce the risk of VEE infection (Weaver et al., 2004).

11.2.11 La Crosse encephalitis

This disease is caused by La Crosse encephalitis virus (LACV) and transmitted by *Ae. triseriatus* (Say). The vertebrate hosts include chipmunks and squirrels. Humans are rarely infected with LACV and are considered dead-end hosts. It is more common in the midwestern United States, but Texas is within the geographic distribution range of its primary vector, *Ae. triseriatus*, a mosquito that develops in tree holes and artificial containers. *Ae. albopictus* has also been known to be naturally infected by LACV (Gerhardt et al., 2001). It is considered rare with only four reported cases of LAC in Texas from 2008 to 2017 (CDC, 2019b).

Symptoms and diagnosis

Its symptoms include fever and headache and can cause severe neuroinvasive disease in children under 16 years of age (CDC, 2019b). The disease can be diagnosed using a serological test or cerebrospinal fluid to detect virus-specific Igm and neutralizing antibodies (CDC, 2019b).

Prevention and treatment

No vaccines or antiviral therapy is available for LAC encephalitis. However, vector control including IMM approaches and use of repellents is known to reduce the risk of LACV infection to humans.

11.2.12 Japanese encephalitis

The JE is a zoonotic disease caused by a member of JE serotype in the genus *Flavivirus* isolated in 1935 from a human brain infected with the virus in Tokyo, Japan. It is the main factor causing encephalitis worldwide, especially in Asia and northern Australia (Solomon et al. 2003; Impoinvil et al., 2012). Wang and Liang (2015) discussed the epidemiology and geographic distribution of JE. Its transmission cycle primarily involves ardeid birds (herons and egrets). It affects other vertebrates including wild and domestic mammals (horses, pigs, etc.) and humans. Cows and other ruminants, humans, and horses are considered dead-end hosts, whereas pigs and wading birds are the main amplifying hosts of the JEV during an epizootic cycle (Impoinvil et al., 2012). It is mainly transmitted by *Cx. tritaeniorhynchus* Giles. In 2010, one of the two imported cases of JE in the United States was in Texas (Chen et al., 2011). *Cx. quinquefasciatus*, a competent vector of the JEV, is ubiquitous in Texas (Huang et al., 2016).

Symptoms and diagnosis

Infection of JE is often asymptomatic, and only <1% of JE infected people develop clinical symptoms (CDC, 2019c). Those who develop JE symptoms usually experience fever, headache, mental disorder, and seizures, particularly in children.

Prevention and treatment

A vaccine using inactivated JE vaccine (manufactured as IXIARO) is licensed and available in the United States (CDC, 2019c). It is recommended that people traveling to JE-endemic Asian countries including China, India, and Korea for a long period of time should get vaccinated with JE vaccine. While there are no specific treatments available for JE patients, supportive care including rest, pain relievers, and use of pain killers may alleviate the illness. The mortality rates in patients who develop encephalitis can reach up to 30% (CDC, 2019c).

11.3 Mosquito-borne protozoan diseases

11.3.1 Malaria

Malaria causes devastating human losses throughout the tropics, mainly in sub-Saharan African countries, Asia, and South America. Human malaria is principally caused by five protozoan species: *Plasmodium falciparum, P. vivax, P. knowlesi, P. malariae,* and *P. ovale* (White, 2008). *P. falciparum* is the most potent pathogen causing malaria in the world. *P. knowlesi* is a natural pathogen of long-tailed and pig-tailed macaques in Southeast Asia and severely infects humans (White, 2008). There are more than 20 *Plasmodium* species, including *P. cynomolgi, P. knowlesi, P. inui, P. schwetzi, P. simium, P. brasilianum,* and *P. rodhani* that recycle in other primates and can naturally or experimentally infect humans (Acha and Szyfres, 1987). The human malaria transmission cycle involves *Anopheles* mosquitoes and human hosts. The mosquito ingests gametocytes from the human blood where the parasite undergoes development and propagation. The mosquito injects the sporozoites into a human host during blood feeding as the parasite undergoes further development and sexual reproduction (Carter et al., 2017).

Malaria was once endemic throughout the United States, including Texas during the late 19th and early 20th centuries before World War II and was eliminated from the United States in the 1950s (Cox, 1944). Several factors including socioeconomic improvements, changes in environmental variables, and later vector control were attributed to the decline of malaria in the United States (Zucker, 1996; Moreno-Madriñán and Turell, 2018).

About 3656 travel and 12 locally transmitted malaria cases were reported in Texas between 1975 and 2017 (Fig. 11.1). Three of the 12 cases of local malaria transmissions caused by *P. vivax* occurred in 1994 in Houston, Texas (CDC, 1995). Mosquitoes in the genus *Anopheles* are responsible for transmitting malaria worldwide, but *Anopheles quadrimaculatus* Say is the main potential vector of malaria in Texas (Table 11.1). Another species, *An. albimanus* Wiedemann, was also a suspected vector of malaria in Rio Grande Valley of Texas in the 1930s (Cox, 1944).

Symptoms and diagnosis

Malaria causes several symptoms with the severity differing between adults and children. It causes fever, respiratory distress, jaundice convulsions, prostration metabolic disorder, renal failure, bleeding, etc. (WHO, 2013). Malaria diagnosis involves microscopic detection of the parasites by staining Giemsa buffer, rapid diagnostic tests, nucleic acid detection, and serological and drug resistance tests (CDC, 2018b).

Treatment and prevention

There are no approved vaccines of malaria for public use to date, despite many efforts in developing effective vaccines (Carter et al., 2017). However, there are several antimalarial drugs including quinine, chloroquine, pyrimethamine, sulphadoxine, primaquine, and artemisinin-based products used for treatment or prevention of malaria infection (CDC, 2013; Carter et al., 2017). Control of *Anopheles* vectors has been the primary method against malaria by using IMM approaches including the use of insecticide impregnated nets, indoor residual spraying, larviciding in mosquito immature developmental habitats, and use of personal protective measures (repellents and wear of proper clothing) to minimize mosquito bites in malaria prone areas.

11.4 Mosquito-borne filarial nematode diseases

Human lymphatic filariasis caused by three nematode species, *Brugia malayi, B. timori,* and *W. bancrofti,* is transmitted by *Culex* spp., *Anopheles* spp., *Aedes* spp., and *Mansonia* spp. *Cx. quinquefasciatus* is the most common species known to transmit these parasitic nematodes throughout the world (Cox, 1944). It is widespread in tropical and subtropical regions including parts of Africa, South America, and Asia. It was also present in some parts of the United States mainly in South Carolina until the 1930s when it was eradicated (Foster and Walker, 2009). Several species of the genus *Dirofilaria* cause other diseases and can also infect humans. These species include *Dirofilaria repens* Railliet and Henry, *D. tenuis* Chandler, *D. immitis* (Leidy), and *D. ursi* Yamaguti, and *D. striata* (Molin) are also known to infect humans (CDC, 2017). *D. immitis* is the most prevalent dog heartworm.

11.4.1 Dirofilariasis

Dirofilariasis caused by *D. immitis* originated in Asia and is now considered one of the most serious diseases of dogs and cats in North America. Although dogs are primarily its main hosts, it can also infect humans incidentally (Theis, 2005). Cases of human infection with *D. immitis* known as human dirofilariasis have been reported in Texas (Skidmore et al., 2000). It is

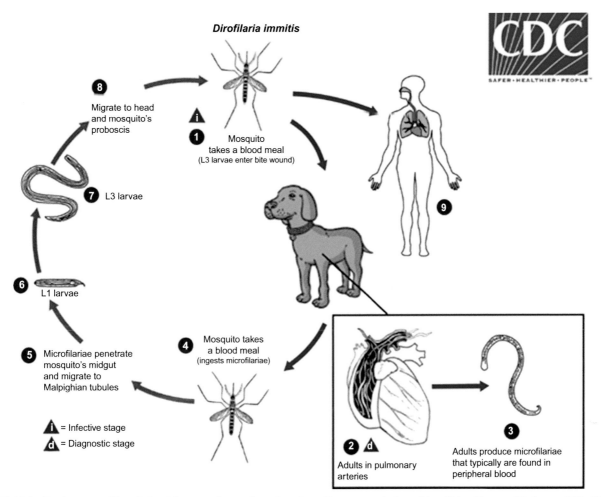

Dirofilaria immitis

8 Migrate to head and mosquito's proboscis

i 1 Mosquito takes a blood meal (L3 larvae enter bite wound)

7 L3 larvae

6 L1 larvae

5 Microfilariae penetrate mosquito's midgut and migrate to Malpighian tubules

4 Mosquito takes a blood meal (ingests microfilariae)

i = Infective stage

d = Diagnostic stage

9

2 d Adults in pulmonary arteries

3 Adults produce microfilariae that typically are found in peripheral blood

FIG. 11.4   Dog heartworm life cycle involving mosquitoes and vertebrate hosts (dogs, cats, and other wild canids) and humans. The larvae of *Dirofiloria immitis* affects the heart of the infected host and can stay in the host for up to 10 years. *(Source: Centers for Disease Control and Prevention (CDC).)*

transmitted by several mosquito species in the genera *Aedes, Culex, Anopheles,* and *Mansonia.* The nematode requires mosquitoes, dogs, and other wild canids (foxes, wolves, coyotes, and raccoons) to complete its life cycle (Fig. 11.4). While the infection of humans with dog heartworm is rare, it is common in cats and dogs as well as wild canine species. Texas is one of the southern states where the incidence of dog heartworm infection increased from 2013 to 2016 (Drake and Wiseman, 2018).

Symptoms and diagnosis

Dirofilariasis can often remain asymptomatic, but it is also associated with inflammation caused by dying nematodes in pulmonary arteries, coughing, chest pain, fever, and pleural effusion (CDC, 2017). Results of coin lesions seen in the human lungs following the death of the filarial nematodes can be detected using an x-ray (Theis, 2005).

Prevention and treatment

Canine dirofilariasis can be prevented in dogs through administration of diethylcarbamazine (DCE), macrocyclic lactones (ML), and milbemycin oxime drugs. However, IMM approaches that reduce mosquito populations will minimize the risk of mosquito-borne filarial infection. More preventive methods of dirofilariasis in dogs and cats are described in Nayar and Rutledge (1998).

11.5   Conclusion

In this chapter, we highlighted the major mosquito-borne pathogens that are of public health importance to the estimated 28.7 million residents of Texas. Despite improved vector control strategies, mosquito-borne diseases continue to be a concern in Texas due to several factors, including the introduction of new pathogens, its proximity to other disease endemic countries, and presence of conducive environment for mosquito vector populations. In addition, only a few counties have a well-equipped and organized mosquito control operations to combat mosquitoes and prevent mosquito-borne diseases. WNF, SLE, and dengue are considered endemic in Texas, while chikungunya and Zika are newly emerging diseases of serious concern. For many of mosquito-borne diseases, adequate mosquito control is the main method of preventing them. Strengthening the mosquito control operations in Texas is critical to combat resurgence of existing mosquito-borne diseases and prevent introduction of new mosquito-borne pathogens.

Acknowledgments

We thank Sepehr Arshadmansab from Zoonosis Control Branch at Texas Department of State Health Services for providing historical data on mosquito-borne diseases. We also thank Heather Tooker-Blue from the University of Texas Health Science Center, School of Public Health, for her valuable suggestions and comments on the medical economic estimate of WNV infection in Texas.

References and further reading

Acha, P.N., Szyfres, B., 1987. Zoonoses and Communicable Diseases Common to Man and Animals. Pan American Health Organization, Washington, DC. 963 p.

Adams, A.P., Navarro-Lopez, R., Ramirez-Aguilar, F.J., Lopez-Gonzalez, I., Leal, G., et al., 2012. Venezuelan equine encephalitis virus activity in the Gulf Coast region of Mexico, 2003–2010. PLoS Negl. Trop. Dis. 6 (11), e1875.

Ali, A., Abbas, A., Debboun, M., 2017. Zika Virus: epidemiology, vector and sexual transmission, neurological disorders and vector management—a review. Int. J. Cur. Res. 9, 58721–58737.

Andreadis, T., Capotosto, P., Shope, R., Tirrell, S., 1994. Mosquito and arbovirus surveillance in Connecticut, 1991–1992. J. Am. Mosq. Control Assoc. 10, 556–564.

Andreadis, T., Anderson, J., Tirrell-Peck, S., 1998. Multiple isolations of eastern equine encephalitis and highlands J viruses from mosquitoes (Diptera: Culicidae) during a 1996 epizootic in southeastern Connecticut. J. Med. Entomol. 35, 296–302.

Andreadis, T.G., Armstrong, P.M., Anderson, J.F., Main, A.J., 2014. Spatial-temporal analysis of Cache Valley virus (Bunyaviridae: Orthobunyavirus) infection in anopheline and culicine mosquitoes (Diptera: Culicidae) in the northeastern United States, 1997–2012. Vector Borne Zoonotic Dis. 14, 763–773.

Armstrong, P.M., Andreadis, T.G., 2010. Eastern equine encephalitis virus in mosquitoes and their role as bridge vectors. Emerg. Infect. Dis. 16, 1869–1874.

Arnell, J.H., 1976. Mosquito studies (Diptera, Culicidae). XXXIII. A revision of the *Scapularis* group of *Aedes* (*Ochlerotatus*). Contrib. Am. Entomol. Inst. 13, 1–144.

Barber, L.M., Schleier, J.J., Peterson, R.K., 2010. Economic cost analysis of West Nile virus outbreak, Sacramento County, California, USA, 2005. Emerg. Infect. Dis 16, 480–486.

Barrett, A.D., Higgs, S., 2007. Yellow fever: a disease that has yet to be conquered. Annu. Rev. Entomol. 52, 209–229.

Beard, C.B., Eisen, R.J., Barker, C.M., Garofalo, J.F., Hahn, M., Hayden, M., Monaghan, A.J., Ogden, N.H., Schramm, P.J., 2016. Ch. 5: Vectorborne diseases. In: The Impacts of Climate Change on Human Health in the United States: A Scientific Assessment. U.S. Global Change Research Program, Washington, DC, pp. 129–156. https://doi.org/10.7930/J0765C7V. (Accessed 18 March 2019).

Bell, R.L., Christensen, B., Holguin, A., Smith, O., 1981. St. Louis encephalitis: a comparison of two epidemics in Harris County, Texas. Am. J. Public Health 71 (2), 168–170.

Bowen, G.S., Fashinell, T.R., Dean, P.B., Gregg, M.B., 1976. Clinical aspects of human Venezuelan equine encephalitis in Texas. Bull. Pan Am. Health Organ. 10 (1), 46.

Carpenter, S.J., LaCasse, W.J., 1955. Mosquitoes of North America (North of Mexico). University of California Press, Berkeley, CA. 495 pp.

Carter, K.H., Escalada, R.P., Singh, P., 2017. Malaria. In: Marcondes, C. (Ed.), Arthropod Borne Diseases. Springer, Cham, pp. 325–346.

Centers for Disease Control and Prevention (CDC), 1995. Local transmission of *Plasmodium vivax* malaria—Houston, Texas, 1994. Morb. Mortal. Wkly. Rep. 44 (15), 295–301.

Centers for Disease Control and Prevention (CDC), 2012. Mosquito Species in Which West Nile Virus Has Been Detected, United States, 1999–2012 (Reported to ArboNET). Centers for Disease Control and Prevention, Atlanta, GA. http://www.cdc.gov/westnile/resources/pdfs/MosquitoSpecies1999-2012.pdf. (Accessed February 20, 2015).

Centers for Disease Control and Prevention (CDC), 2013. Guidelines for Treatment of Malaria in the United States. Centers for Disease Control and Prevention, Atlanta, GA.

Centers for Disease Control and Prevention (CDC), 2015. Arbovirus Catalog. Centers for Disease Control and Prevention, Atlanta, GA. https://wwwn.cdc.gov/Arbocat/Default.aspx/. (Accessed February 20, 2015).

Centers for Disease Control and Prevention (CDC), 2017. Dirofilariasis. Global Health, Division of Parasitic Diseases and Malaria. Centers for Disease Control and Prevention, Global Health, Division of Parasitic Diseases and Malaria, Atlanta, GA.

Centers for Disease Control and Prevention (CDC), 2018a. West Nile Virus. Centers for Disease Control and Prevention, National Center for Emerging and Zoonotic Infectious Diseases (NCEZID), Division of Vector-Borne Diseases (DVBD), Atlanta, GA. https://www.cdc.gov/westnile/index.html. (Accessed February 20, 2019).

Centers for Disease Control and Prevention (CDC), 2018b. Malaria Diagnosis (United States). Centers for Disease Control and Prevention, Global Health, Division of Parasitic Diseases and Malaria, Atlanta, GA. https://www.cdc.gov/malaria/diagnosis_treatment/diagnosis.html. (Accessed February 20, 2019).

Centers for Disease Control and Prevention (CDC), 2019a. Eastern Equine Encephalitis. Centers for Disease Control and Prevention, Global Health, Division of Parasitic Diseases and Malaria, Atlanta, GA. https://www.cdc.gov/easternequineencephalitis/index.html. (Accessed March 18, 2019).

Centers for Disease Control and Prevention (CDC), 2019b. La Crosse Encephalitis. Centers for Disease Control and Prevention, Global Health, Division of Parasitic Diseases and Malaria, Atlanta, GA. https://www.cdc.gov/lac/tech/epi.html. (Accessed March 21, 2019).

Centers for Disease Control and Prevention (CDC), Japanese Encephalitis, 2019c. Centers for Disease Control and Prevention, Global Health, Division of Parasitic Diseases and Malaria, Atlanta, GA. (Accessed 04 June 2019). https://www.cdc.gov/japaneseencephalitis/index.html.

Chen, L., Peek, M., Stokich, D., Todd, R., Anderson, M., Murphy, F.K., Hoffman, R., Evans, A., Jordan-Villegas, A., McCracken Jr., G., Chung, W.M., 2011. Japanese encephalitis in two children in United States, 2010. Morb. Mortal. Wkly. Rep. 60 (9), 276–278.

Christophers, S.R., 1960. *Aedes aegypti* (L.), the Yellow Fever Mosquito: Its Life History, Bionomics, and Structure. Cambridge University Press, Cambridge, UK. 739 pp.

Chungue, E., Marche, G., Plichart, R., Boutin, J.P., Roux, J., 1989. Comparison of immunoglobulin G enzyme-linked immunosorbent assay (IgG-ELISA) and haemagglutination inhibition (HI) test for the detection of dengue antibodies. Prevalence of dengue IgG-ELISA antibodies in Tahiti. Trans. R. Soc. Trop. Med. Hyg. 83 (5), 708–711.

Colpitts, T.M., Conway, M.J., Montgomery, R.R., Fikrig, E., 2012. West Nile virus: biology, transmission, and human infection. Clin. Microbiol. Rev. 25 (4), 635–648.

Cox, G.W., 1944. The Mosquitoes of Texas. Texas State Health Department, Austin, Texas.

Day, J.F., 2001. Predicting St. Louis encephalitis virus epidemics: lessons from recent, and not so recent, outbreaks. Annu. Rev. Entomol. 46 (1), 111–138.

Dennett, J.A., Bala, A., Wuithiranyagool, T., Randle, Y., Sargent, C.B., Guzman, H., Sirrin, M., Hassan, H.K., Reyna-Nava, M., Unnasch, T.R., Tesh, R.B., 2007. Associations between two mosquito populations and West Nile virus in Harris County, Texas, 2003–2006. J. Am. Mosq. Control Assoc. 23 (3), 264–275.

Drake, J., Wiseman, S., 2018. Increasing incidence of *Dirofilaria immitis* in dogs in USA with focus on the southeast region 2013–2016. Parasit. Vectors 11 (1), 39.

Fontenille, D., Traore-Lamizana, M., Diallo, M., Thonnon, J., Digoutte, J., Zeller, H., 1998. New vectors of Rift Valley fever in West Africa. Emerg. Infect. Dis. 4, 289–293.

Foster, W.A., Walker, E.D., 2009. Chapter 14: Mosquitoes (Culicidae). In: Mullen, G.R., Durden, L.A. (Eds.), Medical and Veterinary Entomology. Academic Press, Elsevier, Amsterdam, Netherlands, pp. 207–260.

Gargan, T.P., Clark, G.G., Dohm, D.J., Turell, M.J., Bailey, C.L., 1988. Vector potential of selected north American mosquito species for Rift Valley fever virus. Am. J. Trop. Med. Hyg. 38, 440–446.

Gerhardt, R.R., Gottfried, K.L., Apperson, C.S., Davis, B.S., Erwin, P.C., Smith, A.B., Panella, N.A., Powell, E.E., Nasci, R.S., 2001. First isolation of La Crosse virus from naturally infected *Aedes albopictus*. Emerg. Infect. Dis. 7 (5), 807.

Gilyard, R.T., 1944. Mosquito transmission of Venezuelan equine encephalitis virus in Trinidad. Bull. U.S. Army Med. Dept. 75, 96–107.

Gubler, D.J., 1998. Resurgent vector-borne diseases as a global health problem. Emerg. Infect. Dis. 4 (3), 442.

Gubler, D.J., 2009. Vector-borne diseases. Rev. Sci. Tech. 28 (2), 583.

Hall, N.B., Broussard, K., Evert, N., Canfield, M., 2017. Notes from the field: Zika virus-associated neonatal birth defects surveillance—Texas, January 2016–July 2017. Morb. Mortal. Wkly. Rep. 66 (31), 835–836.

Hayes, E.B., Gubler, D.J., 2006. West Nile virus: epidemiology and clinical features of an emerging epidemic in the United States. Annu. Rev. Med. 57, 181–194.

Hayes, E.B., Komar, N., Nasci, R.S., Montgomery, S.P., O'Leary, D.R., Campbell, G.L., 2005. Epidemiology and transmission dynamics of West Nile virus disease. Emerg. Infect. Dis. 11 (8), 1167–1173. https://doi.org/10.3201/eid1108.050289a.

Hopkins, C.C., Hollinger, F.B., Johnson, R.F., Dewlett, H.J., Newhouse, V.F., Chamberlain, R.W., 1975. The epidemiology of St. Louis encephalitis in Dallas, Texas, 1966. Am. J. Epidemiol. 102 (1), 1–15.

Horsfall, W.R., 1955. Mosquitoes—Their Bionomics and Relation to Disease. Hafner Publishing Company, New York. 723 pp.

Hotez, P.J., 2016. Zika in the United States of America and a fateful 1969 decision. PLoS Negl. Trop. Dis. 10 (5), e0004765.

Hotez, P.J., 2018. The rise of neglected tropical diseases in the "new Texas". PLoS Negl. Trop. Dis. 12 (1), e0005581. https://doi.org/10.1371/journal.pntd.0005581.

Huang, Y.M., 1972. Contributions to the mosquito fauna of Southeast Asia. XIV. The subgenus *Stegomyia* of *Aedes* in Southeast Asia. I—The scutellaris group of species. Contrib. Am. Entomol. Inst. 9 (1), 1–109.

Huang, Y.J.S., Hettenbach, S.M., Park, S.L., Higgs, S., Barrett, A.D., Hsu, W.W., Harbin, J.N., Cohnstaedt, L.W., Vanlandingham, D.L., 2016. Differential infectivities among different Japanese encephalitis virus genotypes in *Culex. quinquefasciatus* mosquitoes. PLoS Negl. Trop. Dis. 10 (10), e0005038.

Impoinvil, D.E., Baylis, M., Solomon, T., 2012. Japanese Encephalitis: on the one health agenda. In: Mackenzie, J., Jeggo, M., Daszak, P., Richt, J. (Eds.), One Health: The Human-Animal-Environment Interfaces in Emerging Infectious Diseases. Current Topics in Microbiology and Immunology. vol. 365. Springer, Berlin, Heidelberg.

Innis, B.L., Nisalak, A., Nimmannitya, S., Kusalerdchariya, S., Chongswasdi, V., Suntayakorn, S., Puttisri, P., Hoke, C.H., 1989. An enzyme-linked immunosorbent assay to characterize dengue infections where dengue and Japanese encephalitis co-circulate. Am. J. Trop. Med. Hyg. 40 (4), 418–427.

Lednicky, J., De Rochars, V.M.B., Elbadry, M., Loeb, J., Telisma, T., Chavannes, S., Anilis, G., Cella, E., Ciccozzi, M., Okech, B., Salemi, M., 2016. Mayaro virus in child with acute febrile illness, Haiti, 2015. Emerg. Infect. Dis. 22 (11), 2000.

Lee, B.Y., Alfaro-Murillo, J.A., Parpia, A.S., Asti, L., Wedlock, P.T., Hotez, P.J., Galvani, A.P., 2017. The potential economic burden of Zika in the continental United States. PLoS Negl. Trop. Dis. 11 (4), e0005531.

Liu, H., Li, M., Zhai, Y., Meng, W., Sun, X., Cao, Y.X., Fu, S.H., Wang, H.Y., Xu, L.H., Tang, Q., Liang, G.D., 2010. Banna virus, China, 1987–2007. Emerg. Infect. Dis. 16, 514–517.

Luby, J.P., Miller, G., Gardner, P., Pigford, C.A., Henderson, B.E., Eddins, D., 1967. The epidemiology of St. Louis encephalitis in Houston, Texas, 1964. Am. J. Epidemiol. 86 (3), 584–597.

Mann, B.R., McMullen, A.R., Swetnam, D.M., Salvato, V., Reyna, M., Guzman, H., Bueno Jr., R., Dennett, J.A., Tesh, R.B., Barrett, A.D., 2013. Continued evolution of West Nile virus, Houston, Texas, USA, 2002–2012. Emerg. Infect. Dis. 19 (9), 1418.

Martinez, D., Murray, K.O., Reyna, M., Arafat, R.R., Gorena, R., Shah, U.A., Debboun, M., 2017. West Nile virus outbreak in Houston and Harris County, Texas, USA, 2014. Emerg. Infect. Dis. 23 (8), 1372.

Mitchell, C.J., Haramis, L.D., Karabatsos, N., Smith, G.C., Starwalt, V.J., 1998. Isolation of La Crosse, Cache Valley, and Potosi viruses from *Aedes* mosquitoes (Diptera: Culicidae) collected at used-tire sites in Illinois during 1994–1995. J. Med. Entomol. 35, 573–577.

Moreno-Madriñán, M.J., Turell, M., 2018. History of mosquito borne diseases in the United States and implications for new pathogens. Emerg. Infect. Dis. 24 (5), 821.

Morse, S.S., 2001. Factors in the emergence of infectious diseases. In: Price-Smith, A.T. (Ed.), Plagues and Politics. Palgrave Macmillan, London, pp. 8–26.

Mota, M.T.D., Ribeiro, M.R., Vedovello, D., Nogueira, M.L., 2015. Mayaro virus: a neglected arbovirus of the Americas. Future Virol. 10 (9), 1109–1122.

Murray, K.O., Rodriguez, L.F., Herrington, E., Kharat, V., Vasilakis, N., Walker, C., Turner, C., Khuwaja, S., Arafat, R., Weaver, S.C., Martinez, D., 2013. Identification of dengue fever cases in Houston, Texas, with evidence of autochthonous transmission between 2003 and 2005. Vector Borne Zoonotic Dis 13 (12), 835–845.

Murray, K.O., Ruktanonchai, D., Hesalroad, D., Fonken, E., Nolan, M.S., 2013. West Nile virus, Texas, USA, 2012. Emerg. Infect. Dis. 19 (11), 1836.

Musso, D., Nhan, T., Robin, E., Roche, C., Bierlaire, D., Zisou, K., Yan, A.S., Cao-Lormeau, V.M., Broult, J., 2014. Potential for Zika virus transmission through blood transfusion demonstrated during an outbreak in French Polynesia, November 2013 to February 2014. Euro Surveill. 19 (14), 20761.

National Institute of Health (NIH), 2018. Guillain-Barré Syndrome Fact Sheet. NINDS, NIH Pub. No. 18-NS-2902.

Nayar, J.K., Rutledge, C.R., 1998. Mosquito-Borne Dog Heartworm Disease. University of Florida Cooperative Extension Service, Institute of Food and Agricultural Sciences, EDIS, Gainesville, FL.

Niebylski, M.L., Mutebi, J.P., Craig, G.B.J., Mulrennan, J.A.J., Hopkins, R.S., 1992. Eastern equine encephalitis virus associated with Aedes albopictus in Florida, 1991. Morb. Mortal. Wkly. Rep. 41, 121.

Nolan, M.S., Schuermann, J., Murray, K.O., 2013. West Nile virus infection among humans, Texas, USA, 2002–2011. Emerg. Infect. Dis. 19 (1), 137.

Pagès, F., Peyrefitte, C.N., Mve, M.T., Jarjaval, F., Brisse, S., Iteman, I., Gravier, P., Nkoghe, D., Grandadam, M., 2009. *Aedes albopictus* mosquito: the main vector of the 2007 Chikungunya outbreak in Gabon. PLoS One 4 (3), e4691.

Pauvolid-Corrêa, A., Kenney, J.L., Couto-Lima, D., Campos, Z.M., Schatzmayr, H.G., Nogueira, R.M., Brault, A.C., Komar, N., 2013. Ilheus virus isolation in the Pantanal, west-central Brazil. PLoS Negl. Trop. Dis. 7, e2318.

Philpott, D.C., Nolan, M.S., Evert, N., Mayes, B., Hesalroad, D., Fonken, E., Murray, K.O., 2019. Acute and delayed deaths after West Nile virus infection, Texas, USA, 2002–2012. Emerg. Infect. Dis. 25 (2), 256.

Polen, K.D., Gilboa, S.M., Hills, S., Oduyebo, T., Kohl, K.S., Brooks, J.T., Adamski, A., Simeone, R.M., Walker, A.T., Kissin, D.M., Petersen, L.R., 2018. Update: interim guidance for preconception counseling and prevention of sexual transmission of Zika Virus for men with possible Zika virus exposure—United States, August 2018. Morb. Mortal. Wkly. Rep. 67 (31), 868.

Ramos, M.M., Mohammed, H., Zielinski-Gutierrez, E., Hayden, M.H., Lopez, J.L.R., Fournier, M., Trujillo, A.R., Burton, R., Brunkard, J.M., Anaya-Lopez, L., Banicki, A.A., 2008. Epidemic dengue and dengue hemorrhagic fever at the Texas–Mexico border: results of a household-based seroepidemiologic survey, December 2005. Am. J. Trop. Med. Hyg. 78 (3), 364–369.

Rawlings, J.A., Hendricks, K.A., Burgess, C.R., Campman, R.M., Clark, G.G., Tabony, L.J., Patterson, M.A., 1998. Dengue surveillance in Texas, 1995. Am. J. Trop. Med. Hyg. 59 (1), 95–99.

Reiter, P., Lathrop, S., Bunning, M., Biggerstaff, B., Singer, D., Tiwari, T., Baber, L., Amador, M., Thirion, J., Hayes, J., Seca, C., 2003. Texas lifestyle limits transmission of dengue virus. Emerg. Infect. Dis. 9 (1), 86.

Sherman, M.B., Weaver, S.C., 2010. Structure of the recombinant *Alphavirus* Western equine encephalitis virus revealed by cryoelectron microscopy. J. Virol. 84 (19), 9775–9782.

Skidmore, J.P., Dooley, P.D., DeWitt, C., 2000. Human extrapulmonary dirofilariasis in Texas. South. Med. J. 93 (10), 1009–1010.

Solomon, T., Ni, H., Beasley, D.W., Ekkelenkamp, M., Cardosa, M.J., Barrett, A.D., 2003. Origin and evolution of Japanese encephalitis virus in southeast Asia. J. Virol. 77 (5), 3091–3098.

Sprenger, D., Wuithiranyagool, T., 1986. The discovery and distribution of Aedes albopictus in Harris County, Texas. J. Am. Mosq. Control. Assoc. 2 (2), 217–219.

Tabachnick, W.J., Powell, J.R., 1979. A world-wide survey of genetic variation in the yellow fever mosquito, *Aedes aegypti*. Genet. Res. 34 (3), 215–229.

Tao, S., Zhang, H., Yang, D., Wang, H., Liu, Q., Zhang, Y.Z., Yang, W.H., Cao, Y.X., Xu, L.H., He, Y., Chen, B.Q., 2003. Investigation of arboviruses in Lancang river down-stream area in Yunnan province. Chin. J. Exp. Clin. Virol. 17, 322–326.

Tesh, R.B., 1982. Arthritides caused by mosquito-borne viruses. Ann. Rev. Med 33 (1), 31–40.

Tesh, R.B., Watts, D.M., Russell, K.L., Damodaran, C., Calampa, C., Cabezas, C., Ramirez, G., Vasquez, B., Hayes, C.G., Rossi, C.A., Powers, A.M., 1999. Mayaro virus disease: an emerging mosquito-borne zoonosis in tropical South America. Clin. Infect. Dis. 28 (1), 67–73.

Texas Department State of Health Services (TDSHS), 2018. Arbovirus Activity in Texas, 2017 Surveillance Report, Austin, Texas.

Theis, J.H., 2005. Public health aspects of dirofilariasis in the United States. Vet. Parasitol. 133 (2–3), 157–180.

Turell, M.J., Dohm, D.J., Sardelis, M.R., Oguinn, M.L., Andreadis, T.G., Blow, J.A., 2005a. An update on the potential of North American mosquitoes (Diptera: Culicidae) to transmit West Nile Virus. J. Med. Entomol. 42 (1), 57–62.

Turell, M.J., Faran, M.E., Cornet, M., Bailey, C.L., 2005b. Vector competence of Sénégalese *Aedes fowleri* (Diptera: Culicidae) for Rift Valley fever virus. J. Med. Entomol. 25, 262–266.

Wang, H., Liang, G., 2015. Epidemiology of Japanese encephalitis: past, present, and future prospects. Ther. Clin. Risk Manag. 11, 435–448. https://doi.org/10.2147/TCRM.S51168.

Wang, Z., An, S., Wang, Y., Han, Y., Guo, J., 2009. A new virus of *Flavivirus*: Chaoyang virus isolated in Liaoning province. Chin. Pub. Health. 25, 769–772.

Weaver, S.C., 2014. Arrival of Chikungunya virus in the new world: prospects for spread and impact on public health. PLoS Negl. Trop. Dis. 8 (6), e2921. https://doi.org/10.1371/journal.pntd.0002921.

Weaver, S.C., Barrett, A.D., 2004. Transmission cycles, host range, evolution and emergence of arboviral disease. Nat. Rev. Microbiol. 2 (10), 789.

Weaver, S.C., Ferro, C., Barrera, R., Boshell, J., Navarro, J.C., 2004. Venezuelan equine encephalitis. Annu. Rev. Entomol. 49 (1), 141–174.

Wellings, F.M., Lewis, A.L., Pierce, L.V., 1972. Agents encountered during arboviral ecological studies: Tampa Bay area, Florida, 1963–1970. Am. Soc. Trop. Med. Hyg. 21, 201–213.

White, N.J., 2008. *Plasmodium knowlesi*: the fifth human malaria parasite. Clin. Infect. Dis. 46 (2), 172–173.

Whitehead, S.S., 2016. Development of TV003/TV005, a single dose, highly immunogenic live attenuated dengue vaccine; what makes this vaccine different from the Sanofi-Pasteur CYD™ vaccine? Expert Rev. Vaccines 15 (4), 509–517.

Wilder-Smith, A., Hombach, J., Ferguson, N., Selgelid, M., O'Brien, K., Vannice, K., Barrett, A., Ferdinand, E., Flasche, S., Guzman, M., Novaes, H.M., 2018. Deliberations of the strategic advisory group of experts on Immunization on the use of CYD-TDV dengue vaccine. Lancet Infect. Dis. 19 (1), e31–e38. https://doi.org/10.1016/S1473-3099(18)30494-8.

World Health Organization (WHO), 2009. Dengue: Guidelines for Diagnosis, Treatment, Prevention and Control: New Edition. World Health Organization, Geneva, Switzerland.

World Health Organization (WHO), 2013. Management of Severe Malaria: A Practical Handbook. WHO Press, Geneva, Switzerland.

World Health Organization (WHO), 2017. World Malaria Report 2017. World Health Organization, Geneva, Switzerland.

World Health Organization (WHO), 2018. Summary of WHO Position Paper on Dengue Vaccines. World Health Organization, Geneva, Switzerland. https://www.who.int/immunization/policy/position_papers/dengue/en/. (Accessed March 17, 2019).

Wozniak, A., Dowda, H.E., Tolson, M.W., Karabatsos, N., Vaughan, D.R., Turner, P.E., Ortiz, D.I., Wills, W., 2001. Arbovirus surveillance in South Carolina, 1996–98. J. Am. Mosq. Contr. Assoc. 17, 73–78.

Zehmer, R.B., Dean, P.B., Sudia, W.D., Calisher, C.H., Sather, G.E., Parker, R.L., 1974. Venezuelan equine encephalitis epidemic in Texas, 1971. Health Serv. Rep. 89 (3), 278.

Zucker, J.R., 1996. Changing patterns of autochthonous malaria transmission in the United States: a review of recent outbreaks. Emerg. Infect. Dis. 2 (1), 37.

Chapter 12

Recent Expansion of Mosquito-Borne Pathogens Into Texas

Scott C. Weaver[a], Alan D.T. Barrett[b]

[a]Institute for Human Infections and Immunity, Western Gulf Center of Excellence for Vector-borne Diseases, University of Texas Medical Branch, Galveston, TX, United States, [b]Sealy Institute for Vaccine Sciences, University of Texas Medical Branch, Galveston, TX, United States

Chapter Outline

Abbreviations

3′-UTR	3′-untranslated region
ADE	antibody-dependent enhancement
CVV	Cache Valley virus
CHIKV	chikungunya virus
CHIKF	chikungunya fever
DENV	dengue virus
DEN	dengue
DENF	dengue fever
DHF	dengue hemorrhagic fever
DSS	dengue shock syndrome
ECSA	East/Central/South African enzootic chikungunya virus lineage
E	envelope
ID_{50}	50% infectious dose
IOL	Indian Ocean chikungunya virus lineage
JE	Japanese encephalitis
JEV	Japanese encephalitis virus
MAYV	Mayaro virus

Mosquitoes, Communities, and Public Health in Texas. https://doi.org/10.1016/B978-0-12-814545-6.00012-2

NHP	nonhuman primate
OROV	Oropouche virus
SLEV	St. Louis encephalitis virus
USUV	Usutu virus
WNV	West Nile virus
YFV	yellow fever virus
ZIKV	Zika virus

12.1 Mosquito-borne viruses

This chapter focuses on mosquito-borne viruses that have expanded recently in Texas. This mainly includes two virus families: *Flaviviridae* and *Togaviridae*. The flaviviruses are probably the most important with the four dengue viruses (DENV-1, DENV-2, DENV-3, and DENV-4), St. Louis encephalitis virus (SLEV), West Nile virus (WNV), and Zika virus (ZIKV). There is only one species of the *Togaviridae* of major current importance and that is the chikungunya virus (CHIKV). The properties of these virus families and summaries of characteristics of the viruses are shown in Table 12.1.

TABLE 12.1 Viruses discussed in this chapter.

Family	Genus	Virus	Serocomplex	Vector	Vertebrate host
Flaviviridae	*Flavivirus*	Dengue-1	Dengue	*Ae. aegypti, Ae. albopictus*	Primates
		Dengue-2			
		Dengue-3			
		Dengue-4			
		St. Louis encephalitis	Japanese encephalitis	*Culex* spp.	Birds
		West Nile	Japanese encephalitis	*Culex* spp.	Birds
		Zika	Not applicable	*Aedes* spp.	Primates
Togaviridae	*Alphavirus*	Chikungunya	Semliki Forest	*Ae. aegypti, Ae. albopictus*	Primates

12.2 Origins, history of spread, and vector transmission

12.2.1 Chikungunya virus

Origins, history of spread, and vector transmission

The CHIKV is an alphavirus (family *Togaviridae*) in the Semliki Forest complex, along with the Mayaro virus (MAYV) discussed later. Like all other arboviruses, it is believed to have originated in a zoonotic cycle, in this case in sub-Saharan Africa in sylvatic habitats involving nonhuman primates (NHPs) and perhaps other amplification hosts and arboreal mosquito vectors (Weaver et al., 2018). These vectors are mainly *Aedes* spp. mosquitoes in subgenera other than *Stegomyia* (the subgenus that includes *Aedes aegypti* (L.) and *Ae. albopictus* (Skuse)). The CHIKV was discovered during a 1952–53 epidemic in present-day Tanzania, where people were afflicted with an acute, febrile illness typically accompanied by rash and arthralgia (now called chikungunya fever, or CHIKF). During this outbreak, *Ae. aegypti* was suspected as the main vector with humans serving as amplification hosts to generate a human-mosquito-human transmission cycle (Weinbren et al., 1958). Subsequently, CHIKV circulation representing the enzootic cycle was detected at several widespread sylvatic sites in sub-Saharan Africa, and evidence of human cases was also detected in many of these locations, due to presumed spillover (directly from enzootic circulation) infections. Additional epidemics with presumed human amplification were later detected in Senegal involving *Ae. aegypti* (Diallo et al., 1999) and in Gabon (Peyrefitte et al., 2008), with *Ae. albopictus* likely serving as an epidemic vector at this location (Paupy et al., 2010).

Although the first direct evidence of epidemic CHIKV emergence was obtained in 1953 (Weinbren et al., 1958), examination of historic medical records suggests outbreaks as early as the 18th century in present-day Indonesia (Carey, 1971) and regular emergences in Asia and the Americas throughout the 19th and 20th centuries (Halstead, 2015). These outbreaks followed the dispersal via sailing ships of both the epidemic vector, *Ae. aegypti*, and the virus through onboard interhuman circulation as has also been hypothesized for the dengue virus (DENV) and yellow fever virus (YFV) (Vasilakis et al., 2011). The most recent CHIKV emergence to be studied using modern phylogenetic and experimental virologic methods occurred sometime before 1958, when a strain now called the Asian lineage, derived up to a century earlier from the East/Central/South African (ECSA) enzootic lineage, was transported to India and Southeast Asia, followed by epidemics in both regions (Powers et al., 2000). Then, in 2004, a CHIKF epidemic along the coast of Kenya (Chretien et al., 2007) spread to islands in the Indian Ocean as well as independently to India, initiating a near pandemic with additional spread to Europe via infected travelers. The etiologic CHIKV strain, also originating from an ECSA enzootic lineage and referred to as the Indian Ocean lineage (IOL), not only circulated efficiently via transmission by *Ae. aegypti* but also adapted for efficient infection of *Ae. albopictus* through a series of adaptive mutations in the envelope glycoprotein genes (reviewed by Tsetsarkin et al., 2016). Unlike *Ae. aegypti*, which was introduced throughout the tropics and subtropics centuries ago following its domestication in sub-Saharan Africa, *Ae. albopictus* had only recently been introduced into many parts of Africa, Southern Europe, and the Americas. These introductions provided an additional opportunity for the CHIKV to circulate in more periurban and rural regions and in temperate climates such as Italy and France where *Ae. aegypti* cannot survive the cold winters to maintain stable populations.

Although the Americas were also considered at high risk for CHIKV introduction during the peak of the 2005–08 IOL epidemic and the large numbers of importations via infected travelers, it was not until 2013 that the virus gained a foothold with autochthonous cases detected in St. Martin, followed by spread throughout the Caribbean, to Central and South America and Mexico, and also to Florida and South Texas (Weaver and Lecuit, 2015). Surprisingly, the etiologic strain was from the older Asian lineage that was imported from somewhere in Southeast Asia or the South Pacific. Another surprise came in 2013 when an ECSA lineage strain was imported directly into Brazil, also initiating sustained epidemic transmission. Both of these epidemic strains (Asian/American and ECSA/American) continue to circulate and cause outbreaks in the Americas. Fortunately, due to epistatic constraints, neither the Asian/American (Tsetsarkin et al., 2011) nor ECSA/American (Tsetsarkin et al., 2009) strains are expected to adapt for efficient transmission by *Ae. albopictus*. This could reduce the risk for outbreaks in northern climates of Texas and the United States (US) where *Ae. aegypti* does not survive cold winters (Weaver, 2014).

12.2.2 Dengue virus

Origins, history of spread, and vector transmission

A clinical syndrome recognizable as DEN has been known for centuries. Indeed, the first detailed description of the disease was written by Benjamin Rush during the epidemic in Philadelphia in 1780.

Currently, DEN is recognized as the most important arboviral disease of humans. It is caused by four genetically and serologically related viruses that are termed DENV-1 to DENV-4. The viruses are often described as serotypes, but are actually four different viruses such that infection by one DENV results in lifelong homotypic immunity to the infecting DENV and short-term heterotypic immunity to the other three DENVs. Therefore, it is possible to acquire four DEN infections, one by each of the four DENVs.

Phylogenetic studies indicate that each of the four DENVs existed in enzootic sylvatic cycles involving *Aedes* spp. mosquitoes and NHPs in Asia with sylvatic isolates of DENV-1, DENV-2, and DENV-4 made in Malaysia and DENV-2 in West Africa (Wang et al., 2000). Over several hundred years, each of the four sylvatic DENVs evolved independently into the human viruses we recognize today. The DENVs are found in endemic and epidemic urban cycles that involve peridomestic *Aedes* spp., especially *Ae. aegypti* as the principal mosquito host and humans as the vertebrate host. Phylogenetic studies indicate that DENV-4 is the oldest and DENV-1 and DENV-3 are the most recently derived from a progenitor virus (Vasilakis and Weaver, 2008).

The four DENVs spread after the second World War, initially into various parts of Asia and then the Americas and other parts of the World where *Ae. aegypti* is found. The usual course was introduction of one DENV into a country followed shortly afterward by the other DENVs. Since humans have high viremias following infection with DENVs, travelers have proved to be a very efficient host for taking the viruses long distances, particularly as air travel expanded in the late 20th century (Guzman et al., 2010).

Currently, DEN is endemic in at least 110 countries, and over 4 billion people (or 60% of the World's population) are at risk from DEN. It is estimated that there are approximately 400 million infections each year, of which 50–100 million are clinical infections, including over 2 million cases of severe DEN and DEN hemorrhagic fever (DHF) (Bhatt et al., 2013). Although morbidity is high, mortality is relatively low at 10,000–24,000 individuals per year, and DEN fever (DENF) is endemic in these areas. Infection by one virus usually causes DENF, a febrile illness that is not life-threatening and appears to lead to lifelong protective immunity against the infecting DENV. However, as stated earlier, individuals who are infected by one DENV remain susceptible to infection by the other three DENVs. Subsequent infection by one of the other DENVs increases the risk of severe DEN, including DHF and DEN shock syndrome (DSS), which are life-threatening diseases (Halstead et al., 2010). A number of hypotheses have been proposed to explain DHF. The most compelling is antibody-dependent enhancement (ADE) where envelope (E) protein DEN cross-reactive epitopes that either elicit nonneutralizing antibodies or are at subneutralizing concentrations induced by the primary DEN infection form virus-antibody complexes in secondary infections that are taken up into macrophages via Fc receptors, thus enhancing virus infection (Halstead et al., 2010). In addition to ADE, cellular immunity and proinflammatory cytokines likely to contribute to the pathogenesis of DHF (Katzelnick et al., 2017).

12.2.3 West Nile virus

Origins and history of spread

WNV is a flavivirus (family Flaviviridae) in the Japanese encephalitis (JE) complex, along with SLEV, discussed later. The genome is approximately 11,000 nucleotides in length with 5′ and 3′ noncoding regions (5′NCR and 3′NCR) with a single open reading frame encoding a polyprotein that encodes 10 proteins: three structural proteins, capsid [C], premembrane/membrane [prM/M] and envelope [E], and seven nonstructural (NS) proteins, NS1, NS2A, NS2B, NS3, NS4A, NS4B, and NS5. The gene order is 5′-C-prM/M-*E*-NS1-NS2A-NS2B-NS3-NS4A-NS4B-NS5-3′. The virus was first isolated from a woman in the West Nile province of Uganda in 1937 who had a mild febrile infection (Smithburn et al., 1940). As such, WNV was not considered a major public health problem at that time. Subsequently, outbreaks of human febrile disease in Egypt and Israel in the early 1950s showed that the virus could be a public health problem. Seroepidemiologic studies showed that humans in many parts of Africa (Uganda, Kenya, Congo, Sudan, South Africa, Mozambique, Namibia, and Botswana) had antibodies to the virus, indicating that the virus was endemic in Africa (Jupp, 2001). The virus has been isolated in Malaysia and Australia and is known as the Kunjin virus and is described in Australia as a subtype of WNV. Periodic outbreaks of febrile disease were reported in Southern and Central Europe, the Middle East, and Northern Africa, but these were not regarded as a major public health problem. However, in 1996 an epidemic of fever and encephalitis due to WNV was reported in Romania (Tsai et al., 1998). Notably, this was the first outbreak involving neurologic disease. The high rate of neurologic disease in this outbreak has set the precedent of subsequent outbreaks where it has been the major clinical disease of concern, not only in the United States (see later) but also in Europe where high levels of neurologic disease and fatalities have been reported since the early 2000s (https://ecdc.europa.eu/en/west-nile-fever/surveillance-and-disease-data). Multiple factors have contributed to the emergence of WNV as a major public health concern, including viral, genetic, host, and environmental factors. However, an important factor was an amino acid substitution in the NS3 protein at residue 249 (NS3-249) that is associated with avian virulence and allowed the virus to multiply to high titers in avian species (Brault et al., 2007). Recent studies have shown that the NS1-NS2A-NS2B region also contributes to avian virulence (Dietrich et al., 2016), indicating that the molecular determinants of avian virulence are multigenic.

The virus emerged in New York in the summer of 1999 and rapidly spread throughout the United States and into areas of Canada, Mexico, the Caribbean, and Central and South America. Overall, WNV has been isolated in five continents: Africa, Asia, Australasia, Europe, and the Americas. Extensive phylogenetic studies have been undertaken, and six genotypes have been described. These include lineages I (found in five continents and consist of at least six genetic clusters of which cluster four contains isolates from the Americas), II (sub-Saharan Africa, Madagascar, and more recently Europe), III (one strain: Rabensburg virus), IV (Russia), and V (India); Koutango, a flavivirus from West Africa, is sometimes classified as lineage VI (May et al., 2011). The phylogeny indicates the virus originated in Africa and spread to Asia, the Middle East, and Europe. Lineage I is the only lineage found in the Americas, and phylogenetic studies suggest the virus that emerged in New York originated in the Middle East or North Africa (Lanciotti et al., 1999; May et al., 2011). The major lineages associated with clinical disease are I and II. Lineage I is associated with the NS3-P249T substitution, which is involved in increased avian virulence. No genetic changes have been associated with an increased human neuroinvasive disease, although it is clear that lineages I and II are evolving into a more neuroinvasive phenotype (Beasley et al., 2004; McMullen et al., 2013).

It is not known how WNV entered the United States in 1999, but there are multiple hypotheses. The most likely routes of introduction are either via migratory birds or importation of vagrant or surreptitiously imported exotic birds that are viremic, although importation of virus-infected mosquitoes on airplanes cannot be excluded. Local mosquitoes probably fed on the virus-infected birds, and the epidemic started. In 2000 and 2001 WNV spread down the eastern seaboard of the United

States and in 2002 spread across many areas of the United States with a concomitant epidemic of West Nile neuroinvasive disease (WNND). By 2004 the virus was isolated in all 48 states except Hawaii, and there has been WNV activity in the United States every year since 1999.

WNV transmission by mosquitoes

WNV is maintained in an enzootic transmission cycle with birds as the principal amplifying hosts and mosquitoes, particularly *Culex* spp., serving as the principal vector. Mammals, including horses and humans, are "dead-end" hosts due to insufficient peak viremia to infect a mosquito upon ingestion of an infected blood meal.

Ornithophilic (bird-seeking) *Culex* spp. mosquitoes are the principal vectors for both epizootic and epidemic transmission with spillover opportunistic feeding on mammals. The actual *Culex* spp. vary by geographic area. For example, in the northern, northeastern, southern, and western regions of the United States, they are as follows: *Cx. pipiens* Linnaeus, *Cx. salinarius* Coquillett, *Cx. pipiens quinquefasciatus* Say, and *Cx. tarsalis* Coquillett, respectively. In addition, other *Culex* species (*Cx. restuans* Theobald, *Cx. nigripalpus* Theobald, and *Cx. stigmatosoma* Dyar) are potential bridge vectors in the enzootic maintenance and amplification of WNV early in the annual transmission season. Other mosquito species can act as vectors for WNV, including *Aedes* spp., *Anopheles* spp., and *Culiseta* spp. To date, at least 62 mosquito species have been implicated as vectors of WNV in the United States with varying efficiency but are not considered as important as *Culex* spp.

Various avian species act as vertebrate hosts for WNV, and to date, 326 species have been implicated as hosts, but their efficiency as amplifying hosts varies greatly. Passeriformes, notably corvids, act as the principal WNV-competent avian hosts in the United States. The American crow, *Corvus brachyrhynchos*, and blue jay, *Cyanocitta cristata*, have viremias $>10^9$ pfu/mL and are excellent amplifying hosts but succumb to WNV infection. Other Passeriformes, including house sparrows, house finches, and American robins, have been implicated as amplification hosts for both local spread and long-distance migration of WNV in the United States. Overall, unlike most mosquito-borne viruses, WNV is able to multiply in many mosquito and avian hosts. However, it should be noted that the related Japanese encephalitis virus (JEV) shows a similar broad range of hosts.

12.2.4 Zika virus

Origins and history of spread

The ZIKV is a flavivirus (*Flaviviridae: Flavivirus*) first discovered in 1947 during field studies of enzootic, sylvatic YFV circulation in the Ziika Forest of Uganda (Dick, 1952; Dick et al., 1952). The virus was isolated from a sentinel macaque placed in the forest canopy that became febrile; inoculation of its serum into mice resulted in ZIKV isolation, and subsequent serologic tests showed that it was a group B arbovirus (now known as the genus *Flavivirus*) related to DENV, YFV, and others. Later, infection of a laboratory worker showed that the ZIKV caused a mild febrile illness (Simpson, 1964); human cases reported from Nigeria in 1954 were later attributed to the closely related Spondweni virus. Virus isolations from *Aedes* (*Stegomyia*) *africanus* (Theobald) collected in the Ziika Forest confirmed mosquito-borne transmission (Dick, 1952, Haddow et al., 1964).

Following its discovery and characterization in Uganda, very little research was done on the ZIKV, and only a handful of human cases was identified over many decades. However, serosurveys and a few virus detections indicated that the virus was widely distributed in sub-Saharan Africa and in Southeast Asia. The first suggestion that the ZIKV could be transmitted in urban settings came from its isolation from *Ae. aegypti* in Malaysia in 1966 (Marchette et al., 1969). Then, in 2007, the first two outbreaks attributed to the ZIKV were detected in Gabon, apparently transmitted by the recently introduced *Ae. albopictus* (Grard et al., 2014), and in Yap Island in the Federated States of Micronesia, transmitted by *Ae.* (*Stegomyia*) *hensilii* Farner (Lanciotti et al., 2008; Duffy et al., 2009). Retrospective epidemiologic studies in Yap indicated that the vast majority of infections were probably asymptomatic and the febrile illness in symptomatic cases closely resembled the mild disease described in Africa, with fever and rash comprising the most common symptoms and signs. Later phylogenetic studies of the ZIKV indicated an African origin, with spread to Asia at least many decades ago (Haddow et al., 2012), with African strains causing the Gabon outbreak (Grard et al., 2014) and a likely importation of the virus to Yap from Southeast Asia. However, these outbreaks involving hundreds and thousands of infections in Gabon and Yap did not foretell the near-pandemic spread that was to occur only a few years later.

The first major step in the major Zika epidemic in the Americas was its introduction into French Polynesia in 2013, probably from somewhere in Southeast Asia followed by rapid spread through the South Pacific, presumably via infected air travelers (Musso et al., 2014). In French Polynesia, the first association of ZIKV infection with a severe disease, Guillain-Barré syndrome, was identified based on a rise in incidence coincident with the outbreak, which probably affected about 100,000 people (Oehler et al., 2014). Perinatal transmission was also documented, but severe congenital manifestations were only recognized in French Polynesia retrospectively after their detection in 2015 in Brazil (Besnard et al., 2016; Cauchemez et al., 2016).

The order-of-magnitude increase in the size of the French Polynesian ZIKV outbreak compared to prior epidemics undoubtedly resulted in a corresponding increase in infected air travelers, and the arrival of the ZIKV in Brazil, first detected in 2015 (Campos et al., 2015) in the northeastern region, but later dated to as early as April 2013 (Passos et al., 2017), probably involved a traveler from the South Pacific. The more than 2-year ZIKV presence in Brazil without its detection was probably the result of a lack of diagnostics or even consideration of the ZIKV in the differential diagnosis of DEN-like disease, which is widespread and common in many areas of the country. However, soon after the epidemic was first recognized in the late spring of 2015, physicians in the northeast noticed a coincident sharp increase in the incidence of fetal microcephaly (Brasil et al., 2016), and the detection of the ZIKV in the brains of fetuses with severe microcephaly (Mlakar et al., 2016) strongly suggested a ZIKV etiology. Subsequently, other surprising findings included the detection of the ZIKV in many body fluids of infected people, including saliva, urine, and semen; the latter, which can result in sexual transmission, first detected in a scientist infected in Africa (Foy et al., 2011) and later confirmed in many infected travelers (Mansuy et al., 2016).

ZIKV transmission by mosquitoes

As the ZIKV outbreak rapidly spread throughout Latin America and the Caribbean, transmission was assumed to occur principally via *Ae. aegypti*. Although few direct detections of infected mosquitoes were made during the outbreaks (Ferreira-de-Brito et al., 2016; Guerbois et al., 2016; Grubaugh et al., 2017), laboratory vector competence studies by many groups showed that this species is susceptible to oral infection and capable of transmission, although minimal or 50% infectious doses ($ID_{50}s$) are much higher than for other *Ae. aegypti*-borne viruses such as DENV, CHIKV, and YFV (Weaver et al., 2018). These findings suggested the possibility that other vectors participate to a significant degree in transmission, such as *Ae. albopictus*, which is similarly susceptible (Wong et al., 2013; Chouin-Carneiro et al., 2016; Di Luca et al., 2016; Azar et al., 2017; Liu et al., 2017) and was incriminated in the 2007 Gabon outbreak (Grard et al., 2014). However, there remains no direct evidence (natural infection during outbreaks) of a role for this species in ZIKV transmission in the Americas. *Culex quinquefasciatus* mosquitoes were also suspected of transmission based on their abundance in tropical cities and association with transmission of the closely related West Nile flavivirus. Although a few groups reported laboratory transmission competence, the only field detections involved ZIKV titers far too low to be associated with transmission competence, and the vast majority laboratory studies of this and the closely related *Cx. pipiens* showed them to be completely refractory to oral infection (Roundy et al., 2017a,b). In contrast, *Ae. aegypti* titers measured during a Mexican outbreak were much higher and consistent with transmission-competence (Azar et al., 2019).

A remaining enigma regarding ZIKV transmission by mosquitoes is that the viremia titers estimated from human blood samples are generally lower than the minimum infectious doses of *Ae. aegypti* and other mosquitoes (Table 12.2). When compared to DENV, YFV, and CHIKV that produce peak viremia titers approximately 10 times higher than the minimum infectious dose for *Ae. aegypti* or *Ae. albopictus*, ZIKV stands out as poorly infectious for these species. Possible explanations include that ZIKV viremia has been underestimated because infected people with mild disease generally seek diagnostics later than for the abovementioned viruses that cause more severe disease, or that sexual transmission contributes more to transmission and spread than has been recognized.

TABLE 12.2 Approximate viremia titers associated with DENV, YFV, CHIKV, and ZIKV human infections and 50% infectious doses (ID_{50}) for *Aedes aegypti* mosquitoes.

Virus	Approximate ID_{50} for *Ae. aegypti* oral infection (infectious units or equivalent)	Approximate peak human viremia titers (infectious units or equivalent)	References
Dengue	10^7	10^8	Vaughn et al. (2000), Whitehorn et al. (2015)
Yellow fever	10^6	10^7	Macnamara (1957), Dickson et al. (2014)
Chikungunya	$10^{3.5-5.5}$	10^6	Thiberville et al. (2013), Tsetsarkin et al. (2016)
Zika	10^{4-7}	10^3	Ciota et al. (2017), Musso et al. (2017), Roundy et al. (2017a,b)

12.3 Dengue, chikungunya, and Zika virus transmission in Texas

12.3.1 Chikungunya virus

In Texas, a laboratory-confirmed, autochthonous CHIKF case occurred in Cameron County in November 2015 (https://www.dshs.texas.gov/news/releases/2016/20160531.aspx). The CHIKV had been reported in Southern Mexico by late 2014 (Kautz et al., 2015), so northward spread through Mexico and across the border into the Rio Grande Valley of Texas was not unexpected considering the history of DEN in this region (Hafkin et al., 1982). Also, as with DEN, the number of CHIKF cases detected in Texas was relatively small, presumably reflecting less efficient transmission due to cultural differences (window and door screening, air conditioning, and more time spent indoors) that reduce human contact with *Ae. aegypti* on the US side of the border (Reiter et al., 2003). The CHIKV circulation in Texas since 2017 has not been detected, presumably because of local extinction due to inefficient epidemic transmission. However, CHIKV introductions and spread to other parts of Texas were and remain risks due to the uncertain level of circulation in Mexico coupled with the large numbers of people who regularly cross the border, as well as the immigration of air travelers from regions of Africa, Asia, and the Americas with ongoing endemic or epidemic transmission. Based on the numbers of such travelers and the populations of *Ae. aegypti*, the highest-risk regions in Texas for CHIKV, ZIKV, and DENV circulation include Brownsville and surrounding communities in the lower Rio Grande Valley, Houston, El Paso, Laredo, Dallas-Fort Worth, and San Antonio.

12.3.2 Dengue virus

Among the 48 contiguous states, nearly all DEN activity is due to travelers to endemic areas importing cases of clinical infections into the United States, including Texas (https://www.cdc.gov/dengue/index.html). The exceptions are Florida and Texas where investigations have shown that, although most clinical cases are due to travelers, there has been identification of a few laboratory-confirmed autochthonous cases and these are described later. Table 12.3 shows local transmission of DENVs over time in Texas.

The majority of eastern Texas has both *Ae. aegypti* and *Ae. albopictus*. Thus, there are opportunities for DENVs to be transmitted to humans. The DEN activity in Texas occurs in the lower Rio Grande Valley region that borders Mexico, and as suggested earlier for the CHIKV, this is in part due to the cultural differences in the region (Reiter et al., 2003).

Following the Second World War to 1980, there was no indigenous DEN activity. However, 1980 saw the first indigenous DEN activity in the United States for 35 years with 27 laboratory-confirmed cases of DENV-1 infection who had no evidence of travel outside Texas with a further nine locally transmitted cases in 1986. Four cases were reported from Brownsville, three from Corpus Christi, and two from Laredo, which were also likely to be DENV-1 infections (Rigau-Perez et al., 1994).

Subsequently, there was another DEN outbreak in 1995, of which seven cases were locally acquired in Texas (four from Hidalgo County and three from Cameron County) and were due to either DENV-2 or DENV-4 (Rawlings et al., 1998).

Ten years later, there was an epidemic of DENV-2 in Mexico that spilled over into Texas in Cameron County. There were 24 laboratory-confirmed cases in Texas, of which two were locally acquired (Centers for Disease Control and Prevention (CDC), 2007).

TABLE 12.3 Local transmission of DENVs in Texas since 1980 (Rigau-Perez et al., 1994).

Year	Number of cases	Infecting DENV
1980	27	DENV-1
1986	9	DENV-1
1995	7	DENV-2 and DENV-4
2005	2	DENV-2
2013	26	DENV-1 and DENV-3

DEN re-emerged in South Texas in 2013 with 53 laboratory-confirmed cases due to either DENV-1 or DENV-3. Of these, 26 were shown to be acquired by local transmission, and 6 had evidence of transmission of members within households (Thomas et al., 2016). Nucleotide sequencing showed the local transmission was due to strains that were circulating in Mexico at that time.

12.3.3 St. Louis encephalitis virus

SLEV is a mosquito-borne flavivirus, which like WNV is a member of the JE serocomplex, and has been endemic in Texas for a very long time. In particular, large epidemics were seen in Texas in 1964 and 1976 (Bell et al., 1981). Mosquito surveillance studies showed that SLEV was endemic in Harris County, Texas, with isolates obtained each year from 1986 to 2003 (Chandler et al., 2001; May et al., 2008). However, the ecology and epidemiology of SLEV have changed in recent years due to the introduction of the related WNV. Numbers of SLEV cases have decreased since WNV arrived in the United States. This is exemplified by the annual data for 2017 from the Centers for Diseases Control and Prevention where WNV was responsible for 2097 (92%) of all arbovirus infections in the United States, while SLEV was only responsible for 11 cases. None of the cases came from Texas (Curren et al., 2018). Overall, during 2008–17, there were only a total of 14 cases of clinical SLEV infections reported from Texas and the last cases were in 2014 when there were four cases (https://www.cdc.gov/sle/technical/epi.html).

Studies in Southern California from 2003 to 2006 suggested that WNV transmission may be displacing SLEV (Reisen et al., 2008); however, similar comparative studies have not been undertaken in Texas. Interestingly, there was a relatively small epidemic of SLE cases in California and Arizona in 2015 (Venkat et al., 2015; White et al., 2016), but no cases were seen in Texas in that year. It remains to be seen if SLEV will re-emerge in Texas again.

12.3.4 West Nile virus

The first WNV isolates in Texas were made in Harris County/Houston and Orange County (Beaumont) in June 2002 (Beasley et al., 2003; Davis et al., 2003). The virus genotype that entered New York in 1999 was named the NY99 genotype. However, the virus isolates from Texas in 2002 were not NY99 genotype. Rather, two new genotypes were identified. In Houston, the genotype was named "North America" or "WN02" (NA/WN02) (Davis et al., 2003; Ebel et al., 2004) and was characterized by 13 nucleotide changes and one amino acid substitution in the E protein (E-V159A), and phenotypically, the genotype had a shorter intrinsic incubation period in *Culex* mosquitoes than the NY99 genotype (Ebel et al., 2004). The NA/WN02 genotype has remained the dominant genotype in the United States since 2002. In Beaumont, the genotype was named "South East Coastal Texas" (SECT) due to its geographic location. Only eight isolates of this genotype were identified and characterized by five amino acid substitutions (E-T76A, NS1-E94G, NS2A-V138I, NS4B-V173I, and NS5-T526I), but not the E-V159A substitution seen in the NA/WN02 genotype. All SECT isolates were from 2002, suggesting that it became extinct because it did not have a selective advantage. Interestingly, this genotype had a naturally attenuated phenotype in mice, birds, and mosquitoes (Davis et al., 2004; Vanlandingham et al., 2008; Brault et al., 2011), which probably contributed to it not having a selective advantage over the NA/WN02 genotype.

In 2003, another genotype was found to be circulating in Texas, which was named the Southwest/WN2003 (SW/WN03) genotype as it was first identified in the Southwestern United States (Arizona, New Mexico, and Texas), and subsequently, most isolates are found in this geographic region plus California. It is characterized by 13 nucleotide changes and two amino acid substitutions (NS4A-A85T and NS5-K314R) (McMullen et al., 2011). To date, there is no information on an advantageous phenotype of this genotype. The SW/WN03 genotype still circulates, but is not dominant like the NA/02 genotype.

WNV has been isolated in Texas every year since 2002. Geographic locations with virus activity vary by year, but it is difficult to provide precise data as some areas survey for virus in mosquitoes every year, while others are based on clinical cases (see Section 12.4.1). Nonetheless, WNND cases provide the relative amount of virus activity in Texas by year. WNV is a typical neurotropic virus where approximately 80% of infections are asymptomatic, 20% cause a febrile fever, 1% are neurologic cases (WNND), and 0.1% cases are fatal. Thus, 100×WNND cases provide an idea of the number of human infections for a particular year. Fig. 12.1 shows the number of WNND cases by year in Texas. As shown, although there is WNV activity every year in Texas, epidemics are cyclic, which is typical of an arbovirus, with peak activity in 2002 and 2012. Both peak years were associated with the NA/WN02 genotype (Beasley et al., 2003; Duggal et al., 2013; Mann et al., 2013). The 2012 epidemic has been the largest to date in Texas, and studies showed that most cases were in Dallas-Fort Worth, which represented >50% of all national WNND cases in 2012. Nucleotide sequencing showed that the virus isolates in Dallas were identical to those in Houston suggesting that environmental factors contributed to more cases in Dallas. However, there was evidence of positive selection in isolates from Texas that year, in particular NS2A-I52T, demonstrating that WNV continues to evolve in the United States.

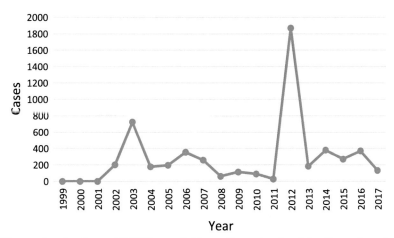

FIG. 12.1   Annual West Nile neuroinvasive disease (WNND) cases in Texas. *(Data taken from https://www.cdc.gov/westnile/statsmaps/cumMaps-Data.html.)*

12.3.5   Zika virus

Mosquito-borne ZIKV transmission in Texas was detected in both 2016 and 2017 in the lower Rio Grande Valley, with six autochthonous cases not associated with travel or sexual contact with an infected person diagnosed in late 2016 (https://www.cdc.gov/zika/reporting/2016-case-counts.html) and five cases in the summer of 2017 (https://www.cdc.gov/zika/reporting/2017-case-counts.html). Most of these occurred in Brownsville, with one possibly acquired in adjacent Hidalgo County. Several of these cases were detected following enhanced surveillance following the initial diagnosis. It is certainly possible, if not likely, that many additional local infections occurred because detailed studies of ZIKV introductions into Florida (Grubaugh et al., 2017) indicate that most result in short-lived transmission chains due to the very low R_0 values (numbers of secondary infections resulting from a primary infection, a measure of transmission efficiency where $R_0 > 1$ is needed for sustained transmission) in regions where *Ae. aegypti* do not have ready access to human hosts where air conditioners and screens are used.

12.4   Potential vectors of dengue, chikungunya, and Zika viruses in Texas and future risk

Experimental vector competence studies and historical evidence suggest that the only mosquitoes that play a major role in CHIKV, DENV, or ZIKV transmission (outside of their ancestral, enzootic cycles) in endemic or epidemic settings are *Ae. aegypti* and *Ae. albopictus*. Unfortunately, both of these species are common in many regions of Texas, especially along the Gulf Coast and the Rio Grande Valley where most transmission of these viruses has been detected. Furthermore, climate change could allow *Ae. aegypti* to become better established in warming temperate regions in northern parts of Texas. Other effects of climate change such as increased flooding and droughts could result in less predictable changes in their populations.

For the CHIKV, prior to 2005, there was no direct evidence of *Ae. albopictus* involvement despite its native presence in areas of Asia with long histories of endemicity or outbreaks. This situation changed when the CHIKV arrived in La Reunion in the Indian Ocean in 2005 following its emergence in Eastern Kenya. There, where *Ae. albopictus* is far more abundant than *Ae. aegypti*, epidemic transmission was accompanied by a genetic shift in the CHIKV population from an Alanine at position 226 of the E1 envelope glycoprotein gene to a Valine. This substitution, which later occurred convergently in Asia, was found to greatly enhance infection of *Ae. albopictus* with little impact on infection of *Ae. aegypti* (Tsetsarkin et al., 2007; Vazeille et al., 2007) and was generally, but not always, associated with virus strains detected in areas of large *Ae. albopictus* populations. Subsequently in Asia, additional mutations associated with slightly lower but still major, highly significant impacts on infection of *Ae. albopictus* were detected during continued evolution of IOL strains (Tsetsarkin and Weaver, 2011; Tsetsarkin et al., 2014). Surprisingly, none of these mutations was detected in CHIKV strains of the Asian lineage despite their circulation in *Ae. albopictus*-native regions of Asia for many decades. This was eventually explained by an epistatic mutation that accompanied the establishment of the Asian lineage from Africa, which prevents the A226V mutation from affecting vector infection in this genetic background (Tsetsarkin et al., 2011). Thus, when the Asian CHIKV lineage arrived in the Americas, it was possible to predict that *Ae. albopictus* would not be a major vector and that spread to temperate climates beyond the range of *Ae. aegypti* would not be extensive (Weaver, 2014). Similarly, the ECSA strain that arrived in Brazil in 2013 was predicted to be unable to adapt efficiently to *Ae. albopictus* due to a different epistatic mutation in some African strains (Tsetsarkin et al., 2009) (Fig. 12.2).

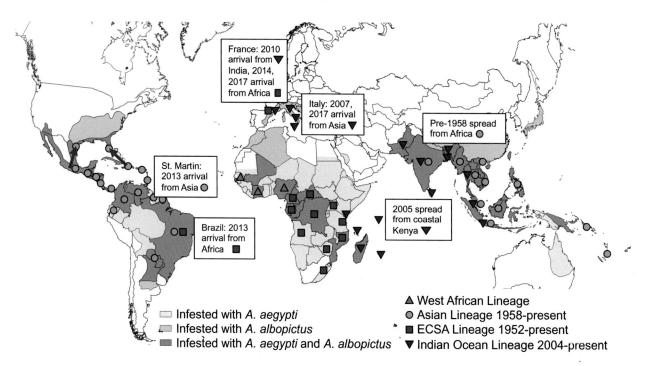

FIG. 12.2 World map with countries where autochthonous (locally initiated) chains of chikungunya virus transmission have been identified. *Red arrows* show routes of introduction that initiated transmission in the United States. *Modified from Rezza, G., Weaver, S.C., 2019. Chikungunya as a paradigm for emerging viral diseases: evaluating disease impact and hurdles to vaccine development. PLoS Negl Trop Dis. 13. e0006919. PMCID: PMCPMC6336248.*

Thus far, these predictions of the predominant role of *Ae. aegypti* in CHIKV transmission in the Americas have been supported by the lack of the adaptive mutations in sequences of both the Asian/American and ECSA/American strains found in GenBank. Furthermore, the only direct incrimination of a vector in transmission involved *Ae. aegypti,* but not *Ae. albopictus* in Southern Mexico (Diaz-Gonzalez et al., 2015). In the United States, the CHIKV has not been isolated from any mosquitoes during outbreaks in Florida or Texas, but it is assumed that *Ae. aegypti*, which was abundant in both locations, was the principal vector. Although other mosquitoes may be competent vectors based on laboratory studies, their vectorial capacity is unlikely to be as high because *Ae. aegypti* exhibits nearly ideal behaviors for transmission of a human-amplified arbovirus including endophilic and anthropophagic behaviors as well as the nearly exclusive use of peridomestic, artificial water containers for its larval habitats (Weaver, 2013).

For DENV, the vast majority of transmission is also believed to occur via *Ae. aegypti*; although the vector competence of *Ae. albopictus* is similar (Paupy et al., 2010; Whitehorn et al., 2015), the vectorial capacity of *Ae. aegypti* is believed to be higher for the same reasons described above related to its behavioral and ecological traits. Furthermore, no *Ae. albopictus*–adapted DENV strains have been described. However, a few DENV outbreaks in regions where either *Ae. albopictus* is more abundant or *Ae. aegypti* is absent have been described, suggesting that it can be the principal vector under some circumstances (Xu et al., 2007).

As for DENV, the available evidence suggests that *Ae. aegypti* is by far the most important vector of the ZIKV outside of the enzootic cycle. Aside from Gabon, there remains no direct evidence of *Ae. albopictus* transmission of the ZIKV, although relatively little mosquito sampling has occurred during outbreaks. *Aedes hensilii* is probably the other main exception, with both field (Lanciotti et al., 2008) and experimental (Ledermann et al., 2014) evidences for its principal role in the Yap outbreak.

12.4.1 Transmission control and disease prevention for dengue, chikungunya, and Zika viruses

In Texas as well as other regions of epidemic risk for human-amplified arboviruses, current control approaches remain focused on killing *Ae. aegypti* and *Ae. albopictus* as well as limiting human exposure to these mosquito vectors. However, their control is especially challenging due to their tendencies to enter and remain inside homes, to exploit artificial containers where they widely distribute their eggs, resulting in highly dispersed, low-density larval populations, and the development of resistance to insecticides, especially by *Ae. aegypti* (Gubler, 2011). Also, mosquito control resources and capabilities vary tremendously in the United States, ranging from longstanding, effective programs in Harris County/Houston to limited, seasonal capabilities in less affluent counties and municipalities, including some of the highest-risk regions of Texas in the lower Rio Grande Valley. Mosquito control in the United States including Texas is mostly funded at the local level with most state and the federal governments playing mainly supporting roles. Detection of arbovirus transmission mainly involves passive surveillance that relies on healthcare providers' familiarity with the signs and symptoms of infections and the availability of diagnostic testing. There is no systematic mechanism to educate healthcare providers about the risks and differential diagnoses of arboviral diseases, and diagnostics are mainly limited to the Texas Department of State Health Services, with some local testing when commercial assays are available. Some testing of mosquitoes for the presence of arboviruses is also performed by the Texas Department of State Health Services. Overall, our limited surveillance and public health capacity leaves the country, including Texas, vulnerable because arboviral outbreaks can amplify rapidly in the absence of early detection and diagnoses in locations with limited capabilities and then spread to threaten even the best prepared, permissive locations. In the future, our ability to control outbreaks of CHIKF as well as other human-amplified arboviruses (DENV, ZIKV, and possibly even YFV) will depend on strengthening vector control programs and increasing public engagement to eliminate larval sources of *Ae. aegypti* and *Ae. albopictus* (mainly artificial containers, especially for the former species) and reduce exposure through wearing appropriate clothing and using repellants in locations of risk.

12.4.2 Future vector control approaches

Unfortunately, the vector control approaches outlined earlier did not have a successful history of controlling the spread and incidence of DENF and also recently to control the CHIKV and ZIKV. Therefore, newer strategies that do not depend on insecticides are being developed to more specifically target *Ae. aegypti*, which overall is the most important vector of the human-amplified arboviruses. These include lethal gravid traps that have been shown to reduce *Ae. aegypti* populations and CHIKV infection in Puerto Rico (Barrera et al., 2017), genetically modified mosquitoes that, when released into the wild as adult males, mate with wild females to produce offspring that die in the larval stages to the toxic product of the transgene. This strategy, which required the sustained release of male mosquitoes that are produced in factory-like settings, has succeeded in dramatically reducing *Ae. aegypti* populations in several locations (Paes de Andrade et al., 2016). Another approach under rapid development is the use of *Wolbachia* bacteria, natural symbionts of many insects including some mosquitoes. These bacteria, including a strain adapted for infection of *Ae. aegypti*, can spread through wild populations via a mechanism termed "cytoplasmic incompatibility," whereby mating of infected males with uninfected females produce nonviable offspring. Some strains of *Wolbachia* also reduce the ability of *Ae. aegypti* to transmit DENV, CHIKV, ZIKV, and other arboviruses, so introductions are underway with the goal of interrupting transmission rather than controlling the vector. Field releases of *Wolbachia*-infected mosquitoes for suppression of arbovirus transmission, which began in Australia, have now been implemented in 10 additional countries (O'Neill, 2018). Releasing infected male *Ae. aegypti* (without females, to avoid persistence of the *Wolbachia*) reared in a laboratory setting can also be used to reduce overall wild populations (Bourtzis et al., 2014). Additional detail on these approaches is found in Chapter 8.

12.5 Vaccine development for CHIKF, DENV, WNV, and ZIKV

12.5.1 Dengue

Due to the phenomenon of ADE, the challenge for DEN vaccine development has been to generate a tetravalent vaccine that induces protective immunity against DENV-1, DENV-2, DENV-3, and DENV-4, simultaneously. This has proved to be a difficult goal due to the four DEN virus immunogens interfering with each other such that a balanced immune response has proved difficult.

DEN vaccine development started following the identification of the viruses in the 1940s. Over the next 70 years, many vaccine candidates were investigated, including formalin-inactivated virus, live attenuated, recombinant proteins, and virus vectored. However, in the 2000s chimeric recombinant live-attenuated vaccine platforms started to show promise as vaccine candidates. The rationale is that the attenuated nonstructural protein gene backbone provides attenuating mutations to help the attenuated phenotype of the chimeric virus with differences based on the particular DENV prM/E genes incorporated into the backbone. Three platforms based on recombinant DNA-derived viruses advanced into phase II and III clinical trials. One utilizes the live-attenuated yellow fever 17D vaccine virus as the backbone (termed "ChimeriVax" technology)

(Guy et al., 2010a,b). The second utilizes a DENV-2 backbone based on strain 16681 primary dog kidney passage 53 (termed DENVax technology) (Osorio et al., 2011). The third utilizes a DENV-4 backbone with a 30-nucleotide deletion in the 3′-untranslated region (3′UTR) (Whitehead, 2016).

Following phase III testing of the ChimeriVax vaccine known as Dengvaxia, it was approved in December 2015 and subsequently licensed in 20 endemic countries. However, in 2017, the vaccine producer noted that a few children who were DENV seronegative when given the vaccine had an increased risk of a hospitalized DENV infection when infected by wild-type DENV, whereas the vaccine was protective if the vaccinees were DENV seropositive at time of immunization (Rosenbaum, 2018; Sridhar et al., 2018). Consequently, the vaccine producer voluntarily removed the vaccine from use unless vaccinees could be shown to be DENV seropositive at time of immunization. In 2018 the vaccine producer applied to have the vaccine licensed in the United State and the FDA gave limited approval in 2019 for those aged 9–16 years who had serological evidence of prior DENV infection.

The other two vaccines are currently being evaluated in phase 3 clinical trials.

12.5.2 Chikungunya

Development of vaccines to prevent CHIKF began in the 1980s and accelerated after the emergence of the IOL in 2004 and the arrival of the CHIKV in the Americas in 2013–14. A wide variety of platforms and approaches was developed and tested in preclinical trials, and three have entered clinical trials: (1) a measles virus–vectored vaccine that expresses the CHIKV structural proteins. A phase 1 clinical trial indicates that this vaccine is safe and highly immunogenic after two doses (Ramsauer et al., 2015) and the candidate vaccine progressed to phase 2 clinical trials; (2) a virus-like particle vaccine made by expressing CHIKV structural proteins in vertebrate cells, where they self-assemble into particles that are identical in external structure to wild-type virus. This vaccine has also performed well in phase 1 trials with demonstrated safety and immunogenicity after two doses (Chang et al., 2014). Most recently, a phase 1 trial (clinical trials.gov NCT03382964) began with a live-attenuated CHIKV strain derived by deletion of a portion of the nonstructural protein 3 gene (Hallengard et al., 2014). The licensure of any of these vaccines will depend on phase 3 clinical trials demonstrating efficacy in protecting against CHIKF. This will be a major challenge considering the difficulty in identifying a site(s) with high enough incidence of disease to adequately power the trial and the cost, typically hundreds-of-millions of US dollars. Further information on CHIKF vaccines is reviewed elsewhere (Erasmus et al., 2016; Powers, 2018; Rezza and Weaver, 2019).

12.5.3 West Nile virus

There are a number of licensed veterinary vaccines that are used to protect horses in the United States. Most are based on formalin-inactivated WNV with a two-dose regimen given 21 days apart and require annual booster doses (Ng et al., 2003). In addition, there is a canarypox-vectored vaccine that incorporates prM/E genes of WNV (Minke et al., 2004). There is no licensed human vaccine, but there are a number of candidate vaccines in various levels of development. Some are based on formalin-inactivated virus, and there are also DNA and chimeric live-attenuated vaccines. The DNA vaccine is based on the prM/E genes and has progressed to phase 1 clinical trials (Martin et al., 2007; Ledgerwood et al., 2011). The live-attenuated yellow fever 17D vaccine virus has been used as a backbone to develop a chimeric virus containing the prM and E protein genes of WNV and the nonstructural protein genes of 17D virus (Guy et al., 2010a,b; Dayan et al., 2013). The rationale is that the attenuated YF nonstructural protein backbone provides attenuating mutations to help the attenuated phenotype of the chimeric virus. This ChimeriVax platform has been used previously to develop candidate vaccines for DEN and JE that are licensed as Dengvaxia and Imojev, respectively. The candidate ChimeriVax-WN vaccine has completed phase 2 clinical trials (Biedenbender et al., 2011). Similar technology has been used to develop a live DENV-4 chimeric virus with the WNV prM and E protein genes, which is based on the live DEN-4 vaccine backbone (Pletnev et al., 2002). This candidate vaccine has completed phase 1 clinical trials (Durbin et al., 2013; Pierce et al., 2017). No vaccine candidate has progressed to phase 3 clinical trials.

12.5.4 Zika virus

The rapid emergence of Zika as a major public health concern in 2015 resulted in many approaches being taken to developing a Zika vaccine. Overall, nearly 50 vaccine candidates have been in discovery phase; of these, approximately 24 progressed to preclinical development in mouse and NHP models, and 9–12 have progressed to clinical evaluation. Those technologies in phase 1 clinical trials include formalin-inactivated virus, DNA, RNA, and live attenuated. To date, phase 1 clinical trial data have been published on the DNA and formalin-activated candidates (Gaudinski et al., 2018;

Modjarrad et al., 2018). All appear safe, and preliminary data indicate they are immunogenic; however, whether or not one or more will induce protective immunity in humans is not known. For a review, see Barrett (2018).

12.6 Viruses with future emergence potential in Texas

Table 12.4 shows a list of arboviruses with potential to emerge in Texas in the future. Owing to its subtropical climate and diversity of vectors including *Ae. aegypti* and *Ae. albopictus*, Texas has been one of the two highest-risk states for the invasion of tropical, urban arboviruses as outlined earlier for DENV, YFV, CHIKV, and ZIKV. The first three have probably been introduced repeatedly into coastal regions of Texas by ships arriving from the Caribbean and Africa for centuries (Halstead, 2015). Unless major advances are made in controlling their spread through vaccination or vector control, Texas will remain vulnerable to future invasions, even for YFV if major urban outbreaks return to Latin America or the Caribbean. The risk of YFV was demonstrated in 2002 when an unvaccinated traveler visited Manaus, Brazil, was infected with YFV and had clinical disease when he returned to Corpus Christi, Texas. There were no secondary cases from the YFV-infected traveler (CDC, 2002).

TABLE 12.4 Partial list of arboviruses with potential to emerge in Texas in the future.

Family	Genus	Virus	Serocomplex	Vector	Vertebrate host
Flaviviridae	*Flavivirus*	Dengue-1	Dengue	*Ae. aegypti, Ae. albopictus*	Primates
		Dengue-2			
		Dengue-3			
		Dengue-4			
		Japanese encephalitis	Japanese encephalitis	*Culex* spp.[a]	Birds
		Usutu	Japanese encephalitis	*Culex* spp.[a]	Birds
		Yellow fever	Not applicable	*Ae. aegypti*	Primates
		Zika	Not applicable	*Ae. aegypti, Ae. albopictus*	Primates
Peribunyaviridae	*Orthobunyavirus*	Cache Valley	Bunyamwera	*Aedes* spp.	Ruminants
				Anopheles spp.	
				Coquillettidia spp.	
				Culiseta spp.	
		Oropouche	Simbu	*Coquillettidia* spp.[a]	Primates and birds
				Cx. quinquefasciatus	
				Ae. serratus	
Phleboviridae	*Phlebovirus*	Rift Valley fever		*Aedes* spp.[a]	Wildlife and livestock
Togaviridae	*Alphavirus*	Chikungunya	Semliki Forest	*Ae. aegypti, Ae. albopictus*	Primates
		Mayaro	Semliki Forest	*Ae. aegypti*[a]	Primates

[a] *Potential vectors in Texas.*

To our current knowledge, these four human-amplified, mosquito-borne viruses are the only ones with evidence of sustained urban transmission. However, ZIKV and its lack of a history of urban transmission before 2007 underscores that possibility plus other viruses with similar potential remain understudied. Recent attempts to examine this possibility have focused on close relatives of the ZIKV such as Spondweni, which was originally confused with the ZIKV due to its close antigenic relationship and presence in sub-Saharan Africa (Haddow and Woodall, 2016). One approach to examining the emergence potential of such viruses is to simply examine their ability to orally infect *Ae. aegypti* and *Ae. albopictus*.

Fortunately, such a study with Spondweni virus demonstrated that they are highly refractory, suggesting little potential for urbanization (Haddow et al., 2016). Additional studies should be undertaken with close relatives of the four human-amplified arboviruses discussed earlier.

A close relative of the CHIKV with a high risk for establishment of an urban, *Ae. aegypti*-amplified transmission cycle is the MAYV. Like DENV, YFV, CHIKV, and ZIKV, the MAYV is believed to have originated in a sylvatic, NHP-amplified cycle, but in this case, it originated in the neotropics where it continues to circulate. Spillover infections have been recognized since the 1950s (Causey and Maroja, 1957), affecting up to hundreds of people living near or working in neotropical forested habitats (LeDuc et al., 1981; Auguste et al., 2015). Although there is no evidence or urban transmission in regions without direct exposure to the NHP-sylvatic mosquito cycle, humans develop viremia levels that may be sufficient for infection of *Ae. aegypti* (Long et al., 2011). The recent identification of MAYV in an infected child in a Haiti (Lednicky et al., 2016) suggests its possible geographic expansion into a region without wild NHPs, possibly indicating that an urban transmission cycle is underway or that other nonhuman amplification hosts can maintain enzootic circulation.

Additional arboviral risks to Texas include other arboviruses that can be introduced and remain in permanent enzootic cycles, as exemplified by WNV. Once present, North American avian hosts and mosquito vectors were highly competent to initiate and maintain enzootic transmission with nearly continuous spillover infections of people.

Other obvious threats in this category include JEV, which can use a wide variety of shorebirds as enzootic hosts as well as several species of *Culex* mosquitoes as vectors. This virus was first detected outside of Asia in 2016, when a coinfection with YFV was diagnosed during the Angola yellow fever epidemic (Simon-Loriere et al., 2017). Although human and veterinary vaccines are available, this virus could be particularly devastating to both public health and agriculture if introduced into Texas because vaccine coverage is minimal (mainly travelers to Asia) and swine, domesticated or feral could serve as amplifying hosts. Human infection can lead to fatal encephalitis, particularly in children.

The JE serocomplex of the *Flavivirus* genus contains 10 viruses, including WNV, SLEV, and JEV. Another emerging pathogen in this serocomplex is the Usutu virus (USUV). As with WNV, the USUV is an African *Flavivirus* that has spread to Europe. The virus was first isolated in South Africa in 1959 and then subsequently emerged in Austria in 2001 where it caused a significant die-off of Eurasian blackbirds, *Turdus merula*. Since 2001 the virus has increased its geographic range in Europe with either serological evidence or virus isolates made in Italy, Germany, Spain, Hungary, Switzerland, Poland, the United Kingdom, the Czech Republic, Greece, and Belgium from USUV-infected birds. In 2009 the first human cases were reported from Italy, albeit they were immunocompromised individuals. However, the clinical cases were similar to WNND. As with other members of the JEV serocomplex, the USUV utilizes avian species as the vertebrate hosts and *Culex* spp. as mosquito hosts, but like JEV and WNV, it can also infect many other vertebrate and mosquito species. Among the mosquito species, *Ae. albopictus*, *Ae. caspius* (Pallas), *Anopheles maculipennis* Meigen, *Coquillettidia aurites* (Theobald), and *Mansonia africana* (Theobald) have been shown to be vectors for the virus (Ashraf et al., 2015; Cheng et al., 2018). Significantly, a phylogeographic study suggests that the major migratory bird flyways may facilitate dispersal of the USUV, including long-distance intercontinental spread as seen with WNV (Engel et al., 2016). Given the availability of mosquito and avian hosts for the USUV in North America, it is possible the virus may emerge in North America and spread to Texas.

Another example is Rift Valley fever, a bunyavirus that has been shown to be capable of transmission by North America mosquito species (Turell et al., 2008) and is highly pathogenic to both people and livestock. There is no licensed vaccine for either human or veterinary use, and both public health and agriculture could be severely damaged. Two other bunyaviruses that should be considered are Cache Valley virus (CVV) and Oropouche virus (OROV), which are both in the genus *Orthobunyavirus* (family Peribunyaviridae, order Bunyavirales). The transmission cycle of CVV involves various mosquito species, including *Ae. albopictus*, and ruminants as the vertebrate hosts. The virus was first identified in 1956 from *Culiseta inornata* (Williston) mosquitoes in Cache Valley, Utah, and is found throughout the United States. It is a veterinary pathogen that infects sheep, is teratogenic, and causes abortion; there have been very few human cases with the first case described in 2005. There were outbreaks of CVV in Texas in 1981 and 1987 involving congenital malformations of sheep (Chung et al., 1991). Since then, there have been no reported outbreaks in Texas or anywhere else in the United States. Thus, the potential for re-emergence of CVV exists. The OROV was first isolated in Trinidad in 1955 from humans and *Cq. venezuelensis* (Theobald). It is the second most prevalent arbovirus in tropical South America after DEN virus and is found in Trinidad, Panama, and tropical South America (Travassos da Rosa et al., 2017). Clinically, ORO fever causes an acute febrile illness characterized by headache, chills, dizziness, photophobia, myalgia, nausea, and vomiting of 3–7 days duration. It is difficult to differentiate ORO fever from classical DEN. A few patients with this disease also develop aseptic meningitis, although there has been no reported mortality. The sylvatic cycle is poorly understood, but available information suggests the three-toed sloth, *Bradypus tridactylus*, and *Callithrix* spp. (marmosets) are the vertebrate hosts, although bird species have shown to have antibodies to the virus. The arthropod host is equally unclear with *Culicoides paraensis* (Goeldi), *Ae. serratus* (Theobald), *Cq. venezuelensis*, and *Cx. p. quinquefasciatus* all considered possibly arthropod vectors.

The urban epidemic cycle involves *C. paraensis* and *Cx. quinquefasciatus* with humans as the vertebrate host. Significantly, *C. paraensis* is widespread in the Americas, including as far north as Wisconsin. Thus, there is potential for the virus to become established in Texas.

Finally, the threat from currently unknown arboviruses is difficult to gauge. While the rate of arbovirus discovery has declined from its peak during the 20th century, there are undoubtedly many more remaining to be discovered, especially those that do not announce their presence by causing cytopathic effects on cells typically used for screening clinical samples or vector collections, such as Vero or mosquito cells, or those that defy detection based on PCR primers designed to react to sequences designed from known virus groups. Next-generation sequencing is generating vast new databases of virus sequences from organisms rarely sampled in the past (Geoghegan and Holmes, 2017). Data to date suggest that no new virus families have been discovered, but the lack of phenotypic characterization of these putative viruses limits assessment of their potential risk to people and other animals. Sequence-based predictions of host range and pathogenicity remain in their infancy and the resources to isolate and study more than a small minority of these viruses will limit progress for the foreseeable future.

12.7 Conclusion

Texas has a long history with mosquito-borne arbovirus introductions, from yellow fever, dengue, and chikungunya in past centuries and the latter two continuing in recent years to WNV and ZIKV since 1999. In addition, tick-borne viruses discovered recently cause an unknown burden of disease. Improved surveillance, education of the public, and vector control offer the best near-term approaches to controlling these vector-borne diseases, while vaccines are ultimately needed for several of these viruses.

Acknowledgments

Research on arboviruses by the authors is supported by NIH grants AI121452 and AI120942 to SCW and AI129844 to ADTB and by the Western Gulf Center of Excellence for Vector-borne Diseases (U01-CK00512, Centers for Disease Control and Prevention).

References and further reading

Ashraf, U., Ye, J., Ruan, X., Wan, S., Zhu, B., Cao, S., 2015. Usutu virus: an emerging flavivirus in Europe. Viruses 7 (1), 219–238.

Auguste, A.J., Liria, J., Forrester, N.L., Giambalvo, D., Moncada, M., Long, K.C., Moron, D., de Manzione, N., Tesh, R.B., Halsey, E.S., Kochel, T.J., Hernandez, R., Navarro, J.C., Weaver, S.C., 2015. Evolutionary and ecological characterization of Mayaro virus strains isolated during an outbreak, Venezuela, 2010. Emerg. Infect. Dis. 21 (10), 1742–1750. PMCID: PMCPMC4593426.

Azar, S.R., Roundy, C.M., Rossi, S.L., Huang, J.H., Leal, G., Yun, R., Fernandez-Salas, I., Vitek, C.J., Paploski, I.A.D., Stark, P.M., Vela, J., Debboun, M., Reyna, M., Kitron, U., Ribeiro, G.S., Hanley, K.A., Vasilakis, N., Weaver, S.C., 2017. Differential vector competency of *Aedes albopictus* populations from the Americas for Zika virus. Am. J. Trop. Med. Hyg. 97 (2), 330–339.

Azar, S.R., Diaz-Gonzalez, E.E., Danis-Lonzano, R., Fernandez-Salas, I., Weaver, S.C., 2019. Naturally infected *Aedes aegypti* collected during a Zika virus outbreak have viral titres consistent with transmission. Emerg. Microbes. Infect. 8, 242–244.

Barrera, R., Acevedo, V., Felix, G.E., Hemme, R.R., Vazquez, J., Munoz, J.L., Amador, M., 2017. Impact of Autocidal Gravid Ovitraps on chikungunya virus incidence in *Aedes aegypti* (Diptera: Culicidae) in areas with and without traps. J. Med. Entomol. 54 (2), 387–395.

Barrett, A.D.T., 2018. Current status of Zika vaccine development: Zika vaccines advance into clinical evaluation. NPJ Vaccines 3, 24.

Beasley, D.W., Davis, C.T., Guzman, H., Vanlandingham, D.L., Travassos da Rosa, A.P., Parsons, R.E., Higgs, S., Tesh, R.B., Barrett, A.D., 2003. Limited evolution of West Nile virus has occurred during its southwesterly spread in the United States. Virology 309 (2), 190–195.

Beasley, D.W., Davis, C.T., Estrada-Franco, J., Navarro-Lopez, R., Campomanes-Cortes, A., Tesh, R.B., Weaver, S.C., Barrett, A.D., 2004. Genome sequence and attenuating mutations in West Nile virus isolate from Mexico. Emerg. Infect. Dis. 10 (12), 2221–2224.

Bell, R.L., Christensen, B., Holguin, A., Smith, O., 1981. St. Louis encephalitis: a comparison of two epidemics in Harris County, Texas. Am. J. Public Health 71 (2), 168–170.

Besnard, M., Eyrolle-Guignot, D., Guillemette-Artur, P., Lastere, S., Bost-Bezeaud, F., Marcelis, L., Abadie, V., Garel, C., Moutard, M.L., Jouannic, J.M., Rozenberg, F., Leparc-Goffart, I., Mallet, H.P., 2016. Congenital cerebral malformations and dysfunction in fetuses and newborns following the 2013 to 2014 Zika virus epidemic in French Polynesia. Euro Surveill. 21 (13), 1–9.

Bhatt, S., Gething, P.W., Brady, O.J., Messina, J.P., Farlow, A.W., Moyes, C.L., Drake, J.M., Brownstein, J.S., Hoen, A.G., Sankoh, O., Myers, M.F., George, D.B., Jaenisch, T., Wint, G.R., Simmons, C.P., Scott, T.W., Farrar, J.J., Hay, S.I., 2013. The global distribution and burden of dengue. Nature 496 (7446), 504–507.

Biedenbender, R., Bevilacqua, J., Gregg, A.M., Watson, M., Dayan, G., 2011. Phase II, randomized, double-blind, placebo-controlled, multicenter study to investigate the immunogenicity and safety of a West Nile virus vaccine in healthy adults. J. Infect. Dis. 203 (1), 75–84.

Bourtzis, K., Dobson, S.L., Xi, Z., Rasgon, J.L., Calvitti, M., Moreira, L.A., Bossin, H.C., Moretti, R., Baton, L.A., Hughes, G.L., Mavingui, P., Gilles, J.R., 2014. Harnessing mosquito-*Wolbachia* symbiosis for vector and disease control. Acta Trop. 132 (Suppl), S150–S163.

Brasil, P., Pereira Jr., J.P., Moreira, M.E., Ribeiro Nogueira, R.M., Damasceno, L., Wakimoto, M., Rabello, R.S., Valderramos, S.G., Halai, U.A., Salles, T.S., Zin, A.A., Horovitz, D., Daltro, P., Boechat, M., Raja Gabaglia, C., Carvalho de Sequeira, P., Pilotto, J.H., Medialdea-Carrera, R., Cotrim da Cunha, D., Abreu de Carvalho, L.M., Pone, M., Machado Siqueira, A., Calvet, G.A., Rodrigues Baiao, A.E., Neves, E.S., Nassar de Carvalho, P.R., Hasue, R.H., Marschik, P.B., Einspieler, C., Janzen, C., Cherry, J.D., Bispo de Filippis, A.M., Nielsen-Saines, K., 2016. Zika virus infection in pregnant women in Rio de Janeiro. N. Engl. J. Med. 375 (24), 2321–2334.

Brault, A.C., Huang, C.Y., Langevin, S.A., Kinney, R.M., Bowen, R.A., Ramey, W.N., Panella, N.A., Holmes, E.C., Powers, A.M., Miller, B.R., 2007. A single positively selected West Nile viral mutation confers increased virogenesis in American crows. Nat. Genet. 39 (9), 1162–1166.

Brault, A.C., Langevin, S.A., Ramey, W.N., Fang, Y., Beasley, D.W., Barker, C.M., Sanders, T.A., Reisen, W.K., Barrett, A.D., Bowen, R.A., 2011. Reduced avian virulence and viremia of West Nile virus isolates from Mexico and Texas. Am. J. Trop. Med. Hyg. 85 (4), 758–767.

Campos, G.S., Bandeira, A.C., Sardi, S.I., 2015. Zika Virus Outbreak, Bahia, Brazil. Emerg. Infect. Dis. 21 (10), 1885–1886.

Carey, D.E., 1971. Chikungunya and dengue: a case of mistaken identity? J. Hist. Med. Allied Sci. 26 (3), 243–262.

Cauchemez, S., Besnard, M., Bompard, P., Dub, T., Guillemette-Artur, P., Eyrolle-Guignot, D., Salje, H., Van Kerkhove, M.D., Abadie, V., Garel, C., Fontanet, A., Mallet, H.P., 2016. Association between Zika virus and microcephaly in French Polynesia, 2013–15: a retrospective study. Lancet 387 (10033), 2125–2132.

Causey, O.R., Maroja, O.M., 1957. Mayaro virus: a new human disease agent. III. Investigation of an epidemic of acute febrile illness on the river Guama in Para, Brazil, and isolation of Mayaro virus as causative agent. Am. J. Trop. Med. Hyg. 6 (6), 1017–1023.

CDC, 2002. Fatal yellow fever in a traveler returning from Amazonas, Brazil, 2002. MMWR Morb. Mortal. Wkly Rep. 51 (15), 324–325.

CDC, 2007. Dengue hemorrhagic fever—U.S.-Mexico border, 2005. MMWR Morb. Mortal. Wkly Rep. 56 (31), 785–789.

Chandler, L.J., Parsons, R., Randle, Y., 2001. Multiple genotypes of St. Louis encephalitis virus (Flaviviridae: Flavivirus) circulate in Harris County, Texas. Am. J. Trop. Med. Hyg. 64 (1–2), 12–19.

Chang, L.J., Dowd, K.A., Mendoza, F.H., Saunders, J.G., Sitar, S., Plummer, S.H., Yamshchikov, G., Sarwar, U.N., Hu, Z., Enama, M.E., Bailer, R.T., Koup, R.A., Schwartz, R.M., Akahata, W., Nabel, G.J., Mascola, J.R., Pierson, T.C., Graham, B.S., Ledgerwood, J.E., The V. R. C. S. T, 2014. Safety and tolerability of chikungunya virus-like particle vaccine in healthy adults: a phase 1 dose-escalation trial. Lancet 384 (9959), 2046–2052.

Cheng, Y., Tjaden, N.B., Jaeschke, A., Luhken, R., Ziegler, U., Thomas, S.M., Beierkuhnlein, C., 2018. Evaluating the risk for Usutu virus circulation in Europe: comparison of environmental niche models and epidemiological models. Int. J. Health Geogr. 17 (1), 35.

Chouin-Carneiro, T., Vega-Rua, A., Vazeille, M., Yebakima, A., Girod, R., Goindin, D., Dupont-Rouzeyrol, M., Lourenco-de-Oliveira, R., Failloux, A.B., 2016. Differential susceptibilities of *Aedes aegypti* and *Aedes albopictus* from the Americas to Zika virus. PLoS Negl. Trop. Dis. 10 (3), e0004543.

Chretien, J.P., Anyamba, A., Bedno, S.A., Breiman, R.F., Sang, R., Sergon, K., Powers, A.M., Onyango, C.O., Small, J., Tucker, C.J., Linthicum, K.J., 2007. Drought-associated chikungunya emergence along coastal East Africa. Am. J. Trop. Med. Hyg. 76 (3), 405–407.

Chung, S.I., Livingston Jr., C.W., Jones, C.W., Collisson, E.W., 1991. Cache Valley virus infection in Texas sheep flocks. J. Am. Vet. Med. Assoc. 199 (3), 337–340.

Ciota, A.T., Bialosuknia, S.M., Zink, S.D., Brecher, M., Ehrbar, D.J., Morrissette, M.N., Kramer, L.D., 2017. Effects of Zika virus strain and *Aedes* Mosquito species on vector competence. Emerg. Infect. Dis. 23 (7), 1110–1117.

Curren, E.J., Lindsey, N.P., Fischer, M., Hills, S.L., 2018. St. Louis encephalitis virus disease in the United States, 2003–2017. Am. J. Trop. Med. Hyg. 99 (4), 1074–1079.

Davis, C.T., Beasley, D.W., Guzman, H., Raj, R., D'Anton, M., Novak, R.J., Unnasch, T.R., Tesh, R.B., Barrett, A.D., 2003. Genetic variation among temporally and geographically distinct West Nile virus isolates, United States, 2001, 2002. Emerg. Infect. Dis. 9 (11), 1423–1429.

Davis, C.T., Beasley, D.W., Guzman, H., Siirin, M., Parsons, R.E., Tesh, R.B., Barrett, A.D., 2004. Emergence of attenuated West Nile virus variants in Texas, 2003. Virology 330 (1), 342–350.

Dayan, G.H., Pugachev, K., Bevilacqua, J., Lang, J., Monath, T.P., 2013. Preclinical and clinical development of a YFV 17 D-based chimeric vaccine against West Nile virus. Viruses 5 (12), 3048–3070.

Di Luca, M., Severini, F., Toma, L., Boccolini, D., Romi, R., Remoli, M.E., Sabbatucci, M., Rizzo, C., Venturi, G., Rezza, G., Fortuna, C., 2016. Experimental studies of susceptibility of Italian *Aedes albopictus* to Zika virus. Euro Surveill. 21 (18), 6–9.

Diallo, M., Thonnon, J., Traore-Lamizana, M., Fontenille, D., 1999. Vectors of chikungunya virus in Senegal: current data and transmission cycles. Am. J. Trop. Med. Hyg. 60 (2), 281–286.

Diaz-Gonzalez, E.E., Kautz, T.F., Dorantes-Delgado, A., Malo-Garcia, I.R., Laguna-Aguilar, M., Langsjoen, R.M., Chen, R., Auguste, D.I., Sanchez-Casas, R.M., Danis-Lozano, R., Weaver, S.C., Fernandez-Salas, I., 2015. First report of *Aedes aegypti* transmission of chikungunya virus in the Americas. Am. J. Trop. Med. Hyg. 93 (6), 1325–1329.

Dick, G.W., 1952. Zika virus. II. Pathogenicity and physical properties. Trans. R. Soc. Trop. Med. Hyg. 46 (5), 521–534.

Dick, G.W., Kitchen, S.F., Haddow, A.J., 1952. Zika virus. I. Isolations and serological specificity. Trans. R. Soc. Trop. Med. Hyg. 46 (5), 509–520.

Dickson, L.B., Sanchez-Vargas, I., Sylla, M., Fleming, K., Black 4th., W.C., 2014. Vector competence in West African *Aedes aegypti* Is Flavivirus species and genotype dependent. PLoS Negl. Trop. Dis. 8 (10). e3153.

Dietrich, E.A., Langevin, S.A., Huang, C.Y., Maharaj, P.D., Delorey, M.J., Bowen, R.A., Kinney, R.M., Brault, A.C., 2016. West Nile virus temperature sensitivity and avian virulence are modulated by NS1-2B polymorphisms. PLoS Negl. Trop. Dis. 10 (8), e0004938.

Duffy, M.R., Chen, T.H., Hancock, W.T., Powers, A.M., Kool, J.L., Lanciotti, R.S., Pretrick, M., Marfel, M., Holzbauer, S., Dubray, C., Guillaumot, L., Griggs, A., Bel, M., Lambert, A.J., Laven, J., Kosoy, O., Panella, A., Biggerstaff, B.J., Fischer, M., Hayes, E.B., 2009. Zika virus outbreak on Yap Island, Federated States of Micronesia. N. Engl. J. Med. 360 (24), 2536–2543.

Duggal, N.K., D'Anton, M., Xiang, J., Seiferth, R., Day, J., Nasci, R., Brault, A.C., 2013. Sequence analyses of 2012 West Nile virus isolates from Texas fail to associate viral genetic factors with outbreak magnitude. Am. J. Trop. Med. Hyg. 89 (2), 205–210.

Durbin, A.P., Wright, P.F., Cox, A., Kagucia, W., Elwood, D., Henderson, S., Wanionek, K., Speicher, J., Whitehead, S.S., Pletnev, A.G., 2013. The live attenuated chimeric vaccine rWN/DEN4Delta30 is well-tolerated and immunogenic in healthy flavivirus-naive adult volunteers. Vaccine 31 (48), 5772–5777.

Ebel, G.D., Carricaburu, J., Young, D., Bernard, K.A., Kramer, L.D., 2004. Genetic and phenotypic variation of West Nile virus in New York, 2000–2003. Am. J. Trop. Med. Hyg. 71 (4), 493–500.

Engel, D., Jost, H., Wink, M., Borstler, J., Bosch, S., Garigliany, M.M., Jost, A., Czajka, C., Luhken, R., Ziegler, U., Groschup, M.H., Pfeffer, M., Becker, N., Cadar, D., Schmidt-Chanasit, J., 2016. Reconstruction of the evolutionary history and dispersal of Usutu virus, a neglected emerging arbovirus in Europe and Africa. MBio 7 (1). e01938-15.

Erasmus, J.H., Rossi, S.L., Weaver, S.C., 2016. Development of vaccines for chikungunya fever. J. Infect. Dis. 214 (Suppl. 5), S488–S496.

Ferreira-de-Brito, A., Ribeiro, I.P., Miranda, R.M., Fernandes, R.S., Campos, S.S., Silva, K.A., Castro, M.G., Bonaldo, M.C., Brasil, P., Lourenco-de-Oliveira, R., 2016. First detection of natural infection of *Aedes aegypti* with Zika virus in Brazil and throughout South America. Mem. Inst. Oswaldo Cruz 111 (10), 655–658.

Foy, B.D., Kobylinski, K.C., Chilson Foy, J.L., Blitvich, B.J., Travassos da Rosa, A., Haddow, A.D., Lanciotti, R.S., Tesh, R.B., 2011. Probable non-vector-borne transmission of Zika virus, Colorado, USA. Emerg. Infect. Dis. 17 (5), 880–882.

Gaudinski, M.R., Houser, K.V., Morabito, K.M., Hu, Z., Yamshchikov, G., Rothwell, R.S., Berkowitz, N., Mendoza, F., Saunders, J.G., Novik, L., Hendel, C.S., Holman, L.A., Gordon, I.J., Cox, J.H., Edupuganti, S., McArthur, M.A., Rouphael, N.G., Lyke, K.E., Cummings, G.E., Sitar, S., Bailer, R.T., Foreman, B.M., Burgomaster, K., Pelc, R.S., Gordon, D.N., DeMaso, C.R., Dowd, K.A., Laurencot, C., Schwartz, R.M., Mascola, J.R., Graham, B.S., Pierson, T.C., Ledgerwood, J.E., Chen, G.L., VRC Teams, 2018. Safety, tolerability, and immunogenicity of two Zika virus DNA vaccine candidates in healthy adults: randomised, open-label, phase 1 clinical trials. Lancet 391 (10120), 552–562.

Geoghegan, J.L., Holmes, E.C., 2017. Predicting virus emergence amid evolutionary noise. Open Biol. 7 (10), 1–9.

Grard, G., Caron, M., Mombo, I.M., Nkoghe, D., Mboui Ondo, S., Jiolle, D., Fontenille, D., Paupy, C., Leroy, E.M., 2014. Zika virus in Gabon (Central Africa)—2007: a new threat from *Aedes albopictus*? PLoS Negl. Trop. Dis. 8 (2), e2681.

Grubaugh, N.D., Ladner, J.T., Kraemer, M.U.G., Dudas, G., Tan, A.L., Gangavarapu, K., Wiley, M.R., White, S., Theze, J., Magnani, D.M., Prieto, K., Reyes, D., Bingham, A.M., Paul, L.M., Robles-Sikisaka, R., Oliveira, G., Pronty, D., Barcellona, C.M., Metsky, H.C., Baniecki, M.L., Barnes, K.G., Chak, B., Freije, C.A., Gladden-Young, A., Gnirke, A., Luo, C., MacInnis, B., Matranga, C.B., Park, D.J., Qu, J., Schaffner, S.F., Tomkins-Tinch, C., West, K.L., Winnicki, S.M., Wohl, S., Yozwiak, N.L., Quick, J., Fauver, J.R., Khan, K., Brent, S.E., Reiner Jr., R.C., Lichtenberger, P.N., Ricciardi, M.J., Bailey, V.K., Watkins, D.I., Cone, M.R., Kopp 4th, E.W., Hogan, K.N., Cannons, A.C., Jean, R., Monaghan, A.J., Garry, R.F., Loman, N.J., Faria, N.R., Porcelli, M.C., Vasquez, C., Nagle, E.R., Cummings, D.A.T., Stanek, D., Rambaut, A., Sanchez-Lockhart, M., Sabeti, P.C., Gillis, L.D., Michael, S.F., Bedford, T., Pybus, O.G., Isern, S., Palacios, G., Andersen, K.G., 2017. Genomic epidemiology reveals multiple introductions of Zika virus into the United States. Nature 546 (7658), 401–405.

Gubler, D.J., 2011. Dengue, urbanization and globalization: the unholy trinity of the 21(st) century. Trop. Med. Health 39 (4), 3–11. Suppl.

Guerbois, M., Fernandez-Salas, I., Azar, S.R., Danis-Lozano, R., Alpuche-Aranda, C.M., Leal, G., Garcia-Malo, I.R., Diaz-Gonzalez, E.E., Casas-Martinez, M., Rossi, S.L., Del Rio-Galvan, S.L., Sanchez-Casas, R.M., Roundy, C.M., Wood, T.G., Widen, S.G., Vasilakis, N., Weaver, S.C., 2016. Outbreak of Zika virus infection, Chiapas State, Mexico, 2015, and first confirmed transmission by *Aedes aegypti* mosquitoes in the Americas. J. Infect. Dis. 214 (9), 1349–1356.

Guy, B., Guirakhoo, F., Barban, V., Higgs, S., Monath, T.P., Lang, J., 2010a. Preclinical and clinical development of YFV 17D-based chimeric vaccines against dengue, West Nile and Japanese encephalitis viruses. Vaccine 28 (3), 632–649.

Guy, B., Saville, M., Lang, J., 2010b. Development of Sanofi Pasteur tetravalent dengue vaccine. Hum. Vaccin. 6 (9), 696–705.

Guzman, M.G., Halstead, S.B., Artsob, H., Buchy, P., Farrar, J., Gubler, D.J., Hunsperger, E., Kroeger, A., Margolis, H.S., Martinez, E., Nathan, M.B., Pelegrino, J.L., Simmons, C., Yoksan, S., Peeling, R.W., 2010. Dengue: a continuing global threat. Nat. Rev. Microbiol. 8 (12 Suppl), S7–16.

Haddow, A.D., Woodall, J.P., 2016. Distinguishing between Zika and Spondweni viruses. Bull. World Health Organ. 94 (10), 711-A.

Haddow, A.J., Williams, M.C., Woodall, J.P., Simpson, D.I., Goma, L.K., 1964. Twelve isolations of Zika virus from *Aedes (Stegomyia) africanus* (Theobald) taken in and above a Uganda Forest. Bull. World Health Organ. 31, 57–69.

Haddow, A.D., Schuh, A.J., Yasuda, C.Y., Kasper, M.R., Heang, V., Huy, R., Guzman, H., Tesh, R.B., Weaver, S.C., 2012. Genetic characterization of zika virus strains: geographic expansion of the Asian lineage. PLoS Negl. Trop. Dis. 6 (2), e1477.

Haddow, A.D., Nasar, F., Guzman, H., Ponlawat, A., Jarman, R.G., Tesh, R.B., Weaver, S.C., 2016. Genetic characterization of Spondweni and Zika viruses and susceptibility of geographically distinct strains of *Aedes aegypti, Aedes albopictus* and *Culex quinquefasciatus* (Diptera: Culicidae) to Spondweni virus. PLoS Negl. Trop. Dis. 10 (10), e0005083.

Hafkin, B., Kaplan, J.E., Reed, C., Elliott, L.B., Fontaine, R., Sather, G.E., Kappus, K., 1982. Reintroduction of dengue fever into the continental United States. I. Dengue surveillance in Texas, 1980. Am. J. Trop. Med. Hyg. 31 (6), 1222–1228.

Hallengard, D., Kakoulidou, M., Lulla, A., Kummerer, B.M., Johansson, D.X., Mutso, M., Lulla, V., Fazakerley, J.K., Roques, P., Le Grand, R., Merits, A., Liljestrom, P., 2014. Novel attenuated chikungunya vaccine candidates elicit protective immunity in C57BL/6 mice. J. Virol. 88 (5), 2858–2866.

Halstead, S.B., 2015. Reappearance of chikungunya, formerly called dengue, in the Americas. Emerg. Infect. Dis. 21 (4), 557–561.

Halstead, S.B., Mahalingam, S., Marovich, M.A., Ubol, S., Mosser, D.M., 2010. Intrinsic antibody-dependent enhancement of microbial infection in macrophages: disease regulation by immune complexes. Lancet Infect. Dis. 10 (10), 712–722.

Jupp, P.G., 2001. The ecology of West Nile virus in South Africa and the occurrence of outbreaks in humans. Ann. N. Y. Acad. Sci. 951, 143–152.

Katzelnick, L.C., Gresh, L., Halloran, M.E., Mercado, J.C., Kuan, G., Gordon, A., Balmaseda, A., Harris, E., 2017. Antibody-dependent enhancement of severe dengue disease in humans. Science 358 (6365), 929–932.

Kautz, T.F., Diaz-Gonzalez, E.E., Erasmus, J.H., Malo-Garcia, I.R., Langsjoen, R.M., Patterson, E.I., Auguste, D.I., Forrester, N.L., Sanchez-Casas, R.M., Hernandez-Avila, M., Alpuche-Aranda, C.M., Weaver, S.C., Fernandez-Salas, I., 2015. Chikungunya virus as cause of febrile illness outbreak, Chiapas, Mexico, 2014. Emerg. Infect. Dis. 21 (11), 2070–2073.

Lanciotti, R.S., Roehrig, J.T., Deubel, V., Smith, J., Parker, M., Steele, K., Crise, B., Volpe, K.E., Crabtree, M.B., Scherret, J.H., Hall, R.A., MacKenzie, J.S., Cropp, C.B., Panigrahy, B., Ostlund, E., Schmitt, B., Malkinson, M., Banet, C., Weissman, J., Komar, N., Savage, H.M., Stone, W., McNamara, T., Gubler, D.J., 1999. Origin of the West Nile virus responsible for an outbreak of encephalitis in the northeastern United States. Science 286 (5448), 2333–2337.

Lanciotti, R.S., Kosoy, O.L., Laven, J.J., Velez, J.O., Lambert, A.J., Johnson, A.J., Stanfield, S.M., Duffy, M.R., 2008. Genetic and serologic properties of Zika virus associated with an epidemic, Yap State, Micronesia, 2007. Emerg. Infect. Dis. 14 (8), 1232–1239.

Ledermann, J.P., Guillaumot, L., Yug, L., Saweyog, S.C., Tided, M., Machieng, P., Pretrick, M., Marfel, M., Griggs, A., Bel, M., Duffy, M.R., Hancock, W.T., Ho-Chen, T., Powers, A.M., 2014. *Aedes hensilli* as a potential vector of chikungunya and Zika viruses. PLoS Negl. Trop. Dis. 8 (10), e3188.

Ledgerwood, J.E., Pierson, T.C., Hubka, S.A., Desai, N., Rucker, S., Gordon, I.J., Enama, M.E., Nelson, S., Nason, M., Gu, W., Bundrant, N., Koup, R.A., Bailer, R.T., Mascola, J.R., Nabel, G.J., Graham, B.S., Team, V.R.C.S., 2011. A West Nile virus DNA vaccine utilizing a modified promoter induces neutralizing antibody in younger and older healthy adults in a phase I clinical trial. J. Infect. Dis. 203 (10), 1396–1404.

Lednicky, J., De Rochars, V.M., Elbadry, M., Loeb, J., Telisma, T., Chavannes, S., Anilis, G., Cella, E., Ciccozzi, M., Okech, B., Salemi, M., Morris Jr., J.G., 2016. Mayaro virus in child with acute febrile illness, Haiti, 2015. Emerg. Infect. Dis. 22 (11), 2000–2002.

LeDuc, J.W., Pinheiro, F.P., Travassos da Rosa, A.P., 1981. An outbreak of Mayaro virus disease in Belterra, Brazil. II. Epidemiology. Am. J. Trop. Med. Hyg. 30 (3), 682–688.

Liu, Z., Zhou, T., Lai, Z., Zhang, Z., Jia, Z., Zhou, G., Williams, T., Xu, J., Gu, J., Zhou, X., Lin, L., Yan, G., Chen, X.G., 2017. Competence of *Aedes aegypti*, *Ae. albopictus*, and *Culex quinquefasciatus* mosquitoes as Zika virus vectors, China. Emerg. Infect. Dis. 23 (7), 1085–1091.

Long, K.C., Ziegler, S.A., Thangamani, S., Hausser, N.L., Kochel, T.J., Higgs, S., Tesh, R.B., 2011. Experimental transmission of Mayaro virus by *Aedes aegypti*. Am. J. Trop. Med. Hyg. 85 (4), 750–757.

Macnamara, F.N., 1957. A clinico-pathological study of yellow fever in Nigeria. West Afr. Med. J. 6 (4), 137–146.

Mann, B.R., McMullen, A.R., Swetnam, D.M., Salvato, V., Reyna, M., Guzman, H., Bueno Jr., R., Dennett, J.A., Tesh, R.B., Barrett, A.D., 2013. Continued evolution of West Nile virus, Houston, Texas, USA, 2002–2012. Emerg. Infect. Dis. 19 (9), 1418–1427.

Mansuy, J.M., Dutertre, M., Mengelle, C., Fourcade, C., Marchou, B., Delobel, P., Izopet, J., Martin-Blondel, G., 2016. Zika virus: high infectious viral load in semen, a new sexually transmitted pathogen? Lancet Infect. Dis. 16 (4), 405.

Marchette, N.J., Garcia, R., Rudnick, A., 1969. Isolation of Zika virus from *Aedes aegypti* mosquitoes in Malaysia. Am. J. Trop. Med. Hyg. 18 (3), 411–415.

Martin, J.E., Pierson, T.C., Hubka, S., Rucker, S., Gordon, I.J., Enama, M.E., Andrews, C.A., Xu, Q., Davis, B.S., Nason, M., Fay, M., Koup, R.A., Roederer, M., Bailer, R.T., Gomez, P.L., Mascola, J.R., Chang, G.J., Nabel, G.J., Graham, B.S., 2007. A West Nile virus DNA vaccine induces neutralizing antibody in healthy adults during a phase 1 clinical trial. J. Infect. Dis. 196 (12), 1732–1740.

May, F.J., Li, L., Zhang, S., Guzman, H., Beasley, D.W., Tesh, R.B., Higgs, S., Raj, P., Bueno Jr., R., Randle, Y., Chandler, L., Barrett, A.D., 2008. Genetic variation of St. Louis encephalitis virus. J. Gen. Virol. 89, 1901–1910. Pt 8.

May, F.J., Davis, C.T., Tesh, R.B., Barrett, A.D., 2011. Phylogeography of West Nile virus: from the cradle of evolution in Africa to Eurasia, Australia, and the Americas. J. Virol. 85 (6), 2964–2974.

McMullen, A.R., May, F.J., Li, L., Guzman, H., Bueno Jr., R., Dennett, J.A., Tesh, R.B., Barrett, A.D., 2011. Evolution of new genotype of West Nile virus in North America. Emerg. Infect. Dis. 17 (5), 785–793.

McMullen, A.R., Albayrak, H., May, F.J., Davis, C.T., Beasley, D.W., Barrett, A.D., 2013. Molecular evolution of lineage 2 West Nile virus. J. Gen. Virol. 94 (Pt 2), 318–325.

Minke, J.M., Siger, L., Karaca, K., Austgen, L., Gordy, P., Bowen, R., Renshaw, R.W., Loosmore, S., Audonnet, J.C., Nordgren, B., 2004. Recombinant canarypoxvirus vaccine carrying the prM/E genes of West Nile virus protects horses against a West Nile virus-mosquito challenge. Arch. Virol. Suppl. 221–230. (18).

Mlakar, J., Korva, M., Tul, N., Popovic, M., Poljsak-Prijatelj, M., Mraz, J., Kolenc, M., Resman Rus, K., Vesnaver Vipotnik, T., Fabjan Vodusek, V., Vizjak, A., Pizem, J., Petrovec, M., Avsic Zupanc, T., 2016. Zika virus associated with microcephaly. N. Engl. J. Med. 374 (10), 951–958.

Modjarrad, K., Lin, L., George, S.L., Stephenson, K.E., Eckels, K.H., De La Barrera, R.A., Jarman, R.G., Sondergaard, E., Tennant, J., Ansel, J.L., Mills, K., Koren, M., Robb, M.L., Barrett, J., Thompson, J., Kosel, A.E., Dawson, P., Hale, A., Tan, C.S., Walsh, S.R., Meyer, K.E., Brien, J., Crowell, T.A., Blazevic, A., Mosby, K., Larocca, R.A., Abbink, P., Boyd, M., Bricault, C.A., Seaman, M.S., Basil, A., Walsh, M., Tonwe, V., Hoft, D.F., Thomas, S.J., Barouch, D.H., Michael, N.L., 2018. Preliminary aggregate safety and immunogenicity results from three trials of a purified inactivated Zika virus vaccine candidate: phase 1, randomised, double-blind, placebo-controlled clinical trials. Lancet 391 (10120), 563–571.

Musso, D., Nilles, E.J., Cao-Lormeau, V.M., 2014. Rapid spread of emerging Zika virus in the Pacific area. Clin. Microbiol. Infect. 20 (10). O595-6.

Musso, D., Rouault, E., Teissier, A., Lanteri, M.C., Zisou, K., Broult, J., Grange, E., Nhan, T.X., Aubry, M., 2017. Molecular detection of Zika virus in blood and RNA load determination during the French Polynesian outbreak. J. Med. Virol. 89 (9), 1505–1510.

Ng, T., Hathaway, D., Jennings, N., Champ, D., Chiang, Y.W., Chu, H.J., 2003. Equine vaccine for West Nile virus. Dev. Biol. (Basel) 114, 221–227.

Oehler, E., Watrin, L., Larre, P., Leparc-Goffart, I., Lastere, S., Valour, F., Baudouin, L., Mallet, H., Musso, D., Ghawche, F., 2014. Zika virus infection complicated by Guillain-Barre syndrome—case report, French Polynesia, December 2013. Euro Surveill. 19 (9). 1–3.

O'Neill, S.L., 2018. The use of Wolbachia by the world mosquito program to interrupt transmission of *Aedes aegypti* transmitted viruses. Adv. Exp. Med. Biol. 1062, 355–360.

Osorio, J.E., Huang, C.Y., Kinney, R.M., Stinchcomb, D.T., 2011. Development of DENVax: a chimeric dengue-2 PDK-53-based tetravalent vaccine for protection against dengue fever. Vaccine 29 (42), 7251–7260.

Paes de Andrade, P., Aragao, F.J., Colli, W., Dellagostin, O.A., Finardi-Filho, F., Hirata, M.H., Lira-Neto, A.C., Almeida de Melo, M., Nepomuceno, A.L., da, G., Nobrega, F., Delfino de Sousa, G., Valicente, F.H., Zanettini, M.H., 2016. Use of transgenic *Aedes aegypti* in Brazil: risk perception and assessment. Bull. World Health Organ. 94 (10), 766–771.

Passos, S.R.L., Borges Dos Santos, M.A., Cerbino-Neto, J., Buonora, S.N., Souza, T.M.L., de Oliveira, R.V.C., Vizzoni, A., Barbosa-Lima, G., Vieira, Y.R., Silva de Lima, M., Hokerberg, Y.H.M., 2017. Detection of Zika virus in April 2013 patient samples, Rio de Janeiro, Brazil. Emerg. Infect. Dis. 23 (12), 2120–2121.

Paupy, C., Ollomo, B., Kamgang, B., Moutailler, S., Rousset, D., Demanou, M., Herve, J.P., Leroy, E., Simard, F., 2010. Comparative role of *Aedes albopictus* and *Aedes aegypti* in the emergence of dengue and chikungunya in Central Africa. Vector Borne Zoonotic Dis. 10 (3), 259–266.

Peyrefitte, C.N., Bessaud, M., Pastorino, B.A., Gravier, P., Plumet, S., Merle, O.L., Moltini, I., Coppin, E., Tock, F., Daries, W., Ollivier, L., Pages, F., Martin, R., Boniface, F., Tolou, H.J., Grandadam, M., 2008. Circulation of chikungunya virus in Gabon, 2006–2007. J. Med. Virol. 80 (3), 430–433.

Pierce, K.K., Whitehead, S.S., Kirkpatrick, B.D., Grier, P.L., Jarvis, A., Kenney, H., Carmolli, M.P., Reynolds, C., Tibery, C.M., Lovchik, J., Janiak, A., Luke, C.J., Durbin, A.P., Pletnev, A.G., 2017. A live attenuated chimeric West Nile virus vaccine, rWN/DEN4Delta30, is well tolerated and immunogenic in flavivirus-naive older adult volunteers. J. Infect. Dis. 215 (1), 52–55.

Pletnev, A.G., Putnak, R., Speicher, J., Wagar, E.J., Vaughn, D.W., 2002. West Nile virus/dengue type 4 virus chimeras that are reduced in neurovirulence and peripheral virulence without loss of immunogenicity or protective efficacy. Proc. Natl. Acad. Sci. U. S. A. 99 (5), 3036–3041.

Powers, A.M., 2018. Vaccine and therapeutic options to control chikungunya virus. Clin. Microbiol. Rev. 31 (1), 1–29.

Powers, A.M., Brault, A.C., Tesh, R.B., Weaver, S.C., 2000. Re-emergence of chikungunya and O'nyong-nyong viruses: evidence for distinct geographical lineages and distant evolutionary relationships. J. Gen. Virol. 81 (Pt 2), 471–479.

Ramsauer, K., Schwameis, M., Firbas, C., Mullner, M., Putnak, R.J., Thomas, S.J., Despres, P., Tauber, E., Jilma, B., Tangy, F., 2015. Immunogenicity, safety, and tolerability of a recombinant measles-virus-based chikungunya vaccine: a randomised, double-blind, placebo-controlled, active-comparator, first-in-man trial. Lancet Infect. Dis. 15, 519–527.

Rawlings, J.A., Hendricks, K.A., Burgess, C.R., Campman, R.M., Clark, G.G., Tabony, L.J., Patterson, M.A., 1998. Dengue surveillance in Texas, 1995. Am. J. Trop. Med. Hyg. 59 (1), 95–99.

Reisen, W.K., Lothrop, H.D., Wheeler, S.S., Kennsington, M., Gutierrez, A., Fang, Y., Garcia, S., Lothrop, B., 2008. Persistent West Nile virus transmission and the apparent displacement St. Louis encephalitis virus in southeastern California, 2003–2006. J. Med. Entomol. 45 (3), 494–508.

Reiter, P., Lathrop, S., Bunning, M., Biggerstaff, B., Singer, D., Tiwari, T., Baber, L., Amador, M., Thirion, J., Hayes, J., Seca, C., Mendez, J., Ramirez, B., Robinson, J., Rawlings, J., Vorndam, V., Waterman, S., Gubler, D., Clark, G., Hayes, E., 2003. Texas lifestyle limits transmission of dengue virus. Emerg. Infect. Dis. 9 (1), 86–89.

Rezza, G., Weaver, S.C., 2019. Chikungunya as a paradigm for emerging viral diseases: evaluating disease impact and hurdles to vaccine development. PLoS Negl. Trop. Dis. 13. e0006919.

Rigau-Perez, J.G., Gubler, D.J., Vorndam, A.V., Clark, G.G., 1994. Dengue surveillance—United States, 1986–1992. MMWR CDC Surveill Summ. 43 (2), 7–19.

Rosenbaum, L., 2018. Trolleyology and the dengue vaccine dilemma. N. Engl. J. Med. 379 (4), 305–307.

Roundy, C.M., Azar, S.R., Brault, A.C., Ebel, G.D., Failloux, A.B., Fernandez-Salas, I., Kitron, U., Kramer, L.D., Lourenco-de-Oliveira, R., Osorio, J.E., Paploski, I.D., Vazquez-Prokopec, G.M., Ribeiro, G.S., Ritchie, S.A., Tauro, L.B., Vasilakis, N., Weaver, S.C., 2017a. Lack of evidence for Zika virus transmission by *Culex* mosquitoes. Emerg. Microbes Infect. 6 (10). e90.

Roundy, C.M., Azar, S.R., Rossi, S.L., Huang, J.H., Leal, G., Yun, R., Fernandez-Salas, I., Vitek, C.J., Paploski, I.A., Kitron, U., Ribeiro, G.S., Hanley, K.A., Weaver, S.C., Vasilakis, N., 2017b. Variation in *Aedes aegypti* mosquito competence for Zika virus transmission. Emerg. Infect. Dis. 23 (4), 625–632.

Simon-Loriere, E., Faye, O., Prot, M., Casademont, I., Fall, G., Fernandez-Garcia, M.D., Diagne, M.M., Kipela, J.M., Fall, I.S., Holmes, E.C., Sakuntabhai, A., Sall, A.A., 2017. Autochthonous Japanese encephalitis with yellow fever coinfection in Africa. N. Engl. J. Med. 376 (15), 1483–1485.

Simpson, D.I., 1964. Zika virus infection in man. Trans. R. Soc. Trop. Med. Hyg. 58, 335–338.

Smithburn, K.C., Hughes, T.P., Burke, A.W., Paul, J.H., 1940. A neurotropic virus isolated from the blood of a native of Uganda. Am. J. Trop. Med. Hyg. s1-20 (4), 471–492.

Sridhar, S., Luedtke, A., Langevin, E., Zhu, M., Bonaparte, M., Machabert, T., Savarino, S., Zambrano, B., Moureau, A., Khromava, A., Moodie, Z., Westling, T., Mascarenas, C., Frago, C., Cortes, M., Chansinghakul, D., Noriega, F., Bouckenooghe, A., Chen, J., Ng, S.P., Gilbert, P.B., Gurunathan, S., DiazGranados, C.A., 2018. Effect of dengue serostatus on dengue vaccine safety and efficacy. N. Engl. J. Med. 379 (4), 327–340.

Thiberville, S.D., Boisson, V., Gaudart, J., Simon, F., Flahault, A., de Lamballerie, X., 2013. Chikungunya fever: A clinical and virological investigation of outpatients on Reunion Island, South-West Indian Ocean. PLoS Negl. Trop. Dis. 7 (1). e2004.

Thomas, D.L., Santiago, G.A., Abeyta, R., Hinojosa, S., Torres-Velasquez, B., Adam, J.K., Evert, N., Caraballo, E., Hunsperger, E., Munoz-Jordan, J.L., Smith, B., Banicki, A., Tomashek, K.M., Gaul, L., Sharp, T.M., 2016. Reemergence of dengue in Southern Texas, 2013. Emerg. Infect. Dis. 22 (6), 1002–1007.

Travassos da Rosa, J.F., de Souza, W.M., Pinheiro, F.P., Figueiredo, M.L., Cardoso, J.F., Acrani, G.O., Nunes, M.R.T., 2017. Oropouche virus: clinical, epidemiological, and molecular aspects of a neglected Orthobunyavirus. Am. J. Trop. Med. Hyg. 96 (5), 1019–1030.

Tsai, T.F., Popovici, F., Cernescu, C., Campbell, G.L., Nedelcu, N.I., 1998. West Nile encephalitis epidemic in southeastern Romania. Lancet 352 (9130), 767–771.

Tsetsarkin, K.A., Weaver, S.C., 2011. Sequential adaptive mutations enhance efficient vector switching by chikungunya virus and its epidemic emergence. PLoS Pathog. 7 (12). e1002412.

Tsetsarkin, K.A., Vanlandingham, D.L., McGee, C.E., Higgs, S., 2007. A single mutation in chikungunya virus affects vector specificity and epidemic potential. PLoS Pathog. 3 (12). e201.

Tsetsarkin, K.A., McGee, C.E., Volk, S.M., Vanlandingham, D.L., Weaver, S.C., Higgs, S., 2009. Epistatic roles of E2 glycoprotein mutations in adaption of chikungunya virus to *Aedes albopictus* and *Ae. aegypti* mosquitoes. PLoS One 4 (8). e6835.

Tsetsarkin, K.A., Chen, R., Leal, G., Forrester, N., Higgs, S., Huang, J., Weaver, S.C., 2011. Chikungunya virus emergence is constrained in Asia by lineage-specific adaptive landscapes. Proc. Natl. Acad. Sci. U. S. A. 108 (19), 7872–7877.

Tsetsarkin, K.A., Chen, R., Yun, R., Rossi, S.L., Plante, K.S., Guerbois, M., Forrester, N., Perng, G.C., Sreekumar, E., Leal, G., Huang, J., Mukhopadhyay, S., Weaver, S.C., 2014. Multi-peaked adaptive landscape for chikungunya virus evolution predicts continued fitness optimization in *Aedes albopictus* mosquitoes. Nat. Commun. 5, 4084.

Tsetsarkin, K.A., Chen, R., Weaver, S.C., 2016. Interspecies transmission and chikungunya virus emergence. Curr. Opin. Virol. 16, 143–150.

Turell, M.J., Dohm, D.J., Mores, C.N., Terracina, L., Wallette Jr., D.L., Hribar, L.J., Pecor, J.E., Blow, J.A., 2008. Potential for North American mosquitoes to transmit Rift Valley fever virus. J. Am. Mosq. Control Assoc. 24 (4), 502–507.

Vanlandingham, D.L., McGee, C.E., Klingler, K.A., Galbraith, S.E., Barrett, A.D., Higgs, S., 2008. Short report: comparison of oral infectious dose of West Nile virus isolates representing three distinct genotypes in *Culex quinquefasciatus*. Am. J. Trop. Med. Hyg. 79 (6), 951–954.

Vasilakis, N., Weaver, S.C., 2008. The history and evolution of human dengue emergence. Adv. Virus Res. 72, 1–76.

Vasilakis, N., Cardosa, J., Hanley, K.A., Holmes, E.C., Weaver, S.C., 2011. Fever from the forest: prospects for the continued emergence of sylvatic dengue virus and its impact on public health. Nat. Rev. Microbiol. 9 (7), 532–541.

Vaughn, D.W., Green, S., Kalayanarooj, S., Innis, B.L., Nimmannitya, S., Suntayakorn, S., Endy, T.P., Raengsakulrach, B., Rothman, A.L., Ennis, F.A., Nisalak, A., 2000. Dengue viremia titer, antibody response pattern, and virus serotype correlate with disease severity. J. Infect. Dis. 181 (1), 2–9.

Vazeille, M., Moutailler, S., Coudrier, D., Rousseaux, C., Khun, H., Huerre, M., Thiria, J., Dehecq, J.S., Fontenille, D., Schuffenecker, I., Despres, P., Failloux, A.B., 2007. Two chikungunya isolates from the outbreak of La Reunion (Indian Ocean) exhibit different patterns of infection in the mosquito, *Aedes albopictus*. PLoS One 2 (11). e1168.

Venkat, H., Krow-Lucal, E., Hennessey, M., Jones, J., Adams, L., Fischer, M., Sylvester, T., Levy, C., Smith, K., Plante, L., Komatsu, K., Staples, J.E., Hills, S., 2015. Concurrent outbreaks of St. Louis encephalitis virus and West Nile virus disease—Arizona, 2015. MMWR Morb. Mortal. Wkly Rep. 64 (48), 1349–1350.

Wang, E., Ni, H., Xu, R., Barrett, A.D., Watowich, S.J., Gubler, D.J., Weaver, S.C., 2000. Evolutionary relationships of endemic/epidemic and sylvatic dengue viruses. J. Virol. 74 (7), 3227–3234.

Weaver, S.C., 2013. Urbanization and geographic expansion of zoonotic arboviral diseases: mechanisms and potential strategies for prevention. Trends Microbiol. 21 (8), 360–363.

Weaver, S.C., 2014. Arrival of chikungunya virus in the new world: prospects for spread and impact on public health. PLoS Negl. Trop. Dis. 8 (6), e2921.

Weaver, S.C., Lecuit, M., 2015. Chikungunya virus and the global spread of a mosquito-borne disease. N. Engl. J. Med. 372 (13), 1231–1239.

Weaver, S.C., Charlier, C., Vasilakis, N., Lecuit, M., 2018. Zika, chikungunya, and other emerging vector-borne viral diseases. Annu. Rev. Med. 69, 395–408.

Weinbren, M.P., Haddow, A.J., Williams, M.C., 1958. The occurrence of chikungunya virus in Uganda. I. Isolation from mosquitoes. Trans. R. Soc. Trop. Med. Hyg. 52 (3), 253–257.

White, G.S., Symmes, K., Sun, P., Fang, Y., Garcia, S., Steiner, C., Smith, K., Reisen, W.K., Coffey, L.L., 2016. Reemergence of St. Louis encephalitis virus, California, 2015. Emerg. Infect. Dis. 22 (12), 2185–2188.

Whitehead, S.S., 2016. Development of TV003/TV005, a single dose, highly immunogenic live attenuated dengue vaccine; what makes this vaccine different from the Sanofi-Pasteur CYD vaccine? Expert Rev. Vaccines 15 (4), 509–517.

Whitehorn, J., Kien, D.T., Nguyen, N.M., Nguyen, H.L., Kyrylos, P.P., Carrington, L.B., Tran, C.N., Quyen, N.T., Thi, L.V., Le Thi, D., Truong, N.T., Luong, T.T., Nguyen, C.V., Wills, B., Wolbers, M., Simmons, C.P., 2015. Comparative susceptibility of *Aedes albopictus* and *Aedes aegypti* to dengue virus infection after feeding on blood of viremic humans: implications for public health. J. Infect. Dis. 212 (8), 1182–1190.

Wong, P.S., Li, M.Z., Chong, C.S., Ng, L.C., Tan, C.H., 2013. Aedes (Stegomyia) albopictus (Skuse): a potential vector of Zika virus in Singapore. PLoS Negl. Trop. Dis. 7 (8). e2348.

Xu, G., Dong, H., Shi, N., Liu, S., Zhou, A., Cheng, Z., Chen, G., Liu, J., Fang, T., Zhang, H., Gu, C., Tan, X., Ye, J., Xie, S., Cao, G., 2007. An outbreak of dengue virus serotype 1 infection in Cixi, Ningbo, People's Republic of China, 2004, associated with a traveler from Thailand and high density of *Aedes albopictus*. Am. J. Trop. Med. Hyg. 76 (6), 1182–1188.

Chapter 13

Functional Relationship Between Public Health and Mosquito Abatement

Robert Vanderslice[a], Nicholas Porter[b], Samantha Williams[b], Abraham Kulungara[b]

[a]Environmental Health Consultant, The Vanderslice Group, North Kingstown, RI, United States, [b]Association of State and Territorial Health Officials, Arlington, VA, United States

Chapter Outline

Abbreviations

APHA	American Public Health Association
ArboNET	a national arboviral surveillance system managed by the Centers for Disease Control and Prevention and the states
ASTHO	Association of State and Territorial Health Officials
CDC	Centers for Disease Control and Prevention
C-FERST	Community-Focused Exposure and Risk Screening Tool
CHM	Community Health Maps
EEE	eastern equine encephalitis
EPA	Environmental Protection Agency
EPHT	Environmental Public Health Tracking
HIA	health impact assessment
HiAP	health in all policies
IMM	integrated mosquito management
MosquitoNET	a national mosquito surveillance system managed by the Centers for Disease Control and Prevention and the states
NASA	National Aeronautics and Space Administration

Mosquitoes, Communities, and Public Health in Texas. https://doi.org/10.1016/B978-0-12-814545-6.00013-4

NOAA	National Oceanic and Atmospheric Association
SLE	St. Louis encephalitis
WNV	West Nile virus
ZIKV	Zika virus

13.1 Introduction

Mosquito-borne diseases pose a continual threat to the nation's health and cause hundreds of thousands of deaths worldwide. In the United States, the prevention and control of this threat to public health relies on the activities of multiple public health partners at each level of government. This chapter examines the functional relationship between public health and mosquito abatement. Understanding this complex relationship requires a foundational knowledge of the mosquito abatement issues found in earlier chapters of this book as well as an appreciation for how the nation's public health infrastructure works to prevent and control vector-borne risks to health.

An appreciation for the nation's public health infrastructure begins with an understanding of the role of public health and how it differs from our system of medical care, also referred to as our system of health care. The US Centers for Disease Control and Prevention (CDC) defines public health as … *the science of protecting and improving the health of people and their communities. This work is achieved by promoting healthy lifestyles, researching disease and injury prevention, and detecting, preventing and responding to infectious diseases* (CDC Foundation, 2018). Public health can also be defined by its mission, core values, and essential services, as described in detail later. In contrast to public health programs, our system of medical/health care focuses on the provision of medical services to prevent or treat specific illness or injury, including diseases transmitted by mosquitoes. Working together, public health programs, systems of medical care, and other programs that benefit public health, like mosquito abatement programs, can be considered part of a larger, public health system (CDC, 2018d).

The United States has successfully prevented/controlled many of the mosquito-borne disease epidemics that affect other parts of the world. This success is due to an effective public health infrastructure, comprehensive mosquito abatement programs, and the ability of these two programs to align their activities. Or could other factors be, at least in part, responsible for this success? Evaluating successful public health prevention programs is always a challenge—it's impossible to count the number of disease cases that did not occur. Evaluating successful mosquito control programs faces similar challenges. Variations in mosquito populations may reflect the effectiveness of a mosquito abatement activity or simply be the result of natural factors, that is, rainfall or temperature that affect population dynamics in the absence of an effective abatement program. Our ability to understand whether public health and mosquito abatement programs coordinate their activities effectively is limited by the challenges we face in evaluating the individual success of public health and mosquito abatement programs.

This relationship between mosquito abatement and public health programs is defined, in part, by how these programs are organized at the local, state, and federal level. Most routine mosquito abatement activities occur at the local level. Some local mosquito control programs have authority to raise funds for their activities through levies or taxes. These mosquito control districts may represent multiple localities to form a regional abatement program. Oversight and support from state programs can supplement these local/regional mosquito abatement activities. Federal support for local/district/state mosquito abatement programs is reactive, with funds appropriated in response to disasters (hurricanes and floods expand mosquito breeding areas) or mosquito disease threats, like Zika virus (Dennis and Sun, 2016). In contrast, state and local public health programs rely heavily on federal funding and support, with the CDC playing a critical role in developing policy, offering technical expertise, and providing the funds for state and local public health vector control and surveillance programs. For issues related to global mosquito-borne disease risks, the CDC serves as a liaison to international public health activities. Federal surveillance networks, like ArboNET (CDC, 2015), MosquitoNET (CDC, 2016b), the National Malaria Surveillance System (Filler et al., 2006) and the US Zika Pregnancy and Infant Registry (CDC, 2018e), begin with case identification by local programs, advance to state programs that collect and manage these case reports, and submit the data to CDC for inclusion into a national surveillance system. In contrast to this strong federal presence in public health, there is no one federal agency responsible for supporting or funding local mosquito abatement activities.

Mosquito abatement program activities routinely include application of chemical pesticides, an activity that raises public health and environmental concerns. Planning processes that are inclusive and address the needs of broad segments of a community can be effective forums to address these concerns. Integrated Mosquito Management (IMM) provides a structure for balancing the needs for effective mosquito abatement with a community's concerns about the public health and environmental risks posed by pesticides.

Technological advancements in collecting and mapping data on mosquito populations and other mosquito-borne disease risk factors hold promise for future improvements in the control of vector-borne diseases. These advancements may also prove to be useful tools for evaluating the effectiveness of both mosquito and public health programs and how well they work together (Braks et al., 2014).

13.2   Burden of mosquito-borne diseases

In 2017 the World Health Organization estimated that vector-borne diseases accounted for more than 17% of all infectious diseases and cause 700,000 deaths annually, the majority caused by mosquitoes (WHO, 2017). Mapping the burden of disease highlights the variable success and failure of public health and mosquito abatement programs (Fig. 13.1).

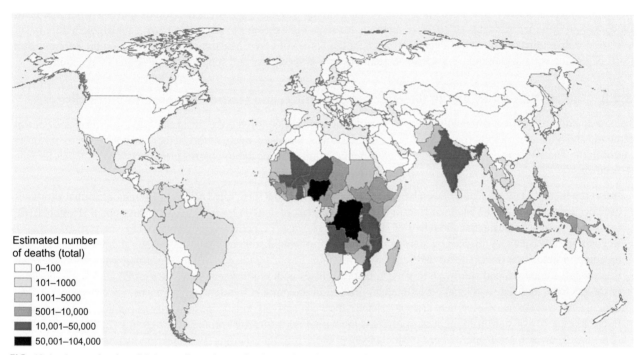

FIG. 13.1   A map showing global mortality estimates for the number of deaths attributed to three major vector-borne diseases: malaria, dengue, and yellow fever. Estimates are based on the latest available national census sources and WHO programs (WHO, 2018b).

For most of the world's population, the primary mosquito-borne diseases of concern are malaria, chikungunya, dengue, and yellow fever. The areas of the world with the highest rates of mosquito-borne diseases often correspond to areas characterized by poor living conditions. Poverty associated with the lack of safe housing, drinking water, and sanitation makes the residents more vulnerable to mosquito-borne diseases. Once infected with a pathogen, residents miss work and lose wages, factors that keep them in poverty and at increased vulnerability to mosquito-borne diseases. The limited capacity of public health and mosquito abatement programs of the world's poorest nations contribute to this burden of disease (WHO, 2014).

In contrast to these areas of endemic mosquito-borne diseases, the local transmission of malaria, chikungunya, dengue, and yellow fever is largely absent in the continental United States. This absence of disease occurs despite the presence of the *Aedes* and *Anopheles* mosquitoes that transmit these diseases. The nation's public health and mosquito abatement infrastructures contribute to this success. The effectiveness of this infrastructure can be largely taken for granted until a new threat of local transmission of these diseases occurs. At these times the public health infrastructure is mobilized to ensure effective identification and follow-up of cases to conduct public education/outreach campaigns and align these activities with local mosquito abatement efforts.

13.2.1   Prevention and control of malaria and other infectious diseases in the United States

The elimination of widespread local transmission of malaria in the United States demonstrates successful public health and mosquito control program collaboration. Local transmission is possible in the United States because three species of *Anopheles* mosquitoes capable of transmitting malaria are prevalent here. Travelers to the United States from areas where malaria is endemic account for about 1700 reported cases of malaria each year. Local transmission occurs when *Anopheles* mosquitoes become infected from biting one of the infected travelers, then spreading the *Plasmodium* parasite to other local residents. Although local outbreaks in the United States have occurred 63 times between 1957 and 2015, all were effectively contained (Filler et al., 2006).

The public health response to a malaria outbreak of eight locally transmitted cases in Palm County, Florida, in 2003 exemplified how malaria control was achieved. The local and state public health response included an aggressive public outreach campaign, active surveillance, and coordination with the local medical community and mosquito control programs. Physician alerts raised awareness about the potential for malaria cases. Case identification included interviews with each of the cases regarding other possible modes of acquiring the disease (transfusion or travel to an endemic area) and laboratory confirmation of the diagnosis. Environmental investigations were conducted at the homes of confirmed cases and at a campsite frequented by one of the individuals who was homeless. Mosquito trapping was conducted within a mile radius of each home, and pesticides were applied via ground/aerial spraying within 3 miles of each case. An aggressive public education campaign informed residents of effective methods for reducing exposure to mosquito bites, campaigns that included door-to-door dissemination of educational materials near the homes of cases. Malaria control focuses on the public health efforts beginning with case finding. Mosquito abatement serves a secondary and supportive role (Filler et al., 2006).

13.2.2 Factors that contribute to the effective prevention and control of mosquito-borne diseases

The coordination of public health and mosquito control programs has contributed to the successful prevention and control of malaria and other mosquito-borne diseases in the United States. However, many of the factors that contribute to mosquito control in the United States are independent of both public health and mosquito control programs. These factors include the following:

- Insect repellents: They are readily available and affordable throughout the United States, although occasional shortages have occurred, that is, on the island of Hawaii during an outbreak of dengue fever (Hawaii Department of Health, 2016). Consumer surveys indicate that about half of US residents use insect repellents (Statistica, 2018).
- Regulation of wastewater discharges, stormwater, and other potential sources of standing water that become mosquito breeding grounds: the management and regulatory oversight of our nation's infrastructure for sewage disposal, other wastewater discharges, and stormwater management originates with requirements of the Clean Water Act, the purview of the US EPA, and state and local environmental protection programs. Oversight of stormwater requirements for residential and commercial development and for transportation projects fall outside the authority of public health programs. Even the majority of state drinking water programs are within environmental protection programs (ASDWA, 2018).
- Housing: In areas of endemic malaria, residents sleep under mosquito netting as protection against mosquitoes that have entered their homes. In contrast, US housing stock features construction methods and barriers, like screens, that prevent mosquitoes from entering living areas. Although the nation's housing stock is far from uniform, with regulatory standards and building codes primarily enforced at the local level, many state and local housing codes require screens on windows used for ventilation and federal home loan programs require appraisals that include documentation of the presence and conditions of screens (Fannie Mae, 2005). In situations where mosquito-borne disease outbreaks occur and housing enforcement is lax, public nuisance ordinances empower local health departments to require landlords to repair screens and address conditions that breed mosquitoes (ASTHO, 2018). In addition, most US homes have air conditioning (U.S. Department of Energy, n.d.). In Texas and other places with robust mosquito populations, almost all homes have air conditioning with more than 80% of Texas homes having central air conditioning. This housing infrastructure reduces contact with peridomestic mosquitoes that live in and around human homes.

13.2.3 Cross border control of vector-borne diseases

Mosquitoes don't respect international borders, and the variability of how mosquito abatement programs are organized in the United States and other countries such as Mexico has an impact on the transmission of vector-borne diseases. While Mexico does have a coordinated nationally managed vector control and surveillance program, there are several other issues that have been identified as gaps and/or challenges in their programs that have implications for the cross border transmission of vector-borne diseases. The Mexican population is very mobile, and while health departments have been able to determine cases of arboviral diseases such as yellow fever and dengue, they have not been able to adequately identify the location of transmission. The inability to pinpoint the location of transmission has made it difficult to identify high-risk areas for interventions and public education. Additionally, Mexico and the United States have had a difficult time enticing the public into action. Even though the Mexican population was informed about Zika and other arboviruses, anecdotal reports indicate that this information did not translate into action to prevent infection.

Bilateral partnerships and agreements with Mexico have brought together representatives from the border states to discuss mosquito-borne diseases of concern to both countries and review and improve communication and mechanisms to detect, prevent, and control mosquito-borne infectious disease outbreaks along the border.

(i) The United States-México Border Health Commission (BIIC or Commission) was created as a binational health commission in July 2000 with the signing of an agreement by the Secretary of Health and Human Services of the United States and the Secretary of Health of México. The primary mission of the commission is to provide international leadership to optimize health and quality of life along the United States-Mexican border that includes four US states (Arizona, California, New Mexico, and Texas) and six Mexican states (Baja California, Sonora, Chihuahua, Coahuila, Nuevo León, and Tamaulipas) (US Department of Health & Human Services, 2017).

(ii) Examples of successful binational programs include (a) Ovitrap-based entomological surveillance program in collaboration with CDC, state health of Tamaulipas (Reynosa jurisdiction of Mexico), and Hidalgo county. This pilot project examined if the data collected from the Mexican side of the border could be used to make inferences about mosquito populations on the United States side of the border (FUMEC, 2017). (b) Binational communication and outbreak investigation in Sonora (Mexico) and Arizona. The governors and health department officials from both states signed cooperative agreements to share and coordinate epidemiologic surveillance. Outbreak investigations where Arizona and Sonora, Mexico, have successfully collaborated include Dengue (2013–14), Chikungunya (2015), and Zika (2016) (AZDHS, 2015; Jones et al., 2016).

13.2.4 Relationship between public health and mosquito abatement varies with vector and disease

In lieu of the prevention and control of vector-borne diseases like malaria in the United States and the ability to reach across international borders on emerging mosquito-borne diseases, why haven't public health and mosquito abatement programs managed to eradicate all mosquito-borne diseases in the United States? Insights into the relationship between public health and mosquito abatement can be gained by evaluating the rapid spread of WNV across the United States in just a few years. The first case report originated from New York City in 1999. Three years later, in 2002, 39 states and the District of Columbia reported over 4000 cases (confirmed and probable) and 284 deaths (CDC, 2018f). The primary strategies for controlling the local transmission of malaria are ineffective in controlling WNV. The control of malaria begins with aggressive case finding and preventing the identified cases from being bitten by mosquitoes, which can then spread the disease. For WNV, mosquitoes do not become infected when they bite humans. Infected birds are the reservoir for WNV, with mosquitoes that bite both birds and humans are responsible for spreading the disease. Surveillance programs provide the data needed to make informed decisions about WNV control. This includes (1) surveillance of mosquito populations, especially *Culex* species that spread WNV to humans, (2) surveillance of trapped mosquitoes for the presence of virus, and (3) epidemiologic surveillance of humans with cases of disease. Public health education and outreach activities focus on personal protective measures to prevent mosquito bites (use of repellents, protective clothing, and mosquito netting and use of window/door screens, air conditioning, and eliminating standing water from uncovered inside and outside containers that hold water, which serve as mosquito breeding sites) and ways to support local mosquito abatement programs to reduce mosquito populations. The control of WNV outbreak relies more heavily on mosquito abatement strategies than the control of malaria, but for both diseases, control is enhanced by access to good surveillance data and coordination of public health and mosquito abatement activities (CDC, 2018b).

Outbreaks of other mosquito-borne diseases, including dengue, chikungunya, Eastern equine encephalitis (EEE), St. Louis encephalitis (SLE), and Zika virus (ZIKV), occur sporadically in the United States. The relationship between public health and mosquito abatement in the prevention and control of these diseases depends on many factors including the way they are transmitted. For diseases in which birds are their reservoir (WNV, EEE, and SLE), mosquito abatement is a primary strategy for disease control. For diseases, that is, malaria, dengue, chikungunya, and ZIKV, in which mosquito infection results from biting humans, disease control strategies focus more on public health activities and case identification (Braks et al., 2014).

Mosquito abatement programs do face controversy. Conflict can arise when there is a desire for greater control of nuisance mosquitoes, a particularly vexing problem in areas like national parks where policies restrict pesticide application (Dillon, 2000). A more common source of public protest arises from concerns about the potential health and environmental impacts of pesticide application. Opposition to pesticide use to control adult mosquitoes has occurred even in situations where the risks of mosquito-borne diseases are high (Flechas, 2016). In these cases, adherence to IMM techniques can reduce reliance on pesticides to control adult mosquitoes, reduce pesticide and mosquito exposures to a concerned public, and, by testing the efficacy of a pesticide before its application, address the growing problem of pesticide resistance in mosquitoes. The IMM builds on surveillance data and promotes coordination between the public health and mosquito abatement programs (American Mosquito Control Association (AMCA), 2017).

The CDC provides technical support and public education materials to assist public health officials in developing effective IMM programs (CDC, 2018b), while more detailed and comprehensive guidance manuals that include best management practices for mosquito control operators are available from trade associations like the AMCA (AMCA, 2017). Contrast these best practices to the all too common "spray and pray" strategies, devoid of the surveillance and monitoring practices needed to prevent adult mosquito populations from becoming resistant to the pesticides.

13.2.5 Current surveillance systems and opportunities for future technological advances

Future threats of emerging diseases present opportunities to use new technologies for mosquito surveillance to better manage the control of these threats. The CDC's National Environmental Public Health Tracking (EPHT) network and the EPA's Community-Focused Exposure and Risk Screening Tool (C-FERST) are two examples of tools that have demonstrated utility for managing environmental health problems through linking public health and environmental data sets. The potential to expand these systems to include novel data sets, for example, those related to electronic health records (EHRs) or information collected from citizen monitoring initiatives, holds promise for advancing mosquito-borne disease surveillance (CDC, 2018c).

In order to detect, respond, and reverse the increase of a vector-borne disease that is being observed nationally, existing surveillance techniques must be strengthened. Fortunately, significant advancements have been made across geospatial technologies, with a variety of data collection and geospatial platforms available to aid agencies in the tracking and monitoring of vectors in real time. While standardized surveillance is still not a routine for all vector control programs, many local and state health agencies are benefiting from the use of this technology with improved operation efficiencies and reduced costs. Agencies can develop their own unique surveillance workflows using a variety of software, allowing their staff to delegate data collection and visualization tasks, resulting in more coordinated and efficient investigation and outbreak management processes. Real-time data capture also helps guide more timely public health actions, such as delivering the appropriate services to at-risk communities during emergency outbreaks (Geraghty, 2017). Fig. 13.2 shows how staff from the Volusia County Public Works Department (FL) can access maps and work orders in real time while conducting activities from the field. Through web-based apps, employees can record surveillance findings and document the time and location of their mitigation activities.

FIG. 13.2   Staff from the Volusia County Public Works Department (FL) access maps and work orders from the field, record surveillance findings, and document their mitigation activities. Using GIS to address the mosquito threat begins with surveillance activities (Geraghty, 2017).

As surveillance technologies have gained popularity, they are becoming more accessible and easier to use. This process, often referred to as the "democratization" of geospatial solutions, has resulted in a wider range of options than ever before becoming available to private and public users. Free and low-cost data management tools and geospatial software are now widely available at a dramatically lower price, with sophisticated capabilities that are comparable with top of the line products. The trend toward low-cost surveillance solutions is gaining swift traction, with initiatives such as the National Library of Medicine's Community Health Maps (CHM) group, which provides free surveillance training to university groups, community organizations, and health agencies. Groups like CHM are putting the power of surveillance into the hands of community groups, nonprofit organizations, and resource-limited public health agencies, with the goal of expanding surveillance capabilities across the nation (US National Library of Medicine, 2018). The success of these initiatives is rooted in developing institutional capacity by educating nongeospatial users with simple tools and workflows to develop and manage their community's data and surveillance activities.

As vector surveillance capabilities continue to expand, several voices from public health have begun to call for a national vector-borne surveillance system (Yong, 2017). Currently, states are responsible for collecting vector-borne diseases via patient case reports, with local mosquito control agencies responsible for collecting mosquito species data. There are over 1000 separate agencies that collect mosquito data (at least one in every county/jurisdiction), but only 152 of these agencies make data available to the public. Additionally, CDC has developed two mosquito-related databases, ArboNET and MosquitoNET that collect county level data for virus infection in mosquito pools and detailed mosquito data by location and insecticide resistance testing results (CDC, 2015, 2016b). However public access and modeling data for local agencies are limited. As a result of this challenge, many agencies use weather data from federal agencies, that is, NASA and NOAA, to forecast the presence of mosquitoes (Porter, 2016). One way to overcome these challenges is to combine these disparate data sets into a single national surveillance system. The CDC's National EPHT network could be used as a model. This network provides a platform for storing and evaluating data related to environmental hazards, exposures, and health outcomes from the portals of 25 different states and one city. The network enhances the ability to identify trends and link exposures to health impacts (CDC, 2018c).

This local, ground-based information collected by the app about mosquito habitats will help NASA by providing information that supports satellite-based research of environmental conditions that support outbreaks of mosquitoes.

Finally, local, state, and federal initiatives are now exploring ways to engage the public in data collection through citizen science activities that are defined as the collection and analysis of data by the general public and are being explored from

FIG. 13.3 The GIS dashboard's, like this one used by Santa Cruz County's Mosquito Abatement and Vector Control District distributes site inspection status information via its Operations Map App. A similar workflow can be used for public health surveillance to monitor mosquito populations and inform outreach to at-risk populations. *(http://www.esri.com/esri-news/arcnews/summer16articles/arcgis-platform-curtails-mosquitos-in-coastal-california [Accessed 15 October 2018].)*

the local to global scale. For example, in Germany, a citizen science project was started by scientists at the Leibniz Centre for Agricultural Landscape Research and the German Federal Research Institute for Animal Health in 2012 to monitor the presence of invasive mosquito species after an increase in WNV, chikungunya, and dengue were observed across the continent (Ricciuti, 2017). Data from volunteers were used to identify at-risk populations to direct investigations, and in 3 years the *Mueckenatlas* received over 7300 submissions on more than 29,000 mosquitoes, its success largely attributed to extensive public relation campaigns. A mosquito atlas was established with the data to display the volunteer submissions (Ricciuti, 2017). In the United States, AccuWeather has developed a Mosquito Zika Risk Index on its online portal and apps, providing the public with a 90-day forecast of their Zika exposure risk levels based on weather data. By continuing to promote mosquito surveillance among the public, researchers can continue supplementing their research with ground-based observations, communities will be better educated about how to protect themselves from the risks of vector-borne diseases, and better health outcomes can be secured for all (AccuWeather, 2017) (Fig. 13.3).

13.3   Foundations of public health and mosquito abatement programs

Public health and mosquito abatement programs operate best when built on a foundation of consistent funding and public support. Communication about the risks of mosquito-borne diseases and the importance of mosquito abatement activities can be challenging, but effective communication is essential to maintaining public support for these programs. The success of public health programs to effectively control vector-borne diseases ultimately depends on the public having an appreciation for the value of public health programs and a willingness to support funding for these programs. While most members of the public cannot be expected to know the CDC definition of public health as *the science of protecting and improving the health of people and their communities* (CDC Foundation, 2018), a general understanding of the role of public health can be necessary for garnering public support for programs to control mosquito-borne diseases (ASTHO, 2009). Misconceptions about public health can be a barrier to responding to public health emergencies, such as funding programs to address emerging mosquito-borne disease threats. This came to light during attempts to secure funding for the control of Zika virus.

In 2016, President Obama proposed a $1.9 billion emergency appropriation for addressing the threat posed by the mosquito-borne Zika virus (Epstein and Lister, 2016). This proposal was controversial. Unlike some political issues, debate did not pit "pro-Zika" against "anti-Zika" factions—the nation was unified in its concern about stopping the virus from gaining a foothold on the continental United States. Debate focused primarily on the best way to fund Zika control measures and how much funding was needed. Congress questioned the need for an immediate appropriation and whether adequate control could be accomplished with a lower level of funding by repurposing funding for the Ebola crisis. Many public health advocates and advocacy organizations such as the American Public Health Association (APHA) and the Association of State and Territorial Health Officials (ASTHO) supported full and immediate funding of the $1.9 billion proposal (APHA, 2016). After 7 months of debate, the Congress ultimately approved measures for a Zika appropriation of $1.1 billion. During discussions and debates of this proposal, advocates for Zika funding identified the lack of a common understanding of the nature of public health programs to be a barrier to their advocacy. Gaps in understanding of public health were an obstacle to gaining congressional support for the funding proposal. Both congressional supporters for greater funding for public health and their opponents displayed misunderstandings about public health practice. Many found it confusing that the resources requested in earlier years to fight Ebola virus were different from the resources needed to fight a mosquito-borne disease such as Zika.

The misunderstandings about public health practice that surfaced during the Zika virus appropriation debates were consistent with the results of a study of how the general public views the term "public health." Partly in response to a long-term trend of decreasing financial support for public health services, the Pew Charitable Trusts commissioned a survey to better understand public opinion about three topics: (1) public health, in general; (2) the role of environmental health in disease; and (3) public health infrastructure. A national telephone survey of 1234 registered voters found that survey participants had trouble defining public health. Most (57%) failed to identify the roles of public health, either in protecting the population from disease or in developing the policies and programs that promote healthy living conditions for everyone. These data, together with the experience of advocates for the federal Zika appropriation, indicate that the lack of a common understanding of public health can be a barrier to those seeking community input and collaboration on public health programs. This includes public health officials seeking to develop relationships with pesticide applicators and others involved in mosquito control (Hearne et al., 2000; Helfrich et al., 2009).

The mission statements and goals of public health agencies provide another way to gain a general understanding of public health. Important differences exist between the missions of international, national, state, and local agencies, but all share common features regarding promoting the health of populations and preventing/controlling diseases.

The international World Health Organization's goal is *to build a better, healthier future for people all over the world. Working through offices in more than 150 countries, WHO staff work side by side with governments and other partners to ensure the highest attainable level of health for all people. Together we strive to combat diseases—infectious diseases like influenza and HIV and noncommunicable ones like cancer and heart disease. We help mothers and children survive and thrive so they can look forward to a healthy old age. We ensure the safety of the air people breathe, the food they eat, the water they drink—and the medicines and vaccines they need* (WHO, 2018a).

The mission of the nation's primary public health agency is *CDC which works 24/7 to protect America from health, safety and security threats, both foreign and in the United States. Whether diseases start at home or abroad, are chronic or acute, curable or preventable, human error or deliberate attack, CDC fights disease and supports communities and citizens to do the same. CDC increases the health security of our nation. As the nation's health protection agency, CDC saves lives and protects people from health threats. To accomplish our mission, CDC conducts critical science and provides health information that protects our nation against expensive and dangerous health threats, and responds when these arise* (CDC, 2018a).

The Texas Department of State Health Services has a stated mission *to improve the health, safety, and well-being of Texans through good stewardship of public resources, and a focus on core public health functions* (Texas Department of State Health Services, 2016).

At the local level, Harris County Public Health has a three-part mission of … *Promoting a healthy and safe community, preventing Illness and Injury and protecting you.* Within Harris County is the city of Houston, and the mission of the Houston Health Department *is to work in partnership with the community to promote and protect the health and social well-being of all Houstonians* (Harris County Public Health, 2018).

The CDC's definition of public health and the mission statements of relevant public health agencies provide an overview for a general understanding of public health.

13.3.1 The core functions and essential services of public health

To move to an understanding of how specific activities like mosquito control fit into the larger context of public health, CDC uses the diagram pictured in Fig. 13.4 (CDC, 2018a). This figure illustrates how public health can be divided into three core functions, 10 essential services and the overarching principles of good management that include planning, evaluation, and communication all with a solid foundation based in research.

FIG. 13.4 The 10 essential public health services. *(Available at: https://phil.cdc.gov/Details.aspx?pid=22747 [Accessed 1 November 2018].)*

The core functions of public health include assessment, policy development, and assurance. Assessment includes the activities needed to understand the health status of a community. Activities that enable a community to respond to information on its health status are included in policy development. Assurance includes activities for ensuring that the policies, medical services, and public health practices are effective and accessible.

13.3.2 The 10 essential public health services

The 10 essential public health services describe the public health activities that all communities should undertake. Examples of how mosquito control activities fit within these services are as follows:

The first two essential public health services are within the core function of assessment:

1. Monitor health status to identify and solve community health problems.
2. Diagnose and investigate health problems and health hazards in the community.

Mosquito control activities relevant to the essential public health services of assessment include activities related to the surveillance of mosquitoes and the diseases they carry and the diagnosis and surveillance of human cases of mosquito-borne diseases. For example, as part of its efforts to prevent and control ZIKV, CDC developed maps of the ranges of the *Aedes* mosquitoes capable of spreading the disease. These maps relied on data from local and state programs. Many local mosquito abatement programs lacked resources to conduct this surveillance, relying on public health programs and laboratories to support field surveys and trapping of mosquitoes, their identification, and testing pools of collected mosquitoes for pathogens. These pathogens can include those that cause human diseases and pathogens that cause diseases in animals and may indicate a greater potential for the spread of human diseases.

Human disease surveillance includes maintenance of disease registries. CDC maintains several vector-borne disease registries including ArboNET, MosquitoNet, the National Malaria Surveillance System, and the US Zika Pregnancy and Infant Registry. These surveillance systems are used to track the geographical and temporal trends of mosquito-borne diseases. Investigation of health hazards also involves the development and improvement of laboratory tests for identifying pathogens in clinical or environmental samples. This was particularly important when ZIKV was first becoming a risk in the United States. Health care providers and their patients had problems getting timely and definitive information. Improved laboratory methods and expanded capacity to analyze samples were critical to making informed decisions about management of this disease.

Effective public health and mosquito abatement programs rely on the data collection activities that fall under these two essential public health services. Vector-borne disease surveillance systems provide a basis for coordination between public health and mosquito abatement programs.

Policy development includes three essential services that focus on community interaction:

3. Inform, educate, and empower people about health issues.
4. Mobilize community partnerships and action to identify and solve health problems.
5. Develop policies and plans that support individual and community health efforts.

Mosquito control activities that relate to policy development include federal, state, and local educational efforts related to how individuals can protect themselves from mosquito bites, making their homes more effective barriers to mosquitoes and their homes and yards less hospitable to mosquito breeding. Public educational efforts benefit from consistency and having public health and mosquito abatement programs agree on public messaging. Policy development is enhanced through creating partnerships that are critical to public education campaigns and getting information to vulnerable populations that may be hard to reach through traditional channels. Processes for creating community-based mosquito control plan benefit from input from diverse partners and promote collaboration between public health and mosquito abatement programs with other agencies whose work touches on mosquito control issues. For example, agencies responsible for planning and development activities, wastewater and stormwater management, and emergency response can provide valuable perspective on the control of mosquito breeding areas. Plans also need to ensure that community-based mosquito control plans are mindful of public health, limit community exposure to pesticides, and are part of emergency response planning efforts. Promoting the use of IMM exemplifies the core public health function of policy development.

The core function of assurance includes a broad range of activities that focus on issues of access to services and ensuring that services are of high quality, including.

6. Enforcing laws and regulations that protect health and ensure safety
7. Linking people to needed personal health services and assure the provision of health care when otherwise unavailable
8. Assuring competent public and personal health care workforce
9. Evaluating effectiveness, accessibility, and quality of personal and population-based health services

These essential public health services include a diverse range of mosquito control activities of which some are part of the environmental follow-up to the identification of cases of malaria or Zika virus. Assurance activities can include door-to-door property inspections in areas of local transmission of these diseases and the enforcement of nuisance laws to prevent careless homeowners from harboring mosquito breeding areas. Assurance activities ensure that people who have contracted mosquito-borne diseases have access to appropriate medical testing and treatment services and that their medical care providers have access to accurate and timely testing of clinical samples. Especially when the nation is faced with emerging infectious mosquito-borne diseases, public health programs develop guidelines and training materials to assure that medical providers are prepared to diagnose and treat their patients exposed to mosquito-borne diseases. These assurance efforts depend on good science and thus include activities related to the development of protocols that ensure mosquito surveillance systems are reliable sources of accurate and timely information.

The final essential public health service is ingrained in each of the core functions:

10. Research for new insights and innovative solutions to health problems

Research activities include investigations into the basic science of disease transmission and expression and the applied science of the surveillance of mosquito populations, surveillance of mosquito-borne diseases, and effectiveness of control strategies (Table 13.1).

TABLE 13.1 Vector control roles of federal, state/territorial, and local government with respect to the core public health functions.

Core public health functions	Vector control specific functions	Mosquito control activities at the local, state/territorial, and federal levels
Assessment	Monitor and investigate mosquitoes	Local programs collect data on mosquitoes and virus (data needed for IMM) with support from states, which compile these data and submit to federal data systems
	Disease surveillance	Initial investigation of cases is performed at the local level, with state support for investigation and response. Federal CDC experts support activities related to outbreaks and emergencies
Policy development	Inform and mobilize communities	Federal agencies develop nationally consistent messaging and web-based materials from which state-specific materials can be developed. Local programs meet with and inform communities
	Develop policies	Federal programs are critical to developing travel alerts and other policies for controlling the global spread of disease and policies, which can be adapted for state/local mosquito control programs
Assurance	Enforce pesticide laws	Federal agencies delegate enforcement authority to state programs. State certification sets standards for pesticide applicators whose compliance is monitored by local officials
	Assure access to medical care	Federal experts provide guidance on mosquito-borne disease diagnosis and treatment. States assure access to laboratory and medical services, which are delivered at the local level
Systems management of crosscutting issues	Planning, especially related to emergencies	Local planning is critical to effective mosquito control requiring coordination with stakeholders at all levels. State plans include strategies for surveillance and coordination with federal agencies, which provide emergency/outbreak response oversight

13.3.3 Health in all policies and health impact assessments

As described in Section 13.2.2, effective mosquito control is due in part to activities that are outside public health and mosquito abatement programs. The greatest public health advances can occur when public health is considered in developing transportation, housing, or public utilities for the control of wastewater/stormwater. This broad consideration of public health issues is referred to as a health in all policies (HiAP) approach (Rudolph et al., 2013). Using the transportation as an example, a HiAP approach to promote mosquito control could include designing and maintaining drainage systems and retention/detention basins in a way to prevent mosquito breeding, roadside plantings to discourage mosquitoes and encourage mosquito predators, and providing insect repellents for highway workers.

The health impact assessments (HIAs) provide a structured way to consider the health impacts of proposed projects. Modeled on environmental impact statements, HIAs could be used to provide a structured way to consider the potential impacts of a proposed project on mosquito populations and mosquito-borne diseases. Massachusetts requires certain transportation projects to use HIAs to measure the potential impacts of proposed projects, a requirement that encourages the public to contribute to the planning process in a way to minimize unintended consequences of development. The HIAs have the potential to bring together public health and mosquito control programs to consider the impact proposed development projects can have on exposure to mosquito-borne diseases (ASTHO, 2017b).

13.3.4 Current vector control and mosquito abatement capacity across the United States

In contrast to public health programs that have close ties to the CDC and rely on federal funding and technical support, mosquito abatement programs are the primary responsibility of state, local, and/or regional programs, receiving relatively little federal support. The roles of federal agencies focus on broader human health and environmental issues (Table 13.2). Some local programs form mosquito control district with the authority to raise funds for their activities through levies or taxes (ASTHO, 2018). Oversight and support from state programs can supplement these local/regional mosquito abatement activities. Federal support for local/district/state mosquito abatement programs is reactive, with funds appropriated in response to disasters (hurricanes and floods expand mosquito breeding areas) or emerging mosquito-borne disease threats.

In 2017 a National Assessment of Mosquito Surveillance and Control was conducted by National Association of County and City Health Officials (NACCHO) to capture the competencies of state vector control programs (NACCHO, 2017). The assessment identified five core competencies that are required by a fully capable mosquito control program:

1. Routine mosquito surveillance through standardized trapping and species identification;
2. treatment decisions using surveillance data;
3. larviciding, adulticiding, or both;
4. routine vector control activities;
5. pesticide resistance testing.

The outcomes of the study showed that just 33 states had at least one fully capable and competent vector control program. These were mostly local vector control programs, which represented over half of the survey respondents. Nationwide an alarming 84% of vector control programs were identified as needing improvement in one or more core competency. These national assessments are important for identifying gaps in vector control and surveillance and for making the critical case for providing funding support to these essential public health services.

Note that the five core competencies of mosquito abatement programs are the primary responsibilities of state, local, and/or regional programs. Although federal agencies can be involved in mosquito control activities (Table 13.2), their roles are generally to support state and local efforts.

TABLE 13.2 Federal agencies with roles in mosquito control.

Agency	Primary responsibility	Mosquito control activities
Centers for Disease Control and Prevention	Human disease	Mosquito management, surveillance, and response
Federal Emergency Management Agency	Disaster preparedness and response	Resources for mosquito control following floods or other disasters (Federal Emergency Management Agency, 2016)
US Department of Agriculture	Livestock health	Zoonoses prevention and surveillance; research on management of human and animal pests
Environmental Protection Agency	Pesticide impacts on human health and the environment	Pesticide registration and review; Clean Water Act permits for discharge of pesticides into waters of the United States
Department of Defense	Vector impacts on military personnel	Mosquito control at military installations; technical assistance to adjacent communities

13.3.5 Mosquito abatement programs as defined by their missions

Recent surveys of state and local governments regarding mosquito abatement and control activities demonstrate the lack of uniformity in how mosquito abatement programs are organized and managed across the nation (NACCHO, 2017; ASTHO, 2018). Mosquito abatement activities are primarily conducted at the local level. Local agencies responsible for mosquito abatement can include independent local mosquito control boards, health departments, public works departments, and independent mosquito abatement districts. In a few states, such as Delaware, state agencies assume primary control of mosquito abatement (ASTHO, 2018).

A mosquito abatement program mission is influenced by its enabling authority and organizational structure. Not surprisingly, programs within a health agency have missions that speak to the control of mosquito-borne diseases, while independent mosquito control districts were often established in response to a need to control nuisance mosquitoes. Three examples are provided in the succeeding text:

The Harris County Public Health Mosquito and Vector Control Division *protects the health and well-being of county residents through surveillance, control, education, research, and technology to prevent and control mosquito-borne diseases, including St. Louis encephalitis (SLE) and West Nile virus (WNV) encephalitis* (Harris County Public Health, 2017).

In Minnesota, mosquito abatement activities are conducted by a seven-county mosquito control district funded through property taxes. The Metropolitan Mosquito Control District's mission is to promote ... *health and well-being by protecting the public from disease and annoyance caused by mosquitoes, black flies, and ticks in an environmentally sensitive manner* (Metropolitan Mosquito Control District, 2017).

In New Jersey, (NJ), mosquito abatement programs include the state Mosquito Control Commission and New Jersey's local/county mosquito control programs. The control of nuisance mosquitoes is of primary importance and is mentioned first in mission statements, with disease control an important but secondary function. New Jersey's Mosquito Control Commission lists among its primary goals, "to protect the general public from nuisance mosquitoes and the threat of mosquito-borne disease" and maintain a "... safe environmentally sound Mosquito Aerial Application Operation (NJ DEP, 2016). The NJ's local mosquito control programs begin their mission statements with an emphasis on the control of nuisance mosquitoes. For example, the Camden County Mosquito Commission states its objective is "... to eliminate and control mosquitoes which cause great suffering, economic loss and disease transmission ... Mosquitoes are well known throughout the area as a nuisance. Uncontrolled mosquito populations can still ruin outdoor activities and interfere with many elements of business. A less common but even more serious problem with the blood sucking of mosquitoes is the spread of disease" (Camden County Mosquito Commission, 2017).

Despite their differences, these programs have much in common. All promote the safety and efficacy of their activities. All include measures to educate the public about effective methods to avoid bites and control mosquito breeding sites. All programs use multiple strategies to reduce mosquito populations that include measures to modify mosquito habitat and the use of chemical/biological agents to reduce populations of mosquito larvae and adults. Other strategies may also be effective but are less common.

Ideally, mosquito abatement strategies reflect a planning process that includes a broad segment of local interests. Mosquito abatement programs have ready access to guidance for creating an effective plan that reflects local community concerns (AMCA, 2017; ASTHO, 2018). In addition, mosquito abatement can be an essential element within emergency preparedness and response plans, for two very different scenarios. Hurricanes and other causes of flooding can alter mosquito habitat, greatly expanding the areas of mosquito breeding sites resulting in significant increases in mosquito populations. Strategies for responding to the threats of emerging mosquito-borne diseases warrant inclusion in emergency preparedness and response plans.

Most routine mosquito abatement activities occur at the local level. Some local programs have authority to raise funds for their activities through levies or taxes. Oversight and support from state programs can supplement these local/regional mosquito abatement activities. Federal support for local/district/state mosquito abatement programs is reactive, with funds appropriated in response to disasters (hurricanes and floods expand mosquito breeding areas) or mosquito disease threats, that is, ZIKV (ASTHO, 2018).

The relationship between public health and mosquito abatement is complex, in part, because public health and mosquito abatement activities are not confined to public programs. As described in Section 13.2.2, housing programs and other factors outside the traditional realm of public health can influence the risk of mosquito-borne diseases. Similarly, not all mosquito abatements are conducted by public programs. Fee-based services to homeowners by private pest control companies are big business. These services face a different set of expectations and constraints from those that define public programs. In general, clients do not want to pay for surveillance or long-term strategies to manage mosquito populations. Clients expect immediate results and relief, as well as fees as low as possible. Clients may have little knowledge or concern about issues of public health or environmental concern and therefore are unlikely to pay extra to ensure these concerns are addressed (Robinson, 2018). Planning activities, mosquito control, and pest control operator professional/trade organizations offer opportunities for improved public-private mosquito control coordination.

13.4 Factors that affect the relationship between public health and mosquito abatement

Since the deadly cholera epidemics of the late 19th century, health officials have been working together to control infectious diseases and protect the health of the public. The idea of an association of health officials was first conceived in 1879 at a meeting of the Sanitary Council of the Mississippi Valley. At a meeting of 19 health officials 5 years later in Washington DC, the National Conference of State Boards of Health was established. Less than two decades later, the US surgeon general and state and territorial health officials began meeting annually to discuss cost-effective public health strategies for disease control leading to the establishment of the Association of State and Territorial Health Officials (ASTHO) in 1942. Together with other national organizations like the National Association of Vector-Borne Disease Control Officials, the National Environmental Health Association, the NACCHO and the American Public Health Association's, ASTHO continue to provide state public health leaders with formal networks to discuss vector control issues, exchange information, and advocate for public health priorities (Maddox, 2017).

However, in spite of public health leadership's commitment to cooperation, a chronically underfunded public health system has left millions of Americans exposed to preventable diseases (Segal and Martin, 2017). For the past two decades, vector control capabilities have been experiencing a steady decline in funding support, and even large cities like San Antonio are undergoing budget cuts to public health surveillance capacity as high as 75% (Kofler, 2016). The reemergence and rapid spread of WNV in 1999 was attributed to the severe fund reductions in vector control capabilities, and WNV remains the leading cause of arboviral encephalitis in the United States. Emergency outbreaks are a matter of when, not if, and without adequate funding, public health's response will continue to remain reactive, inadequate, and unsustainable. These are grave concerns at a time when zoonotic diseases account for 75% of recently emerging infectious diseases (CDC, 2018g).

In addition to reduced funding for vector-borne disease control programs, affected communities face a double burden when they are faced with health care costs associated with treating the infections, which includes medication and missed time at work. At a broader scale an outbreak also negatively impacts a community's overall human, animal, and economic health. The biggest impacts are felt by local industry such as tourism, outdoor recreation, and agriculture, and ultimately, it falls to many local and state programs to incur the costs for abatement, public education campaigns, and vaccinations associated with an outbreak. Across the United States, many states at risk of mosquito-borne disease outbreaks rely greatly on the income generated from local tourism and industries impacted by vector-borne diseases. For example, 3 months after the 1990 SLE outbreak, Florida witnessed a 15% decline in tourism revenue (MosquitoMagnet, 2018). National losses in beef and milk production associated with infected cattle were estimated to be as high as $61 million (Lloyd et al., 2018).

The following section will examine the role of finance as a driver for vector-borne disease control across state and local programs and explore the cost-effectiveness of prevention and the need for sustained investment.

13.4.1 Funding state and local programs

The national landscape of public health is composed of a variety of models that operate at the federal, state, and local level. The relationships that exist between federal and state agencies are as different as those that exist between a state agency and its local counterparts. As described earlier in this chapter's introduction, most mosquito abatement program activities occur at the local level. In some cases, these local programs have the authority to raise funds through levies or tax mechanisms (ASTHO, 2018). State agency oversight and support are often provided to address the broader needs of mosquito control programs in the form of surveillance, technical, and laboratory assistance. There is no official federal agency to provide regular assistance to local and state programs, and when federal assistance is provided, it is often in response to an emergency such as the Zika virus outbreak. In spite of this, local and state programs still rely heavily on federal funding sources and technical assistance and expertise from the CDC.

Although local health agencies are responsible for much of a state's routine mosquito abatement activities, they are often contingent on the relationship that exists between a local and state health agency. State public health agencies are distinguished by three governance structures that can influence the delivery of public health services (including vector control) and are based on governmental authority as well as where a state public health agency is situated, that is, it is part of a larger agency. This affects the budgeting process, decision-making, as well as programmatic responsibility (ASTHO, 2017a). The three governance structures are centralized, decentralized, and hybrid. In centralized states, local health units are essentially operated by state health agency staff and funds; in decentralized states, local health departments are largely independent of the state; and in hybrid or mixed states, no single structure predominates between state and local staff to varying degrees. These structures inform the extent to which a local health agency relies on state or federal funding or what kinds of self-financing options are available for vector control and surveillance, ultimately impacting the sustainability and strength of a program.

The ASTHO conducted a recent analysis of select state statutory and regulatory mosquito control authorities to understand the structure of state mosquito control, the role of state and local authorities, financing, and enforcement. The states examined were those where the structure of mosquito control responsibilities and activities could be ascertained from express statutory provisions addressing mosquito control at a state or local level. It was found that majority of states with mosquito control responsibilities are characterized by a hybrid or mixed structure, followed by a decentralized structure, and lastly a centralized structure. Roles and responsibilities of state and local entities were captured for enforcement and financing and illustrated the range and the complexity of options that are available to public health staff who support vector control activities. Across all governance structures, it was found that most financing for a local health agency is largely based on needs.

Hybrid public health agencies

In states governed by a hybrid structure, both state and local authorities share responsibilities for mosquito control. For example, in Colorado, the Department of Agriculture is the authority for state mosquito control, and the Department of Public Health and the Environment has authority to "investigate and control the causes of epidemic and communicable diseases." In addition, the commissioner of agriculture is the responsible individual for advancing funds for pest control. Coordination and relationship building are necessary components to fund both state and local programs, including those within mosquito control districts (as earlier mentioned, a district may represent multiple localities to form a regional program).

Across other hybrid states, mosquito control districts are used to support statewide activities and are financed by county budgets, tax rates (e.g., property and water bills), or matching state grants. A district can be established through a petition and ballot and financed by such a property tax. In Florida the funding of mosquito control districts is made through matching state grants, and at the local level, special taxing districts are established for mosquito control. These districts may levy a tax on real and personal property (needs based) and not to exceed $1 for every $100 in assessed value. Mosquito control authority is shared across the Department of Agriculture, the Florida Coordinating Council on Mosquito Control, and the state health officer.

Decentralized public health agencies

States that are governed by a decentralized structure see responsibilities shared at the local level. There is little difference in the use and choice of funding mechanisms that are available to hybrid states. For example, in Texas, Nevada, North Dakota, and Arkansas, vector control districts are also established by petition (often needs based) and financed by property taxes, county budgets, and annual assessments. In Nebraska, counties and cities have the authority to undertake mosquito control activities and appropriate necessary funds to do so.

Centralized public health agencies

Few states are governed by a centralized governance structure, but for those that are primary, responsibilities and activities are concentrated mainly at the state level. There are few mosquito control laws and statutes that allow for the establishment of a mosquito control district. The primary authorities have the right to conduct abatement activities at the property owner's expense. For example, in the US Virgin Islands, the Commissioner of Health can charge private property owners reasonable fees for pest abatement. A more structured approach is used in the District of Columbia where a vector-borne disease control fund is financed through fines, civil penalties, and judgments received from violations from rules relating to vector control.

13.4.2 Need for sustained investment

As the previously mentioned governance structures demonstrate, funding for vector control programs varies considerably. Local programs can be highly dependent on the support of their state public health agency or needs based and self-financed using a variety of taxing mechanisms. It can also be a blend of both. This variability can present significant challenges in securing funds for mosquito control, in addition to limited data on the national capacity of vector control. While federal support is limited to public health surveillance activities, funding is often made available to support public health activities during an outbreak. For example, during the 2016 Zika virus outbreak, it became evident that many local and state programs across the nation relied heavily on these emergency funds to revive their vector control programs, with delays in the appropriation of emergency funds ultimately delayed the outbreak response activities. Across high-risk states such as Texas and Florida, it was found that 68% were lacking in competency for mosquito control and surveillance (Segal and Martin, 2017).

Also demonstrated by the Zika crisis was the reality that many vector control programs were encountered: the need to mobilize their resources and develop more sustainable funding models. Maintaining well-trained staff and adequate levels of supplies are necessary to ensure an outbreak can be managed while maintaining surveillance capabilities to prevent them in the first place (NACCHO, 2017). Across the nation, many local and state mosquito control programs are combining their resources to cut costs, and programs are exploring how to share service agreements, equipment, regional districts, and standard contracts for services to maximize their funding impact and remain in operation. Programs are also beginning to train nonmosquito control employees to use abatement equipment and techniques and developing stronger relationships with their state public health laboratories to secure access to testing facilities and surveillance platforms during an outbreak. In the case where local and state health agencies are unable to provide these essential services, private mosquito control companies are also being contracted.

Sustaining robust vector control programs within a local or state public health agency is a challenging task. Public health leaders are crucial players in developing messages to communicate the importance of investments in these critical public health services to national leaders and emphasize the importance of the role of prevention to the public to control vector-borne diseases. During the Zika crisis, state leaders advocated for increased funding on behalf of their local and regional programs, calling on the Senate Appropriations Committee to request additional funding for vector-borne disease line at CDC's National Center for Emerging and Zoonotic Infectious Diseases (NCEZID) to support research and laboratory capabilities (Bennet, 2016). Nationwide the CDC awarded $184 million to states, territories, local jurisdictions, and universities to protect Americans from the Zika virus, which was part of the $350 million in funding provided to CDC under the Zika Response and Preparedness Appropriations Act of 2016 (CDC, 2016a). The CDC continues to be called on to expand its state and local capacity support to prevent and detect mosquito-borne illnesses (Segal and Martin, 2017).

It is critical that public health builds on the momentum produced from the Zika emergency to reduce the loss of vector control capacity. While emergency funds can help alleviate gaps in capacity in the short-term, more long-term solutions and investments that work upstream to prevent an outbreak in the first place are key to safeguarding health and preventing the resurgence of vector-borne disease in a nation that has fought hard to respond to these threats.

Case study: Astronomical medical costs of Zika

In 2015, health officials first confirmed the presence of Zika virus in Brazil. The first confirmed case of microcephaly and other public health impacts soon followed, and in less than a year, the first US case was confirmed (Mestrovic, 2017). The rapid spread of the virus sparked a nationwide emergency response, prompting Congress to determine about how much to invest in prevention, control, and response. $1.9 billion was proposed to support Zika response programs, and ultimately $1.1 billion was allocated (Washington Post Editorial Board, 2016). While it is difficult to determine the exact economic burden of the disease to date, studies have made estimates for total medical costs and costs associated with workforce productivity. One study focused on six at-risk states, Alabama, Florida, Georgia, Louisiana, Mississippi, and Texas, and estimated that over a 230-day epidemic period, medical costs ranged from $223.9 million to over $2 billion (Lee et al., 2017). Texas was estimated to have incurred direct medical costs between $54 million and $99.8 million. While these figures are still estimates, they put a spotlight on the need for sustained public health investments to prevent vector-borne disease outbreaks to secure human and economic health.

Environmental impacts of mosquito control can also influence the mosquito abatement/public health relationship. Negative environmental impacts can turn public sentiment against mosquito abatement programs, making it more difficult to implement these programs and protect the public from certain vector-borne diseases. The IMM is a promising mosquito control approach that can be used to minimize negative environmental impacts due to the many tools included in IPM principles.

13.4.3 Adverse impacts to the environment from mosquito abatement activities

Mosquito abatement, specifically when applying pesticides to kill mosquito adults and larvae, can have adverse environmental impacts on air, water, and land. One only has to consider the devastating effects of DDT to know that pesticides can have unintended consequences on the environment. While DDT was a potent and useful insecticide initially, further research and observations uncovered its unintended toxicity in birds, fish, and other animals (NPIC, 1999). Because of this, EPA banned DDT in 1972 (NPIC, 1999). Despite the well-documented impacts on different organisms and the environmental perseverance of DDT (NPIC, 1999), as well as more recent research on its human health impacts, the use of DDT for the control of malaria continues to be debated. Public health advocates can be found on both sides of the debate. Even for a chemical such as DDT, balancing the negative impacts of pesticides on the environment and human health with the need to control disease-carrying mosquitoes is a challenge for both practitioners in public health and mosquito abatement. Completely eliminating risks to human health and the environment from pesticide applications is nearly impossible. However, steps can be taken to minimize the chance of directly exposing people and animals, such as applying pesticides at night when pollinators and people aren't as active or instituting Integrated Pest Management (IPM) practices to decrease overall pesticide usage.

Pesticides sprayed into the atmosphere can disperse, or "drift" from the originally selected location (NPIC, 2017a). This movement can cause the small liquid beads from aerial spraying to deposit into waterways and land outside of the original spray location (NPIC, 2017a). The California Department of Food and Agriculture studied the potential drift of malathion, a popular mosquito pesticide, during a 1990 spraying event to eliminate fruit fly populations and found that airborne concentrations were elevated outside of the administration zone (Newhart, 2006). In this case the area sampled outside of the administration zone was an endangered species area. This pesticide drift can be an issue if it causes the pesticide to settle on land or in water that is outside of the designated spray area (NPIC, 2017a). It can interact with and negatively affect organisms in and around those nondesignated spray areas (NPIC, 2017a). For example, bees are very susceptible to malathion's neurotoxicity (Gervais et al., 2009). However, there are multiple ways to decrease the risk of drift as noted by US EPA's Drift Reduction Technology Program, including the use of "spray shields" and "drift-reducing adjuvant chemicals" (U.S. EPA, 2018a).

Pesticides can have detrimental effects on freshwater and saltwater environments when they settle in waterways or get washed in through rain runoff (Helfrich et al., 2009). Pyrethroids, organophosphates, and carbamates are classes of insecticides that can all present some level of toxicity to aquatic life (Helfrich et al., 2009). Awareness of the environmental fate in water for the pesticide being applied and familiarity with the water bodies in or near designated spraying areas are critical to taking appropriate preventive measures to protect waterways from unnecessary insecticide exposures.

Pesticides can also settle or be sprayed directly onto the ground and soil. Different types of pesticides have different environmental fates in soil (NPIC, 2016). Pesticides sprayed on soil can revolatilize and drift further away (NPIC, 2017a) or be washed into bodies of water after heavy rains (Helfrich et al., 2009). Wildlife can be exposed to pesticide residues by ingesting plants growing in contaminated soils.

These potential adverse impacts to the environment need to be taken into consideration when spraying pesticides for the purpose of mosquito abatement. An IPM approach to mosquito abatement can help decrease the amount of pesticides used by implementing other precautionary methods, such as the elimination of stagnant or temporary bodies of water (U.S. EPA, 2017a). By following an IPM approach to mosquito abatement, environmental and human health risks from pesticides are minimized while simultaneously decreasing the risk of mosquito-borne disease infections in people (U.S. EPA, 2017a).

Challenges to implementing IPM/IMM

The IPM is a method of managing mosquitoes, or "pests," in a way that allows for decreased usage of pesticides that could cause environmental or human harm (U.S. EPA, 2017a). The US EPA outlines four distinct actions that lead to a successful IPM program: "identify pests and monitor progress," "set action thresholds," "prevent," and "control" (U.S. EPA, 2017a).

The first two actions are generally straightforward with regard to a mosquito control context. It is necessary to "identify" the target mosquitoes for control and "monitor progress" in controlling them, keeping in mind that changes may need to be made to the IPM plan if mosquito control is not proceeding satisfactorily (U.S. EPA, 2017a). "Action thresholds" are related to the overall goal of the mosquito control program; when mosquito numbers begin to pose a "health hazard" to humans or an "economic threat" to a specific industry, action thresholds have been surpassed (U.S. EPA, 2017a).

Action three is often difficult due to the significant amount of public input necessary. It requires combined efforts from local, state, and private mosquito abatement entities, as well as homeowners, renters, and home inhabitants. "Prevention" is aimed at "removing conditions that attract pests, such as food, water, and shelter" (U.S. EPA, 2017a). In order to successfully decrease the prevalence of mosquitos without the use of pesticides, breeding habitats and sustenance sources need to be taken out of the environment (U.S. EPA, 2017a). This means limiting the amount of stagnant and/or temporary bodies of water in the environment, as the larva requires these environments to facilitate growth and maturation. However, getting rid of stagnant and/or temporary bodies of water can be difficult for several reasons. Often times, people can have numerous stagnant water sites on their property that they are aware or unaware of, facilitating mosquito proliferation. Also, rain storms can produce flooding that leads to large, temporary bodies of stagnant water. The recent flooding in Houston from

Hurricane Harvey is a perfect example (Nutt, 2017). The U.S. EPA also lists other prevention activities appropriate for an IPM approach, including preserving a tidy living space and "weatherizing" buildings (U.S. EPA, 2017a). Even with IPM programs in place, applying pesticides will sometimes be necessary, raising the likelihood of exposing the environment, and public, to potentially harmful chemicals (U.S. EPA, 2017a). Any IPM program that plans to use pesticides as a mosquito control measure should take care to evaluate the potential environmental and health risks. Public health professionals and environmental scientists should be consulted before any pesticide spraying event to minimize these impacts.

Sensitive environments

Depending on where the pesticides are applied, different environments can have different levels of susceptibility to the spraying. For example, in the previously mentioned example from the California Environmental Protection Agency (Newhart, 2006), spray drift near an endangered species reserve could have negative outcomes, especially if the endangered animals in question were especially susceptible to the sprayed pesticide. Sensitive environments can also include those where children live and play. Pesticides pose a greater chance of contributing to adverse health outcomes in children due to the susceptibility of their developing organ systems (NPIC, 2018). Care should be taken to limit their exposures, especially in areas where pesticide spraying recently occurred (NPIC, 2018). Similarly, pregnant women should also exercise caution to avoid exposures to their developing fetus (NPIC, 2017b).

IPM and its links to positive environmental impacts

Green infrastructure can greatly reduce the amount of suitable habitat for mosquito proliferation, that is, stagnant and/or temporary bodies of water. It allows for rainwater to be absorbed into the ground, as opposed to collecting in areas with impermeable surfaces, such as streets and sidewalks (U.S. EPA, 2017c). Examples of green infrastructure include rain gardens, planter boxes, bioswales, green roofs, and permeable pavement (U.S. EPA, 2017c). Green infrastructure also has benefits of groundwater regeneration, decreased flooding, and dissipation of heat islands—all issues relevant to public health (U.S. EPA, 2016, 2017b, 2018b).

13.5 Future directions and conclusions

What does the future hold for mosquito-borne diseases and how will our public health and mosquito abatement programs respond to these future changes? Clearly, there is a potential for increase in mosquito-borne diseases due to a variety of factors. For many vectors, ranges are increasing. Environmental conditions associated with climate change, for example, increased temperatures, severe weather events, flooding, and changes in rainfall, can contribute to the changes in the distribution and range of mosquitoes. However, as can be seen with the distribution of malaria and other vector-borne diseases, weather and climate impacts on the transmission of disease are not the only factors to influence disease transmission. Social factors can be as or more important related to disease transmission. Increased international travel and the movement of refugee populations can increase the risk of travel-related cases of mosquito-borne diseases and their sequelae (USGCRP, 2016).

Despite the threats of increased risks of mosquito-borne diseases, advances in technology related to both disease surveillance and mosquito abatement, opportunities for mobilizing public health advocates, and enlisting support from partners in sectors like emergency planning and response provide reason for optimism that the future burden of mosquito-borne diseases will be minimized. Surveillance of mosquitoes and mosquito-borne disease is the foundation for informed public health and mosquito abatement programs. From the promise of citizen monitoring (Ricciuti, 2017) to the development of comprehensive environmental health tracking networks, public health advocates have shown a willingness to lobby for funding mosquito control programs at the federal, state, and local levels. Recognition of the advantages to using a comprehensive planning approach to implementing mosquito abatement initiatives increases the likelihood of sharing resources and gaining broad public support for these initiatives (ASTHO, 2009).

Mosquito abatement is a multidiscipline effort that can and should involve many agencies and organizations at the local, state, and federal level. Public health may not be the primary agency for mosquito control, but public health data, alerts, warnings, and information often drive the political will to reduce mosquito populations and habitats. The functional relationship between public health and mosquito abatement is similar to the relationship between public health and other environmental programs. Mosquito abatement programs are often outside of public health agencies. Similarly, most drinking water and sewage treatment programs are within public works/environmental protection programs. Emergency planning and response activities provide a forum for public health and environmental program leaders to interact and integrate their activities. Local planning efforts around mosquito abatement, emergency preparedness and response, community/economic development, hurricane readiness, etc., all provide opportunities for building constructive relationships between public health and mosquito abatement programs. Continued progress in developing this public health/mosquito abatement program partnership is the only rational response to the severity of the risks of existing and emerging mosquito-borne diseases. As history demonstrates, the mighty mosquito always returns and frequently with a previously unknown and unpredictable disease threat. Public health has a responsibility and an opportunity to be part of a comprehensive and thoughtful approach to continue mosquito control through partnerships and teamwork at all levels of government.

References and further reading

AccuWeather, 2017. AccuWeather Launches New Mosquito Zika Risk Index for the Contiguous U.S. on AccuWeather.com and iOS App. Available from: https://www.accuweather.com/en/press/70023866. (Accessed May 7, 2018).

American Mosquito Control Association (AMCA), 2017. AMCA Best Practices for Integrated Mosquito Management: A Focused Update. American Mosquito Control Association, Mt. Laurel, NJ. Available from: https://www.naccho.org/uploads/downloadable-resources/amca-guidelines-final_pdf.pdf. (Accessed May 7, 2018).

American Public Health Association (APHA), 2016. Organizational Letter in Support of the President's Request for $1.9 Billion in Emergency Funding to Prepare for and Respond to the Zika Crisis. Available from: https://www.apha.org/-/media/files/pdf/advocacy/letters/2016/160223_zika_supplemental.ashx?la=en&hash=4E356978F72AB67FDFC5EAA4E3E93FC2E1501AC3. (Accessed May 21, 2018).

Arizona Department of Health Services (AZDHS), 2015. Binational infectious disease cases along the US—Mexico border. Annual Report 2015. Available from: https://azdhs.gov/documents/director/border-health/bids/binational-cases-reports/annual%20reports/2015-AZ-Binational-Cases-Report.pdf. (Accessed May 24, 2018).

Association of State and Territorial Health Officials (ASTHO), 2009. Communicating About Effective Mosquito Control: Tailor Made for Our Community. Available from: http://www.astho.org/Programs/Environmental-Health/Natural-Environment/AsthoMosquitoCommGuide011509/. (Accessed May 21, 2018).

Association of State and Territorial Health Officials (ASTHO), 2017a. Profile of State and Territorial Public Health (Volume 4). Available from: http://www.astho.org/Profile/Volume-Four/2016a-ASTHO-Profile-of-State-and-Territorial-Public-Health/. (Accessed May 24, 2018).

Association of State and Territorial Health Officials (ASTHO), 2017b. Using Health Impact Assessments to Enhance the Environmental Regulatory Process: Case Studies and Key Messages. Available from: http://www.astho.org/Environmental-Health/Using-HIA-to-Enhance-the-Environmental-Regulatory-Process/. (Accessed May 3, 2018).

Association of State and Territorial Health Officials (ASTHO), 2018. Analysis of Express Legal Authorities for Mosquito Control in the U.S. States, the District of Columbia and Puerto Rico. Available from: http://www.astho.org/ASTHOReports/Analysis-of-Express-Legal-Authorities-for-Mosquito-Control-in-the-US-DC-and-PR/10-12-18/. (Accessed October 22, 2018).

Association of State Drinking Water Administrators (ASDWA), 2018. About ASDWA. Available from: https://www.asdwa.org/about-asdwa/. (Accessed May 3, 2018).

Bennet, M., 2016. Bennet Calls for Increased Funding for Mosquito-Control Programs to Fight the Spread of Zika. Available from: https://www.bennet.senate.gov/?p=release&id=3637. (Accessed May 7, 2018).

Braks, M., Medlock, J., Hubalek, A., et al., 2014. Vector-borne disease intelligence: strategies to deal with disease burden and threats. Front. Public Health 280. Available from: https://www.ncbi.nlm.nih.gov/pmc/articles/PMC4273637/. (Accessed May 7, 2018).

Camden County Mosquito Commission, 2017. Mosquito Commission. Available from: http://www.camdencounty.com/service/mosquito-commission/. (Accessed May 24, 2018).

Centers for Disease Control and Prevention (CDC), 2015. Surveillance Resources. Available from: https://www.cdc.gov/westnile/resourcepages/survresources.html. (Accessed March 30, 2018).

Centers for Disease Control and Prevention (CDC), 2016a. CDC Awards Nearly $184 Million to Continue the Fight Against Zika. Available from: https://www.cdc.gov/media/releases/2016/p1222-zika-funding.html. (Accessed May 7, 2018).

Centers for Disease Control and Prevention (CDC), 2016b. Surveillance and Control of Aedes aegypti and Aedes Albopictus in the United States. Available from: https://www.cdc.gov/zika/vector/vector-control.html. (Accessed March 30, 2018).

Centers for Disease Control and Prevention (CDC), 2018a. About CDC 24-7. Available from: https://www.cdc.gov/about/organization/mission.htm. (Accessed May 14, 2018).

Centers for Disease Control and Prevention (CDC), 2018b. Integrated Mosquito Management. Available from: https://www.cdc.gov/westnile/vectorcontrol/integrated_mosquito_management.html. (Accessed March 30, 2018).

Centers for Disease Control and Prevention (CDC), 2018c. National Environmental Public Health Tracking Network. Available from: https://ephtracking.cdc.gov/showHome.action. (Accessed May 24, 2018).

Centers for Disease Control and Prevention (CDC), 2018d. The Public Health System & the 10 Essential Public Health Services. Available from: https://www.cdc.gov/stltpublichealth/publichealthservices/essentialhealthservices.html. (Accessed March 28, 2018).

Centers for Disease Control and Prevention (CDC), 2018e. US Zika Pregnancy and Infant Registry. Available from: https://www.cdc.gov/pregnancy/zika/research/registry.html. (Accessed March 30, 2018).

Centers for Disease Control and Prevention (CDC), 2018f. West Nile virus. Available from: https://www.cdc.gov/westnile/index.html. (Accessed May 3, 2018).

Centers for Disease Control and Prevention (CDC), 2018g. Zoonotic Diseases. Available from: https://www.cdc.gov/onehealth/basics/zoonotic-diseases.html. (Accessed October 29, 2018).

Centers for Disease Control and Prevention (CDC) Foundation, 2018. What Is Public Health? Available from: https://www.cdcfoundation.org/what-public-health. (Accessed March 28, 2018).

Dennis, B., Sun, L., 2016. Without Federal Funding, Counties Brace to Confront Zika on Their Own. The Washington Post. Available from: https://www.washingtonpost.com/national/health-science/without-federal-help-counties-brace-to-confront-zika-on-their-own/2016/06/24/6591f1e2-36f2-11e6-a254-2b336e293a3c_story.html?noredirect=on&utm_term=.c3460ea90e01. (Accessed May 1, 2018).

Dillon, C., 2000. Mosquitoes and Public Health: Protecting a Resource in the Face of Public Fear. George Wright Forum 63–72. Available from: http://www.georgewright.org/174dillon.pdf. (Accessed May 7, 2018).

Epstein, S., Lister, S., 2016. Zika Response Funding: Request and Congressional Action. Congressional Research Service. Available from: https://fas.org/sgp/crs/misc/R44460.pdf. (Accessed May 21, 2018).

Fannie Mae, 2005. Uniform Property Appraisal Form (Form 1004). Available from: https://www.fanniemae.com/content/guide_form/1004.pdf. (Accessed May 21, 2018).

Federal Emergency Management Agency, 2016. Public Assistance Program and Policy Guide—Appendix G: Mosquito Abatement. Available from: https://www.fema.gov/media-library-data/1456167739485-75a028890345c6921d8d6ae473fbc8b3/PA_Program_and_Policy_Guide_2-21-2016_Fixes.pdf. (Accessed October 29, 2018).

Filler, S., MacArthur, J., Parise, M., et al., 2006. Locally acquired mosquito transmitted infection: a guide for investigation in the United States. MMWR 55 (13), 1–9. Available from: https://www.cdc.gov/mmwr/preview/mmwrhtml/rr5513a1.htm. (Accessed May 7, 2018).

Flechas, J., 2016. Tempers flare as public protests spraying for Zika mosquitoes in South Beach. Miami Herald. 14 September. Available from: http://www.miamiherald.com/news/health-care/article101823792.html. (Accessed May 7, 2018).

Geraghty, E., 2017. Infectious Mosquitoes Pose Increasing Health Threat. ESRI Blog. Available from: https://www.esri.com/about/newsroom/blog/gis-aids-mosquito-control/. (Accessed May 24, 2018).

Gervais, J., Luukinen, B., Buhl, K., et al., 2009. Malathion General Fact Sheet. National Pesticide Information Center, Oregon State University Extension Services. Available from: http://npic.orst.edu/factsheets/malagen.html. (Accessed May 17, 2018).

Harris County Public Health, 2017. Mosquito & Vector Control. Available from: https://publichealth.harriscountytx.gov/About/Organization-Offices/Mosquito-and-Vector-Control. (Accessed May 24, 2018).

Harris County Public Health, 2018. Mission, Vision and Values. Available from: http://publichealth.harriscountytx.gov/About/Mission-Vision-Values. (Accessed May 14, 2018).

Hawaii Department of Health, 2016. If You Are Traveling to the Big Island of Hawai'i. Available from: https://health.hawaii.gov/docd/files/2015/12/Travelers-to-HI_2015rev.pdf. (Accessed March 29, 2018).

Hearne, S., Locke, P., Mellman, M., et al., 2000. Public opinion about public health—United States 1999. MMWR Weekly 49 (12), 258–260. Available from: https://www.cdc.gov/mmwr/preview/mmwrhtml/mm4912a4.htm. (Accessed May 7, 2018).

Helfrich, L., Weigman, D., Hipkins, P., et al., 2009. Pesticides and Aquatic Animals: A Guide to Reducing Impacts on Aquatic Systems. Available from: https://vtechworks.lib.vt.edu/bitstream/handle/10919/48060/420-013_pdf.pdf?sequence=1&isAllowed=y. (Accessed May 17, 2018).

Jones, J., Lopez, B., Adams, L., et al., 2016. Binational dengue outbreak along the United States—Mexico border—Yuma County, Arizona, and Sonora, Mexico, 2014. MMWR 65 (19), 495–499. Available from: https://www.cdc.gov/mmwr/volumes/65/wr/mm6519a3.htm. (Accessed May 7, 2018).

Kofler, S., 2016. San Antonio Slashed Its Mosquito Control Budget-Could It Respond to Zika? Available from: http://www.tpr.org/post/san-antonio-slashed-its-mosquito-control-budget-could-it-respond-zika. (Accessed October 29, 2018).

Kriete, B.P.M., 2017. ArcGIS Platform Curtails Mosquitoes in Coastal California. ARC News. Available from: http://www.esri.com/esri-news/arcnews/summer16articles/arcgis-platform-curtails-mosquitos-in-coastal-california. (Accessed 30 March 2018).

Lee, B., Alfara-Murrilo, J., Parpia, A., et al., 2017. The potential economic burden of Zika in the continental United States. PLOS Negl. Trop. Dis. Available from: http://journals.plos.org/plosntds/article/file?id=10.1371/journal.pntd.0005531&type=printable. (Accessed May 16, 2018).

Lloyd, A.M., Connelly, C.R., Carlson, D.B., 2018. Florida Mosquito Control: The State of the Mission as Defined by Mosquito Controllers, Regulators, and Environmental Managers. Available from: https://fmel.ifas.ufl.edu/media/fmelifasufledu/7-15-2018-white-paper.pdf. (Accessed May 7, 2018).

Maddox, N., 2017. Celebrating 75 years of ASTHO: milestones in public health leadership. J. Public Health Manag. Pract. 23 (5), 524–530. Available from: https://journals.lww.com/jphmp/Fulltext/2017/09000/Celebrating_75_Years_of_ASTHO___Milestones_in.16.aspx. (Accessed May 16, 2018).

Mestrovic, T., 2017. Zika virus history. News Medical. Available from: https://www.news-medical.net/health/Zika-Virus-History.aspx. (Accessed May 16, 2018).

Metropolitan Mosquito Control District, 2017. Budget in Brief for the Fiscal Year Beginning January 1, 2018. Available from: http://www.mmcd.org/wp-content/uploads/2017/12/BudgetBrief2018dft.pdf. (Accessed May 24, 2018).

MosquitoMagnet, 2018. The Economic Cost of Mosquito-Borne Disease. Available from: http://www.mosquitomagnet.com/articles/the-economic-cost-of-mosquito-borne-diseases. (Accessed May 7, 2018).

National Association of County and City Health Officials (NACCHO), 2017. Mosquito Control Capabilities of the U.S. Available from: https://www.naccho.org/uploads/downloadable-resources/Mosquito-control-in-the-U.S.-Report.pdf. (Accessed May 17, 2018).

National Pesticide Information Center (NPIC), 1999. DDT. Available from: http://npic.orst.edu/factsheets/ddtgen.pdf. (Accessed May 17, 2018).

National Pesticide Information Center (NPIC), 2016. Soil and Pesticides. Available from: http://npic.orst.edu/envir/soil.html. (Accessed May 17, 2018).

National Pesticide Information Center (NPIC), 2017a. Pesticide Drift. Available from: http://npic.orst.edu/reg/drift.html. (Accessed August 20, 2018).

National Pesticide Information Center (NPIC), 2017b. Pesticides and Pregnancy. Available from: http://npic.orst.edu/health/preg.html. (Accessed August 20, 2018).

National Pesticide Information Center (NPIC), 2018. Pesticides and Children. Available from: http://npic.orst.edu/health/child.html. (Accessed May 17, 2018).

New Jersey Department of Environmental Protection (NJ DEP), 2016. Mosquito Control & West Nile Virus. Available from: http://www.nj.gov/dep/mosquito/. (Accessed May 24, 2018).

Newhart, K., 2006. Environmental Fate of Malathion. California Environmental Protection Agency. Available from: http://www.cdpr.ca.gov/docs/emon/pubs/fatememo/efate_malathion.pdf. (Accessed May 24, 2018).

Nutt, A., 2017. Houston's next big storm: mosquitoes. Washington Post. Available from: https://www.washingtonpost.com/news/to-your-health/wp/2017/09/07/houstons-next-big-storm-mosquitoes/?utm_term=.b26b3f2f98c2. (Accessed May 17, 2018).

Porter, M., 2016. NASA Helps Forecast Zika Risk. Available from: https://www.nasa.gov/centers/marshall/news/news/releases/2016/nasa-helps-forecast-zika-risk.html. (Accessed May 7, 2018).

Ricciuti, E., 2017. Citizen scientists collect 29,000 mosquitoes in Germany and help detect spreading population of invasive species. Entomology Today. 25 September. Available from: https://entomologytoday.org/2017/09/25/citizen-scientists-collect-29000-mosquitoes-in-germany-and-help-detect-spreading-populations-of-invasive-species/. (Accessed May 16, 2018).

Robinson, K., 2018. Challenges between public health and pest management. In: Presented at the Enhancing Environmental Health: Vectors & Public Health Pests Virtual Conference 2018. National Environmental Health Association. (Accessed 7 May 2018).

Rudolph, L., Caplan, J., Ben-Moshe, K., Dillon, L., 2013. Health in All Policies: A Guide for State and Local Governments. Available from: http://www.phi.org/uploads/files/Health_in_All_Policies-A_Guide_for_State_and_Local_Governments.pdf. (Accessed May 7, 2018).

Segal, L., Martin, A., 2017. A Funding Crisis for Public Health Safety: State-by-State Public Health Funding and Key Health Facts. Trust for American's Health, Washington, DC. Available from: https://www.tfah.org/report-details/a-funding-crisis-for-public-health-and-safety-state-by-state-public-health-funding-and-key-health-facts-2017/. (Accessed May 16, 2018).

Statistica, 2018. U.S. Population: Which Brand of Insect Repellent Do You Use Most Often? Available from: https://www.statista.com/statistics/275079/us-households-most-used-brands-of-insect-repellents. (Accessed March 29, 2018).

Texas Department of State Health Services, 2016. Vision and Mission. Available from: http://www.dshs.texas.gov/visionmission.shtm. (Accessed May 14, 2018).

The United States-Mexico Foundation for Science (FUMEC), 2017. Mexico Foments Actions Against Zika, Dengue and Chikungunya. Available from: https://fumec.org.mx/v6/index.php?option=com_content&view=article&id=752:healt-actions&catid=98&Itemid=442&lang=en. (Accessed May 24, 2018).

U.S. Department of Energy, n.d. Air Conditioning. Available from: https://energy.gov/energysaver/home-cooling-systems/air-conditioning [Accessed 29 March 2018].

U.S. Department of Health & Human Services, 2017. U.S.- Mexico Border Health Commission: Who we are. Available from: https://www.hhs.gov/about/agencies/oga/about-oga/what-we-do/international-relations-division/americas/border-health-commission/index.html. (Accessed May 24, 2018).

U.S. Environmental Protection Agency (U.S. EPA), 2016. Manage Flood Risk. Available from: https://www.epa.gov/green-infrastructure/manage-flood-risk. (Accessed August 20, 2018).

U.S. Environmental Protection Agency (U.S. EPA), 2017a. Introduction to Integrated Pest Management. Available from: https://www.epa.gov/managing-pests-schools/introduction-integrated-pest-management. (Accessed May 17, 2018).

U.S. Environmental Protection Agency (U.S. EPA), 2017b. Reduce Urban Heat Island Effect. Available from: https://www.epa.gov/green-infrastructure/reduce-urban-heat-island-effect. (Accessed August 20, 2018).

U.S. Environmental Protection Agency (U.S. EPA), 2017c. What Is Green Infrastructure? Available from: https://www.epa.gov/green-infrastructure/what-green-infrastructure. (Accessed May 17, 2018).

U.S. Environmental Protection Agency (U.S. EPA), 2018a. About the Drift Reduction Technology Program. Available from: https://www.epa.gov/reducing-pesticide-drift/about-drift-reduction-technology-program. (Accessed May 17, 2018).

U.S. Environmental Protection Agency (U.S. EPA), 2018b. Green Infrastructure and Ground Water Impacts. Available from: https://www.epa.gov/green-infrastructure/green-infrastructure-and-ground-water-impacts. (Accessed August 20, 2018).

U.S. Global Change Research Program (USGCRP), 2016. The Impacts of Climate Change on Human Health in the United States: A Scientific Assessment. Available from: https://health2016.globalchange.gov/vectorborne-diseases. (Accessed May 24, 2018).

U.S. National Library of Medicine, 2018. Community Health Maps. Available from: https://communityhealthmaps.nlm.nih.gov/. (Accessed May 7, 2018).

Washington Post Editorial Board, 2016. It's us sgainst Zika—whose side is Congress on? Washington Post. 23 June. Available from: https://www.washingtonpost.com/opinions/its-us-against-zika--whose-side-is-congress-on/2016/06/23/51715eec-3971-11e6-9ccd-d6005beac8b3_story.html?utm_term=.1a8631e83910. (Accessed May 16, 2018).

World Health Organization (WHO), 2014. A Global Brief on Vector-Borne Diseases. Available from: http://apps.who.int/iris/bitstream/10665/111008/1/WHO_DCO_WHD_2014.1_eng.pdf. (Accessed March 28, 2018).

World Health Organization (WHO), 2017. Vector-Borne Diseases. Available from: http://www.who.int/mediacentre/factsheets/fs387/en/. (Accessed March 30, 2018).

World Health Organization (WHO), 2018a. About WHO. Available from: http://www.who.int/about/en/. (Accessed May 14, 2018).

World Health Organization (WHO), 2018b. Global Health Estimates 2016: Deaths by Cause, Age, Sex, by Country and by Region, 2000–2016. Available from: http://www.who.int/healthinfo/global_burden_disease/estimates/en/. (Accessed November 1, 2018).

Yong, E., 2017. The case for sharing all of American's data on mosquitoes. The Atlantic. 24 August 2017. Available from: https://www.theatlantic.com/science/archive/2017/08/mosquito-data/537735/. (Accessed May 7, 2018).

Chapter 14

Vaccines for Mosquito-Borne Human Viruses Affecting Texas

Peter J. Hotez

Texas Children's Hospital Center for Vaccine Development, National School of Tropical Medicine, Baylor College of Medicine, Houston, TX, United States

14.1 Introduction

Together with Florida, the state of Texas has historically led the United States in terms of numbers of arbovirus epidemics, as well as the high endemicity of neglected tropical diseases (Hotez, 2018). In the 21st century alone, Texas has experienced outbreaks of several important arbovirus infections. They include epidemics of dengue fever in Brownsville and elsewhere in Cameron county on the border with Mexico (Adalja et al., 2012) and one in Houston in the early 2000s (Murray et al., 2013a). In addition, both chikungunya virus and Zika virus infection have emerged in South Texas (Texas Department of State Health Services, 2016a,b). Texas has also experienced some of the worst West Nile virus (WNV) infection epidemics in American history, including one in 2012 that especially affected North Texas, with thousands of cases and dozens of deaths (Murray et al., 2013b; Cervantes et al., 2015). In addition to their public health impact, WNV, dengue, and Zika virus infections also adversely affect the Texas economy.

The underlying reasons for the susceptibility of human populations living in Texas to arbovirus infections and other neglected tropical diseases have been recently reviewed (Hotez, 2018). They include rapid urbanization and expansion of a large (and nonimmune) population such that the cities in Texas now rank among the fastest growing in America (Hotez, 2018). Adding to this mix are frequent human immigrations from arbovirus-endemic regions of Latin America and the Caribbean, Africa, and Asia, with the potential for multiple virus introductions to susceptible populations (Hotez, 2018). A similar situation led to the 2016 emergence of Zika virus infection in Miami, following multiple introductions of the virus through infected individuals emigrating or visiting from the Caribbean.

Still another key factor is extreme poverty, which is widespread both in South Texas and in the major cities of Texas (Hotez, 2018). Urbanization and poverty combine to produce urban slums, which are especially vulnerable to arbovirus outbreaks, as both mosquitoes and arbovirus infections flourish in the setting of inadequate housing with no air-conditioning, together with crowding and poor sanitation. Climate change and global warming, which are believed to disproportionately affect Texas and the adjoining Gulf Coast region more than other areas in the United States, also must be considered (Hotez, 2018; Petkova et al., 2015).

Many of the risk factors for human arbovirus infections in Texas will likely remain in place in the coming years and decades and therefore become even more problematic. Indeed, if such trends continue, there is a high likelihood that existing and new arbovirus vaccines will be required to protect the human population of Texas, especially in the absence of enhanced mosquito control efforts. Adding to this challenge are findings that today Texas has emerged as a new epicenter of a growing antivaccine movement in America (Hotez, 2016).

Mosquitoes, Communities, and Public Health in Texas. https://doi.org/10.1016/B978-0-12-814545-6.00014-6

Briefly summarized here are some of the current and anticipated vaccines for human arbovirus infections that may need to be used in Texas in the near future and a discussion of both scientific and socioeconomic challenges to new vaccine introduction.

14.2 Vaccines for *Aedes*-transmitted human virus infections

Aedes aegypti mosquitoes are widespread in Texas, especially in the Gulf coastal regions and major urban centers. Accordingly, there is potential risk during the Texas summers for outbreaks of *A. aegypti*–transmitted virus infections, including yellow fever, dengue, chikungunya, and Zika virus infection. Several vaccines for these diseases are under development, while two vaccines have so far reached licensure (Table 14.1).

TABLE 14.1 Approved vaccines or vaccine in advanced stages of clinical development (phase 2 and higher) for mosquito-borne arboviruses currently circulating in the Western Hemisphere.

Disease	Vaccine	Stage of development	Manufacturer	Comments and references
Yellow fever	YF-Vax	Licensed in the United States and many other countries	Sanofi Pasteur + several other developing country vaccine manufacturers	Not recommended for infants less than 6 months of age (Staples et al., 2018). Prevents infection and useful to interrupt transmission
Dengue	Dengvaxia	Licensed in at least 10 countries, including the United States and Mexico	Sanofi Pasteur	Initial indication for ages 9–45 years of age (Hadinegoro et al., 2015; Simmons, 2015; Aguiar et al., 2016). Decreases disease but effect on transmission unknown
Dengue	TV003/005	Phase 2	NIAID-NIH/Merck & Co. (United States)/Butantan (Brazil)/Panacea (India)/Vabiotech (Vietnam)	Protective in human challenge studies (Halstead and Thomas, 2018; Kirkpatrick et al., 2016; Vannice et al., 2016)
Dengue	DenVax	Phase 2	Takeda	Vannice et al. (2016)
Chikungunya	MV-CHIK	Phase 2	Themis Bioscience	Lyon (2017)
	VLP	Phase 2	NIAID-NIH	Lyon (2017)
West Nile virus infection	ChimeriVax-WN02	Phase 2	Sanofi Pasteur	Vaccine development appears to have halted, with no plans to proceed towards licensure (Dayan et al., 2012, 2013)

14.2.1 Yellow fever

According to the Texas State Historical Association, both Galveston and Houston experienced multiple yellow fever outbreaks during the middle of the 19th century, including several that affected thousands of residents and caused hundreds of deaths (Burns, n.d.). The last Galveston epidemic, which occurred in the summer of 1867, killed an estimated 725 residents (Burns, n.d.). There is an approved yellow fever vaccine, known as the 17D vaccine, which is a live vaccine first developed during the 1930s by Max Theiler and his associates at the Rockefeller Foundation laboratories in New York. To develop the vaccine, live yellow fever virus was first passaged in mouse embryo tissues and then whole chick embryos, before finally passing the virus through eggs (Staples et al., 2018). The vaccine was first introduced into Brazil between 1938 and 1941 (Staples et al., 2018). In the United States, the 17D vaccine is manufactured by Sanofi Pasteur based in Swiftwater, Pennsylvania, that produces it under the trade name of YF-Vax (Staples et al., 2018). The vaccine is administered by a subcutaneous route, with host neutralizing antibodies peaking roughly 1 month afterwards (Staples et al., 2018). It is considered extremely safe, with some estimates suggesting that more than 500 million people have been vaccinated over the last century (Staples et al., 2018). However, the vaccine can be reactogenic and result in fever, chills, malaise, headaches, and myalgias (Staples et al., 2018). In terms of serious adverse events, allergic hypersensitivity reactions, possibly to egg or chicken proteins, are considered rare (Staples et al., 2018). Another rare side effect is neurotropic disease resulting from live virus invasion in the human central nervous system, while sometimes fatal viscerotropic infection from the 17D virus has been reported in approximately 100 cases worldwide (Staples et al., 2018).

In 2017 an outbreak of rural (jungle) yellow fever (transmitted by mosquitoes of the genus *Sabethes* or *Haemagogus*) emerged in Brazil, and there is potential for this or future epidemics to spread to Texas or the Gulf Coast where the urban vector, *A. aegypti*, is found (Paules and Fauci, 2017). However, there is not sufficient vaccine available if the epidemic was to spread widely across the Americas. If a yellow fever outbreak was to occur in Texas, it would likely be recommended to implement mass vaccinations, except for infants younger than 6 months (because this age group is considered more susceptible to neurologic disease), while infants aged 6–8 months and adults age 60 and older should be vaccinated with a "precaution" because these populations are also at greater risk of neurologic disease (and viscerotropic disease in the elderly) (Staples et al., 2018). To reduce the demand for vaccine doses, potentially fractional dosing could be implemented or even ring vaccination if it was combined with adequate vector control. Newer-generation vaccines are also under development.

14.2.2 Dengue

As noted earlier, Texas has experienced modern dengue outbreaks, both in South Texas and in Houston during the 2000s (Adalja et al., 2012; Murray et al., 2013a). An unanswered question is whether dengue has now become endemic to South Texas especially in Cameron County on the Texas border where a high seroprevalence to dengue virus was noted among randomly selected adults or in poor communities lacking basic city services and adequate housing (Hotez, 2018; Bouri et al., 2012; Vitek et al., 2014). In such areas, it would be highly instructive to conduct cost-effectiveness analyses for administering dengue vaccines. Approved in Mexico and nine other countries, including Brazil, Costa Rica, El Salvador, Guatemala, Indonesia, Paraguay, the Philippines, Singapore, Thailand, and most recently the United States, is a new dengue vaccine, which is also manufactured by Sanofi Pasteur and known as Dengvaxia (World Health Organization, n.d.). The Sanofi Pasteur vaccine is a chimeric vaccine using the 17D yellow fever backbone, with dengue preM (membrane protein) and E (envelope glycoprotein) genes inserted (Halstead and Thomas, 2018). Currently the World Health Organization recommends that the Sanofi Pasteur vaccine should be used only in geographic areas where there is evidence for a high dengue disease burden. In Mexico, the vaccine was licensed at the end of 2015 for individuals between the ages of 9 and 45 years living in endemic areas.

Currently, there is not sufficient disease burden or cost-effectiveness data to justify the introduction of the Sanofi Pasteur vaccine into South Texas, but this situation could change. There remain questions, however, regarding the safety of Dengvaxia, especially among certain age groups. Particularly in young children, it has been noted that an excess of hospitalizations occurred during clinical trials in Asia (Hadinegoro et al., 2015; Simmons, 2015). In addition, the efficacy of the vaccine was lower for recipients who never had previous dengue and were seronegative prior to receiving the vaccine (Hadinegoro et al., 2015; Simmons, 2015). The basis for these observations is still under investigation but may reflect partly the fact that there are four dengue virus serotypes that may either cross-react or exacerbate disease through an immune enhancement phenomenon. Some investigators have therefore recommended that the vaccine should only be administered to individuals who have been first immunologically screened and show evidence of previous exposure to at least one of the four dengue serotypes (Aguiar et al., 2016). However, the practicality of taking this approach remains unclear. Still another concern is whether widespread vaccination with Dengvaxia would interrupt transmission and prevent a dengue outbreak.

Additional dengue vaccines are under development. For instance, scientists at the National Institute of Allergy and Infectious Diseases (NIAID) of the National Institutes of Health (NIH) have developed a tetravalent formulation of four dengue virus strains with deletions in their untranslated regions to produce an attenuated virus that result in lower levels of viremia (CV003/CV005) (Halstead and Thomas, 2018). In turn the NIH vaccine has been licensed to Merck & Co. in the United States, as well as several developing country vaccine manufacturers (Halstead and Thomas, 2018). In human challenge studies with live virus of a dengue virus type 2 strain, this vaccine was shown to elicit complete protection (Kirkpatrick et al., 2016). A live recombinant virus vaccine known as DENVax is also under development by Takeda, a Japanese-based pharmaceutical company, and consists of an attenuated dengue type 2 virus strain, together with types 1, 3, and 4 using the dengue type 2 background. This vaccine has also undergone phase 2 testing (Vannice et al., 2016). Finally, there are numerous other vaccine constructs in earlier stages (phase 1 and preclinical) of development (Vannice et al., 2016).

14.3 Other vaccines: Chikungunya and Zika virus infection

Both chikungunya and Zika virus infections have emerged in South Texas, especially Cameron County on the border with Mexico, but the level of autochthonous transmission has so far remained low and would probably not justify beginning a vaccine program (Texas Department of State Health Services, 2016a,b). Neither a chikungunya vaccine nor Zika virus vaccine is currently in advanced clinical development. However, early-stage phase 1/2 trials of a recombinant live-attenuated measles vaccine that expresses chikungunya antigens (MV-CHIKV) are in development, as is a viruslike particle vaccine containing a chikungunya envelope protein (Lyon, 2017). A live-attenuated virus vaccine has also reached phase 2 trials but was shown to induce arthritis in some human volunteers (Erasmus et al., 2016). Several experimental vaccines for Zika virus infection are also in early-stage clinical development, but as of this writing, none have yet reached phase 2 trials. Both the chikungunya and Zika virus vaccines will face considerable challenges if they are to advance to licensure. With regard to the former, it has been pointed out that field-based testing could be daunting because of the clinical similarities between chikungunya and dengue and other arbovirus infections that are often coendemic (Erasmus et al., 2016). For Zika virus, there are several candidate vaccines now in phase 1 trials (Thomas, 2017). However, it will be essential to show that they do not induce Guillan-Barré syndrome, a serious illness possibly caused by autoimmunity, as does the Zika virus. Another complication is that a likely target product profile for the Zika virus vaccine will include women of reproductive age and pregnant women, which presents a very high safety bar. Adding further to the complexities for the clinical testing of a Zika vaccine is finding a field setting suitable for a phase 3 trial, given the fact that (as of this writing) the level of Zika virus transmission in the Western Hemisphere is significantly down compared with previous years. Overall a challenge for both the chikungunya and Zika vaccines is that the diseases they target are highly periodic, so that these vaccines might not be as cost-effective as dengue vaccines.

14.4 Vaccines for *Culex*-transmitted human virus infections

West Nile virus (WNV) infection is the dominant human disease transmitted by *Culex* mosquitoes in Texas, as well as the state's most important arbovirus in terms of both health and economic impact (Hotez, 2018). The 2012 epidemic was devastating in terms of acute and chronic illness (Murray et al., 2013b; Cervantes et al., 2015), but in addition, WNV outbreaks have returned to Texas each year, with the worst epidemics noted to occur in three-year cycles (Murray et al., 2013b). Another feature of WNV (and one that is not commonly appreciated) is the ability of the virus to invade the central nervous system to produce a range of short-term and long-term neurologic sequelae, including depression and losses in cognition (Murray et al., 2014). Studies conducted at the National School of Tropical Medicine at Baylor College of Medicine have provided evidence that WNV can enter a chronic and persistent state, which is also linked to long-term kidney disease and even renal failure (Nolan et al., 2012). Such findings need to be incorporated in cost-effectiveness estimates to justify the development of WNV vaccines. In addition, WNV utilizes an avian reservoir host, which rapidly spreads virus infections among human populations; thus, together with *Culex* vector control, a vaccine would be necessary to contain newly emerging outbreaks.

Several candidate vaccines for WNV have been developed, but so far none have progressed to licensure. The one furthest along in terms of clinical development is ChimeriVax-WN02, a chimeric vaccine using the yellow fever 17D vaccine backbone with genes encoding WNV preM and E proteins inserted (Dayan et al., 2012, 2013). In a phase 2 study of healthy adult volunteers 50 years of age and older, the vaccine was found to be both safe and immunogenic (Dayan et al., 2012). Despite these findings the vaccine has not been developed further. In terms of other WNV vaccines, a second chimeric WNV vaccine on an attenuated dengue virus type 4 background has been developed (Pierce et al., 2017), as has a formalin-inactivated WNV vaccine (Posadas-Herrera et al., 2010), and a recombinant protein vaccine (Hawaii Biotech) encoding the WNV E protein (WN-80E) (Van Hoeven et al., 2016).

A key point for WNV vaccine development is the feasibility of taking several different approaches, with each vaccine seeming to show the ability to elicit neutralizing antibodies. Despite such promising results in both the laboratory and the clinic, none of these WNV vaccine candidates have advanced to licensure and introduction. The reasons for such failures need to be investigated further, but possibly key to this aspect is missing investor enthusiasm due to perceived absences for a medical need or promising lucrative commercial markets. However, such analyses may not adequately consider the long-term neurologic and renal sequelae of WNV due to chronic virus persistence and, therefore, need to be reevaluated.

14.5 Barriers to vaccine introduction

Currently, WNV causes the highest disease burden in Texas relative to any other arbovirus infection. However, there are no immediate prospects for advancing a WNV to licensure and introduction. As highlighted earlier, there is the perception that there is not a readily available market for such a vaccine, despite the susceptibility of older populations to severe neurologic disease and the emerging concerns that WNV is an important cause of kidney disease in the state of Texas. In terms of the arboviruses transmitted by *A. aegypti* mosquitoes, there is not yet a sufficient level of their transmission in Texas to warrant vaccine introduction, but given the regular appearance of dengue and now possibly Zika virus infection, some thought might be given to begin clinical testing among selected populations in Texas.

Still another important hurdle for arbovirus vaccine introduction is the rise of a vaccine hesitancy movement in Texas, with estimates that up to 50,000 children or more in Texas schools are not receiving their routine vaccines (Hotez, 2016). Among the reasons for widespread vaccine hesitancy is a well-organized local antivaccine movement having the ability to aggressively lobby the state legislature (Hotez, 2016). Should both the need arise and a licensed arbovirus vaccine becomes available, the Texas antivaccine lobby could become a formidable obstacle to vaccine introduction.

14.6   Conclusion

There are no immediate prospects for introducing an arbovirus vaccine for the state of Texas. The need is possibly greatest to introduce a WNV vaccine, but none so far have gone past phase 2 clinical trials. Both yellow fever and dengue vaccines are now available should outbreaks of either infection become significant, especially during the summer months corresponding to "mosquito season" in Texas. However, a strong and vigorous antivaccine lobby in the state would likely complicate this process.

Conflict of interest statement

The author is a principal investigator on a grant from the Chao Family Foundation to develop a therapeutic West Nile virus vaccine, in collaboration with Hawaii Biotech.

References and further reading

Adalja, A.A., Sell, T.K., Bouri, N., Franco, C., 2012. Lessons learned during dengue outbreaks in the United States, 2001–2011. Emerg. Infect. Dis. 18, 608–614.

Aguiar, M., Stollenwerk, N., Halstead, S.B., 2016. The impact of the newly licensed dengue vaccine in endemic countries. PLoS Negl. Trop. Dis. 10, e0005179.

Bouri, N., Sell, T.K., Franco, C., Adalja, A.A., et al., 2012. Return of epidemic dengue in the United States: implications for the public health practitioner. Public Health Rep. 127, 259–266.

Burns C.R. Epidemic Diseases [Online]. Texas State Historical Association. Available from: https://dshs.texas.gov/news/releases/2016/20161222.aspx [Accessed 7 December 2018].

Cervantes, D.T., Chen, S., Sutor, L.J., Stonecipher, S., et al., 2015. West Nile virus infection incidence based on donated blood samples and neuroinvasive disease reports, Northern Texas, USA, 2012. Emerg. Infect. Dis. 21, 681–683.

Dayan, G.H., Bevilacqua, J., Coleman, D., Buldo, A., et al., 2012. Phase II, dose ranging study of the safety and immunogenicity of single dose West Nile vaccine in healthy adults ≥ 50 years of age. Vaccine 30, 6656–6664.

Dayan, G.H., Pugachev, K., Bevilacqua, J., Lang, J., et al., 2013. Preclinical and clinical development of a YFV 17 D-based chimeric vaccine against West Nile virus. Viruses 5, 3048–3070.

Erasmus, J.H., Rossi, S.L., Weaver, S.C., 2016. Development of vaccines for chikungunya fever. J. Infect. Dis. 214, S488–S496.

Hadinegoro, S.R., Arredondo-Garcia, J.L., Capeding, M.R., Deseda, C., et al., 2015. Efficacy and long-term safety of a dengue vaccine in regions of endemic disease. N. Engl. J. Med. 373, 1195–1206.

Halstead, S.B., Thomas, S.J., 2018. Dengue vaccines. In: Plotkin, S.A., Worenstein, W.A., Offit, P.A., Edwards, K.M. (Eds.), Plotkin's Vaccines. Elsevier.

Hotez, P.J., 2016. Texas and its measles epidemics. PLoS Med. 13, e1002153.

Hotez, P.J., 2018. The rise of neglected tropical diseases in the "new Texas". PLoS Negl. Trop. Dis. 12, e0005581.

Kirkpatrick, B.D., Whitehead, S.S., Pierce, K.K., Tibery, C.M., et al., 2016. The live attenuated dengue vaccine TV003 elicits complete protection against dengue in a human challenge model. Sci. Transl. Med. 8, 330ra36.

Lyon, J., 2017. Chikungunya vaccine trials begin. JAMA 318, 322.

Murray, K.O., Rodriguez, L.F., Herrington, E., Kharat, V., et al., 2013a. Identification of dengue fever cases in Houston, Texas, with evidence of autochthonous transmission between 2003 and 2005. Vector Borne Zoonotic Dis. 13, 835–845.

Murray, K.O., Ruktanochai, D., Hesalroad, D., Fonken, E., Nolan, M.S., 2013b. West Nile Virus, Texas, USA, 2012. Emerg. Infect. Dis. 19 (11), 1836–1838.

Murray, K.O., Garcia, M.N., Rahbar, M.H., Martinez, D., et al., 2014. Survival analysis, long-term outcomes, and percentage of recovery up to 8 years post-infection among the Houston West Nile virus cohort. PLoS One 9, e102953.

Nolan, M.S., Podoll, A.S., Hause, A.M., Akers, K.M., et al., 2012. Prevalence of chronic kidney disease and progression of disease over time among patients enrolled in the Houston West Nile virus cohort. PLoS One 7, e40374.

Paules, C.I., Fauci, A.S., 2017. Yellow fever—once again on the radar screen in the Americas. N. Engl. J. Med. 376, 1397–1399.

Petkova, E.P., Ebi, K.L., Culp, D., Redlener, I., 2015. Climate change and health on the U.S. Gulf Coast: public health adaptation is needed to address future risks. Int. J. Environ. Res. Public Health 12, 9342–9356.

Pierce, K.K., Whitehead, S.S., Kirkpatrick, B.D., Grier, P.L., et al., 2017. A live attenuated chimeric West Nile virus vaccine, rWN/DEN4Delta30, is well tolerated and immunogenic in flavivirus-naive older adult volunteers. J. Infect. Dis. 215, 52–55.

Posadas-Herrera, G., Inoue, S., Fuke, I., Muraki, Y., et al., 2010. Development and evaluation of a formalin-inactivated West Nile Virus vaccine (WN-VAX) for a human vaccine candidate. Vaccine 28, 7939–7946.

Simmons, C.P., 2015. A candidate dengue vaccine walks a tightrope. N. Engl. J. Med. 373, 1263–1264.

Staples, J.E., Monath, T.P., Gershman, M.D., Barrett, A.D.T., 2018. Yellow fever vaccines. In: Plotkin, S.A., Worenstein, W.A., Offit, P.A., Edwards, K.M. (Eds.), Plotkin's Vaccines. seventh ed. Elsevier.

Texas Department of State Health Services, 2016a. Additional Locally-Acquired Zika Case in Cameron County.

Texas Department of State Health Services, 2016b. DSHS Announces First Texas-Acquired Chikungunya Case.

Thomas, S.J., 2017. Zika virus vaccines—a full field and looking for the closers. N. Engl. J. Med. 376, 1883–1886.

Van Hoeven, N., Joshi, S.W., Nana, G.I., Bosco-Lauth, A., et al., 2016. A novel synthetic TLR-4 agonist adjuvant increases the protective response to a clinical-stage West Nile virus vaccine antigen in multiple formulations. PLoS One 11, e0149610.

Vannice, K.S., Durbin, A., Hombach, J., 2016. Status of vaccine research and development of vaccines for dengue. Vaccine 34, 2934–2938.

Vitek, C.J., Gutierrez, J.A., Dirrigl Jr., F.J., 2014. Dengue vectors, human activity, and dengue virus transmission potential in the lower Rio Grande Valley, Texas, United States. J. Med. Entomol. 51, 1019–1028.

World Health Organization, Immunization, Vaccines and Biologicals. Questions and Answers on Dengue Vaccines [Online]. Available from: https://www.who.int/immunization/research/development/dengue_q_and_a/en/. Accessed 7 December 2018.

Chapter 15

Personal Protective Measures Against Mosquitoes

Stephen P. Frances[a], Mustapha Debboun[b]

[a]Australian Defence Force Malaria & Infectious Disease Institute, Enoggera, QLD, Australia, [b]Mosquito and Vector Control Division, Harris County Public Health, Houston, TX, United States

Chapter Outline

Abbreviations

BDU	Battle Dress Uniform
DDT	dichlorodiphenyltrichloroethane
DPCU	Disruptive Pattern Combat Uniforms
IDA	Individual Dynamic Absorption
PMD	*p*-menthane-3,8 diol
PPM	personal protective measures
USEPA	US Environmental Protection Agency
WHO	World Health Organization

15.1 Introduction

The use of insect repellents is part of an integrated approach to preventing bites from nuisance and vector mosquitoes and other biting arthropods (Debboun and Strickman, 2013). This chapter will review some of the strategies that have been used for personal protective measures (PPM) against vectors, especially mosquitoes. It will discuss basic engineering, repellents in combination with impregnated clothing, insecticide-treated nets, and chemical barriers to arthropod vectors.

Personal protective measures are the first line of defense against vectors of infectious diseases. Although PPM can reduce the incidence of vector-borne diseases, they can rarely eliminate the risk because not all individuals in a community use the measures (Debboun and Strickman, 2013). In this chapter the strategies used to reduce nuisance and the potential reduction in the transmission of vector-borne diseases will be discussed. Readers will note that none of the methods or combination of methods is 100% effective due to inconsistency of repellent application and not all individuals in a community adhere to PPM.

15.2 Basic engineering

Humans live in dwellings or homes made of a variety of materials for shelter from the elements of weather, security, and safety from others. In areas where mosquitoes are a problem, methods have been used in construction of homes to minimize the effects of mosquito vectors. At the same time, the homes constructed by humans are sometimes used by mosquitoes for shelter and rest areas, as well as completion of the larval stages in stagnant water that humans also need for their lifestyle.

In Texas, modern homes have been constructed and provide a physical barrier to the entry of mosquitoes. The use of window screens will improve barrier protection against mosquitoes and provide fresh and cool air to enter homes. In areas where mosquitoes are a significant problem, homeowners can use bed nets to protect themselves while sleeping. Homeowners can also minimize the habitats of mosquito larvae by eliminating standing water in containers around the home, including water that settles in roof gutters, plant pots, pet water containers, and a variety of other water-holding containers that may be left outdoors in the front or back yards.

Work in Africa, where limited funds are available for vector control and personal protection, has focused on low-cost community methods such as bed nets. The use of bed nets will be discussed later in this chapter. Earlier studies showed that houses and huts in many areas in the African continent were made with local materials and generally not mosquito-proof. During the 1950s, the World Health Organization (WHO) had enthusiastically pursued the use of dichlorodiphenyltrichloroethane (DDT) to treat the inside walls of houses to control *Anopheles* vectors of malaria. The rationale was that *Anopheles* entered the house, bit the sleeping occupants, and then rested on the interior walls of the house. Applying DDT to the inside of these houses resulted in the mortality of resting mosquitoes. The method was used in many countries with financial support from the WHO; however, due to the development of resistance in many mosquito species to DDT, the practice ceased in many countries during the 1980s.

Some studies in the late 1980s investigated the use of permethrin-treated materials as barriers to the entry of *Anopheles* mosquitoes to huts and houses. In some countries, a fabric called sisal was treated with permethrin and placed as insecticide-treated barrier in windows and other areas where *Anopheles* mosquitoes entered homes.

Impregnated curtains were also used with some success. In Tanzania, permethrin-treated curtains placed over eaves reduced the number of *Anopheles* mosquitoes that fed and survived after entering houses (Lines et al., 1987). A study in Burkina Faso in West Africa between July 1987 and May 1989 showed that the use of permethrin-treated curtains reduced the number of malaria episodes, parasitaemia, and splenomegaly during a 22-month period for children 6 months to 6 years old (Procacci et al., 1991). Although these methods were effective, limited funds were used to supply bed nets and insecticide chemicals to communities throughout the world.

In the last decade, placing barriers in homes in Africa has had renewed research interest. Early studies showed that the fungus *Beauveria bassiana* (Bals.) Vuill and *Metarhizium anisopliae* (Metschn.) Sorokin had promise in the control of adult *Anopheles* (Scholte et al., 2004). The studies showed that the fungus provided variable mortality against mosquitoes when applied to dark fabric swatches, which were then hung over gaps in the roofs and walls where mosquitoes entered houses. The mosquitoes contacting the treated cloth became infected and died from the fungal infection. A study also showed that strains of insecticide-resistant mosquitoes were susceptible to fungal infection (Kikankie et al., 2010). The *M. anisopliae* was also used to treat fabrics that were placed in traps to kill adult mosquitoes. The use of mosquito coils, vaporizer mats, and emanators have been shown to provide protection against mosquitoes, especially within homes (Fig. 15.1). Coils that burn pyrethroid insecticides inhibit biting and feeding of vector mosquitoes, causing knockdown and mortality of some (Ogama et al., 2012).

FIG. 15.1 Mosquito coils are burned within homes to reduce mosquito biting. *(Courtesy of Modeha [CC BY-SA 3.0 (https://creativecommons.org/ licenses/by-sa/3.0)].)*

15.3 Repellents

The use of topical arthropod repellents applied to exposed skin is the most important personal intervention against vectors of vector-borne diseases, and numerous studies have been undertaken to enumerate the protection against biting mosquitoes. Some of the main active ingredients used in repellent formulations for mosquitoes include *N,N*-diethyl-3-methyl benzamide (DEET), 2-(2-hydroxyethyl)-1-piperidinecarboxylic acid 1-methylpropyl ester (picaridin); ethyl butylacetylaminopropionate (IR3535); lemon-scented eucalyptus oil or *p*-menthane-3,8-diol (PMD), and 2-undecanone, Bite Blocker BioUD, which are discussed in more detail later in the chapter.

15.4 DEET

This active ingredient (also known as *N,N*-diethyl-*m*-toluamide) was synthesized by the US Department of Agriculture and became widely available in 1956. Since its introduction, DEET has become the most widely used active ingredient in topical repellents to protect against a variety of biting flies, especially mosquitoes, and other arthropods such as chiggers, ticks, and leeches. DEET has a broad spectrum of activity and has become the gold standard for comparison of novel active ingredients for other arthropod repellent formulations (Fig. 15.2).

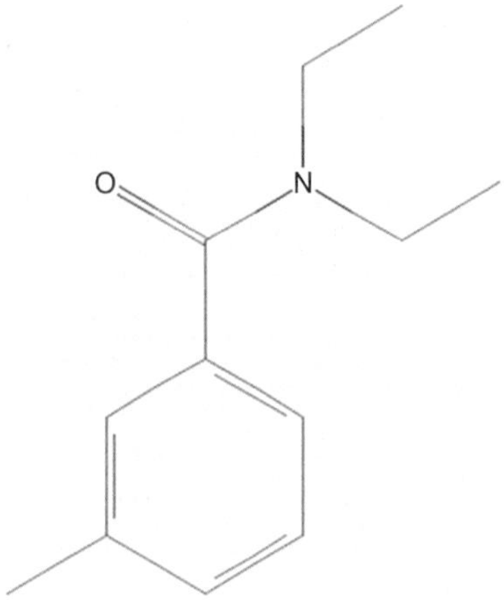

FIG. 15.2   DEET.

Numerous formulations containing DEET have been evaluated in field and laboratory settings since the 1940s (Frances, 2007a). DEET provided good protection against all genera of mosquitoes, including *Culex* spp. and *Aedes* spp. Field studies over the last 20 years in Africa, Australia, Papua New Guinea, and Thailand have shown that protection provided against *Anopheles* spp. was less than provided against Culicine mosquitoes. The response of different mosquito species to DEET is variable. Field tests of repellent formulations containing DEET against biting *Culex* spp., *Aedes* spp., *Mansonia* spp., and *Verralina* spp. have shown longer protection against these mosquitoes compared with *Anopheles* spp. (Frances et al., 1998).

There have been a number of reviews of the safety of DEET (Goodyer and Behrens, 1998; Sudakin et al., 2003), which have attested to its generally acceptable safety profile. There are a few reports of systemic toxicity in adults following dermal application. The use of DEET was shown to be safe in the second and third trimester of pregnancy (McCready et al., 2001), and animal models do not indicate any teratogenic effects (Schoenig et al., 1994). The scientific evidence and continued use for more than 60 years has shown that DEET is the best broad spectrum repellent available for minimizing the bites of mosquitoes and other biting arthropods.

Some of the formulations that contain DEET and are available in the United States are shown in Table 15.1.

TABLE 15.1 Mosquito repellent formulations available in Texas.

Product brand name	Active ingredient	Formulation	% active ingredient
Off! Deep Woods Insect Repellent	DEET	Aerosol	100
Off! Deep Woods Unscented	DEET	Aerosol	30
Off! Deep Woods Insect Repellent	DEET	Aerosol	25
Coleman 100 Max Insect Repellent	DEET	Aerosol	98.1
Off Skintastic	DEET	Lotion	7
Off! Deep Woods Insect Repellent Towelettes	DEET	Towelettes	25
Sawyer	DEET	Controlled release	19
Off! Active	DEET	Aerosol	15
Repel Family Formula	DEET	Lotion	10
Repel Sportsmen	DEET	Aerosol	29
Repel Insect Repellent	DEET	Aerosol	40
Muskol	DEET	Lotion	29
Cutter All Family Mosquito Wipes	DEET	Towelettes	7.15
Cutter All Family	DEET	Aerosol	10
Cutter Outdoorsman	DEET	Aerosol	30
Cutter Backwoods Dry Insect Repellent	DEET	Aerosol	25
Cutter Skinsations Insect Repellent	DEET	Liquid	7
Cutter Dry Insect Repellent	DEET	Aerosol	10
Cutter Natural Outdoor Fogger	Lemongrass oil	Aerosol	3
Fit Organic Mosquito Repellent	Lemongrass oil	Aerosol	10
Maggie's Farm Natural Insect Repellent	Lemongrass oil Geraniol Citronella oil Clove oil	Aerosol	1.7 1.75 0.95 0.95
Coleman Insect Repellent Skin Smart	IR3535	Aerosol	20
Avon Skin So Soft	IR3535	Cream	7.5
Autan	Picaridin	Aerosol	10
Cutter Citro Guard	Citronella oil	Candle	3
Off! Triple Wick Citronella Candle	Citronella oil	Candle	0.5
Bite Me Not Insect Repelling Wristband	Geraniol Citronella oil	Wristband	10 2
Repel Natural Insect Repellent	Geraniol	Aerosol	5
Cutter Natural Insect Repellent	Geraniol	Aerosol	5
Tiki BiteFighter	Cedar oil Citronella oil	Torch fuel canister	0.8 0.2
Off! Mosquito Lamp	Metofluthrin	Candle and heat	9
Off! Mosquito Coil IV	Metofluthrin	Vapor	0.02
Off! Clip-On	Metofluthrin	Spatial	31.2
Sawyer Permethrin Insect Repellent	Permethrin	Aerosol	0.5
Thermacell Mosquito Repellent	*d-cis*/transallethrin	Mats	21.97

See Xue (2007). Commercially available insect repellents and criteria for their use. In Debboun, Frances, and Strickman (eds.). Insect Repellents: Principles, Methods and Uses, Chapter 25.

15.5 Picaridin

This active ingredient was developed by Bayer AG in the 1990s and by 2005 was registered in over 50 countries throughout the world as a 7% formulation referred to as Cutter Advanced. A number of picaridin formulations have been evaluated throughout the world (Frances, 2007b). In comparative studies conducted to date, picaridin has been shown to be as good as or better than comparable formulations of DEET. In initial field tests in Malaysia (Yap et al., 1998, 2000) against *Aedes albopictus* and *Culex quinquefasciatus*, picaridin provided similar protection to DEET. In subsequent field studies in the United States (Barnard et al., 2002) and Australia (Frances et al., 2002), picaridin provided extended protection against mosquitoes. In a study in Burkina Faso against *Anopheles gambiae*, the main vector of malaria in Africa, picaridin provided the best protection compared with DEET and IR3535 (Constantini et al., 2004) (Fig. 15.3).

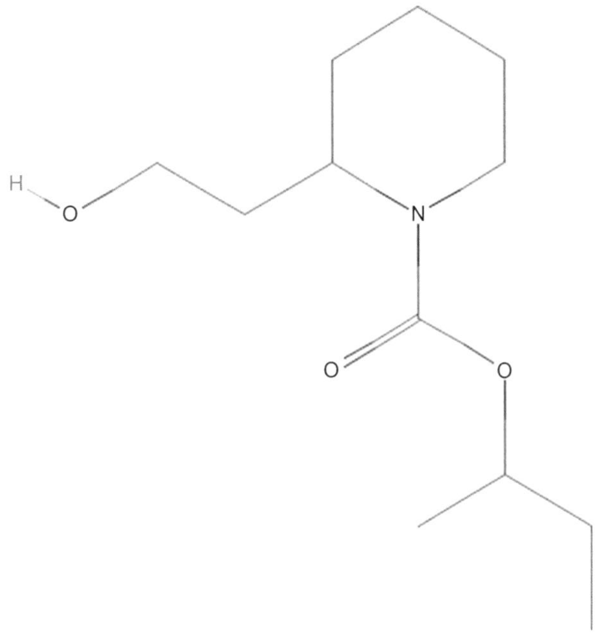

FIG. 15.3 Picaridin.

15.6 PMD

The principal repellent component of oil of lemon eucalyptus extract is PMD, which is the main by-product after hydrodistillation. The repellent was known initially as Quwenling when used in China. The first field study in the United States of PMD provided poor results (Schreck and Leonhardt, 1991), possibly because a Chinese commercial formulation was used. However, subsequent studies in the laboratory and field have shown PMD to have equal efficacy as DEET (Goodyer et al., 2010). Formulations containing this active ingredient were evaluated against anopheline mosquitoes in Tanzania (Trigg, 1996). It was found that 50% PMD provided 6–8-h protection against *An. gambiae* compared with 7-h protection provided by 50% DEET. Carroll and Loye (2006) showed that PMD provided excellent protection against *Ae. melanimon* and *Ae. vexans* in field trials in California (Fig. 15.4).

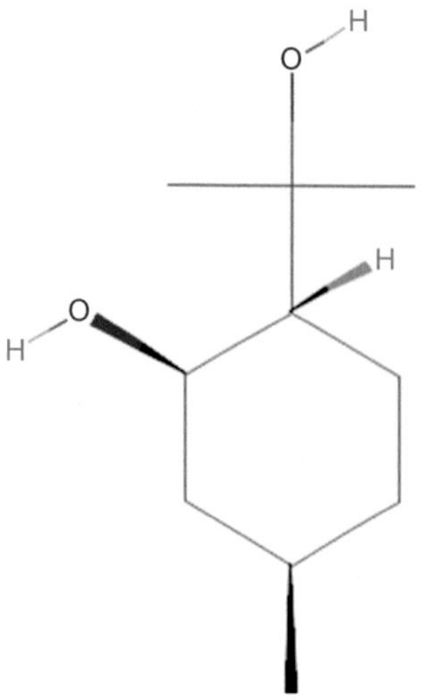

FIG. 15.4 PMD.

15.7 IR3535

The efficacy of formulations containing this active ingredient is reviewed by Puccetti (2007). Ethyl butylacetylamino-propionate is a synthetic molecule derived from a natural amino acid, β-alanine. It was developed in the early 1970s and is currently in more than 150 consumer products worldwide. This chemical has been used in commercial products since 1999. Field tests conducted in Florida, showed 25% IR3535 in ethanol provided protection of 3–4 h, compared with 5 h for 25% DEET and 7 h for 25% picaridin (Barnard et al., 2002). Field trials in Burkina Faso showed that higher concentrations of IR3535 were needed to provide similar protection as DEET and picaridin (Constantini et al., 2004). In field trials, IR3535 provided good protection against mosquitoes in Southeast Asia (Thavara et al., 2001; Liu et al., 2003) (Fig. 15.5).

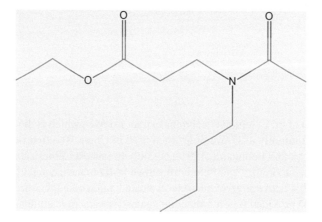

FIG. 15.5 IR3535.

15.8 2-Undecanone

The repellent compound, 2-undecanone (methyl nonyl ketone) is a US Environmental Protection Agency (EPA)-registered natural compound from leaves and stems of the wild tomato plant, *Lycopersicon hirsutum* (Farrar and Kennedy, 1987). The insect repellent, Bite Blocker with BioUD formula, is an EPA-registered product containing the active ingredient 2-undecanone that was registered in 2007 and classified as a biopesticide product. The Bite Blocker BioUD formulation is nonflammable and not a plasticizer. It is safe for use on pregnant women, children, and elderly and can be reapplied as often as needed. Witting-Bissinger et al. (2008) evaluated 7.75% 2-undecanone (Bite Blocker BioUD) in comparison to 30% DEET in North Carolina against field populations of *Psorophora ferox* and found that BioUD provided similar repellency as 30% DEET and a 90%

reduction in mosquito bites for up to 6h. In another study conducted by Qualls et al. (2011), they found that Bite Blocker BioUD insect repellent provided 140min protection against the floodwater mosquito, *Ps. columbiae* in St. Johns County, Florida. In addition, BioUD was found to be an effective repellent against ticks (Witting-Bissinger et al., 2008; Bissinger et al., 2009a,b) (Fig. 15.6).

FIG. 15.6   2-Undecanone.

An important consideration in the use of topical repellents by individuals within communities is the cost of obtaining them (Frances and Wirtz, 2005). This is a limiting factor in the use of repellents in developing countries, where vector-borne diseases may be endemic and personal incomes are low. This has led to the development of lower-cost repellent formulations. For example, a repellent soap formulation containing 20% DEET and 0.5% permethrin was developed in Australia in 1985 (Simmons, 1985). It was prepared in small blocks (approximate weight, 70g) and packed in relatively low-cost grease proof paper and was initially priced at $US 0.25 per piece. This formulation was applied to wet skin, lathered, and the residue was left on the skin surface to dry. Several field trials showed that the formulation provided satisfactory protection against mosquitoes in Malaysia (Yap, 1986), Papua New Guinea (Charlwood and Dagoro, 1987), Australia (Frances, 1987), India (Mani et al., 1991), and Ecuador and Peru (Kroeger et al., 1997). This formulation was commercialized and marketed as "Mosbar" in Southeast Asia. A survey of PPM used by inhabitants of East Honiara, Solomon Islands, showed a variety of measures were used by people to protect themselves from potential vectors of malaria (Bell et al., 1997). The survey showed 10.3% of respondents used Mosbar and 8.4% used unidentified repellent formulations. The study showed that only respondents who used prophylactic drugs or Mortein (pyrethroid aerosol) had increased protection against malaria (Bell et al., 1997).

Another important consideration in the use of topical repellents and other personal PPM is the user acceptability of the formulation. A number of studies have been undertaken to determine what factors affect the use of repellents by individuals and groups (Frances and Debboun, 2007).

The availability of PPM against mosquitoes is variable within communities and differences are due primarily to socioeconomic reasons. In most communities the seasonal increase in the density of mosquitoes is expected to result in an increase in the use of mosquito control and protection products. In Tanzania, Chavasse et al. (1996) showed a relationship between mosquito densities obtained from trap collections and the sale of mosquito coils in the shops of Mikocheni, Dar es Salaam. In Gambia, districts with higher mosquito densities had higher rates of bed net usage (Thomson et al., 1994). Conversely, a study in Southern Tanzania showed that mosquitoes were diverted away from households where the occupants used repellents to households who did not (Maia et al., 2012). These researchers noted that policy makers should take into consideration results showing vectors diverted from privileged families to those less privileged who may be exposed to nuisance and vector mosquitoes.

The evaluation of new active ingredients and formulations continues. Many studies have tested plant extracts as topical repellents against mosquitoes. The success of these active ingredients to repel mosquitoes has been variable, and as yet active ingredients to replace those mentioned earlier, namely, DEET, picaridin, PMD, IR3535, and 2-undecanone, have not become widely available. It is possible that new and potentially safer active ingredients may be developed in the future (Nentwig et al., 2017; Tisgratog et al., 2018).

15.9   Spatial repellents

In the last decade, entomologists have investigated the use of highly volatile chemicals that were dispersed around people using candles, coils, and battery-powered dispersing devices to protect against mosquito bites by providing spatial repellency.

A number of recent studies have evaluated the use of battery-powered devices dispersing the chemical metofluthrin. The trial by Lucas et al. (2007) showed that these devises reduced landing rates of mosquitoes by 85% (5.5 landings per minute) to 100% protection against *Ae. canadensis* in Pennsylvania in 2003. Bibbs and Xue (2016) evaluated a metofluthrin emanator against caged *Ae. aegypti* in the field and showed high mortality of mosquitoes at 0.3m from the emanator, thereby showing that mortality could be provided outdoors. A battery-powered chemical emanator called Off Clip-On was evaluated in field trials in Northeastern Florida by Xue et al. (2012). They showed that the emanator provided 79% protection against *Ae. albopictus* and *Ae. taeniorhynchus* in the field and 79% protection against *Ae. taeniorhynchus* for 3 h after it had remained open for more than a week. In contrast, Lucas et al. (2007) showed that devices used for 40 h or more would be unlikely to provide good protection in the field. This was contradicted by the findings of Kawada et al. (2004a,b) who demonstrated that impregnated papers remained active for a 4-week period. Recent studies have shown the potential of battery-powered devices, which can be attached to clothing at times when biting mosquitoes are active. These devices may be preferred to the application of repellents topically on the skin. The overall effectiveness of these chemical emanators has been variable, but it is likely that better emanators and active ingredients will be developed in the future.

15.10 Clothing

A number of studies evaluating the effectiveness of combinations of repellents and insecticide-treated clothing have been undertaken and are discussed here. The protection of an individual is enhanced by reducing the amount of skin exposed to biting insects and arthropods. In most human communities, clothes are worn and provide a physical barrier to arthropod bites. In many tropical countries, fewer clothes may be worn, as the use of long sleeves and long trousers may not be necessary, allowing more skin to be exposed to arthropods.

The recommendation to wear long sleeve shirts and long trousers to protect a person against mosquitoes is moderated by the discomfort of wearing such clothing in warm tropical environments. For short-term visitors, they may decide to protect themselves against vectors, while longer-term residents are less likely to wear/need long sleeve clothing.

The use of insecticide treatment of military clothing with synthetic pyrethroid insecticides, usually permethrin was first suggested in the 1970s. Earlier studies showed that dipping clothes into a water emulsion containing 0.6% permethrin was toxic to flying and crawling insects and ticks (Schreck et al., 1978). Schreck and colleagues showed the optimal concentration of permethrin in fabric to protect against mosquitoes, ticks, and chiggers. They also showed that the insecticide was lost from the fabric following washing, whereas relatively little was lost when clothes were worn normally. Abrasion of clothing also resulted in limited loss of permethrin from fabric (Schreck et al., 1980).

Early studies with freshly treated and unwashed clothing in combination with DEET repellent on the skin provided the best protection against mosquitoes in a rainforest habitat in Northern Queensland, Australia (Gupta et al., 1987). This method was shown to provide the best protection against mosquitoes in field trials in Thailand (Harbach et al., 1990) and Alaska, United States (Lillie et al., 1988).

The effect of washing fabric on the persistence of permethrin has been shown in Thailand (Eamsila et al., 1994), Australia (Gupta et al., 1987), United States (Gupta et al., 1990), and recently Europe (Faulde et al., 2003). Permethrin treatment of military uniforms has been used by United States and coalition forces, including Australia, for the last two decades. The first field evaluations showed freshly treated unwashed uniforms in combination with the application of DEET to the exposed skin provided the best protection against mosquitoes (Gupta et al. 1987). Treatment of US Battle Dress Uniforms (BDU) and Australian Disruptive Pattern Combat Uniforms (DPCU) was achieved initially by dipping or spraying in a water/permethrin emulsion (Frances and Cooper, 2007). The US BDUs are also treated by the Individual Dynamic Absorption (IDA) kit. Recently, fabrics have been treated with permethrin by a number of methods in the factory prior to manufacturing them into uniforms (Faulde et al., 2003, 2006). The loss of permethrin from fabric is primarily due to washing of the uniform during its use by soldiers (Schreck et al., 1982). The rate of loss of permethrin from the fabric has been shown to be variable, depending on the method of treatment, number of washes, and method of entomological evaluation.

Previous studies have evaluated persistence of permethrin in fabric by exposing mosquitoes to treated fabric for the amount of time needed to kill a proportion or all of the arthropods exposed (Faulde et al., 2003, 2006; Faulde and Uedelhoven, 2006). However, mosquitoes that are seeking a blood meal spend a relatively short time on surfaces such as clothing. When fabrics have been freshly treated with permethrin, mosquitoes spend only a few seconds on treated fabric and quickly fly off (Frances 1987; Miller et al., 2004).

The method of application of permethrin to clothing and other fabrics was developed during the last 20 years. Initially, clothes were treated by placing them into an emulsion of permethrin and water, allowing the clothes to become saturated, and then allowing the fabric to air-dry. On many occasions, military uniforms were initially treated with permethrin and worn by soldiers who were working in areas where vector-borne diseases occurred. However, the rate of retreatment of uniforms was variable and protection against vectors was compromised (Frances et al., 2003; Coleman et al., 2006). Methods of treating uniforms with permethrin were developed to increase the persistence of permethrin against washing. During the last 10 years, the treatment of fabrics has been conducted in factories prior to making the uniforms. The United States and European countries have used uniforms treated in this way in recent wars in Iraq and Afghanistan (Coleman et al., 2006). The main benefit to soldiers is that they are able to carry out their duties with reduced concern about vectors of vector-borne diseases. Education briefs advise that they should continue to apply DEET repellent to exposed skin in areas where vector-borne disease is a primary concern. The developmental improvements in application methods should allow for fabrics to be treated with permethrin and other pyrethroids to increase the protection of civilian workers and people living in malarious areas.

The use of clothes and fabrics treated with synthetic pyrethroids in civilian communities has been investigated in the last decade. A community-wide study in a refugee camp located in Northeastern Kenya on the use of personal clothes (Diraa, Saria, Jalabaa, Ma'awia, and shirts) and bedding (sheets and blankets) treated with permethrin was compared with people with untreated clothes and bedding on the rate of malaria infection (Kimani et al., 2006). The study showed that the use of permethrin-treated clothes and bedding reduced the rate of malaria infection and the mosquito biting rate in huts inhabited by people with treated cloths and bedding. It also showed that the concept of treating clothes was well accepted by the participants who had no side effects due to insecticide treatment of clothes or bedding over a 6-month period. The use of permethrin treatment of clothing was shown to provide protection from the black-legged tick, *Ixodes scapularis* Say, in laboratory studies. The authors concluded that do-it-yourself treatment of summer clothing (T-shirts, shorts, socks, and sneakers) significantly reduced tick bites and tick-borne pathogen transmission (Miller et al., 2011).

A field assessment of the effects of wearing permethrin-treated uniforms on the incidence of malaria in Thai soldiers was conducted in 1992. The results showed no decrease in malaria over a 6-month period (Eamsila et al., 1994). In this trial, soldiers were not required to wear their uniforms after hours, and although they were given topical repellent, they were not required to apply it in the evening or at night. In Columbia the efficacy of permethrin-treated uniforms for the prevention of malaria and leishmaniasis over a 4-week period was studied (Sota et al., 1995). The study showed that soldiers wearing the treated uniforms had an increased protection against both diseases, and the authors concluded that permethrin-treated clothing was recommended for exposure to both diseases for a period of 1–2 months (Sota et al., 1995). In contrast, a subsequent study in Iran showed that permethrin-impregnated uniforms were not effective for protection of cutaneous leishmaniasis in Iranian troops (Asilan et al., 2003). In 2000, 22,000 French troops were deployed on military service to Ivory Coast. A study showed that industrial impregnation of permethrin offered some protection from mosquitoes, but not enough to reduce significantly the incidence of malaria among nonimmune troops (Deparis et al., 2004). The contrast in the results shown in these trials may be due to the period of time that the use of permethrin-treated uniforms were monitored. In the trial in Columbia, good protection was shown in a 1–2-month period, whereas, in other studies, permethrin-treated uniforms were not effective against malaria in Thailand over 6 months and in Ivory Coast over 1–2 months.

15.11 Barriers

Using physical barriers as protection against mosquitoes has been reported to have occurred in ancient times (Lindsay and Gibson, 1988). Fishermen working in the Mediterranean Sea sometimes slept under fishing nets to protect themselves from mosquitoes. Mosquito bed nets have been used in the tropics for a long time. During World War II, soldiers operating in the Southwest Pacific theater used bed nets to protect themselves from malaria mosquito vectors. In the 1980s, the treatment of bed nets with pyrethroid insecticides gained impetus. The synthetic pyrethroid, permethrin, became available in 1975. In many malarious countries, untreated mosquito bed nets were used despite the presence of holes, which allowed mosquitoes to enter the net and bite the sleeping person. Sometimes, people were awoken to find many engorged mosquitoes enclosed in the bed net. A number of studies showed that treating the net with synthetic pyrethroids reduced the ability of *Anopheles* mosquitoes to fly through holes in the bed nets, thereby increasing the protection provided by nets with holes. Bed nets became damaged, and due to the cost of replacement, they were continued to be used to provide protection against vectors. In the late 1980s the use of permethrin-treated bed nets was adopted by the WHO as a major intervention to reduce the transmission of malaria in Africa and Asia.

Models of host-seeking processes in the context of local human host availability and elucidating the impacts and mechanisms of pyrethroid treated bed nets were undertaken (Killeen and Smith, 2007). The modeling showed the excitorepellency of pyrethroid chemicals increased the exposure of untreated humans by concentrating mosquito biting on this vulnerable (untreated) group. The model predicted that nets would have a significant impact on transmission of pathogens among users of bed nets. These results are consistent with the outcomes of many controlled randomized trials. For example, a study in India in military camps showed that deltamethrin-treated bed nets provided an 87% reduction in malaria cases over a 2-year period where three treatments of nets with deltamethrin were undertaken (Joshi et al., 2003).

An important consideration in the use of bed nets is that *Anopheles* vectors of malaria bite between dusk and dawn, and biting activity usually begins before people have gone to bed to be protected by bed nets and screens in their homes. The use of PPM during this time before going to bed has been investigated, with the use of repellents and protective fabrics. In a study in Cambodia, Sochantha et al. (2010) investigated protection provided by permethrin-treated hammocks using Olyset technology against exophagic mosquito vectors. The study showed that only 46% protection was provided by the insecticide-treated hammocks against *An. minimus* Theobald, and poor protection was provided against *An. dirus* Peyton and Harrison, *An. maculatus* Theobald, and Culicine mosquitoes. The researchers suggested that despite the poor results, the insecticide-treated hammocks could prove effective in protecting forest workers and villagers before sleep time, and it is a valuable tool in areas where artemisinin-resistant malaria parasites are emerging. The use of topical repellents in malaria endemic areas has been suggested as a method to further reduce the risk of exposure to biting *Anopheles* mosquitoes. A study in Pakistan suggested that a repellent containing DEET was popular among Afghan refugees and provided protection from malaria in the early evening (Rowland et al., 2004). A subsequent study in Bolivia showed that the incidence of malaria in adults was reduced when people used a repellent formulation containing a natural active ingredient (30% PMD) in combination with sleeping under permethrin-treated bed nets (Hill et al., 2007). In South America, a low-cost repellent containing PMD and lemongrass oil provided comparable protection to DEET against malaria vectors in Guatemala and Peru (Moore et al., 2007).

A possible impediment to the success of bed nets is the development of resistance in mosquito vectors to the pyrethroid insecticides used to treat bed nets. Resistance of malaria vectors to a range of pyrethroid active ingredients has been shown by some workers in Africa. John et al. (2008) showed a significant increase in the knockdown time and mean mortality of *An. gambiae* Giles to net material treated with resmethrin, cyfluthrin, and cypermethrin. The study showed the development of resistance to pyrethroids in *An. gambiae* in Western Uganda (John et al., 2008). Other workers have shown that treating nets with a combination of active ingredients is beneficial and can potentially decrease the likelihood of the rapid development of resistance in malaria vectors (N'Guessan et al., 2006).

The use of indoor residual sprays was used successfully to control *Anopheles* mosquitoes and reduce the transmission of malaria. The use of DDT to treat indoor wall and ceiling surfaces of houses has been advocated since the 1950s, when WHO had hoped to eliminate malaria. Due primarily to the development of resistance, decreased allowance by householders to allow their homes to be treated, and decreased political will among governments who funded the intervention, the use of indoor residual spraying has decreased dramatically since the 1980s. Despite this, Graham et al. (2002) used insecticide-treated tarpaulins for the control of malaria in refugee camps in Afghanistan. More recently, Diabate et al. (2006) and Chandre et al. (2010) have investigated the use of insecticide-treated plastic sheeting, known as durable lining, as a lining for interior walls and ceilings. The plastic lining is placed into the interior of houses and is treated with a contact irritant to prevent mosquitoes from resting inside homes and potentially causing death among mosquitoes that spend too much time on the plastic surface. Chandre et al. (2010) noted that the plastic sheet was pretreated with insecticide in the factory, and retreatment was not required. The use of insecticide-treated plastic sheeting as both a shelter and protection against malaria was conducted in refugee camps in Sierra Leone (Burns et al., 2012). This study showed a protective efficacy of 61% under fully lined insecticide-treated plastic sheeting and only 15% in shelter where only the roof was lined. The authors reported improvements in anemia rates when both methods were used and considered as convenient, safe, and long-lasting methods of malaria control (Burns et al., 2012). The use of insecticide-treated plastic sheeting to line shelters used in refugee camps and during humanitarian aid programs provides emergency shelter as well as improved protection against vector-borne diseases and will be utilized by aid programs into the future.

The application of insecticides and repellents to clothing, other fabrics, and screens has been evaluated and used in a number of places to minimize contact between people and vectors, especially mosquitoes. During World War II, Allied forces operating in the Southwest Pacific region experienced significant morbidity and mortality from scrub typhus, a disease caused by *Orientia tsutsugamushi* (Hayashi) and carried by trombiculid mite larvae, commonly called chiggers (McCulloch, 1946). Studies in Australia and Papua New Guinea during 1943 showed that dibutylphthalate applied externally to military shirts and trousers provided excellent protection against chiggers. A concerted effort was undertaken to show soldiers the best method of application and encouraged its use among the allies, and in 1944, there was an 80% decrease in the incidence of scrub typhus (McCulloch, 1947). This is one of the few examples where the use of a repellent chemical has had a measurable effect on the incidence of disease (Rutledge et al., 1978). This method was subsequently adapted for use in communities after the war as protection against chiggers and scrub typhus.

Bifenthrin [2-methylbifenyl-3-ylmethyl (Z)-(IRS)-*cis*-3 (2-chloro-3, 3, 3trifluroprop-1-enyl)-2, 2-dimethyl cyclopropane carboxylate] is a non-alpha cyano pyrethroid and is used against a range of agricultural pests and as an insecticide treatment for mosquito bed nets (Hougard et al., 2002). The chemical has a relatively low irritant and knockdown effect compared with permethrin and deltamethrin. Bifenthrin causes a higher mortality by allowing mosquitoes to rest on treated surfaces for longer periods (Hewitt et al., 1995). In a study in India, the rate of entry of mosquitoes into rooms containing bifenthrin-treated and lamda-cyhalothrin–treated bed nets was fewer than those entering rooms containing untreated bed nets (McGinn et al., 2008).

Military and civilian tents are commonly used to house refugees, displaced persons, and aid workers following natural disasters and civilian unrest. These tents may be in place for extended periods of time when the occurrence of vector-borne

FIG. 15.7 Habitats in tent folds that can be used by *Aedes aegypti* eggs and larvae.

diseases such as malaria and dengue may increase. Following rain, the folds in these tents provide breeding sites for *Ae. aegypti* and *Ae. albopictus*, the vectors of dengue fever (Fig. 15.7). For example, a survey of Australian military peacekeeping installations in Timor-Leste in 2001 found 41 containers with dengue vectors, 22 (53.7%) of which were from water found in tent folds (Cooper, unpublished). As a control measure, military tent fabrics have been treated with synthetic pyrethroids, such as permethrin and bifenthrin to reduce mosquito entry and mosquito bites (World Health Organization, 2001; Batra et al., 2005). The effects of the insecticide when applied to tent fabric on the eggs and larvae of *Ae. aegypti* showed that the eggs and larvae did not develop and the ability of adults to obtain a blood meal after exposure to treated fabric was significantly reduced (Frances, 2007c) (Fig. 15.8). The treatment of the inner walls of tents with permethrin reduced the nuisance of mosquitoes and probably invasive pests (Batra et al., 2005) and has been shown to provide good protection against vectors of malaria (Schreck, 1991). The treatment of Australian military tents with bifenthrin provided an 81.5% reduction of mosquito entry and 91% reduction in mosquito biting in treated tents over a 10-day period in Northern Territory, Australia (Frances et al., 2008).

The protection of humans from mosquitoes by applying insecticides inside and around dwellings and community areas has been used for many years. Indoor residual spraying with DDT was used to control malaria by reducing the longevity of mosquitoes in many communities during the 1970s (World Health Organization, 1975).

FIG. 15.8 Application of bifenthrin to the inside/outside of Australian military tents as a barrier to entry of mosquitoes.

Permethrin and bifenthrin have also been applied to vegetation as barrier treatment against mosquitoes (Anderson et al., 1991; Perich et al., 1993). A comparative trial of both of these pyrethroids in Kentucky in 2004 showed that application to low-lying vegetation did not properly target adult resting sites of *Culex*, but reduced *Aedes* mosquitoes. The study showed no differences between the two pyrethroids, and both provided increased protection against mosquitoes (Trout et al., 2007). The use of insecticides, especially permethrin, has been shown to enhance the barrier effect of tents in preventing the entry and mosquito bites within and around treated tents. This method was first evaluated with the application of repellents such as DEET on tent fabrics. Sholdt et al. (1977) showed that mosquito bites were reduced in and near tents treated with the repellent DEET. The treatment of the inner walls of tents with permethrin reduced the nuisance of mosquitoes and invasive pests (Batra et al., 2005) and provided good protection against vectors of malaria (Schreck, 1991).

Qualls et al. (2012) compared the large-scale treatment of a golf course with the chemical bifenthrin with an area that only used ULV spray to control floodwater mosquitoes in Florida. Both methods significantly decreased mosquito populations, compared to a control, but the two treatments did not differ. The mosquito population in the golf course treated with bifenthrin decreased by 84% compared to a 52% decrease in an area where ULV was used. The authors showed that the service requests in the barrier-treated area was decreased compared to the ULV treated area, and the overall cost savings were estimated to be $2700. The workers concluded that barrier spraying was an appropriate tool for controlling mosquito populations (Qualls et al., 2012).

15.12 Conclusion

The use of arthropod repellents as personal protection is practiced by military personnel, mining personnel, and tourists. Protection of expatriate miners who fly into mines located in malarious areas and periodically return to their home location where the risk of vector-borne disease is reduced or absent. This may be the case in Africa, Southeast Asia, and Australasia. In Australasia, there are many mines operating in Papua New Guinea and Irian Jaya, where workers are flown in and out of the mine site on a rotational basis. The mining companies need to maintain production and put an important emphasis on protection against vectors of diseases such as malaria and dengue. Many mining companies provide uniforms pretreated with permethrin and advocate the correct wearing of uniforms by staff in the mine site and local areas. Military groups are also encouraged to correctly wear protective clothing, while the use of topical arthropod repellents on exposed skin is advocated by most of the groups of people in this chapter.

The use of PPM against vectors of diseases, especially in developing countries, particularly in Africa and Asia, has focused on methods that are low cost and can be conducted by local communities. In the last decade, more funds have become available to pursue some of the innovative methods of personal protection, and progress to reduce the incidence of vector-borne diseases, especially malaria, were made. The vector control workers in these countries have contributed knowledge and expertise in using and developing new methods.

This chapter discussed the use of innovative PPM, which reduce vector, nuisance biting, and disease transmission among people in Texas and varied locations around the world. Despite a variety of methods, mosquitoes still cause nuisance and transmit pathogens of vector-borne diseases, and it is hoped that the wider use of the methods described in this chapter and those under development will further decrease the burden of vector-borne diseases in Texas and the world. It is also sobering to note that personal protection products are sometimes subject to "false" advertising, where untested and sometimes unrealistic outcomes for success are stated (Revay et al., 2013).

References and further reading

Anderson, A.L., Apperson, C.S., Knake, R., 1991. Effectiveness of mist-blower applications of malathion and permethrin to foliage as barrier sprays for salt marsh mosquitoes. J. Am. Mosq. Control Assoc. 7, 116–117.

Asilan, A., Sadeghinia, A., Shariati, F., Jome, I., Ghoddusi, A., 2003. Efficacy of permethrin-impregnated uniforms in the prevention of cutaneous leishmaniasis in Iranian soldiers. J. Clin. Pharm. Therapeut. 28, 175–178.

Barnard, D.R., Bernier, U.R., Posey, K.H., Xue, R.-D., 2002. Repellency of IR3535, KBR 3023, *para*-menthane -3-diol, and deet to black salt marsh mosquitoes (Diptera: Culicidae) in the Everglades National Park. J. Med. Entomol. 39, 895–899.

Batra, C.P., Raghavendra, K., Adak, T., Singh, O.P., Singh, S.P., Mittal, P.K., Malhotra, M.S., Sharma, R.S., Subbarao, S.K., 2005. Evaluation of bifenthrin treated mosquito nets against anopheline and culicine mosquitoes. Indian J. Med. Res. 121, 55–62.

Bell, D., Bryan, J., Cameron, A., Fernando, M., Leafascia, J., Pholsyna, K., 1997. Malaria in Honiara, Solomon Islands: reasons for presentation and human and environmental factors influencing prevalence. Southeast Asian J. Trop. Med. Public Health 28, 482–488.

Bibbs, C.S., Xue, R.-D., 2016. OFF! Clip on repellent device with metofluthrin tested on *Aedes aegypti* (Diptera: Culicidae) for mortality at different time intervals and distances. J. Med. Entomol. 53, 480–483.

Bissinger, B.W., Apperson, C.S., Sonenshine, D.E., Watson, D.W., Roe, R.M., 2009a. Efficacy of the new repellent BioUD against three species of ixodid ticks. Exp. Appl. Acarol. 48, 239–250.

Bissinger, B.W., Zhu, J., Apperson, C.S., Sonenshine, D.E., Watson, D.W., Roe, R.M., 2009b. Comparative efficacy of BioUD to other commercially available arthropod repellents against the ticks, *Amblyomma americanum* and *Dermacentor variabilis* on cotton cloth. Am. J. Trop. Med. Hyg. 81, 658–690.

Burns, M., Rowland, M., N'Guessan, R., Carneiro, I., Beeche, A., Ruiz, S.S., Kamara, S., Takken, W., Carnevale, P., Allan, R., 2012. Insecticide-treated plastic sheeting for emergency malaria prevention and shelter among displaced populations: an observational cohort study in a refugee setting in Sierra Leone. Am. J. Trop. Med. Hyg. 87, 242–250.

Carroll, S.P., Loye, J., 2006. A registered botanical mosquito repellent with deet-like efficacy. J. Am. Mosq. Control Assoc. 22, 507–514.

Chandre, F., Dabire, R.K., Hougard, J.-M., Djogbenou, L.S., Irish, S.R., Rowland, M., N'Guessan, R., 2010. Field efficacy of pyrethroid treated plastic sheeting (durable lining) in combination with long lasting insecticidal nets against malaria vectors. Parasit. Vectors 3, 65–71.

Charlwood, J.D., Dagoro, H., 1987. Repellent soap for use against malaria vectors in Papua New Guinea. Papua New Guinea Med. J. 30, 301–303.

Chavasse, D.C., Lines, J.D., Ichimori, K., 1996. The relationship between mosquito density and mosquito coil sales in Dar es Salaam. Trans. R. Soc. Trop. Med. Hyg. 90, 493–495.

Coleman, R.E., Burkett, D.A., Putnam, J.L., Sherwood, V., Caci, J.B., Jennings, B.T., Hochberg, L.P., Spradling, S.L., Rowton, E.D., Blount, K., Bloch, J., Hopkins, G., Raymond, J.-L.W., O'Guinn, M.L., Lee, J.S., Weina, P.J., 2006. Impact of phlebotomine sand flies on U.S military operations at Tallil Air Base: 1. Background, military situation, and development of a "Leishmaniasis control program". J. Med. Entomol. 43, 647–662.

Constantini, C., Badolo, A., Iboudo-Sanoo, E., 2004. Field evaluation of the efficacy and persistence of insect repellents deet, IR3535 and KBR 3023 against *Anopheles gambiae* complex and other Afrotropical vector mosquitoes. Trans. R. Soc. Trop. Med. Hyg. 98, 644–652.

Debboun, M., Strickman, D., 2013. Insect repellents and associated personal protection for a reduction in human disease. Med. Vet. Entomol. 27, 1–9.

Deparis, X., Frere, B., Lamizana, M., N'Guessan, R., Leroux, F., Lefevre, P., Finot, L., Hougard, J.-M., Carnevale, P., Gillet, P., Baudon, D., 2004. Efficacy of permethrin-treated uniforms in combination with deet topical repellent for protection of French military troops in Cote d'Ivoire. J. Med. Entomol. 41, 914–921.

Diabate, A., Chandre, F., Rowland, M., N'Guessan, R., Duchon, S., Dabire, K.R., Hougard, J.-M., 2006. The indoor use of plastic sheeting pre-impregnated with insecticide for control of malaria vectors. Trop. Med. Int. Health 11, 597–603.

Eamsila, C., Frances, S.P., Strickman, D., 1994. Evaluation of permethrin-treated military uniforms for personal protection against malaria in northeastern Thailand. J. Am. Mosq. Control Assoc. 10, 515–521.

Farrar, R.R., Kennedy, G.G., 1987. 2-Undecanone, a constituent of the glandular trichomes of *Lycopersicon hirsutum f. glabratum*: effects *on Heliothis zea* and *Manduca sexta* growth and survival. Entomol. Exp. Appl. 43, 17–23.

Faulde, M., Uedelhoven, W., 2006. A new clothing impregnation method for personal protection against ticks and biting insects. Int. J. Med. Microbiol. 296, 225–229.

Faulde, M.K., Uedelhoven, W.M., Robbins, R.G., 2003. Contact toxicity and residual activity of different permethrin-based fabric impregnation methods for *Aedes aegypti* (Diptera: Culicidae), *Ixodes ricinus* (Acari: Ixodidae), and *Lepisma saccharina* (Thysanura: Lepismatidae). J. Med. Entomol. 40, 935–941.

Faulde, M.K., Uedelhoven, W.M., Malerius, M., Robbins, R.G., 2006. Factory-based permethrin impregnation of uniforms: residual activity against *Aedes aegypti* and *Ixodes ricinus* in Battle Dress Uniforms worn under field conditions, and cross-contamination during the laundering and storage process. Mil. Med. 171, 472–477.

Frances, S.P., 1987. Effectiveness of deet and permethrin, alone, and in a soap formulation as skin and clothing protectants against mosquitoes in Australia. J. Am. Mosq. Control Assoc. 3, 648–650.

Frances, S.P., 2007a. Efficacy and safety of repellents containing deet. In: Debboun, M., Frances, S.P., Strickman, D. (Eds.), Insect Repellents: Principles, Methods and Uses. CRC Press, Boca Raton, FL, pp. 311–325.

Frances, S.P., 2007b. Picaridin. In: Debboun, M., Frances, S.P., Strickman, D. (Eds.), Insect Repellents: Principles, Methods and Uses. CRC Press, Boca Raton, FL, pp. 337–340.

Frances, S.P., 2007c. Evaluation of bifenthrin and permethrin as barrier treatments for military tents against mosquitoes in Queensland, Australia. J. Am. Mosq. Control Assoc. 23, 208–212.

Frances, S.P., Cooper, R.D., 2007. Personal protective measures against mosquitoes: insecticide-treated uniforms, bednets and tents. ADF Health 8, 50–56.

Frances, S.P., Debboun, M., 2007. User acceptability: public perceptions of insect repellents. In: Debboun, M., Frances, S.P., Strickman, D. (Eds.), Insect Repellents, Principles, Methods and Uses. CRC Press, Boca Raton, FL, pp. 397–403.

Frances, S.P., Wirtz, R.A., 2005. Repellents: past, present, and future. J. Am. Mosq. Control Assoc. 21, 1–3. Suppl.

Frances, S.P., Cooper, R.D., Sweeney, A.W., 1998. Laboratory and field evaluation of the repellents deet, CIC-4 and AI3-37220, against *Anopheles farauti* (Diptera: Culicidae) in Australia. J. Med. Entomol. 35, 690–693.

Frances, S.P., Dung, N.V., Beebe, N.W., Debboun, M., 2002. Field evaluation of repellent formulations against daytime and nighttime biting mosquitoes in a tropical rainforest in northern Australia. J. Med. Entomol. 39, 541–544.

Frances, S.P., Auliff, A.M., Edstein, M.D., Cooper, R.D., 2003. Survey of personal protection measures against mosquitoes among Australian Defense Force personnel deployed to East Timor. Mil. Med. 168, 227–230.

Frances, S.P., Huggins, R.L., Cooper, R.D., 2008. Evaluation of the inhibition of egg laying, larvicidal effects and bloodfeeding success of *Aedes aegypti* exposed to permethrin- and bifenthrin-treated military tent fabric. J. Am. Mosq. Control Assoc. 24, 598–600.

Goodyer, L., Behrens, R.H., 1998. Short report: the safety and toxicity of insect repellents. Am. J. Trop. Med. Hyg. 59, 323–324.

Goodyer, L.I., Croft, A.M., Frances, S.P., Hill, N., Moore, S.J., Sangoro, P., Onyango, P., Debboun, M., 2010. Expert review of the evidence base for arthropod bite avoidance. J. Travel Med. 17, 182–192.

Graham, K., Mohammed, N., Rehman, H., Nazari, A., Ahmad, M., Skovmand, O., Guillet, P., Allan, R., Zaim, M., Yates, A., Lines, J., Rowland, M., 2002. Insecticide-treated plastic tarpaulins for control of malaria vectors in refugee camps. Med. Vet. Entomol. 16, 404–408.

Gupta, R.K., Sweeney, A.W., Rutledge, L.C., Cooper, R.D., Frances, S.P., Westrom, D.R., 1987. Effectiveness of controlled-release personal-use arthropod repellents and permethrin-impregnated clothing in the field. J. Am. Mosq. Control Assoc. 3, 556–560.

Gupta, R.K., Rutledge, L.C., Reifenrath, W.G., Gutierrez, G.A., Korte Jr., D.W., 1990. Resistance to weathering in fabrics treated for protection against mosquitoes. J. Med. Entomol. 27, 494–500.

Harbach, R.E., Tang, D.B., Wirtz, R.A., Gingrich, J.B., 1990. Relative repellency of two formulations of N,N-diethyl-3-methylbenzamide (deet) and permethrin-treated clothing against *Culex sitiens* and *Aedes vigilax* in Thailand. J. Am. Mosq. Control Assoc. 6, 641–644.

Hewitt, S., Rowland, M., Muhammad, N., Kamal, M., Kemp, E., 1995. Pyrethroid-sprayed tents for malaria control: an entomological evaluation in Pakistan. Med. Vet. Entomol. 9, 344–352.

Hill, N., Lenglet, A., Arnez, A.M., Caniero, I., 2007. Plant based repellent and insecticide treated bed nets to protect against malaria in areas of early evening biting vectors: double blind randomized placebo controlled clinical trial in the Bolivian Amazon. Br. Med. J. 335, 1023–1027.

Hougard, J.-M., Duchon, S., Zaim, M., Guillet, P., 2002. Bifenthrin: A useful pyrethroid insecticide for treatment of mosquito nets. J. Med. Entomol. 39, 526–533.

John, R., Ephraim, T., Andrew, A., 2008. Reduced susceptibility to pyrethroid insecticide treated nets by the malaria vector *Anopheles gambiae* s.l. in western Uganda. Malar. J. 7, 92.

Joshi, R.M., Ghose, G., Som, T.K., Bala, S., 2003. Study of the impact of deltamethrin impregnated mosquito nets on malaria incidence at a malaria station. MJAFI 59, 12–14.

Kawada, H., Maekawa, Y., Tsuda, Y., Takagi, M., 2004a. Trial of spatial repellency of metofluthrin-impregnated paper strip against *Anopheles* and *Culex* in shelters without walls in Lombok, Indonesia. J. Am. Mosq. Control Assoc. 20, 434–437.

Kawada, H., Maekawa, Y., Tsuda, Y., Takagi, M., 2004b. Laboratory and field evaluation of spatial repellency with metofluthrin-impregnated paper strip against mosquitoes in Lombo Island, Indonesia. J. Am. Mosq. Control Assoc. 20, 292–298.

Kikankie, C.K., Brooke, B.D., Knols, B.G.J., Koekemoer, L.L., Farenhorst, M., Hunts, R.H., Thomas, M.B., Coetzee, M., 2010. The infectivity of the entomopathogenic fungus *Beauveria bassiana* to insecticide-resistant and susceptible *Anopheles arabiensis* mosquitoes at two different temperatures. Malar. J. 9, 71.

Killeen, G.F., Smith, T.A., 2007. Exploring the contributions of bednets, cattle, insecticides and excitorepellency to malaria control: a deterministic model of mosquito host-seeking behaviour and mortality. Trans. R. Soc. Trop. Med. Hyg. 101, 867–880.

Kimani, E.W., Vulule, J.M., Kuria, I.W., Mugisha, F., 2006. Use of insecticide-treated clothes for personal protection against malaria: a community trial. Malar. J. 5, 63–72.

Kroeger, A., Gerhardus, A., Kruger, G., Mancheno, M., Pisse, K., 1997. The contribution of repellent soap to malaria control. Am. J. Trop. Med. Hyg. 56, 580–584.

Lillie, T.H., Schreck, C.E., Rahe, A.J., 1988. Effectiveness of personal protection against mosquitoes in Alaska. J. Med. Entomol. 25, 475–478.

Lindsay, S.W., Gibson, M.E., 1988. Bednets revisited—old idea, new angle. Parasitol. Today 4, 270–272.

Lines, J.D., Myamba, J., Curtis, C.F., 1987. Experimental hut trials of permethrin-impregnated mosquito nets and eave curtains against malaria vectors in Tanzania. Med. Vet. Entomol. 1, 37–51.

Liu, D., Zhang, A., Zhou, M., 2003. Preliminary selective studies on the long lasting effect repellent formulations. Chin. J. Vector Bio. Control 14, 358.

Lucas, J.R., Shono, Y., Iwasaki, T., Ishiwatari, T., Spero, N., Benzon, G., 2007. U.S. laboratory and field trials of metofluthrin (SumiOne®) emanators for reducing mosquito biting outdoors. J. Am. Mosq. Control Assoc. 23, 47–54.

Maia, M., Sangaro, P., Thiele, M., Turner, E., Moore, S., 2012. Do topical repellents divert mosquitoes within a community? Malar. J. 11 (Suppl. 1), 120.

Mani, T.R., Reuben, R., Akiyama, J., 1991. Field efficacy of "Mosbar" repellent soap against vectors of Bancroftian filariasis and Japanese Encephalitis in southern India. J. Am. Mosq. Control Assoc. 7, 565–568.

McCready, R., KA Hamilton, K.A., Simpson, J.A., Cho, T., Luxemberger, C., Edwards, R., Looareesuwan, S., White, N.J., Nosten, F., Lindsay, S.W., 2001. Safety of the insect repellent N,N-diethyl-m-toluamide (deet) in pregnancy. Am. J. Trop. Med. Hyg. 65, 285–289.

McCulloch, R.N., 1946. Studies in the control of scrub typhus. Med. J. Aust. 1, 717–738.

McCulloch, R.N., 1947. The adaption of military scrub typhus mite control to civilian needs. Med. J. Aust. 1, 449–452.

McGinn, D., Frances, S.P., Sweeney, A.W., Brown, M.D., Cooper, R.D., 2008. Evaluation of Bistar 80SC (Bifenthrin) as a tent treatment for protection against mosquitoes in Northern Territory, Australia. J. Med. Entomol. 45, 1087–1091.

Miller, R.J., Wing, J., Cope, S.E., Klavons, J.A., Kline, D.L., 2004. Repellency of permethrin-treated Battle-Dress Uniforms during Operation Tandem Thrust 2001. J. Am. Mosq. Control Assoc. 20, 462–464.

Miller, N.J., Rainone, E.E., Dyer, M.C., Gonzalez, M.L., Mather, T.N., 2011. Tick bite protection with permethrin-treated summer-weight clothing. J. Med. Entomol. 48, 327–333.

Moore, S.J., Darling, S.T., Sihuincha, M., Padilla, N., Devine, G.J., 2007. A low-cost repellent for malaria vectors in the Americas: results of two field trials in Guatemala and Peru. Malar. J. 6, 101.

Nentwig, G., Frohberger, R., Sonneck, R., 2017. Evaluation of clove oil, icaridin, and transfluthrin for spatial repellent effects in Three Tests Systems against the *Aedes aegypti* (Diptera: Culicidae). J. Med. Entomol. 54, 150–158.

N'Guessan, R., Rowland, M., Moumouni, T.-L., Kesse, N.B., Carnevale, P., 2006. Evaluation of synthetic repellents on mosquito nets in experimental huts against insecticide-resistant *Anopheles gambiae* and *Culex quinquefasciatus* mosquitoes. Trans. R. Soc. Trop. Med. Hyg. 100, 1091–1097.

Ogama, S.B., Moore, S.J., Maia, M.F., 2012. A systematic review of mosquito coils and passive emanators: defining recommendations for spatial repellency testing methodologies. Parasit. Vectors 5, 287.

Perich, M.J., Tidwell, M.A., Dobson, S.E., Sardelis, M.R., Zaglul, A., Williams, D.C., 1993. Barrier spraying to control the malaria vector *Anopheles albimanus*: laboratory and field evaluation in the Dominican Republic. Med. Vet. Entomol. 7, 363–368.

Procacci, P.G., Lamizana, L., Kumlein, S., Habluetzel, A., Rotigliano, G., 1991. Permethrin-impregnated curtains in malaria control. Trans. R. Soc. Trop. Med. Hyg. 85, 181–185.

Puccetti, G., 2007. IR3535 (ethyl butylacetylaminopropionate). In: Debboun, M., Frances, S.P., Strickman, D. (Eds.), Insect Repellents: Principles, Methods and Uses. CRC Press, Boca Raton, FL, pp. 353–360.

Qualls, W.A., Xue, R.-D., Holt, J.A., Smith, M.L., Moeller, J.J., 2011. Field evaluation of commercial repellents against the floodwater mosquito *Psorophora columbiae* (Diptera: Culicidae) in St. Johns County, Florida. J. Med. Entomol. 48 (6), 1247–1249.

Qualls, W.A., Smith, M.L., Muller, G.C., Zhao, T.-Y., Xue, R.-D., 2012. Field evaluation of a large-scale barrier application of bifenthrin on a golf course to control floodwater mosquitoes. J. Am. Mosq. Control Assoc. 28, 219–224.

Revay, E.E., Junnila, A., Xue, R.-D., Kline, D.L., Bernier, U.R., Kravchenko, V.D., Qualls, W.A., Ghattas, N., Muller, G.C., 2013. Evaluation of commercial products for personal protection against mosquitoes. Acta Trop. 125, 226–230.

Rowland, M., Downey, G., Rab, A., Freeman, T., Mohammad, N., Rehman, H., Durrani, N., Curtis, C., Lines, J., Fayaz, M., 2004. Deet mosquito repellent provides personal protection against malaria: a household randomized trial in an Afghan refugee camp in Pakistan. Med. Vet. Entomol. 9, 335–342.

Rutledge, L.C., Sofield, R.K., Mussa, M.A., 1978. A bibliography of diethyl toluamide. ESA Bull. 24, 431–439.

Schoenig, G.P., Neeper-Bradley, T.L., Fisher, L.C., Hartnagel Jr., R.E., 1994. Teratological evaluations of deet in rats and rabbits. Fundam. Appl. Toxicol. 23, 63–69.

Scholte, E.-J., Knols, B.G.J., Samson, R.A., Takken, W., 2004. Entomopathogenic fungi for mosquito control: a review. J. Insect Sci. 4, 19.

Schreck, C.E., 1991. Permethrin and dimethylphthalate as a tent fabric treatments against *Aedes aegypti*. J. Am. Mosq. Control Assoc. 7, 533–535.

Schreck, C.E., Leonhardt, B., 1991. Efficacy assessment of Quwenling, a mosquito repellent from China. J. Am. Mosq. Control Assoc. 7, 43–436.

Schreck, C.E., Posey, K., Smith, D., 1978. Durability of permethrin as a potential clothing treatment to protect against blood-feeding arthropods. J. Econ. Entomol. 71, 397–400.

Schreck, C.E., Carlson, D.A., Weidhaas, D.E., Posey, K., Smith, D., 1980. Wear and aging tests with permethrin-treated cotton-polyester fabric. J. Econ. Entomol. 73, 451–453.

Schreck, C.E., Mount, G.A., Carlson, D.A., 1982. Wear and wash persistence of permethrin used as a clothing treatment for personal protection against the lone star tick (Acari: Ixodidae). J. Med. Entomol. 19, 143–146.

Sholdt, L.L., Holloway, M.L., Chandler, J.A., Fontaine, R.E., van Elsen, A., 1977. Dwelling space repellents: Their use on military tentage against mosquitoes in Kenya, East Africa. J. Med. Entomol. 14, 252–253.

Simmons, T.E., 1985. Insect-repellent and insecticidal soap composition. UK Patent No. 2160216A.

Sochantha, T., van Bortel, W., Savannaroth, S., Marchotty, T., Speybroeck, N., Coosemans, M., 2010. Personal protection by long-lasting insecticidal hammocks against the bites of forest malaria vectors. Trop. Med. Int. Health 15, 336–341.

Sota, J., Mendina, F., Dember, N., Berman, J., 1995. Efficacy of permethrin-impregnated uniforms in the prevention of malaria and leishmaniasis in Columbian soldiers. Clin. Infect. Dis. 21, 599–602.

Sudakin, D.L., Wade, R., Trevathan, B.S., 2003. Deet: a review and update of safety and risk I the general population. J. Toxicol. Clin. Toxicol. 41, 831–839.

Thavara, U.A., Tawatsin, A., Chompoosri, J., Suwonkerd, W., Chansuang, U.R., Asavadachanukon, P., 2001. Laboratory and field evaluations of the insect repellent 3535 (ethyl butylacetylaminopropionate) and deet against mosquito vectors in Thailand. J. Am. Mosq. Control Assoc. 17, 190.

Thomson, M.C., Alessandro, U., Bennett, S., Connor, S.J., Langerock, P., Jawara, M., Todd, J., Greenwood, B.C., 1994. Malaria prevalence is inversely related to vector density in The Gambia, West Africa. Trans. R. Soc. Trop. Med. Hyg. 88, 638–643.

Tisgratog, R., Sukkanon, C., Grieco, J.P., Sanguanpong, U., Chauhan, K.R., Coats, J.R., Chareonviriyaphap, T., 2018. Evaluation of the constituents of vetiver oil against *Anopheles minimus* (Diptera: Culicidae), a malaria vector in Thailand. J. Med. Entomol. 55, 193–199.

Trigg, J.K., 1996. Evaluation of a eucalyptus-based repellent against *Anopheles* spp. in Tanzania. J. Am. Mosq. Control Assoc. 12, 243–246.

Trout, R.T., Brown, G.C., Potter, M.F., Hubbard, J.L., 2007. Efficacy of two pyrethroid insecticides applied as barrier treatments for managing mosquito (Diptera: Culicidae) populations in suburban residential properties. J. Med. Entomol. 44, 470–477.

Witting-Bissinger, B.E., Stumpf, C.E., Donohue, K.V., Apperson, C.S., Roe, R.M., 2008. Novel arthropod repellent, BioUD, is an efficacious alternative to Deet. J. Med. Entomol. 45, 891–898.

World Health Organization, 1975. Manual of Practical Entomology in Malaria, Part II. Methods and Techniques. World Health Organization, Geneva, Switzerland.

World Health Organization, 2001. Report of the Fifth WHOPES Working Group Meeting. WHO/CDS/WHOPES/2001.4, World Health Organization, Geneva, Switzerland.

Xue, R.-D., Qualls, W.A., Smith, M.L., Gaines, M.K., Weaver, J.H., Debboun, M., 2012. Field evaluation of the Off! Clip-on mosquito repellent (metofluthrin) against *Aedes albopictus* and *Aedes taeniorhynchus* (Diptera: Culicidae) in northeastern Florida. J. Med. Entomol. 49, 652–655.

Yap, H.H., 1986. Effectiveness of soap formulations containing deet and permethrin as personal protection against outdoor mosquitoes in Malaysia. J. Am. Mosq. Control Assoc. 2, 63–67.

Yap, H.H., Jahangir, S.C., Chong, C., Adanan, C.R., Chong, N.L., Malik, Y.A., Rohaizat, B., 1998. Field efficacy of a new repellent, KBR 3023, against *Aedes albopictus* (Skuse) and mosquitoes in a tropical rainforest. J. Vector Ecol. 23, 62–68.

Yap, H.H., Jahangir, K., Zairi, J., 2000. Field efficacy of four insect repellents against vector mosquitoes in a tropical rainforest. J. Am. Mosq. Control Assoc. 16, 241–244.

Index

Note: Page numbers followed by *f* indicate figures and *t* indicate tables.

Printed in the United States
By Bookmasters